ADVANCES IN MICROBIAL BIOTECHNOLOGY

BIOTECHNOLOGY

Current Trends and Future Prospects

ADVANCES IN MICROBIAL BIOTECHNOLOGY

Current Trends and Future Prospects

Edited by

Pradeep Kumar
Jayanta Kumar Patra
Pranjal Chandra

Apple Academic Press Inc.
3333 Mistwell Crescent
Oakville, ON L6L 0A2 Canada

Apple Academic Press Inc.
9 Spinnaker Way, Waretown,
New Jersey, NJ 08758, USA

Library and Archives Canada Cataloguing in Publication

Advances in microbial biotechnology : current trends and future prospects / edited by Pradeep Kumar, Jayanta Kumar Patra, Pranjal Chandra.
Includes bibliographical references and index.
Issued in print and electronic formats.
ISBN 978-1-77188-667-3 (hardcover).--ISBN 978-1-351-24891-4 (PDF)
1. Microbial biotechnology. I. Chandra, Pranjal, editor II. Kumar, Pradeep, 1982-, editor III. Patra, Jayanta Kumar, editor

TP248.27.M53A38 2018	660.6'2	C2018-903683-4	C2018-903684-2

Library of Congress Cataloging-in-Publication Data

Names: Kumar, Pradeep (Professor of biotechnology), editor. | Patra, Jayanta Kumar, editor. | Chandra, Pranjal, editor.
Title: Advances in microbial biotechnology : current trends and future prospects / editors, Pradeep Kumar, Jayanta Kumar Patra, Pranjal Chandra.
Other titles: Advances in microbial biotechnology (Kumar)
Description: Oakville, ON ; Waretown, NJ : Apple Academic Press, Inc., 2018.
| Includes bibliographical references and index.
Identifiers: LCCN 2018027315 (print) | LCCN 2018028005 (ebook) | ISBN 9781351248914 (ebook) | ISBN 9781771886673 (hardcover : alk. paper)
Subjects: | MESH: Microbiological Phenomena | Biotechnology--trends
Classification: LCC QR41.2 (ebook) | LCC QR41.2 (print) | NLM QW 4 | DDC 579--dc23
LC record available at https://lccn.loc.gov/2018027315

CONTENTS

LIST OF CONTRIBUTORS

Gautam Anand
Department of Biotechnology, D.D.U. Gorakhpur University, Gorakhpur, U.P. 273009, India

Bigneswar Baliyarsingh
School of Life science, Ravenshaw University. Cuttack, Odisha, India

Chandrajit Balomajumder
Professor and Head, Department of Chemical Engineering, Indian Institute of Technology Roorkee, Roorkee 247667, Uttarakhand, India

Riddhi Banerjee
Organelle Biology and Cellular Ageing Lab, Department of Biosciences and Bioengineering, Indian Institute of Technology Guwahati, Guwahati 781039, Assam, India

Anupriya Baranwal
Department of Biosciences and Bioengineering, Indian Institute of Technology Guwahati, Guwahati 781039, Assam, India

Manash P. Borgohain
MS Pharm, Laboratory for Stem Cell Engineering and Regenerative Medicine (SCERM), Department of Biosciences and Bioengineering, Indian Institute of Technology Guwahati (IITG), Guwahati, Assam 781039, India
Mobile: +919854850277, E-mail: m.borgohain@iitg.ernet.in

Pranjal Chandra*
Department of Biosciences and Bioengineering
Indian Institute of Technology Guwahati, Guwahati 781039, Assam, India
E-mail: pranjalmicro13@gmail.com, pchandra13@iitg.ernet.in

Gitishree Das
Research Institute of Biotechnology & Medical Converged Science, Dongguk University Seoul, Ilsandong-gu, Gyeonggi-do 10326, Republic of Korea

Sushrirekha Das
Department of Zoology, North Orissa University, Baripada, Odisha 757003, India

Rachayeeta Deb
Organelle Biology and Cellular Ageing Lab, Department of Biosciences and Bioengineering, Indian Institute of Technology Guwahati, Guwahati-781039, Assam, India

Nayan Moni Deori
Organelle Biology and Cellular Ageing Lab, Department of Biosciences and Bioengineering, Indian Institute of Technology Guwahati, Guwahati-781039, Assam, India

Sagarika Devi*
DTU Bioengineering, Department of Biotechnology and Biomedicine, Technical University of Denmark, 2800 Kgs. Lyngby, Denmark
E-mail: sadevi@dtu.dk, devi.sagarika@gmail.com

Chandrima Dey
M. Sc., Laboratory for Stem Cell Engineering and Regenerative Medicine (SCERM), Department of Biosciences and Bioengineering, Indian Institute of Technology Guwahati (IITG), Guwahati, Assam 781039, India
Mobile: +919163303801, E-mail: dey.chandrima@iitg.ernet.in

Eleftherios H. Drosinos
Laboratory of Food Quality Control and Hygiene, Department of Food Science and Human Nutrition, Agricultural University of Athens, Iera Odos 75, GR-118 55 Athens, Greece

Shubha Dwivedi*
Department of Biotechnology, Uttarakhand Technical University, Dehradun, India
E-mail: shubha.dwivedi@rediffmail.com

Naveen Dwivedi*
Department of Biotechnology, S. D. College of Engineering & Technology, Muzaffarnagar, UP, India
E-mail: naveen.dwivedi@gmail.com

Lekshmi K. Edison
Microbiology Division, Jawaharlal Nehru Tropical Botanic Garden and Research Institute, Palode, Trivandrum, Kerala 695562, India

Sushanto Gouda
Amity Institute of Wildlife Science, Noida 201303, Uttar Pradesh, India

Rout George Kerry
Department of Biotechnology, AMIT College, Khurda 752057, Odisha, India

Kavindra Kumar Kesari
Department of Applied Physics & Department of Bioproduct and Biosystem, Aalto University, Espoo, Finland

Anuj Kumar
School of Chemical Engineering, Yeungnam University, 280 Daehak-Ro, Gyeongsan 38541, South Korea and Department of Nano, Medical and Polymer Materials, College of Engineering, Yeungnam University, 280 Daehak-Ro, Gyeongsan 38541, South Korea

Devendra Kumar*
Centre for Conservation and Utilization of Blue Green Algae, Indian Agricultural Research Institute Pusa New Delhi 110012, India
E-mail: devendra2228@gmail.com

Neeraj Kumar
Department of Microbiology, Kurukshetra University Kurukshetra, Haryana, India

Pradeep Kumar*
Department of Forestry, North Eastern Regional Institute of Science and Technology, Nirjuli, Arunachal Pradesh 791109, India
E-mail: pkbiotech@gmail.com

H. Krishna Kumar
M. Tech., Laboratory for Stem Cell Engineering and Regenerative Medicine (SCERM), Department of Biosciences and Bioengineering, Indian Institute of Technology Guwahati (IITG), Guwahati, Assam 781039, India Mobile: +919962180535, E-mail: hk.kumar@iitg.ernet.in

Ravinder Kumar
Faculty of Engineering Technology, Universiti Malaysia Pahang, Kuantan 26300, Malaysia

Gyana Prakash Mahapatra
Department of Biotechnology, AMIT College, Khurda 752057, Odisha, India

Aseem Mishra
P.G. Department of Biotechnology, Utkal Univeristy, Bhubaneswar, Odisha, India

Nakulananda Mohanty
Department of Zoology, North Orissa University, Baripada, Odisha 757003, India

D. Mohapatra
School of Life Sciences, Ravenshaw University, Cuttack, India

Nihar Ranjan Singh*
School of Life Sciences, Ravenshaw University, Cuttack, India
E-mail: nihar.singh@gmail.com

Prasenjit Mondal
Associate Professor, Department of Chemical Engineering, Indian Institute of Technology Roorkee, Roorkee 247667, Uttarakhand, India

Shirisha Nagotu*
Organelle Biology and Cellular Ageing Lab, Department of Biosciences and Bioengineering, Indian Institute of Technology Guwahati, Guwahati 781039, Assam, India
E-mail: snagotu@iitg.ernet.in

Gloria Narayan
M. Tech., Laboratory for Stem Cell Engineering and Regenerative Medicine (SCERM), Department of Biosciences and Bioengineering, Indian Institute of Technology Guwahati (IITG), Guwahati, Assam 781039, India
Mobile: +919957966856, E-mail: gloria@iitg.ernet.in

Priyanka Nehra
Centre for Conservation and Utilization of Blue Green Algae, Indian Agricultural Research Institute Pusa New Delhi 110012, India

Smita Hasini Panda*
Department of Zoology, North Orissa University, Baripada, Odisha, India
E-mail: panda.smita@gmail.com

Spiros Paramithiotis*
Department of Food Science and Human Nutrition, Agricultural University of Athens, Iera Odos 75, GR-11855, Athens, Greece
E-mail: sdp@aua.gr

Jayanta Kumar Patra*
Research Institute of Biotechnology & Medical Converged Science, Dongguk University-Seoul, Ilsandong-gu, Gyeonggi-do 10326, REPUBLIC OF KOREA
E-mail: jkpatra.cet@gmail.com

N. S. Pradeep*
Microbiology Division, Jawaharlal Nehru Tropical Botanic Garden and Research Institute, Palode, Trivandrum, Kerala 695562, India

Anjan Pradhan*
Virginia Commonwealth University, Richmond, Virginia, USA
E-mail: pradhan.anjan@gmail.com

Saktikanta Rath*
Ravenshaw University, Cuttack, Odisha, India
E-mail: saktirath@gmail.com

Reecha Sahu
Department of Biotechnology, National Institute of Technology, Raipur Chattisgarh 492010, India

Sushil Kumar Sahu
John Hopkins University, Baltimore, Maryland, USA

Dibyaranjan Samal
Department of Biotechnology, AMIT College, Khurda 752057, Odisha, India

S. Shiburaj
Microbiology Division, Jawaharlal Nehru Tropical Botanic Garden and Research Institute, Palode, Trivandrum, Kerala 695562, India

Ananya Srivastav
Department of Chemistry, Indian Institute of Technology Delhi, New Delhi 10016, India

Vishal Kumar Soni*
National Filaria Control Program Unit (Department of Medical and Health, U.P.), Civil lines, Deoria, Uttar Pradesh 274001, India
E-mail: vishalkumars@gmail.com

Aiman Tanveer
Department of Biotechnology, D.D.U. Gorakhpur University, Gorakhpur, U.P. 273009, India

Hrudayanath Thatoi
Department of Biotechnology, Baripada, Odisha 757003, India

Rajkumar P. Thummer*
Assistant Professor, Laboratory for of Stem Cell Engineering and Regenerative Medicine (SCERM), Department of Biosciences and Bioengineering, Indian Institute of Technology Guwahati (IITG), Guwahati, Assam 781039, India.
Mobile: +917086867025, E-mail: rthu@iitg.ernet.in

Pooja Tripathi
Center of Bioinformatics,
IIDS, University of Allahabad, Allahabad, UP, India

Vijay Tripathi*
Department of Molecular and Cellular Engineering, Jacob Institute of Biotechnology and Bioengineering, Sam Higginbottom University of Agriculture, Technology and Sciences, Naini, Allahabad, India
E-mail: vijaytripathi84@gmail.com

Lata S. B. Upadhyay*
Assistant Professor, Department of Biotechnology, National Institute of Technology, Raipur, C.G, India
E-mail: lupadhyay.bt@nitrr.ac.in

Chethala N. Vishnuprasad*
School of Life Sciences, Trans-disciplinary University (TDU) 74/2, Jarakabande Kaval, Post: Attur, Via Yelahanka, Bangalore 560106, Karnataka, India
E-mail: drcnvp@gmail.com

Dinesh Yadav*
Department of Biotechnology, D.D.U. Gorakhpur University, Gorakhpur, U.P. 273009, India
E-mail: dinesh_yad@rediffmail.com

Mukesh Kumar Yadav*
Department of Otorhinolaryngology-Head and Neck Surgery & Institute for Medical Device Clinical Trials, Korea University College of Medicine, Seoul, South Korea
E-mail: mukiyadav@gmail.com

Sangeeta Yadav
Department of Biotechnology, D.D.U. Gorakhpur University, Gorakhpur, U.P. 273009, India

ABOUT THE EDITOR

Pradeep Kumar, PhD, is currently working as Assistant Professor in the Department of Forestry at the North Eastern Regional Institute of Science and Technology (Deemed University-MHRD, Govt. of India), Nirjuli (Itanagar), Arunachal Pradesh, India. Before he joined NERIST, he worked as a Research Professor in the Department of Biotechnology, Yeungnam University, South Korea. He was Postdoctorate Researcher in the Department of Biotechnology Engineering, Ben Gurion University of the Negev, Israel, and was awarded and PBC Outstanding PostDoc Fellowship for more than three years. His area of research and expertise are wide, including microbial biotechnology, plant pathology, bacterial genetics, insect-pest biocontrol, gene expression of cry genes, plant-microbe interaction, and molecular breeding. He has been honored with an international travel grant from Ben Gurion University of Negev, Israel, to attend an international conference. He is the recipient of a best paper presentation and the Narasimhan Award by the Indian Phytopathological Society, India. He has presented several oral and poster presentations at various national and international conferences and has published one book and more than 40 research articles, including original research papers in peer-reviewed journals and several book chapters with international publishers, including Springer, CABI, Bentham, and Apple Academic Press. He provides his service to many journals as a guest editor, associate editor, editorial board member, technical editor, and peer reviewer.

Dr. Jayanta Kumar Patra, MSc PhD, is currently working as assistant professor at Research Institute of Biotechnology and Medical Converged Science, Dongguk University, Ilsandong, Gyeonggido, Republic of Korea. His current research is focussed on nanoparticle synthesis by green technology methods and their potential application in biomedical and agricultural fields. He has about 11 years of research and teaching experience in the field of food, pharmacological and nanobiotechnology. Dr. Patra completed his Ph.D. (Life Sciences) from North Orissa University, India, and PDF (Biotechnology) from Yeungnam University, South Korea. Since 2007, he has more than 90 papers published in various national and international peer-reviewed journals of Elsevier, Springer, Taylor and Francis, Wiley and so forth and around 15 book chapters in different reputed edited books. Dr. Patra has also edited four books, for STUDIUM Press LLC USA, STUDIUM Press India Pvt. Ltd., CRC Press USA, Springer Nature publisher, Singapore and so forth. Besides, he is serving as editorial board member of several international journals and science magazines.

 **Pranjal Chandra, PhD,** is an Assistant Professor and Principal Investigator in the Department of Biosciences and Bioengineering at the Indian Institute of Technology Guwahati, Assam, India. He has published over 70 research articles in reputed journals and three books. He is also a visiting scientist and visiting professor at the Institute of Biomedical Science and Technology (IBST), South Korea. Dr. Pranjal's research contributions are highly interdisciplinary, spanning a wide range in nanobiotechnology, nanobiosensors, lab-on-chip systems for biomedical diagnostics, and nanomedicine. His work has been highlighted in the Chemistry World News of the Royal Society of Chemistry, Cambridge, as "a new system for cancer detection" and also featured as a key scientific article in the Global Medical Discovery News Canada. He is the recipient of many prestigious awards and fellowships such as the Ramanujan Fellowship (Government of India), Early Carrier Research Award (Government of India), BK-21 and NRF fellowship of South Korea, Technion postdoctoral fellowship (Israel), University of Montreal Postdoc fellowship (Canada), NMS Young scientist Award (2016), etc. He is also an editorial board member of a dozen international journals.

Dr. Chandra earned his PhD from Pusan National University, South Korea, and did postdoctoral training at Technion-Israel Institute of Technology, Israel.

PREFACE

In the past few decades, the rapid and vast development of advanced microbial bioresources and metagenomic techniques has completely transformed the area of research across the biological sciences. Microbiology is one of the most influential fields and attracts researchers from all over the globe. Our understanding of microbial diversity, evolutionary biology, and microbial interaction with their animal and plant hosts at molecular level has been revolutionized.

In this volume, various applications of microorganisms have been covered broadly and have been appropriately reflected in depth in different chapters. This book covers four major sections: (i) applied microbiology in agriculture, (ii) microbes in the environment, (iii) microbes and human health, and (iv) microbes in nanotechnology. The book provides insights into the diverse microorganisms that have been explored and exploited in the development of various applications for improvements in agriculture. It also looks at the utilization of microorganisms to improve the desired traits for achieving optimal production of microbials.

Applications of microbes for the removal of pollutants and for the recovery of metals and oils have also been discussed for environmental purposes. The book also includes information on the synthesis and applications of nanoparticles derived from plants and microbes, their uses and applications in agri-food production, pathogen diagnostics, contaminant removal, and the antimicrobial and pharmacological properties. We have also included chapters that discuss nanotechnology and biosensors applications for human health.

The volume provides a holistic look at applications of microorganisms and other multidisciplinary approaches for improvement of agriculture, environment, human health, and biomedicine. The book is written in simple and clear text, and we also used many figures and tables to make the book easier to understand.

This volume is the culmination of the efforts of several researchers, scientists, graduate students, and postdoctoral fellows across the world who are well-known and respected in different frontiers of biotechnology and microbiology research. This book caters to the needs of the graduate and postgraduate students pursuing their study in branches of life sciences, microbiology, health sciences, and environmental biotechnology. It will also be of interest to researchers and scientists working in laboratories and industries involved in the study and research in the fields of microbiology, environmental biotechnology, and allied areas.

We express our appreciation to all of the contributing authors who helped us tremendously with their contributions, time, critical thoughts, and suggestions to put together this peer-reviewed edited volume. The editors are also thankful to Apple Academic Press and their team members for the opportunity to publish this volume. Lastly, we thank our family members for their love, support, encouragement, and patience during the entire period of this work.

Pradeep Kumar
NERIST, Arunachal Pradesh, India

Jayanta Kumar Patra
Dongguk University, Goyang-si, South Korea

Pranjal Chandra
IIT Guwahati, Assam, India

PART I
Applied Microbiology in Agriculture

CHAPTER 1

THE GENOMICS OF MAJOR FOODBORNE PATHOGENS: AN UPDATE

SPIROS PARAMITHIOTIS* and ELEFTHERIOS H. DROSINOS

*Laboratory of Food Quality Control and Hygiene, Department of Food Science and Human Nutrition, Agricultural University of Athens, Iera Odos 75, 11855 Athens, Greece, *E-mail: sdp@aua.gr*

CONTENTS

ABSTRACT

Advances in the field of molecular biology resulted in the development of analytical approaches and tools that allow the detailed characterization of foodborne pathogens. More accurately, a series of platforms and biosensors for rapid and reliable detection, a range of methodological approaches for the effective epidemiological assessment as well as the study of the transcriptomic responses of stress- and virulence-associated genes in food-related environments have improved our understanding regarding their subsistence in them. In this chapter, an update of the recent advances in the genomic studies of the major foodborne pathogens *Listeria monocytogenes*, *Salmonella* Serovars and *Escherichia coli* is offered.

1.1 INTRODUCTION

Listeria monocytogenes, *Salmonella* Serovars and *Escherichia coli* are annually implicated as the causative agents in several foodborne outbreaks. More accurately, each year in the USA, *L. monocytogenes* has been reported to cause 1600 illnesses, 1500 hospitalizations and 250 deaths. *Salmonella enterica* serovar Typhi caused 1800 illnesses, 200 hospitalizations but no deaths and nontyphoidal *Salmonella* 1,000,000 illnesses, 19,000 hospitalizations and 380 deaths.[1] Regarding shiga toxin-producing *E. coli* (STEC), serovar O157 has been reported to cause 63,000 illnesses, 2100 hospitalizations and 20 deaths while non-O157, 110,000 illnesses, 270 hospitalizations and 1 death.[1] On the basis of the estimated annual deaths, nontyphoidal *Salmonella* infections are the most important agents among bacteria, viruses and parasites with *L. monocytogenes* in the third position behind *Toxoplasma gondii*. As far as the trend of foodborne illnesses caused by the above-mentioned microorganisms in the USA was concerned, in 2014 the incidence of STEC O157 was significantly ($P<0.05$) lower while that of STEC non-O175 was significantly higher, compared to 2011–2013. Regarding *L. monocytogenes*, the incidence was significantly lower only compared to 1996–1998, whereas that of *Salmonella* exhibited no significant changes from 1996.[2]

On the other hand, in 2014 in the EU a total of 88,715, 2,161 and 5,955 salmonellosis, listeriosis and STEC infections were reported, respectively, with the hospitalized cases summing up to 9830, 812 and 930 and the fatal one as many as 65, 210 and 7, respectively. Interestingly, on the basis of the fatal cases, listeriosis was the most common cause, salmonellosis ranked second with STEC infections in the fourth place behind campylobacteriosis. Additionally, in 2014 in the EU the decreasing prevalence of salmonellosis and the increasing of listeriosis since 2008 continued, whereas the number of STEC *E. coli* infections slightly decreased compared to 2013.[3]

The genomics of these foodborne pathogens has been in the epicenter of intensive study over the last decades. Research has focused on the accurate, reliable and fast detection, the diversity assessment, mostly from an epidemiological point of view, as well as their response to particular stresses occurring during food manufacturing and storage. As a result, a wealth of information has been generated, leading to a variety of proposed approaches for effective detection, and epidemiological assessment as well as an improvement in the level of our understanding regarding their physiology under several stressful conditions. In the next paragraphs, the recent developments regarding these issues are presented and critically discussed.

1.2 DETECTION OF FOODBORNE PATHOGENS

The classical microbiological approach for the detection of foodborne pathogens necessary involves one or more enrichment steps, cultivation on selective solid media, isolation and purification of the colonies exhibiting 'typical' phenotypic characteristics and performance of a number of tests in order to verify the identity of the pathogen. Depending on the pathogen, this procedure may require as long as one week to be completed, time that in many cases exceeds the shelf-life of the product under examination, especially in the case of fresh-cut fruits and vegetables. Molecular techniques have been introduced in pathogen detection to reduce this time and improve selectivity, sensitivity, specificity and reliability. Their underlying principle is the detection of molecules that are specific for the desired taxonomic level, that is, species, serotype or strain. The nature of the detection, that is the food matrix, the low numbers of the target microorganisms in the presence of another biota in much higher population, along with the presence of compounds that may interfere with the analysis are the challenges that need to be addressed before any detection step may take place. Selective enrichment is the pretreatment that in most of the cases is applied as it increases the population of the pathogen and at the same time dilutes the food matrix. However, it may be applied only when qualitative detection is required. Alternatively, the concentration rather than propagation of the microbial population under examination has been proposed by strategies that include filtration,[4,5] buoyant density centrifugation,[6,7] use of aqueous polymer two-phase systems,[8,9] use of some type of biosorbent matrix,[10–12] bacteriophages[13] and dielectrophoresis.[14,15] However, in most of the cases, additional pretreatments are necessary, since the dominant microbial population components such as fat and protein as well as the food debris interferes with the analysis, thus increasing the processing time and complexity.

The molecular techniques used for detection purposes, as a step in either a culture-dependent or -independent protocol, may be divided into the nucleic acid- and the immuno-based ones. For the former, lysis of the cells is necessary to expose the nucleic acids to the detection chemistry involved, whereas the latter relies on antigens residing on the surface of the bacterial cells.

1.2.1 IMMUNO-BASED DETECTION

Enzyme-linked immunosorbent assay (ELISA) has been extensively employed for the detection of foodborne pathogens. It relies on the specific antibody-antigen binding and concomitant visualization. Sandwich ELISA, the most frequently used ELISA strategy, involves the binding of the target antigen to an immobilized primary antibody and an enzyme-conjugated secondary one. The enzyme is used to achieve visualization, mostly through the conversion of a colorless substrate to a colored one. This approach has been used for the detection of both microbial cells and toxins.[16–21] The ELISA principle has been employed to develop lateral flow immunoassays as well as fully automated detection systems. The former is suitable for on-site detection purposes, a wide range of dipsticks and immunochromatographic strips have been developed, some of which are currently commercially available, such as the Reveal® for *E. coli* O157:H7 (Neogen), the VIP® Gold for EHEC, *Listeria* spp. and *Salmonella* (Biocontrol) and the DuPont™ lateral flow system for *E. coli* O157, *Listeria* spp. and *Salmonella* (DuPont).[22–28] On the other hand, fully automated detection systems, such as the VIDAS (BioMerieux), are more suitable for high throughput quality control installations and for that purpose have been extensively employed.[29–33]

Despite the several advantages that characterize the ELISA-based detection strategies, with small detection time and automation capacities being the most important, they suffer from significant disadvantages as well. Cross-reactivity and low sensitivity are the most important ones. Therefore, research has focused on the improvement of these fields. More accurately, the main strategy that has been employed to address cross-reactivity is the continuous development of new antibodies; on the other hand, sensitivity is improved through the increase of antibody density or coupling with concentration methods.

The use of aptamers as ligands has been extensively considered. Aptamers are single-stranded nucleic acids that may bind to a wide range of non-nucleic acid targets.[34] The selection of the aptamers suitable for the desired purpose is based on a procedure named systematic evolution of ligands by exponential enrichment (SELEX).[35,36] This procedure is based on the repetition of a three-step cycle that involves incubation of the target with a mixture of aptamers, elution of the bound ones and cloning through polymerase chain reaction (PCR) until the desired level of specificity is reached. An improvement to this technique was very recently proposed by Wang et al.[37] In this study, the use of fluorescence-activated cell sorting was proposed to assist screening of aptamers for their affinity and specificity simultaneously instead of sequencially; this approach was named Multiparameter Particle Display (MPPD). Aptamers have been selected for the detection of the major foodborne bacterial pathogens,[38–42] mostly in simplex but also in the multiplex format.[43] All necessary features of aptamer chemistry and use have been recently gathered and critically reviewed by Davydova et al.[44] and Teng et al.[45]

Sensitivity improvement using nanomaterials has been studied to some depth. Carbon nanotubes have become a very promising tool in biodiagnosis mostly due to the large surface to volume ratio and the ease by which new functional groups can be introduced.[46] Using these properties Chunglok et al.[46] employed single-walled carbon nanotubes (SWCNTs) as a labelling platform for antibody and horseradish peroxidase (HRP) co-immobilizing for the development of an immunoassay for the detection of *S. enterica* ser. Typhimurium and reported a 100-times improvement of the detection limit when the ratio of the antibody to the pathogen to HRP was 1:400.

Gold nanoparticles have also been effectively used for the same purpose. Cho and Irudayaraj[47] developed an in situ immuno-AuNP network-based ELISA biosensor coupled with immunomagnetic separation for the detection of foodborne pathogens. The biosensor was tested for the detection of *E. coli* O157:H7 and *S. enterica* ser. Typhimurium in actual food matrices with very promising results. Shen et al.[48] presented a nanoparticle-enhanced enzyme-linked immunosorbent assay (FNP-ELISA) for the detection of *E. coli* O157:H7. Immunomagnetic nanoparticles (IMMPs) were used to capture the target cells, while beacon gold nanoparticles (B-GNPs) were used for the immunoreaction providing significant improvement of the detection limit.

1.2.2 NUCLEIC ACID-BASED DETECTION

DNA was the molecule on which the detection was based for a long time. The major disadvantage of the use of DNA is the ability of the molecule to remain undegraded for extended periods of time that may reach several weeks.[49] However, due to the recent methodological advances, the discrimination of DNA from metabolically active cells or the use of RNA for detection purposes became feasible.

1.2.2.1 DNA-BASED DETECTION

DNA-based detection necessarily involves DNA isolation, PCR amplification and post-PCR visualization of the amplicons. Numerous protocols are currently available in the literature employing any kind of format, that is simplex, multiplex or real-time. The efficacy of this approach is greatly compromised by the nature of the detection as already described. In addition, the ability of DNA to remain undegraded for significant amount of time involves the risk of overestimation. The most important advancement in this field was the use of ethidium monoazide (EMA) or propidium monoazide (PMA) in order to differentiate 'environmental' DNA, that is originating from lysed or cells with compromised cell membrane integrity from DNA of metabolically active cells. Thus, the addition of EMA/PMA treatment seems necessary when microecosystem composition or pathogen detection is performed through genomic approaches. When PMA/EMA is added to the cell suspension, it intercalates into any DNA molecule accessible, that is environmental molecules or molecules residing into membrane-compromised cells. In the effect of strong visible light nitrene radicals are formed leading to a reaction of the dye-DNA complex with any organic molecule in proximity, including other DNA molecules, rendering these molecules inaccessible for PCR reaction. At the same time, any unbound dye turns into hydroxylamine which is not reactive.[50] The addition of this step is becoming more and more common not only in detection protocols but also in microecosystem composition assessment studies. Regarding the latter, the addition of this step before PCR-DGGE has exhibited the usefulness in assessing the microcommunity in complex matrices.[51,52] Propidium monoazide is generally preferred over EMA due to the ability of the latter to penetrate metabolically active cells.[53,54] In addition, this penetration was correlated with the temperature used for the treatment, whereas PMA exhibited no such effect.[55] The effectiveness of PMA treatment depends upon several factors. Among them, the dead-to-viable ratio, the minimum and maximum population of the target microorganism as well as the amplicon size that is important in the case of qPCR applications have drawn special attention. Pan and Breidt[55] reported that for effective *L. monocytogenes* quantification, the ratio of dead to viable cells should not exceed 10^4 CFU/mL while the minimum population of living cells at 10^3 CFU/mL. On the other hand, Zhu et al.,[56] working with *Vibrio parahaemolyticus* and Elizaquivel et al.,[57] working with *E. coli* O157:H7 reported that the higher the cell density the lower the efficiency of PMA treatment. In this sense, the effect of PMA on living cells may not be excluded. Amplicon size plays an important role in qPCR protocols, amplicon sizes of 70–200 bp are generally considered ideal for typical thermocycling conditions because optimum efficiency is achieved. Luo et al.[58] working with *E. coli* O157:H7 reported that PMA treatment may not effectively exclude DNA molecules less than 190 bp., thus amplicons of this size is not suitable for qPCR protocols. Similarly, the lack of concordance between classical microbiological count determination and the respective by PMA-qPCR that was reported by Pacholewicz et al.[59] and was referring to *Campylobacter* detection on broiler chicken carcasses may be attributed to the amplicon size of 287 bp. Banihashemi et al.[60] working with *Campylobacter* concluded that the signal from environmental DNA may not be excluded in the case of amplicons less than 200 bp., on the contrary, when the amplicon size was more than 1.5 kbp the signal from dead cells was completely excluded. The latter was also the case for *E. coli*. Schnetzinger et al.[61] targeting a 1108 bp 16 S rRNA gene fragment of *E. coli* observed a complete suppression of signals from exogenous DNA. On the contrary, Li and Chen[62] reported that an amplicon of 130 bp was optimal for PMA treatment and was selected for further PMA-qPCR assay development for the detection of *Salmonella* spp. living cells in food. An innovative procedure that promises to address the above-mentioned drawbacks was introduced by Soejima et al.[63] This procedure employed the use of platinum compounds. When microorganisms were in vivo exposed to these compounds, the latter penetrate compromised cells and chelated by DNA. Thus, the PCR that follows is exclusively based on the DNA originating from living cells.

Fluorescence in situ hybridization (FISH) has been characterized as a promising approach already from its early days.[64,65] As a matter of fact, it has become a routine method in biomedical applications.[66] However, in applications related to Food Microbiology, although many protocols have been developed for pathogen detection[67–74] it has not met widespread application. The main reason for this is the substantial matrix effect in actual food systems. The matrix effect is so significant that an actual in situ detection is not always feasible. Moreover, additional steps, such as selective enrichment, matrix dilution, etc. constitute necessary interventions for effective use of this technique. However, in that way the in situ advantage of this approach is lost and the information obtained is no different from the respective received through PCR application, which is much simpler and therefore preferred.

1.2.2.2   RNA-BASED DETECTION

The use of RNA for detection purposes offers a better view of the cells with metabolic activity.[75] Isolation of RNA is followed by a reverse transcription step and a sequence characteristic of the desired taxonomic level is amplified by PCR. This approach is gaining interest mostly in microecosystem composition molecular assessment rather than pathogen detection over the last years.[76–82] In the former case, this approach seems, in some cases, necessary to obtain a complete picture of the microbiota involved, whereas in the latter, the nature of the determination as already described seems to inhibit the effective application of this approach. However, it may be considered as an alternative that may be selected according to the needs and capacity of the laboratory.

1.2.3   BIOSENSORS

Biosensors consist of two basic elements: a bio-receptor and a transducing system.[83] The purpose of the former is to capture the target and of the latter to convert this interaction into a measurable signal. The target microorganism may be recognized through the use of antibodies, nucleic acids, aptamers, etc. Then, the transduction of the signal may be electrochemical, magnetoelastic, optical or piezoelectric.[21,84] The main advantage of biosensors is the user-friendly interface and thus trained personnel is not always required. On the other hand, their major drawback is that the time required for the development of a novel approach to commercialization is usually long enough for a proposed improvement to become out of date.

Optical biosensors are the most frequently reported ones in the detection of foodborne pathogens. The main advantages of these biosensors are the ability for real-time monitoring of the target binding reaction and the low cost. Among the several approaches used, for example light absorbance, fluorescence, chemiluminescence and surface plasmon resonance (SPR), the latter is the most commonly employed technique. In addition, commercial biosensors based on SPR are currently available, such as the Biacore (GE Healthcare) and the Spreeta (Sensata Technologies). The former biosensor was effectively used for many studies including the detection of *E. coli* O157:H7 in cucumber and ground beef,[85] *L. monocytogenes,*[86] *Salmonella* groups B, D and E,[87] *E. coli* and *S. enterica* ser. Enteritidis in skim milk[88] with reported detection limits ranging from 10^3 to 10^5 CFU mL^{-1}. Similarly, many studies describe the use of the Spreeta platform for the effective detection of foodborne pathogens, among which the detection of *Campylobacter jejuni* in chicken rinse,[89] *E. coli* O157:H7 in milk, apple juice and ground beef,[90] *S. enterica* ser. Enteritidis in chicken extract,[91] *S. enterica* ser. Typhimurium in chicken[92] with detection limits ranging between 10^2 to 10^6 CFU mL^{-1}.

Several studies have reported the use of magnetoelastic biosensors. Among the most interesting ones are the ones in which filamentous bacteriophages and the lectin concanavalin A were selected as detecting elements. The former was used for the detection of *S. enterica* ser. Typhimurium[93] and spores of *Bacillus anthracis,*[94] even in the sequential format,[95] whereas the latter for the detection of *E. coli* O157:H7.[96] However, only the detection of *S. enterica* ser. Typhimurium has been exhibited in an actual food matrix. Lakshmanan et al.[97] reported a detection limit of 5 10^3 CFU/mL for spiked with *S. enterica* ser. Typhimurium water and fat-free milk. Li et al.[98] studied spiked fresh tomato surface with *S. enterica* ser. Typhimurium and reported a detection limit of 5 10^2 CFU/mL. Finally, Park et al.[99,100] studied spiked with *S. enterica* ser. Typhimurium spinach and cantaloupe surfaces; in the first case a detection limit of 2.17 and 1.94 log CFU/spinach for the adaxial and abaxial surface, respectively was obtained while in the second case a detection limit of 2.47 ± 0.50 log CFU/mm^2 surface of cantaloupe was reported.

Electrochemical biosensors may be further subdivided according to the measured parameter into amperometric, conductometric, impedimetric and potentiometric.[101] Many studies describing the successful detection of foodborne pathogens in various food matrices are currently available. Among these, the development of a disposable amperometric magneto-immunosensor, based on the use of functionalized magnetic beads (MBs) and gold screen-printed electrodes (Au/SPEs), for the specific detection and quantification of *Staphylococcus aureus* with a very low detection limit of 1 CFU/mL of raw milk was reported by Esteban-Fernandez de Avila et al.[102] The development of an electrochemical immunosensor based on the use of a multiwall carbon nanotube-polyallylamine modified screen-printed electrode (MWCNT-PAH/SPE) for antibody immobilization for simultaneous detection of *E. coli* O157:H7, *Campylobacter* spp. and *Salmonella* spp. in milk was described by Viswanathan et al.[103] with a limit of detection at 400 cells/mL for *Salmonella* and *Campylobacter* and 800 cells/mL for *E. coli*. Alfonso et al.[104] described a disposable immunosensor based on the use of a screen-printed carbon electrode (SPCE) permanently incorporating a magnet underneath and the use of gold nanoparticles (AuNPs) as label for electrochemical detection of *S. enterica* ser. Typhimurium. The detection limit was reported at 143 cells/mL. The real-time potentiometric detection of *E. coli* in milk and apple juice samples based on covalently immobilized aptamers as biorecognition elements and SWCNT as transducers was reported by Zelada-Guillen et al.[105] The detection limit of this approach was reported at 6 CFU/mL in milk and 26 CFU/mL in apple juice. Yang et al.[106] reported the real-time detection of staphyloccoccal enterotoxin B based on electrical percolation through a single-walled carbon nanotubes (SWNTs)-antibody complex. The detection of *S. enterica* ser. Typhimurium in milk samples with the use of an electrochemical impedance immunosensor was reported by Dong et al.[107] with anti-*Salmonella* antibodies immobilized onto the gold nanoparticles and poly (amidoamine)-multiwalled carbon nanotubes-chitosan nanocomposite film modified glassy carbon electrode (AuNPs/PAMAM-MWCNT-Chi/GCE). The detection limit was reported at $5.0 \ 10^2$ CFU/mL. Wan et al.[108] developed an impedimetric immunosensor doped with reduced graphene sheets (RGSs) and combined with a controllable electrodeposition technique for the selective detection of marine pathogenic sulphate-reducing bacteria (SRB) in the range of 1.8 10^1–1.8 10^7 CFU/mL. Detection of *E. coli* O157:H7 was studied by Li et al.,[109] Joung et al.[110] and Wang et al.[111] In the first study, an immunosensor based on quartz crystal Au electrode was developed and was able to discriminate between populations ranging from 10^2 to 10^5 CFU/mL. The biosensor was effectively applied to pasteurized milk, ground beef and fresh spinach. Joung et al.[110] introduced a nanoporous membrane-based impedimetric immunosensor and successfully detected the pathogen in whole milk samples with a detection limit of 83.7 CFU/mL and 95% probability. Finally, Wang et al.[111] developed an impedimetric immunosensor based on gold nanoparticles modified free-standing graphene paper electrode. The biosensor was tested on spiked ground beef and cucumber samples and the detection limit was reported at 1.5×10^4 and 1.5×10^3 CFU/mL, respectively.

Regarding piezoelectric biosensors, the development of cantilever-based and quartz crystal microbalance (QCM)-based ones have been described. Regarding the former, they have been used for effective detection of *E. coli* O157:H7 in ground beef with a detection limit of ca. 10 cells/mL[112] *L. monocytogenes* in milk at the infection dose limit of 10^3 CFU/mL[113] and staphylococcal enterotoxin B in apple juice and milk with a limit of detection at 10 and 100 fg, respectively.[114] On the other hand, QCM-based ones were used for the detection of *E. coli* through microcontact imprinting of whole cell[115] and *E. coli* O157:H7 using a thiolated probe specific to *eaeA* gene. The latter approach was tested in apple juice, milk and ground beef with promising results.[116] Finally, Salmain et al.[117] developed a QCR-based immunosensor operating in flow-through mode for assaying staphylococcal enterotoxin A in milk. The reported limit of detection was 0.02 mg/L. However, when the sandwich type assay was utilized, the detection limit dropped to 0.007 mg/L.

1.3   EPIDEMIOLOGICAL ASSESSMENT

The aim of the epidemiological studies is to identify as early as possible the infection source and the transmission route of a foodborne pathogen. A range of approaches is currently available; for the selection of the most appropriate one, their specific characteristics along with the clonality of the microorganism under investigation should be carefully considered;

only then concordance with the epidemiological data is achieved and conclusions of epidemiological nature can be safely reached.

The approaches that are currently available may be divided into the sequence-based ones, such as Multilocus Sequence Typing (MLST) and the profile-based ones, such as Pulsed-Field Gel Electrophoresis (PFGE) and Multilocus Variable number of tandem repeats Analysis (MLVA). The cornerstone of the analysis is the detection of differences in the nucleotide sequence between isolates. In the former ones, these are directly assessed while in the latter ones they are visualized through the genotypic profiles.

1.3.1 PULSED-FIELD GEL ELECTROPHORESIS (PFGE)

Pulsed-Field Gel Electrophoresis relies on DNA cleavage by restriction enzymes and subsequent separation of the fragments with gel electrophoresis in a pulsed electric field. It is characterized by high discriminatory power, remarkable reproducibility and, depending on the clonality of the pathogen under surveillance, adequate epidemiological concordance. The latter may be retained by adjusting the similarity threshold used to infer epidemiological relationships. The disadvantages of this approach are also important: the application is rather expensive and trained personnel is required for both bench-work and interpretation of the results. Especially regarding the latter, and due to the increased importance of this approach in epidemiological studies, guidelines have been published,[118] updated after years of application[119] and strongly recommended to follow.

PFGE has been extensively used both in biodiversity and epidemiological studies. Regarding the latter, PulseNet, an Internet-based platform, has been created to facilitate epidemiological surveillance. In this platform, apart from guidance regarding the application and inter-laboratory standardization and normalization of the results, exchange of data may take place enabling thus the establishment of an international network of laboratories and assessment of transmission routes at the global level.

Effective differentiation of the isolates is obtained with the use of more than one restriction enzyme. Differentiation of indistinguishable genotypic profiles in most of the cases is obtained with the application of a secondary or a tertiary restriction enzyme. However, for the epidemiological surveillance of foodborne pathogens that are characterized by high clonality, the combination of techniques rather than enzymes is most of the times required.

Effective epidemiological surveillance of *L. monocytogenes* is achieved with *Asc*I, *Apa*I and *Xba*I as primary, secondary and tertiary enzymes, respectively. Since the populations of this microorganisms are principally clonal, application of PFGE offers adequate epidemiological concordance. Similar is the case regarding typing of *E. coli* O157:H7 as well as non-O157. Adequate differentiation may be achieved with the use of *Xba*I, *Bln*I and *Spe*I as primary, secondary and tertiary restriction enzymes, respectively. On the contrary, the case of *Salmonella* Serovars is rather different. The use of *Xba*I as primary, *Bin*I/*Avr*II as a secondary and *Spe*I as a tertiary enzyme is recommended by PulseNet. However, PFGE may fail to provide with adequate differentiation of isolates belonging to highly clonal Serovars such as *S. enterica ser.* Typhimurium, *S. enterica* ser. Enteritidis, *S. enterica* ser. Typhi and *S. enterica* ser. Paratyphi A and combination with other techniques may be necessary.

1.3.2   MULTILOCUS VARIABLE NUMBER OF TANDEM REPEATS ANALYSIS (MLVA)

Multilocus Variable number of tandem repeats Analysis is a PCR-based technique that relies on the amplification of specific tandem repeated DNA sequences that occur in many loci. Differentiation between the isolates is thus achieved through the comparison of the number of these repetitions in several loci. Generally, MLVA and PFGE provide with comparable results from an epidemiological perspective. Moreover, the technically less demanding nature of the analysis and the increased processing capacity make this approach an attractive alternative. However, data exchange between laboratories is hindered by the fact that genetic relatedness between the isolates is assessed through PCR-generated genotypic profiles, which are extremely difficult to standardize. Despite that, several MLVA protocols have been proposed for the epidemiological surveillance of major foodborne pathogens, an example of which is presented in Table 1.1.

TABLE 1.1   Examples of Established Multilocus Variable Number of Tandem Repeats Analysis (MLVA) Schemes Used for the Epidemiological Assessment of *Listeria monocytogenes*, *Salmonella* Serovars and *E. coli*

VNTR locus name	Primers (5′-3′)	Reference
Listeria monocytogenes		
LM-TR1	GGCGGAAAATGGGAAGC	[120]
	TGCGATGGTTTGGACTGTTG	
LM-TR2	CCTAGAACAAATCCGCCACCAT	
	TCGCCATTGTAAACATCCCCTATT	
LM-TR 3	GCGTGTATTAGATGCGGTTGAG	
	GCATTCCACTATCCCCTGTTTT	
LM-TR4	TCCGAAAAAGACGAAGAAGTAGCA	
	TGGAACGACGGACGAAATAATAAT	
Lm-TR5	GTTTATGCGAATGGCGAGAT	
	CTGGCTTCATAGGATTTACTGGAT	
LM-TR6	AAAAGCAGCGCCACTAACG	
	TAAAAATCCCAATAACACTCCTGA	
Salmonella enterica ser. Typhimurium		
STTR1	FAM-CAGCAGTACAACCGTCAGCAGGAT	[129]
	CCCCACCGTTAGCGCCCGATGTA	
STTR2	FAM-GGGTTCCCTTCCAGATTTACG	
	TTTACCCGCGCATATTACCACACT	
STTR3	FAM-CCCCCTAAGCCCGATAATGG	
	TGACGCCGTTGCTGAAGGTAATAA	
STTR4	FAM-GGCGCCGCTATGGGTGGTGA	
	CAAAAGCGGTAGCGATGAGC	
STTR5	FAM-ATGGCGAGGCGAGCAGCAGT	
	GGTCAGGCCGAATAGCAGGAT	
STTR6	FAM-TCGGGCATGCGTTGAAA	
	CTGGTGGGGAGAATGACTGG	
STTR7	FAM-CGCGCAGCCGTTCTCACT	
	TGTTCCAGCGCAAAGGTATCTA	
STTR8	FAM-TTATGTCACCACGGCTGTCAAT	
	AAGGCCAAATAGGGGTTCATAAGG	

TABLE 1.1   *(Continued)*

VNTR locus name	Primers (5′-3′)	Reference
Salmonella Enteritidis		
SE1	AGACGTGGCAAGGAACAGTAG	[130]
	CCAGCCATCCATACCAAGAC	
SE2	CTTCGGATTATACCTGGATTG	
	TGGACGGAGGCGATAG	
SE3	CAACAAAACAACAGCAGCAT	
	GGGAAACGGTAATCAGAAAGT	
SE4	ACTTTAGAAAATGCGTTGAC	
	AAGTCAACTGCTCTACCAAC	
SE5	CGGGAAACCACCATCAC	
	CAGGCCGAATAGCAGGAT	
SE6	CCCCTAAGCCCGATAATG	
	GCCGTTGCTGAAGGT	
SE7	GATAATGCTGCCGTTGGTAA	
	ACTGCGTTTGGTTTCTTTTCT	
SE8	TTGCCGCATAGCAGCAGAAGT	
	GCCTGAACACGCTTTTTAATAGGCT	
SE9	CGTAGCCAATCAGATTCATCCCGCG	
	TTTGAAACGGGGTGTGGCGCTG	
SE10	GCTGAGATCGCCAAGCAGATCGTCG	
	ACTGGCGCAACAGCAGCAGCAACAG	
Escherichia coli		
Vhec1	FAM-AGCCCGCAGTTGATACCTACG	[304]
	GATGCCGGATGAAAATGATAAGTT	
Vhec2	FAM-AACCGTTATGAAAGAAAGTCCT	
	TCGCCCAGTAAGTATGAAATC	
Vhec3	FAM-GGCCGGATGGAAACTATGCTATT	
	GCCGCTATTTTTAACCACTGACTA	
Vhec4	FAM-ATCGCCTTCTTCCTCCGTAATG	
	CTCCTCGCGCTCAGACAGTG	
Vhec5	FAM-CTCAGGCGCCGTTAAGGTGTAGC	
	TCCGCGGAGTGCAGAGAAAATAAA	
Vhec6	FAM-ACGTTAAACCCGGAATGGAAAATC	
	GAGCGAAAAATGTCTATCTTGAGG	
Vhec7	FAM-ATGCGCGGTTAGCTACACGACA	
	TGAAAGCCCACACCATGCGATAAT	

VNTR: variable number tandem repeats

As many as 21 loci have been used for subtyping of *L. monocytogenes*. Murphy et al.[120] developed the first MLVA protocol for *L. monocytogenes*. Six tandem repeats (TRs) were used for the discrimination between related and unrelated isolates that were comparable to the one obtained by PFGE. More protocols were reported in 2008. More accurately, Miya et al.[121] developed and applied an MLVA protocol to 60 serotype 4b isolates from a variety of sources and compared the results with the respective obtained from *Apa*I PFGE, *Eco*RI ribotyping and an MLST scheme. The MLVA protocol provided with higher discriminatory power than the other methods, highlighting its efficiency. Coupling with high-resolution capillary gel electrophoresis was performed by Volpe Sperry et al.[122] and Lindstedt et al.[123] In both studies, the MLVA protocol that was developed was capable to accurately differentiate sets of strains isolated from different sources. Since then, many protocols based on six to eleven TRs were developed, combined and effectively applied taking advantage of the reduced time and cost of this approach. In most of the cases, the results obtained were comparable to the ones of PFGE with two enzymes.[124–128]

MLVA is a serotype-specific analysis; therefore, there are no generic schemes for *Salmonella* sp. and *E. coli*. In the first case MLVA protocols have been developed for a wide range of serotypes including Typhimurium, Enteritidis, Typhi, Paratyphi A, Newport, Gallinarum, Heidelberg and 4,5,12, i:-.[129–136] Development of a protocol essentially involves the screening of potential variable number tandem repeats (VNTRs) and the discriminatory power that confer to the analysis. Then, the most suitable ones are selected according to the concordance with the epidemiological data. *S. enterica* ser. Typhi and *S. enterica* ser. Paratyphi A may serve as an example of the discriminatory power of MLVA. These Serovars are highly monomorphic[137] resulting in the inability of PFGE to provide with adequate discrimination. Eventually, as many as 20 suitable loci for the analysis have been reported, along with their variability,[131,133,138–140] revealing the ones that are capable for effective epidemiological assessment.

Development of MLVA protocols for *S. enterica* ser. Typhimurium and *S. enterica* ser. Enteritidis has been the epicenter of intensive study due to their significant contribution in human foodborne salmonellosis[141] as well as the inability of PFGE to provide with adequate discrimination. In both cases, a number of loci have been examined for TRs. More accurately, 28 loci have been examined for *S. enterica* ser. Typhimurium and 16 for *S. enterica* ser. Enteritidis.[129,142–150] In the first case, the 5-loci scheme proposed by Lindstedt et al.[142] seem to be the most commonly applied in Europe.[151] In the case of *S. enterica* ser. Enteritidis surveillance, the protocols currently employed are improvements of the one developed by Cho et al.[143] In both cases, the discriminatory power and concomitantly the epidemiological surveillance have been improved, compared to PFGE.

MLVA protocols have been developed and effectively applied in diversity and epidemiology studies of O157 as well as non-O157 *E. coli*.[152–162] Although the effectiveness of this approach in epidemiological assessment has been exhibited, the complementarity with PFGE and MLST and the accuracy obtained by their combination has been highlighted.[163–171]

1.3.3 MULTILOCUS SEQUENCE TYPING (MLST)

With MLST epidemiological relationships are inferred through the comparison of the nucleotide sequences of a number of loci. The criteria used for the selection of loci are similar to MLVA. Multilocus Sequence Typing schemes have been proposed for many foodborne pathogens and have become an established approach for epidemiological surveillance.[172,173] Examples of established MLST schemes used for the epidemiological assessment of *L. monocytogenes*, *Salmonella* Serovars and *E. coli* are given in Table 1.2.

TABLE 1.2 Examples of Established MLST Schemes Used for the Epidemiological Assessment of *L. monocytogenes*, *Salmonella* Serovars and *E. coli*

Gene	Primers (5′-3′)	Reference
L. monocytogenes		
absZ	TCGCTGCTGCCACTTTTATCCA	[174]
	TCAAGGTCGCCGTTTAGAG	
bglA	GCCGACTTTTTATGGGGTGGAG	
	CGATTAAATACGGTGCGGACATA	
cat	ATTGGCGCATTTTGATAGAGA	
	AGATTGACGATTCCTGCTTTTG	
dapE	CGACTAATGGGCATGAAGAACAAG	
	ATCGAACTATGGGCATTTTTACC	
dat	GAAAGAGAAGATGCCACAGTTGA	
	TGCGTCCATAATACACCATCTTT	
ldh	ATTTTGATCGTATTGGGGTTTT	
	TACTGAATGGATTAGCGAAGATGA	
lhkA	AGAATGCCAACGACGAAACC	
	TGGGAAACATCAGCAATAAAC	
pgm	CCGATGATCAGGAAGAAGAAAT	
	CTGTCAAAATCGCCATCAAA	
sod	GCGGTTGCTGGTCATCCT	
	GCGTTTGTTAGCTTCATCCCAGTT	

TABLE 1.2 *(Continued)*

Gene	Primers (5′-3′)	Reference
Salmonella Serovars		
aroC	GGACTACAGCGCGATTAA	[196]
	ATCATTTCGACTTCTTCACC	
dnaN	TTACCGTTGAACGTGAACAT	
	CGCGGATATTATTACTGC	
hemD	ATGAGTATTCTGATCACCCG	
	ATCAGCGACCTTAATATCTTGCCA	
hisD	GTCTCGTCTGTCGGTCTGTA	
	TGCGCTGGTAATCGCATC	
purE	TTTTAAGACGCATGTCTTCC	
	ATCTCTGCGGTAATGACG	
sucA	CGAAGAGAAACGCTGGAT	
	AGTGGGTTAGAGGTGGTGA	
thrA	GGCCATTACCTTGAATCTAC	
	ATAGAACTCATCCTGCATCG	
gluA (glnA)	AGGATATCCGTTCTGAAATG	
	CTACCTGTGGGATCTCTTT	
manB	ATCAGTGTTAACAATCTCACC	
	CGCCAGTCTGCTGTTGAT	
panB	TACAGCTTCGCTAAGTTATTT	
	AGCCATATACTGCTGTACG	
fimA	GTGTGTAATTCAAGGGAAAT	
	TTCCTTATCATCTGCTATGTT	
E. coli		
aspC	GTTTCGTGCCGATGAACGTC	[200]
	AAACCCTGGTAAGCGAAGTC	
clpX	CTGGCGGTCGCGGTATACAA	
	GACAACCGGCAGACGACCAA	
fadD	GCTGCCGCTGTATCACATTT	
	GCGCAGGAATCCTTCTTCAT	
icdA	CTGCGCCAGGAACTGGATCT	
	ACCGTGGGTGGCTTCAAACA	
lysP	CTTACGCCGTGAATTAAAGG	
	GGTTCCCTGGAAAGAGAAGC	
mdh	GTCGATCTGAGCCATATCCCTAC	
	TACTGACCGTCGCCTTCAAC	

Regarding *L. monocytogenes*, the first attempts to set up an MLST scheme were performed by Salcedo et al.[174] and Revazishvili et al.[175] In the first study *abcZ*, *bglA*, *cat*, *dapE*, *dat*, *ldh* and *lhkA* were utilized resulting in discrimination comparable to PFGE. In the latter study, *actA*, *betL*, *hlyA*, *gyrB*, *pgm* and *recA* were selected resulting in improved discrimination than PFGE. An improved version[176] of the protocol proposed by Salcedo et al.[174] provided with improved epidemiological concordance and therefore was widely accepted and used.[177–181] A variation of MLST, namely multi-virulence-locus sequence typing (MVLST) was proposed by Zhang et al.[182] In that variation three virulence (*prfA*, *inlB* and *inlC*) and three virulence-associated genes (*dal, lisR* and *clpP*) were used to effectively differentiate a set indistinguishable by *Eco*RI-RT and PFGE isolates. Since then, it was applied in many studies.[121,180,183–187]

The importance of the selection of proper loci has been highlighted in the case of *Salmonella* sp. Several studies have been performed that reported the inability of the developed protocols to provide with higher discrimination that PFGE[188–192] mainly due to the improper loci selection. Several alternatives to the classic MLST schemes have been proposed. Among them, the ones proposed by Tankouo-Sandjong et al.,[193] Liu et al.,[194,195] and Singh et al.,[196] are the most interesting. In the first study, the new protocol was designated as MLST-v, for multilocus sequence typing, and was based on two housekeeping genes, namely *gyrD* and *atpD*, and the two flagellin genes *fliC* and *fliB*. This study was important from an evolutionary perspective as well since a recombination that was detected in seven out of 22 serotypes examined in the *fliC* gene indicated the multiple phylogenetic origins of these Serovars that could not be achieved by classical serotyping. Liu et al.[194,195] utilized two virulence genes, namely *fimH* and *sseL* and two CRISPR loci. This approach was efficiently utilized for subtyping of *S. enterica ser.* Typhimurium and *S. enterica* ser. Heidelberg,[197] *S. enterica* ser. Newport[198] and *S. enterica* ser. Enteritidis[199] but was not successful in the case of *S. enterica* ser. Muenchen. The latter could be assigned to an increased evolving speed of the selected CRISPRs of *S. enterica* ser. Muenchen, and thus their reduced applicability for outbreak investigation. Finally, Singh et al.[196] amplified 20 housekeeping and virulence genes using a new hairpin-primed multiplex amplification (HPMA) protocol and designated this approach as MLST-seq. However, as next-generation sequencing (NGS) becomes more available, MLST-seq may become more practical and economically obtainable for strain subtyping.

As far as *E. coli* is concerned, three standard MLST schemes are currently available. The scheme developed by Reid et al.[200] that utilizes the sequences of *aspC, clpX, fadD, icd, lysP, mdh* and *uidA* is currently hosted in http://www.shigatox.net/ecmlst/cgi-bin/index. The one developed by Wirth et al.[201] that uses *adk, icd, fumC, recA, mdh, gyrB* and *purA* is accessible at http://mlst.warwick.ac.uk/mlst/dbs/Ecoli, while the one proposed by Jaureguy et al.[202] polymorphisms in eight house-keeping genes, namely *dinB*, *icdA*, *pabB*, *ploB*, *putP*, *trpA*, *trpB* and *uid* were studied, this scheme is available at http://www.pasteur.fr/recherche/genopole/PF8/mlst/EColi.html. Finally, Weissman et al.[203] proposed a cost-effective scheme suitable for clinical applications. More accurately, the use of only two loci, namely 489-nucleotide internal fragment of *fimH* and a 469-nucleotide internal *fumC* fragment could predict the outcome of standard MLST.

1.4 STUDIES ON VIRULENCE-ASSOCIATED GENOMIC DETERMINANTS

In the last two decades, an increase of the studies assessing genomic determinants of foodborne pathogens was evident. These determinants were either associated with their food subsistence and the mechanisms of stress response and adaptation or with their virulence potential. The initial trend was to detect the genes that were necessary for virulence, in an attempt to assess the pathogenic potential of an isolate. By addressing the virulence potential of a group of isolates, an improvement of the accuracy of risk assessment studies is possible. However, the pathogenic potential may accurately be depicted only through in vivo experiments, despite the several indicators that have been proposed. At the same time, several studies addressing gene expression in actual food matrices took place increasing our level of understanding regarding the physiology of foodborne pathogens.

1.4.1 ESCHERICHIA COLI

The virulence factors of the *E. coli* pathotypes have been studied. These include attaching and effacing (e.g. intimin) that are characteristics of diffusely adherent *E. coli* (DAEC), enteropathogenic *E. coli* (EPEC) and enterohaemorrhagic *E. coli* (EHEC),[204] aggregative adherence fimbriae in enteroaggregative *E. coli* (EAEC),[205] afimbrial adhesins in DAEC,[206] haemolysins in EAEC, EPEC and EHEC[207–209] as well as a wide range of toxins.[210–215]

Six genes, namely *stx1*, *stx2*, *eae*, *hlyA*, *rfbE*, and *fliC* are usually detected through PCR for confirmation purposes of O157 serotype.

[216] Since these genes are significant from a virulence perspective, their presence is very frequently assessed. Their prevalence has been extensively studied and significant variation has been reported. Ateba and Bezuidenhout[217] studied 30 human, cattle and pig isolates and reported the absence of the shiga toxin genes (*stx1* and *stx2*) and a prevalence of 60% and 23.3% of *hlyA* and *eae*, respectively. Moreover, only the 16.7% of the isolates contained both genes. Fifteen O157 strains, isolated from retaining beef products were assessed by Sallam et al.[218] *stx1* gene was detected in 7, *stx2* in 13, *eae* in all and *hlyA* in 14 isolates. All four genes were detected in 7 isolates, while 6 of them were positive for *stx2*, *eae* and *hlyA*. A total of 70 isolates from sheep flock were characterized by Martins et al.[219] It was reported that most of the isolates harbored only *stx1*, with only a few carrying *stx2*. Among the latter, the *stx2b* subtype was the most common. *ehxA* and *eae* were detected in 87.1% and 5.7% of the isolates, respectively. Additionally, genes encoding putative adhesins (*saa*, *iha*, *lpfO113*) and toxins (*subAB* and *cdtV*) were also observed. Jiang et al.[220] studied 115 strains isolated from small and very small beef processing plants and reported that only 13 of them (11.3%) contained *stx1*, *stx2* or *eae*, with none of them harboring all three and only 3 containing 2 of them, namely *stx1* and *eae*. Douellou et al.[221] studied 197 strains, recovered from dairy and meat products, humans and the environment. Dairy and human STEC exhibited similar virulence gene profiles. Regarding O26:H11 isolates, *stx1* was more prevalent among dairy than human isolates (87% vs. 44%) while *stx2* was more prevalent among human isolates (23% vs. 81%). As far as O157:H7 isolates were concerned, *stx1* (0% vs. 39%), *nleF* (40% vs 94%) and Z6065 (40% vs 100%) were more prevalent among human than dairy isolates. In another study, the virulence gene content of 70 strains isolated from chicken farms was studied by Al-Arfaj et al.[222] *uidA* and *cvaC* were detected in all isolates while *iutA* and *iss* in 81.43% and 64.29%, respectively. Ranjbar et al.[223] investigated the virulence gene content of 100 *E. coli* strains isolated from different water sources. The results of the PCR exhibited that 97% of the isolates harbored *ipaH* gene, while *estA*, *eaeA* and *bfpA* genes were detected in 37%, 31% and 3% of the isolates, respectively. A total of 27 isolates obtained from a screening of hen egg shells were studied by Burgos et al.[224] Seven isolates carried both *eae* and *bfpA*, while 14 isolates only carried *eae*. Shiga toxin genes *stx1* and *stx2* were detected in four isolates. Heat stable and heat labile enterotoxin genes as well as *aggR* were also detected. Moreover, efflux pump genes

detected included *acrB* (96.29%), *mdfA* (85.18%) and *oxqA* (37.03%), in addition to quaternary ammonium compound (QAC) resistance genes *qacA/B* (11.11%) and *qacE* (7.40%). Finally, Khan et al.[225] isolated a total of 285 strain along the production and supply chain of pork around Hubei, China, and assessed the occurrence of 13 virulence-related genes, namely *kpsMII, papA, papC, iutA, sfaS, focG, afa, hlyD, fimH, cnf, vat, fyuA* and *ireA*. The most frequent ones among the isolates were *kpsMII, sfaS* and that is, with 69.8%, 69.5% and 68.7%, respectively while the least prevalent were *cnf, hlyD* and *afa* with 24.9%, 22.8% and 12.3%, respectively.

The stress response and resistance strategies of *E. coli* to stresses faced in the food industry, such as acidity, reduced water activity, non-lethal heat shock, oxidative stress, etc. have been extensively studied and reviewed by McClure.[226] However, studies assessing the expression of virulence-associated genes during growth of *E. coli* on an actual food matrix are rather scarce. Carey et al.[227] studied the transcription of *stx1A, stx2A, eaeA, fliC, rpoS* and *sodB* during growth of *E. coli* O157:H7 on Romaine lettuce stored at 4°C and 15°C over a 9-day period. The occasional upregulation that was observed in all cases did not allow the detection of any particular trend. The effect of *Lippia graveolens* and *Haematoxylon brassiletto* extracts as well as carvacrol, brazilin, citral and rifaximin on virulence gene expression in enterohemorrhagic *E. coli* O157:H7, enteroaggregative *E. coli* strain 042 and a strain of O104:H4 serotype was studied by Garcia-Heredia et al.[228] No effect on *pic, rpoS* and *stx2* transcription level was observed; on the contrary upregulation of *aggR* of the *E. coli* strain 042 as a result of rifaximin and *Haematoxylon brassiletto* extracts addition was evident.

1.4.2 LISTERIA MONOCYTOGENES

The Listeria Pathogenicity Island 1 (LIPI-1) is a 9-kb gene cluster in which a series of genes, namely *prfA, plcA, hly, mpl, actA* and *plcB* are physically gathered. The transcription of these genes is necessary for the effective intracellular parasitism of this foodborne pathogen. The physical and transcriptional organization of LIPI-1 has been recently reviewed by Paramithiotis et al.[229] In addition, the transcription of another series of genes encoding for the invasion factors referred to as internalins is equally important. Until now as many as 25 internalins have been described; InlA (encoded by *inlA*) and InlB (encoded by *inlB*) are generally recognized as the most important for internalization of the pathogen into the epithelial cells of the host, regarding the former and hepatocytes, fibroblast and epithelioid cells regarding the latter. Other internalins such as InlC (encoded by *inlC*) and InlJ (encoded by *inlJ*) are considered to contribute during the later stages of infection.[230,231] The structure, variability, transcriptional organization and function of internalins have been recently reviewed by Bierne et al.[232]

The above-mentioned genes were the preferred targets in studies assessing in vitro the pathogenic potential of isolates. *inlA* and *inlB* have been detected in all examined strains that were isolated from human cases, food and environment.[233–237] In addition, *inlA* is probably the most studied internalin with at least 18 distinct mutations been described so far.[234,238–240] *inlJ* and *inlC* were also present in all isolates studied by Liu et al.,[233] Lomonaco et al.,[241] Soni et al.,[237] Jamali et al.[235] and Moreno et al.[236] with only one exception, in which *inlC* was not detected. Presence of the genes clustered in LIPI-1 has also been extensively evaluated in *L. monocytogenes* isolated from a variety of sources. Kaur et al.[242] studied four strains isolated from human patients with spontaneous abortions. Two isolates contained all studied genes, that is *prfA*, *plcA*, *hlyA* and *actA*, and two were lacking *plcA* gene as well as the respective PI-PLC phenotype. The in vivo tests conducted verified that the former strains were pathogenic while the latter were not. Fifty-two isolates obtained from Gorgonzola cheese were studied by Lomonaco et al.[241] The virulence-associated genes assessed were *actA*, *inlC*, *inlJ*, *plcA*, *prfA*, *hlyA* and *iap* and their presence was verified in all isolates but one that lacked *inlC*. Soni et al.[237] isolated 30 strains from vegetables and detected *inlA*, *inlC*, *inlJ*, *plcA*, *prfA*, *actA*, *hlyA* and *iap* in all isolates gene except one that lacked the *plcA* gene. Finally, Moreno et al.[236] studied 29 isolates from pork slaughterhouses, markets and human infection samples and reported that *inlA*, *inlB*, *inlC*, *inlJ*, *hly*, *prfA*, *plcA*, and *plcB* genes were detected in all of them.

The in vitro transcriptomic responses of several *L. monocytogenes* strains to a variety of environmental stimuli including extracellular pH value,[243] heat,[244–248] cold,[249–252] acid,[253–255] osmotic shock,[256–261] carbohydrates,[262–264] high hydrostatic pressure processing,[265] various disinfectants,[266] essential oils[267] and the presence of antimicrobials[268–277] have been extensively studied and reviewed.[278]

On the contrary, a rather limited amount of studies assessing the expression of virulence-associated genes during growth on an actual food matrix have been performed. More accurately, the expression of key virulence genes has been assessed during growth on salmon, fermented sausages, soft cheese, UHT milk, minced meat, standard liver pate, rocket and melon.[279–284] In these studies, the effect of strain variability, growth substrate, storage temperature and time on the transcription of genes including *prfA*, *sigB*, *hly*, *actA*, *inlA*, *inlB*, *inlC*, *inlJ*, *plcA* and *plcB* was assessed. In most of the cases, a mixed response was observed and no particular trend was identified indicating the presence of additional regulatory mechanisms that are yet to be explored, a conclusion that was also drawn by the studies of Williams et al.[285] and Kazmierczak et al.[286]

1.4.3 SALMONELLA SEROVARS

The genomic determinants associated with salmonellae virulence and their chromosomal organization have been extensively studied. Many virulence-associated genes are physically clustered in genomic islands, known as salmonella pathogenicity islands (SPIs). As many as 22 SPIs have been so far described. Some of them seem to be serovar specific, such as the SPI-7 of *S. enterica ser.* Typhi, whereas others are characterized by a wider distribution, such as the SPIs-1 to 6.[287]

Salmonella infection involves invasion of intestinal cells, formation of the *Salmonella*-containing vacuole (SCV), crossing of the intestinal mucosa and entering the bloodstream and the lymphatic system. These may occur through several mechanisms according to *Salmonella* serotype, the host and the type of cell.[288–290] The best-characterized system and, therefore the most important virulence determinant in *Salmonella* is the type three secretion system (T3SS). In addition, the more recently described type six secretion system (T6SS) seem to play an important role in various processes including virulence. Both secretion systems are very widespread among Gram-negative bacteria and ensure translocation of proteins, referred to as effectors, from the bacterial cytoplasm into the host cell cytoplasm; a process that is required for survival, colonization and spread within the host. T3SS may be divided into T3SS1 and T3SS2 that are physically assembled in SPI-1 and SPI-2, respectively, and are sequentially activated during infection. T3SS1 is activated upon contact with host epithelial cells and directs invasion and formation of the SCV.

Then, T3SS2 is activated as a response to the intracellular environment and directs replication of the pathogen. A detailed description of T3SS has been recently provided by Ramos-Morales.[291] Regarding T6SS, five phylogenetically distinct T6SS have been described so far, located in SPI-6, SPI-19, SPI-20, SPI-21 and SPI-22.[292,293] They seem to be differentially distributed among serotypes and their relative contribution to pathogenesis is currently under intensive study.[292,294–296]

Numerous studies are available regarding the antimicrobial resistance and virulence gene content of *Salmonella* isolates. The former is mostly studied through the agar diffusion assay but may as well be assessed through the detection of the respective genes. As an example, the very recent studies by Kuang et al.,[297] Oueslati et al.,[32] Zadernowska and Chajecka-Wierzchowska[298] and Li et al.[299] are mentioned. In the first study, *avrA*, *ssaQ*, *mgtC*, *siiD*, *sopB* and *bcfC* were detected in the majority of the 79 S. Newport isolates assessed. Oueslati et al.[32] isolated 17 strains from beef samples, assigned them to the serotypes *S. enterica* ser. Montevideo, *S. enterica* ser. Anatum, *S. enterica* ser. Minnesota, *S. enterica* ser. Amsterdam, *S. enterica* ser. Kentucky and *S. enterica* ser. Brandenburg and the reported presence of *invA* and absence of *spvC* in all isolates. Only one isolate (*S. enterica* ser. Kentucky) harbored the *hli* virulence gene. Similarly, Zadernowska and Chajecka-Wierzchowska[298] reported the presence of *invA* in all strains isolated from food samples, many of which belong to *S. enterica* ser. Typhimurium and *S. enterica* ser. Enteritidis, and *spvC* in the 18.7% of the isolates. Finally, Li et al. [299] studied a total of 138 *S. enterica* ser. Typhimurium isolates recovered from retail raw chickens and reported that among the 30 virulence genes examined, the top five genes were *pagK* (82.6%), *sodC1* (66.7%), *siiE* (58.7%), *pefA* (58.7%) and *marT* (50.7%). Collectively, such studies highlight the conclusion drawn by Suez et al.[300] that a common core of virulence genes currently exists and is accompanied by a remarkable degree of genetic and phenotypic diversity.

The response strategies to stresses frequently encountered during food manufacturing and storage, such as acid, desiccation, osmotic, oxidative, starvation and thermal stresses have been extensively studied and reviewed in detail by Spector and Kenyon;[301] on the contrary, the studies addressing gene expression in actual food matrices are very few. In the more recent ones, Jakociune et al.[302] studied *S. enterica* ser. Enteritidis gene expression during growth in the whole egg at 20°C and reported that twenty-six genes were significantly upregulated and fifteen were significantly downregulated. The former belonged to amino acid biosynthesis, di/oligo-peptide transport system, biotin synthesis, ferrous iron transport system and type III secretion system whereas the latter were related to formate hydro-genlyase (FHL) and trehalose metabolism. There results suggested that during growth in whole, for example, *S. enterica ser.* Enteritidis starved for amino-acids, biotin and iron. Doulgeraki et al.[303] studied *S. enterica ser.* Typhimurium gene expression during biofilm formation on rocket leaves at 20°C and reported downregulation of *osmY* and *dps*, upregulation of *csgB* and *sspH2* and no effect on the transcription level of *dppA*, *rpoH* and *sdiA*.

1.5   CONCLUSION

Advances in the field of molecular biology have provided with a series of tools and approaches for effective detection and characterization of foodborne pathogens. After years of application, the limitations of each approach have been revealed and should always be considered. Moreover, studies addressing their physiological responses indicate the existence of additional regulatory mechanisms yet to be explored. This is anticipated with excitement in the years to come.

KEYWORDS

- foodborne pathogens
- immuno-based detection
- DNA-based detection
- RNA-based detection
- Biosensors
- epidemiological studies
- PFGE
- MLVA
- MLST
- virulence
- *Escherichia coli*
- *Listeria monocytogenes*
- *Salmonella* Serovars

REFERENCES

1. Scallan, E.; Hoekstra, R. M.; Angulo, F. J.; Tauxe, R. V.; Widdowson, M-A.; Roy, S. L.; Jones, J. L.; Griffin, P. M. Foodborne Illness Acquired in the United States— Major Pathogens. *Emerging Infect. Dis.* **2011,** *17,* 7–15.

2. Centers of Disease Control and Prevention (CDC). Foodborne Diseases Active Surveillance Network FoodNet 2014 Surveillance Report. https://www.cdc.gov/foodnet/reports/annual-reports-2014.html. (Accessed September 2016.)

3. European Food Safety Authority (EFSA). The European Union Summary Report on Trends and Sources of Zoonoses, Zoonotic agents and food-borne outbreaks in 2014. *EFSA J.* 2015, *13,* 4329.

4. Venkateswaran, K.; Kamikoh, Y.; Ohashi, E.; Nakanishi, H. A Simple Filtration Technique to Detect Enterohemorrhagic *Escherichia coli* O157:H7 and Its Toxins in Beef by Multiplex PCR. *Appl. Environ. Microbiol.* **1997,** *63,* 4127–4131.

5. Samkutty, P. J.; Gough, R. H.; Adkinson, R. W.; McGrew, P. Rapid Assessment of the Bacteriological Quality of Raw Milk Using ATP Bioluminescence. *J. Food Prot.* **2001,** *64,* 208–212.

6. Lindqvist, R. Detection of *Shigella spp.* in Food with a Nested PCR Method—Sensitivity and Performance Compared with a Conventional Culture Method. *J. Appl. Microbiol.* **1999,** *86,* 971–978.

7. Lambertz, S. T.; Lindqvist, R.; Ballagi-Pordany, A.; Danielsson-Tham, M. L. A Combined Culture and PCR Method for Detecting Pathogenic *Yersinia enterocolitica* in Food. *Int. J. Food Microbiol.* **2000,** *57,* 63–73.

8. Lantz, P. G.; Knutsson, R.; Blixt, Y.; Al-Soud, W. A.; Borch, E.; Radstrom, P. Detection of Pathogenic *Yersinia enterocolitica* in Enrichment Media and Pork by a Multiplex PCR: A Study of Sample Preparation and PCR-Inhibitory Components. *Int. J. Food Microbiol.* **1998,** *45,* 93–105.

9. Pedersen, L. H.; Skouboe, P.; Rossen, L.; Rasmussen, O. F. Separation of *Listeria monocytogenes* and *Salmonella* Berta from a Complex Food Matrix by Aqueous Polymer Two-Phase Partitioning. *Lett. Appl. Microbiol.* **1998,** *26,* 47–50.

10. Li, X.; Boudjellab, N.; Zhoa, X. Combined PCR and Slot Blot Assay for Detection of *Salmonella* and *Listeria monocytogenes. Int. J. Food Microbiol.* **2000,** *56,* 167–177

11. Tu, S.; Patterson, D.; Uknalis, J.; Irwin, P. Detection of *Escherichia coli* O157:H7 Using Immunomagnetic Capture and Luciferin-Luciferase ATP Measurement. *Food Res. Int.* **2000,** *33,* 375–380.

12. Hudson, J. A.; Lake, R. J; Savill, M. G.; Scholes, P.; McCormick, R. E. Rapid Detection of *Listeria monocytogenes* in Ham Samples Using Immunomagnetic Separation Followed by Polymerase Chain Reaction. *J. Appl. Microbiol.* **2001,** *90,* 614–621

13. Bennett, A. R.; Davids, F. G. C.; Valhodimou, S.; Banls, J. G.; Betts, R. P. The Use of Bacteriophage-Based Systems for the Separation and Concentration of *Salmonella. J. Appl. Microbiol.* **1997,** *83,* 259–265.

14. Betts, W. B. The Potential of Dielectrophoresis for the Real-Time Detection of Microorganisms in Foods. *Trends Food Sci. Technol.* **1995,** *6,* 51–58.

15. Markx, G. H.; Dyda, P. A.; Pethig, R. Dielectrophoretic Separation of Bacteria Using Conductivity Gradient. *J. Biotechnol.* **1996,** *51,* 175–180.

16. Bolton, F. J.; Fritz, E.; Poynton, S. Rapidenzyme-Linkedimmunoassay for the Detection of *Salmonella* in Food and Feed Products: Performance Testing Program. *J. AOAC Int.* **2000,** *83,* 299–304.

17. Aschfalk, A.; Mülller, W. Clostridium Perfringens Toxin Types from Wild Caught Atlantic Cod (*Gadus morhua* L.) Determined by PCR and ELISA. *Can. J. Microbiol.* **2002,** *48,* 365–368.

18. Kumar, B. K.; Raghunath, P.; Devegowda, D.; Deekshit, V. K.; Venugopal, M. N.; Karunasagar, I.; Karunasagar, I. Development of Monoclonal Antibody Based Sandwich ELISA for the Rapid Detection of Pathogenic *Vibrio parahaemolyticus* in Seafood. *Int. J. Food Microbiol.* **2011,** *145,* 244–249.

19. Cavaiuolo, M.; Paramithiotis, S.; Drosinos, E. H.; Ferrante, A. Development and Optimization of an ELISA Based Method to Detect *Listeria monocytogenes* and *Escherichia coli* O157 in Fresh Vegetables. *Anal. Methods* **2013,** *5,* 4622–4627.

20. Shen, Z.; Hou, N.; Jin, M.; Qiu, Z.; Wang, J.; Zhang, B.; Wang, X.; Wang, J.; Zhou, D.; Li, J. A Novel Enzyme-Linked Immunosorbent Assay for Detection of *Escherichia coli* O157:H7 Using Immunomagnetic and Beacon Gold Nanoparticles. *Gut Pathog.* **2014,** *6,* 14.

21. Zhao, X.; Lin, C-W.; Wang, J.; Oh, D. H. Advances in Rapid Detection Methods for Foodborne Pathogens. *J. Microbiol. Biotechnol.* **2014,** *24,* 297–312.

22. Jung, B. Y.; Jung, S. C., Kweon, C. H. Development of a Rapid Immunochromatographic Strip for the Detection of *Escherichia coli* O157. *J. Food Protect* **2005,** *68,* 2140–2143.

23. Kim, S. H.; Kim, J. Y.; Han, W.; Jung, B. Y.; Chuong, P. D.; Joo, H. Development and Evaluation of an Immunochromatographic Assay for *Listeria* Spp. in Pork and Milk. *Food Sci. Biotechnol.* **2007,** *16,* 515–519.

24. Shukla, S.; Leem, H.; Kim, M. Development of a Liposome-Based Immunochromatographic Strip Assay for the Detection of *Salmonella. Anal. Bioanal. Chem.* **2011,** *401,* 2581–2590.

25. Shukla, S.; Leem, H.; Lee, J. S.; Kim, M. Immunochromatographic Strip Assay for the Rapid and Sensitive Detection of *Salmonella* Typhimurium in Artificially Contaminated Tomato Samples. *Can. J. Microbiol.* **2014,** *60,* 399–406.

26. Xu, D.; Wu, X.; Li, B.; Li, P.; Ming, X.; Chen, T.; Wei, H.; Xu, F. Rapid Detection of *Campylobacter jejuni* Using Fluorescent Microspheres as Label for Immune Chromatographic Strip Test. *Food Sci. Biotechnol.* **2013,** *22,* 585–591.

27. Niu, K.; Zheng, X.; Huang, C.; Xu, K.; Zhi, Y.; Shen, H.; Jia, N. A Colloidal Gold Nanoparticle-Based Immunochromatographic Test Strip for Rapid and Convenient Detection of *Staphylococcus aureus. J. Nanosci. Nanotechnol.* **2014,** *14,* 5151–5156.

28. Leem, H.; Shukla, S.; Song, X.; Heu, S.; Kim, M. An Efficient Liposome-Based Immunochromatographic Strip Assay for the Sensitive Detection of *Salmonella* Typhimurium in Pure Culture. *J. Food Saf.* **2014,** *34,* 239–248.

29. Moon, H. W.; Kim, H. N.; Hur, M.; Shim, H. S.; Kim, H.; Yun, Y. M. Comparison of Diagnostic Algorithms for Detecting *Toxigenic Clostridium* Difficile in Routine Practice at a Tertiary Referral Hospital in Korea. *PLoS One* **2016,** *11,* e0161139.

30. Mata, G. M.; Martins, E.; Machado, S. G.; Pinto, M. S.; de Carvalho, A. F.; Vanetti, M. C. Performance of Two Alternative Methods for *Listeria* Detection Throughout Serro Minas Cheese Ripening. *Braz. J. Microbiol.* **2016,** *47,* 749–756.

31. Zeleny, R.; Nia, Y.; Schimmel, H.; Mutel, I.; Hennekinne, J. A.; Emteborg, H.; Charoud-Got, J.; Auvray, F. Certified Reference Materials for Testing of the Presence/Absence of *Staphylococcus aureus* enterotoxin A (SEA) in Cheese. *Anal. Bioanal. Chem.* **2016,** *408,* 5457–5465.

32. Oueslati, W.; Rjeibi, M. R.; Mhadhbi, M.; Jbeli, M.; Zrelli, S.; Ettriqui, A. Prevalence, Virulence and Antibiotic Susceptibility of *Salmonella* Spp. Strains, Isolated from Beef in Greater Tunis (Tunisia). *Meat Sci.* **2016,** *119,* 154–159.

33. Rodrigues, M. J.; Martins, K.; Garcia, D.; Ferreira, S. M.; Gonçalves, S. C.; Mendes, S.; Lemos, M. F. Using the Mini-VIDAS® Easy *Salmonella* Protocol to Assess Contamination in Transitional and Coastal Waters. *Arch. Microbiol.* **2016,** *198,* 483–487.

34. Jayasena, S. D. Aptamers: An Emerging Class of Molecules that Rival Antibodies in Diagnostics. *Clin. Chem.* **1999,** *45,* 1628–1650.

35. Ellington, A. D.; Szostak, J. W. In Vitro Selection of RNA Molecules that Bind Specific Ligands. *Nature* **1990,** *346,* 818–822.

36. Tuerk, C.; Gold, L. Systematic Evolution of Ligands by Exponential Enrichment: RNA Ligands to Bacteriophage T4 DNA Polymerase. *Science* **1990,** *249,* 505–510.

37. Wang, J.; Yu, J.; Yang, Q.; McDermott, J.; Scott, A.; Vukovich, M.; Lagrois, R.; Gong, Q.; Greenleaf, W.; Eisenstein, M.; Ferguson, B. S.; Soh, H. T. Multiparameter Particle Display (MPPD): A Quantitative Screening Method for the Discovery of Highly Specific Aptamers. *Angew. Chem. Int. Ed.* in press. DOI: 10.1002/anie.201608880.

38. Duan, N.; Wu, S. J.; Chen, X. J.; Huang, Y. K.; Xia, Y.; Ma, X. Y.; Wang, Z. Selection and Characterization of Aptamers Against *Salmonella* Typhimurium Using Whole-Bacterium Systemic Evolution of Ligands by Exponential Enrichment (SELEX). *J. Agric. Food Chem.* **2013,** *61,* 3229–3234.

39. Liu, G. Q.; Lian, Y. Q.; Gao, C.; Yu, X. F.; Zhu, M.; Zong, K.; Chen, X-J.; Yan, Y. In Vitro Selection of DNA Aptamers and Fluorescence-Based Recognition for Rapid Detection *Listeria monocytogenes. J. Integr. Agric.* **2014,** *13,* 1121–1129.

40. Lee, S. H.; Ahn, J. Y.; Lee, K. A.; Um, H. J.; Sekhon, S. S.; Sun Park, T.; Min, J.; Kim, Y. H. Analytical bioconjugates Aptamers, Enable Specific Quantitative Detection of *Listeria monocytogenes. Biosens. Bioelectron.* **2015,** *68,* 272–280.

41. Wang, Q. Y.; Kang, Y. J. Bioprobes Based on Aptamer and Silica Fluorescent Nanoparticles for Bacteria *Salmonella* Typhimurium Detection. *Nanoscale Res. Lett.* **2016,** *11,* 150.

42. Wang, L.; Wang, R.; Chen, F.; Jiang, T.; Wang, H.; Slavik, M.; Wei, H.; Li, Y. QCM-Based Aptamer Selection and Detection of *Salmonella* Typhimurium. *Food Chem.* **2017,** *221,* 776–782.

43. Wang, X.; Huang, Y.; Wu, S.; Duan, N.; Xu, B.; Wang, Z. Simultaneous Detection of *Staphylococcus aureus* and *Salmonella* Typhimurium Using Multicolor Time-Resolved Fluorescence Nanoparticles as Labels. *Int. J. Food Microbiol.* **2016,** *237,* 172–179.

44. Davydova, A.; Vorobjeva, M.; Pyshnyi, D.; Altman, S.; Vlassov, V.; Venyaminova, A. Aptamers Against Pathogenic Microorganisms. *Crit. Rev. Microbiol.* **2016,** *42,* 847–865.

45. Teng, J.; Yuan, F.; Ye, Y.; Zheng, L.; Yao, L.; Xue, F.; Chen, W.; Li, B. Aptamer-Based Technologies in Foodborne Pathogen Detection. *Front. Microbiol.* **2016,** *7,* 1426.

46. Chunglok, W.; Wuragil, D. K.; Oaewc, S.; Somasundrum, M.; Surareungchai, W. Immunoassay Based on Carbon Nanotubes-Enhanced ELISA for *Salmonella enterica* Serovar Typhimurium. *Biosens. Bioelectron.* **2011**, *26,* 3584–3589.

47. Cho, I-H.; Irudayaraj, J. In Situ Immuno-gold Nanoparticle Network ELISA Biosensors for Pathogen Detection. *Int. J. Food Microbiol.* **2013**, *164,* 70–75.

48. Shen, Z.; Hou, N.; Jin, M.; Qiu, Z.; Wang, J.; Zhang, B.; Wang, X.; Wang, J.; Zhou, D.; Junwen, L. (A Novel Enzyme-Linked Immunosorbent Assay for Detection of *Escherichia coli* O157:H7 Using Immunomagnetic and Beacon Gold Nanoparticles. *Gut Pathog.* **2014**, *6,* 14.

49. Nielsen, K. M.; Johnsen, P. J.; Bensasson, D.; D'Affonchio, D. Release and Persistence of Extracellular DNA in the Environment. *Environ. Biosafety Res.* **2007**, *6,* 37–53.

50. Zeng, D.; Chen, Z.; Jiang, Y.; Xue, F.; Li, B. Advances and Challenges in Viability Detection of Foodborne Pathogens. *Front. Microbiol.* **2016**, *7,* 1833.

51. Nocker, A.; Sossa-Fernandez, P.; Burr, M. D.; Camper, A. K. Use of Propidium Monoazide for Live/Dead Distinction in Microbial Ecology. *Appl. Environ. Microbiol.* **2007**, *73,* 5111–5117.

52. Zhao, F.; Liu, H.; Zhang, Z.; Xiao, L.; Sun, X.; Xie, J.; Pan, Y.; Zhao, Y. Reducing Bias in Complex Microbial Community Analysis in Shrimp Based on Propidium Monoazide Combined with PCR-DGGE. *Food Control* **2016**, *68,* 139–144.

53. Nocker, A.; Camper, A. K. Selective Removal of DNA from Dead Cells of Mixed Bacterial Communities by Use of Ethidium Monoazide. *Appl. Environ. Microbiol.* **2006**, *72,* 1997–2004.

54. Flekna, G.; Stefanic, P.; Wagner, M.; Smulders, F. J.; Mozina, S. S.; Hein, I. Insufficient Differentiation of Live and Dead *Campylobacter jejuni* and *Listeria monocytogenes* Cells by Ethidium Monoazide (EMA) Compromises EMA/Real-Time PCR. *Res. Microbiol.* **2007**, *158,* 405–412.

55. Pan, Y.; Breidt, F. Jr. Enumeration of Viable *Listeria monocytogenes* Cells by Real-Time PCR with Propidium Monoazide and Ethidium Monoazide in the Presence of Dead Cells. *Appl. Environ. Microbiol.* **2007**, *73,* 8028–8031.

56. Zhu, R. G.; Li, T. P.; Jia, Y. F.; Song, L. F. Quantitative Study of Viable *Vibrio parahaemolyticus* Cells in Raw Seafood Using Propidium Monoazide in Combination with Quantitative PCR. *J. Microbiol. Methods* **2012**, *90,* 262–266.

57. Elizaquivel, P.; Sanchez, G.; Aznar, R. Application of Propidium Monoazide Quantitative PCR for Selective Detection of Live *Escherichia coli* O157:H7 in Vegetables after Inactivation by Essential Oils. *Int. J. Food Microbiol.* **2012**, *159,* 115–121.

58. Luo, J-F.; Lin, W-T.; Guo, Y. Method to Detect Only Viable Cells in Microbial Ecology. *Appl. Microbiol. Biotechnol.* **2010**, *86,* 377–384.

59. Pacholewicz, E.; Swart, A.; Lipman, L. J.; Wagenaar, J. A.; Havelaar, A. H.; Duim, B. Propidium Monoazide does not Fully Inhibit the Detection of Dead Campylobacter on Broiler Chicken Carcasses by qPCR. *J. Microbiol. Methods* **2013**, *95,* 32–38.

60. Banihashemi, A.; Van Dyke, M. I.; Huck, P. M. Long-Amplicon Propidium Monoazide-PCR Enumeration Assay to Detect Viable Campylobacter and *Salmonella*. *J. Appl. Microbiol.* **2012**, *113,* 863–873.

61. Schnetzinger, F.; Pan, Y.; Nocker, A. Use of Propidium Monoazide and Increased Amplicon Length Reduce False-Positive Signals in Quantitative PCR for Bioburden Analysis. *Appl. Microbiol. Biotechnol.* **2013,** *97,* 2153–2162.

62. Li, B.; Chen, J. Q. Development of a Sensitive and Specific qPCR Assay in Conjunction with Propidium Monoazide for Enhanced Detection of Live *Salmonella* Spp. in Food. *BMC Microbiol.* **2013,** *13,* 273.

63. Soejima, T.; Minami, J.; Xiao, J. Z.; Abe, F. Innovative Use of Platinum Compounds to Selectively Detect Live Microorganisms by Polymerase Chain Reaction. *Biotechnol. Bioeng.* **2016,** *113,* 301–310.

64. De Long, E. F.; Wickham, G. S.; Pace, N. R. Phylogenetic Stains: Ribosomal RNA Based Probes for the Identification of Single Cells. *Science* **1989,** *243,* 1360–1363.

65. Amann, R. I.; Krumholz, L.; Stahl, D. A. Fluorescent-Oligonucleotide Probing of Whole Cells for Determinative, Phylogenetic, and Environmental Studies in Microbiology. *J. Bacteriol.* **1990,** *172,* 762–770.

66. Hu, L.; Ru, K.; Zhang, L.; Huang, Y.; Zhu, X.; Liu, H.; Zetterberg, A.; Cheng, T.; Miao, W. Fluorescence in Situ Hybridization (FISH): An Increasingly Demanded Tool for Biomarker Research and Personalized Medicine. *Biomarker. Res.* **2014,** *2,* 3.

67. Schmid, M. W.; Lehner, A.; Stephan, R.; Schleifer, K. H.; Meier, H. Development and Application of Oligonucleotide Probes for in Situ Detection of Thermotolerant Campylobacter in Chicken Faecal and Liver Samples. *Int. J. Food Microbiol.* **2005,** *105,* 245–255.

68. Kitaguchi, A.; Yamaguchi, N.; Nasu, M. Enumeration of Respiring Pseudomonas Spp. in Milk Within 6 Hours by Fluorescence in Situ Hybridization Following Formazan Reduction. *Appl. Environ. Microbiol.* **2005,** *71,* 2748–2752.

69. Laflamme, C.; Gendron, L.; Filion, N.; Ho, J.; Duchaine, C. Rapid Detection of Germinating *Bacillus cereus* Cells Using Fluorescent in Situ Hybridization. *J. Rapid Methods Autom.Microbiol.* **2009,** *17,* 80–102.

70. Fuchizawa, I.; Shimizu, S.; Ootsubo, M.; Kawai, Y.; Yamazaki, K. Specific and Rapid Quantification of Viable *Listeria monocytogenes* Using Fluorescence in Situ Hybridization in Combination with Filter Cultivation. *Microbes. Environ.* **2009,** *24,* 273–275.

71. Angelidis, A. S.; Tirodimos, I.; Bobos, M.; Kalamaki, M. S.; Papageorgiou, D. K.; Arvanitidou, M. Detection of *Helicobacter pylori* in Raw Bovine Milk by Fluorescence in Situ Hybridization (FISH). *Int. J. Food Microbiol.* **2011,** *151,* 252–256.

72. Oliveira, M.; Vieira-Pinto, M.; Martins da Costa, P.; Vilela, C. L.; Martins, C.; Bernardo, F. Occurrence of *Salmonella* Spp. in Samples from Pigs Slaughtered for Consumption: A Comparison Between ISO 6579:2002 and 23S rRNA Fluorescent in Situ Hybridization Method. *Food Res. Int.* **2012,** *45,* 984–988.

73. Almeida, C.; Cerqueira, L.; Azevedo, N. F.; Vieira, M. J. Detection of *Salmonella enterica* Serovar Enteritidis Using Real Time PCR, Immunocapture Assay, PNA FISH and Standard Culture Methods in Different Types of Food Samples. *Int. J. Food Microbiol.* **2013,** *161,* 16–22.

74. Almeida, C.; Sousa, J. M.; Rocha, R.; Cerqueira, L.; Fanning, S.; Azevedo, N. F.; Vieira, M. J. Detection of *Escherichia coli* O157 by Peptide Nucleic Acid Fluorescence

in Situ Hybridization (PNA-FISH) and Comparison to a Standard Culture Method. *Appl. Environ. Microbiol.* **2013,** *79,* 6293–6300.

75. Arraiano, C. M.; Yancey, S. D.; Kushner, S. R. Stabilization of Discrete mRNA Breakdown Products in *ams,* *pnp* and *rnb* Multiple Mutants of *Escherichia coli* K-12. *J. Bacteriol.* **1998,** *170,* 4625–4633.

76. Livezey, K.; Kaplan, S.; Wisniewski, M.; Becker, M. M. A New Generation of Food-Borne Pathogen Detection Based on Ribosomal RNA. *Annu. Rev. Food Sci. Technol.* *2013, 4,* 313–325.

77. Dolci, P.; De Filippis, F.; La Storia, A.; Ercolini, D.; Cocolin, L. rRNA-Based Monitoring of the Microbiota Involved in Fontina PDO Cheese Production in Relation to Different Stages of Cow Lactation. *Int. J. Food Microbiol.* **2014,** *185,* 127–135.

78. Dolci, P.; Zenato, S.; Pramotton, R.; Barmaz, A.; Alessandria, V.; Rantsiou, K.; Cocolin, L. Cheese Surface Microbiota Complexity: RT-PCR-DGGE, a Tool for a Detailed Picture? *Int. J. Food Microbiol.* **2013,** *162,* 8–12.

79. Greppi, A.; Ferrocino, I.; La Storia, A.; Rantsiou. K.; Ercolini, D.; Cocolin, L. Monitoring of the Microbiota of Fermented Sausages by Culture Independent rRNA-based Approaches. *Int. J. Food Microbiol.* **2015,** *212,* 67–75.

80. Wang, C.; Esteve-Zarzoso, B.; Cocolin, L.; Mas, A.; Rantsiou, K. Viable and Culturable Populations of *Saccharomyces cerevisiae, Hanseniaspora uvarum and Starmerella bacillaris (synonym Candida zemplinina)* During Barbera must Fermentation. *Food Res. Int.* **2015,** *78,* 195–200.

81. Alessandria, V.; Ferrocino, I.; De Filippis, F.; Fontana, M.; Rantsiou, K.; Ercolini, D.; Cocolin, L. Microbiota of an Italian Grana-like Cheese During Manufacture and Ripening, Unraveled by 16S rRNA-based Approaches. *Appl. Environ. Microbiol.* **2016,** *82,* 3988–3995.

82. Cravero, F.; Englezos, V.; Rantsiou, K.; Torchio, F.; Giacosa, S.; Segade, S. R.; Gerbi, V.; Rolle, L.; Cocolin, L. Ozone Treatments of Post Harvested Wine Grapes: Impact on Fermentative Yeasts and Wine Chemical Properties. *Food Res. Int.* **2016,** *87,* 134–141.

83. Law, J. W-F.; Mutalib, N-S. A.; Chan, K-G.; Lee, L-H. Rapid Methods for the Detection of Foodborne Bacterial Pathogens: Principles, Applications, Advantages and Limitations. *Front. Microbiol.* **2015,** *5,* 770.

84. Wang, Y.; Salazar, J. K. Culture-Independent Rapid Detection Methods for Bacterial Pathogens and Toxins in Food Matrices. *Comp. Rev. Food Sci. Food Saf.* **2016,** *15,* 183–205.

85. Wang, Y.; Ping, J.; Ye, Z.; Wu, J.; Ying, Y. Impedimetric Immunosensor Based on Gold Nanoparticles mModified Grapheme Paper for Label-Free Detection of *Escherichia coli* O157:H7. *Biosens. Bioelectron.* **2013,** *49,* 492–498.

86. Leonard, P.; Hearty, S.; Quinn, J.; O'Kennedy, R. A Generic Approach for the Detection of Whole *Listeria monocytogenes* Cells in Contaminated Samples Using Surface Plasmon Resonance. *Biosens. Bioelectron.* **2004,** *19,* 1331–1335.

87. Bokken, G. C.; Corbee, R. J.; van Knapen, F; Bergwerff, A. A. Immunochemical Detection of *Salmonella* Group B, D and E Using an Optical Surface Plasmon Resonance Biosensor. *FEMS Microbiol. Lett.* **2003,** *222,* 75–82.

88. Waswa, J. W.; Debroy, C.; Irudayaraj, J. Rapid Detection of *Salmonella* Enteritidis and *Escherichia coli* Using Surface Plasmon Resonance Biosensor. *J. Food Process Eng.* **2006**, *29*, 373–385.

89. Wei, D.; Oyarzabal, O. A.; Huang, T.; Balasubramanian, S.; Sista, S.; Simonian, A. L. Development of a surface plasmon resonance biosensor for the identification of *Campylobacter jejuni. J. Microbiol. Methods* **2007**, *69*, 78–85.

90. Waswa, J.; Irudayaraj, J.; DebRoy, C. Direct Detection of *E. coli* O157:H7 in Selected Food Systems by a Surface Plasmon Resonance Biosensor. *LWT-Food Sci. Technol.* **2007**, *40*, 187–192.

91. Son, J. R.; Kim, G.; Kothapalli, A.; Morgan, M. T.; Ess, D. (2007) Detection of *Salmonella* Enteritidis Using a Miniature Optical Surface Plasmon Resonance Biosensor. *J Phys Conf Ser.* **2007**, *61*, 1086–1090.

92. Lan, Y. B.; Wang, S. Z.; Yin, Y. G.; Hoffmann, W. C.; Zheng, X. Z. Using a Surface Plasmon Resonance Biosensor for Rapid Detection of *Salmonella* Typhimurium in Chicken Carcass. *J. Bionic. Eng.* **2008**, *5*, 239–246.

93. Lakshmanan, R. S.; Guntupalli, R.; Hu, J.; Kim, D-J.; Petrenko, V. A.; Barbaree, J. M.; Chin, B. A. Phage Immobilized Magnetoelastic Sensor for the Detection of *Salmonella* Typhimurium. *J. Microbiol. Methods* **2007**, *71*, 55–60.

94. Wan, J.; Johnson, M. L.; Guntupalli, R.; Petrenko, V. A.; Chin, B. A. Detection of *Bacillus anthracis* Spores in Liquid Using Phage-Based Magnetoelastic Micro-Resonators. *Sens. Actuators, B* **2007**, *127*, 559–566.

95. Huang, S.; Yang, H.; Lakshmanan, R. S.; Johnson, M. L.; Wan, J.; Chen, I-H.; Wikle, H. C. III; Petrenko, V. A.; Barbaree, J. M.; Chin, B. A. Sequential Detection of *Salmonella* Typhimurium and *Bacillus anthracis* Spores Using Magnetoelastic Biosensors. *Biosens. Bioelectron.* **2009**, *24*, 1730–1736.

96. Lu, Q.; Lin, H.; Ge, S.; Luo, S.; Cai, Q.; Grimes, C. A. A Wireless, Remote-Query, and High Sensitivity *Escherichia coli* O157:H7 Biosensor Based on the Recognition Action of Concanavalin A. *Anal. Chem.* **2009**, *81*, 5846–5850.

97. Lakshmanan, R. S.; Guntupalli, R.; Hu, J.; Petrenko, V. A.; Barbaree, J. M.; Chin, B. A. Detection of *Salmonella* Typhimurium in Fat free Milk Using a Phage Immobilized Magnetoelastic Sensor. *Sens. Actuators, B* **2007**, *126*, 544–550.

98. Li, S.; Li, Y.; Chen, H.; Horikawa, S.; Shen, W.; Simonian, A.; Chin, B. A. Direct Detection of *Salmonella* Typhimurium on Fresh Produce Using Phage-Based Magnetoelastic Biosensors. *Biosens. Bioelectron.* **2010**, *26*, 1313–1319.

99. Park, M-K.; Park, J. W.; Wikle, H. C. III; Chin, B. A. Evaluation of Phage-Based Magnetoelastic Biosensors for Direct Detection of *Salmonella* Typhimurium on Spinach Leaves. *Sens. Actuators, B.* **2013**, *176*, 1134–1140.

100. Park, M-K.; Weerakoon, K. A.; Oh, J-H.; Chin, B. A. The Analytical Comparison of Phage-Based Magnetoelastic Biosensor with TaqMan-based Quantitative PCR Method to Detect *Salmonella* Typhimurium on Cantaloupes. *Food Control* **2013**, *33*, 330–336.

101. Velusamy, V.; Arshak, K.; Korostynska, O.; Oliwa, K.; Adley, C. An Overview of Foodborne Pathogen Detection: In the Perspective of Biosensors. *Biotechnol Adv.* **2010**, *28*, 232–254.

102. Esteban-Fernandez de Ávila, B.; Pedrero, M.; Campuzano, S.; Escamilla-Gómez, V.; Pingarrón, J. M. Sensitive and Rapid Amperometric Magnetoimmunosensor for the Determination of *Staphylococcus aureus*. *Anal. Bioanal. Chem.* **2012**, *403*, 917–925.

103. Viswanathan, S.; Rani, C.; Ho, J. A. Electrochemical Immunosensor for Multiplexed Detection of Food-Borne Pathogens Using Nanocrystal Bioconjugates and MWCNT Screen-Printed Electrode. *Talanta* **2012**, *94*, 315–319.

104. Afonso, A. S.; Perez-Lopez, B.; Faria, R. C.; Mattoso, L. H. C.; Hernandez-Herrero, M.; Roig-Sagues, A. X.; Maltez-da Costa, M.; Merkoci, A. Electrochemical Detection of *Salmonella* Using Gold Nanoparticles. *Biosens. Bioelectron.* **2013**, *40*, 121–126.

105. Zelada-Guillen, G. A.; Bhosale, S. V.; Riu, J.; Rius, F. X. Real-Time Potentiometric Detection of Bacteria in Complex Samples. *Anal. Chem.* **2010**, *82*, 9254–9260.

106. Yang, M.; Sun, S.; Bruck, H. A.; Kostov, Y.; Rasooly, A. Electrical Percolation-Based Biosensor for Real-Time Direct Detection of Staphylococcal Enterotoxin B (SEB). *Biosens. Bioelectron.* **2010**, *25*, 2573–2578.

107. Dong, J.; Zhao, H.; Xu, M.; Ma, Q.; Ai, S. A Label-Free Electrochemical Impedance Immunosensor Based on AuNPs/PAMAM-MWCNT-Chi Nanocomposite Modified Glassy Carbon Electrode for Detection of *Salmonella* Typhimurium in Milk. *Food Chem.* **2013**, *141*, 1980–1986.

108. Wan, Y.; Lin, Z.; Zhang, D.; Wang, Y.; Hou, B. Impedimetric Immunosensor Doped with Reduced Graphene sheets Fabricated by Controllable Electrodeposition for the Non-labelled Detection of Bacteria. *Biosens. Bioelectron.* **2011**, *26*, 1959–1964.

109. Li, D.; Feng, Y.; Zhou, L.; Ye, Z.; Wang, J.; Ying, Y.; Ruan, C.; Wang, R.; Li, Y. Label-Free Capacitive Immunosensor Based on Quartz Crystal Au Electrode for Rapid and Sensitive Detection of *Escherichia coli* O157:H7. *Anal. Chim. Acta.* **2011**, *687*, 89–96.

110. Joung, C-K.; Kim, H-N.; Lim, M-C.; Jeon, T-J.; Kim, H-Y.; Kim, Y-R. A Nanoporous Membrane-Based Impedimetric Immunosensor for Label-Free Detection of Pathogenic Bacteria in Whole Milk. *Biosens. Bioelectron.* **2013**, *44*, 210–215.

111. Wang, Y.; Ye, Z.; Si, C.; Ying, Y. Monitoring of *Escherichia coli* O157:H7 in Food Samples Using Lectin Based Surface Plasmon Resonance Biosensor. *Food Chem.* **2013**, *136*, 1303–1308.

112. Maraldo, D.; Mutharasan, R. 10-minute Assay for Detecting *Escherichia coli* O157:H7 in Ground Beef Samples Using Piezoelectric-Excited Millimeter-Size Cantilever Sensors. *J. Food Protect* **2007**, *70*, 1670–1677.

113. Sharma, H.; Mutharasan, R. Rapid and Sensitive Immunodetection of *Listeria monocytogenes* in Milk Using a Novel Piezoelectric Cantilever Sensor. *Biosens. Bioelectron.* **2013**, *45*, 158–162.

114. Maraldo, D.; Mutharasan, R. Detection and Confirmation of Staphylococcal Enterotoxin B in Apple Juice and Milk Using Piezoelectric-Excited Millimeter-Sized Cantilever Sensors at 2.5 fg/mL. *Anal. Chem.* **2007**, *79*, 7636–7643.

115. Yilmaz, E.; Majidi, D.; Ozgur, E.; Denizli, A. Whole Cell Imprinting Based *Escherichia coli* Sensors: A Study for SPR and QCM. *Sens. Actuators, B* **2015**, *209*, 714–721.

116. Chen, S.; Wu, V. C. H.; Chuang, Y.; Lin, C. Using Oligonucleotide-Functionalized Au Nanoparticles to Rapidly Detect Foodborne Pathogens on a Piezoelectric Biosensor. *J. Microbiol. Methods* **2008**, *73*, 7–17.

117. Salmain, M.; Ghasemi, M.; Boujday, S.; Pradier, C. Elaboration of a Reusable Immunosensor for the Detection of Staphylococcal Enterotoxin A (SEA) in Milk with a Quartz Crystal Microbalance. *Sens. Actuators, B* **2012,** *173,* 148–156.

118. Tenover, F. C.; Arbeit, R. D.; Goering, R. V.; Mickelsen, P. A.; Murray, B. E.; Persing, D. H.; Swaminathan, B. Interpreting Chromosomal DNA Restriction Patterns Produced by Pulsed-Field Gel Electrophoresis: Criteria for Bacterial Strain Typing. *J. Clin. Microbiol.* **1995,** *33,* 2233–2239.

119. Barrett, T. J.; Gerne-Smidt, P.; Swaminathan, B. Interpretation of Pulsed-Field Gel Electrophoresis Patterns in Foodborne Disease Investigations and Surveillance. *Foodborne Pathog. Dis.* **2006,** *3,* 20–31.

120. Murphy, M.; Corcoran, D.; Buckley, J. F.; O'Mahony, M.; Whyte, P.; Fanning, S. Development and Application of Multiple-Locus Variable Number of Tandem Repeat Analysis (MLVA) to Subtype a collection of *Listeria monocytogenes. Int. J. Food Microbiol.* **2007,** *115,* 187–194.

121. Miya, S.; Kimura, B.; Sato, M.; Takahashi, H.; Ishikawa, T.; Suda, T.; Takakura, C.; Fujii, T. Wiedmann, M. Development of a Multilocus Variable-Number of Tandem Repeat Typing Method for *Listeria monocytogenes* Serotype 4b Strains. *Int. J. Food Microbiol.* **2008,** *124,* 239–249.

122. Volpe Sperry, K. E.; Kathariou, S.; Edwards, J. S.; Wolf, L. A. Multiple-Locus Variable-Number Tandem-Repeat Analysis as a Tool for Subtyping *Listeria monocytogenes* Strains. *J. Clin. Microbiol.* **2008,** *46,* 1435–1450.

123. Lindstedt, B. A.; Tham, W.; Danielsson-Tham, M. L.; Vardund, T.; Helmersson, S.; Kapperud, G. Multiple-Locus Variable-Number Tandem-Repeats Analysis of *Listeria monocytogenes* Using Multicolour Capillary Electrophoresis and Comparison with Pulsed-Field Gel Electrophoresis Typing. *J. Microbiol. Methods* **2008,** *72,* 141–148.

124. Larsson, J. T.; Roussel, S.; Moller Nielsen, E. *Better and Faster Typing, MLVA – Shall We Play Together?* In ISOPOL XVII Porto, 2010, Portugal.

125. Chen, S.; Li, J.; Saleh-Lakha, S.; Allen, V.; Odumeru, J. Multiple-Locus Variable Number of Tandem Repeat Analysis (MLVA) of *Listeria monocytogenes* Directly in Food Samples. *Int. J. Food Microbiol.* **2011,** *148,* 8–14.

126. Chenal-Francisque, V.; Diancourt, L.; Cantinelli, T.; Passet, V.; Tran-Hykes, C.; Bracq-Dieye, H.; Leclercq, A.; Pourcel, C.; Lecuit, M.; Brissed, S. Optimized Multilocus Variable-Number Tandem-Repeat Analysis Assay and its Complementarity with Pulsed-Field Gel Electrophoresis and Multilocus Sequence Typing for *Listeria monocytogenes* Clone Identification and Surveillance. *J. Clin. Microbiol.* **2013,** *51,* 1868–1880.

127. Saleh-Lakha, S.; Allen, V. G ; Li, J.; Pagotto, F.; Odumeru, J.; Taboada, E.; Lombos, M.; Tabing, K. C.; Blais, B.; Ogunremi, D.; Downing, G.; Lee, S.; Gao, A.; Nadon, C.; Chen, S. Subtyping of a Large Collection of Historical *Listeria monocytogenes* Strains from Ontario, Canada, by an Improved Multilocus Variable-Number Tandem-Repeat Analysis (MLVA). *Appl. Environ. Microbiol.* **2013,** *79,* 6472–6480.

128. Gorski, L.; Walker, S.; Liang, A. S.; Nguyen, K. M.; Govoni, J.; Carychao, D.; Cooley, M. B.; Mandrell, R. E. Comparison of Subtypes of *Listeria monocytogenes* Isolates from Naturally Contaminated Watershed Samples with and Without a Selective Secondary Enrichment. *PLoS ONE* **2014,** *9,* e92467.

129. Lindstedt, B. A.; Heir, E.; Gjernes, E.; Kapperud, G. DNA Fingerprinting of *Salmonella enterica* Subsp. enterica serovar Typhimurium with Emphasis on Phage Type DT104 Based on Variable Number of Tandem Repeat Loci. *J. Clin. Microbiol.* **2003,** *41,* 1469–1479.

130. Boxrud, D.; Pederson-Gulrud, K.; Wotton, J.; Medus, C.; Lyszkowicz, E.; Besser, J.; Bartkus, J. M. Comparison of Multiple-Locus Variable-Number Tandem Repeat Analysis, Pulsed-Field Gel Electrophoresis, and Phage Typing for Subtype Analysis of *Salmonella enterica* Serotype Enteritidis. *J. Clin. Microbiol.* **2007,** **45,** 536–543.

131. Ramisse, V.; Houssu, P.; Hernandez, E.; Denoeud, F.; Hilaire, V.; Lisanti, O.; Ramisse, F.; Cavallo, J. D.; Vergnaud, G. Variable Number of Tandem Repeats in *Salmonella enterica* Subsp. enterica for Typing Purposes. *J. Clin. Microbiol.* **2004,** *42,* 5722–5730.

132. Davis, M. A.; Baker, K. N. K.; Call, D. R.; Warnick, L. D.; Soyer, Y.; Wiedmann, M.; Grohn, Y.; McDonough, P. L.; Hancock ,D. D.; Besser, T. E. Multilocus Variable-Number Tandem-Repeat Method for Typing *Salmonella enterica* Serovar Newport. *J. Clin. Microbiol.* **2009,** *47,* 1934–1938.

133. Tien, Y. Y.; Wang, Y. W.; Tung, S. K.; Liang, S. Y.; Chiou, C. S. Comparison of Multi-locus Variable Number Tandem Repeat Analysis and Pulsed-Field Gel Electropho-resis in Molecular Subtyping of *Salmonella enterica* Serovars Paratyphi A. *Diagn. Microbiol. Infect. Dis.* **2011,** *69,* 1–6.

134. Bergamini, F.; Iori, A.; Massi, P.; Pongolini, S. Multilocus Variable-Number of Tandem-Repeats Analysis of *Salmonella enterica* Serotype Gallinarum and Compar-ison with Pulsed-Field Gel Electrophoresis Genotyping. *Vet. Microbiol.* **2011,** *149,* 430–436.

135. Young, C. C.; Ross, I. L.; Heuzenroeder, M. W. New Methodology for Differentia-tion and Typing of Closely Related *Salmonella enterica* Serovar Heidelberg Isolates. *Curr. Microbiol.* **2012,** *65,* 481–487.

136. Lettini, A. A.; Saccardin, C.; Ramon, E.; Longo, A.; Cortini, E.; Dalla Pozza, M. C.; Barco, L.; Guerra, B.; Luzzi, I.; Ricci, A. Characterization of an Unusual *Salmonella* Phage Type DT7a and Report of a Foodborne Outbreak of Salmonellosis. *Int. J. Food Microbiol.* **2014,** *189,* 11–17.

137. McClelland, M.; Sanderson, K. E.; Clifton, S. W.; Latreille, P.; Porwollik, S.; Sabo, A.; Meyer, R.; Bieri, T.; Ozersky, P.; McLellan, M.; Harkins, C. R.; Wang, C.; Nguyen, C.; Berghoff, A.; Elliott, G.; Kohlberg, S.; Strong, C.; Du, F.; Carter, J.; Kremizki, C. Layman, D.; Leonard, S.; Sun, H.; Fulton, L.; Nash, W.; Miner, T.; Minx, P.; Delehaunty, K.; Fronick, C.; Magrini, V.; Nhan, M.; Warren, W.; Florea, L.; Spieth, J.; Wilson, R. K. Comparison of Genome Degradation in Paratyphi A and Typhi, Human-Restricted Serovars of *Salmonella enterica* that Cause Typhoid. *Nat. Genet.* **2004,** *36,* 1268–1274.

138. Octavia, S.; Lan, R. Multiple-Locus Variable-Number Tandem-Repeat Analysis of *Salmonella enterica* Serovar Typhi. *J. Clin. Microbiol.* **2009,** *47,* 2369–2376.

139. Tien, Y. Y.; Ushijima, H.; Mizuguchi, M.; Liang, S. Y.; Chiou, C. S. Use of Multilocus Variable Number Tandem Repeat Analysis in Molecular Subtyping of *Salmonella enterica* Serovar Typhi Isolates. *J. Med. Microbiol.* **2012,** *61,* 223–232.

140. Liu, Y.; Lee, M. A.; Ooi, E. E.; Mavis, Y.; Tan, A. L; Quek, H. H. Molecular Typing of *Salmonella enterica* Serovar Typhi Isolates from Various countries in Asia by a Multiplex PCR Assay on Variable-Number Tandem Repeats. *J. Clin. Microbiol.* **2003,** *41,* 4388–4394.

141. Paramithiotis, S.; Hadjilouka, A.; Drosinos, E. H. Molecular Typing Schemes of Food-Associated Salmonellae. *Salmonella: Prevalence, Risk Factors and Treatment Options;* Hacket C. B., Ed.; Nova Publishers, NY, USA: 2015; pp 159–178.

142. Lindstedt, B. A.; Vardund, T.; Aas, L.; Kapperud, G. Multiple-Locus Variable-Number Tandem Repeats Analysis of *Salmonella enterica* Subsp. enterica serovar Typhimurium Using PCR Multiplexing and Multicolor Capillary Electrophoresis. *J. Microbiol. Methods* **2004,** *59,* 163–172.

143. Cho, S.; Boxrud, D. J.; Bartkus, J. M.; Whittam, T. S.; Saeed, M. Multiple Locus Variable-Number Tandem Repeat Analysis of *Salmonella* Enteritidis Isolates from Human and Non-human Sources Using a Single Multiplex PCR. *FEMS Microbiol. Lett.* **2007,** *275,* 16–23.

144. Cho, S.; Whittam, T. S.; Boxrud, D. J.; Bartkus, J. M.; Saeed, A. M. Allele Distribution and Genetic Diversity of VNTR Loci in *Salmonella enterica* serotype Enteritidis Isolates from Different Sources. *BMC Microbiol.* **2008,** *8,* 146.

145. Malorny, B.; Junker, E.; Helmuth, R. Multi-Locus Variable-Number Tandem Repeat Analysis for Outbreak Studies of *Salmonella enterica* Serotype Enteritidis. *BMC Microbiol.* **2008,** *8,* 84.

146. Ross, I. L.; Parkinson, I. H.; Heuzenroeder, M. W. A Comparison of Two PCR-Based Typing Methods with Pulsed-Field Gel Electrophoresis in *Salmonella enterica* serovar Enteritidis. *Int. J. Med. Microbiol.* **2009,** *299,* 410–420.

147. Ross, I. L.; Parkinson, I. H.; Heuzenroeder, M. W. The Use of MAPLT and MLVA Analyses of Phenotypically Closely Related Isolates of *Salmonella enterica* Serovar Typhimurium. *Int. J. Med. Microbiol.* **2009,** *299,* 37–41.

148. Beranek, A.; Mikula, C.; Rabold, P.; Arnhold, D.; Berghold, C.; Lederer, I.; Allerberger, F.; Kornschober, C. Multiple-Locus Variable-Number Tandem Repeat Analysis for Subtyping of *Salmonella enterica* Subsp. enterica serovar Enteritidis. *Int. J. Med. Microbiol.* **2009,** *299,* 43–51.

149. Chiou, C. S.; Hung, C. S.; Torpdahl, M.; Watanabe, H.; Tung, S. K.; Terajima, J.; Liang, S. Y.; Wang, Y. W. Development and Evaluation of Multilocus Variable Number Tandem Repeat Analysis for Fine typing and Phylogenetic Analysis of *Salmonella enterica* Serovar Typhimurium. *Int. J. Food Microbiol.* **2010,** *142,* 67–73.

150. Boxrud, D.; Pederson-Gulrud, K.; Wotton, J.; Medus, C.; Lyszkowicz, E.; Besser, J., Bartkus, J. M. Comparison of Multiple-Locus Variable-Number Tandem Repeat Analysis, Pulsed-Field Gel Electrophoresis, and Phage Typing for Subtype Analysis of *Salmonella enterica* Serotype Enteritidis. *J. Clin. Microbiol.* **2007,** *45,* 536–543.

151. Lindstedt, B. A.; Torpdahl, M.; Vergnaud, G.; Le Hello, S.; Weill, F. X.; Tietze, E.; Malorny, B.; Prendergast, D. M.; Ní Ghallchóir, E.; Lista, R. F.; Schouls, L. M.; Söderlund, R.; Börjesson, S.; Åkerström, S. Use of Multilocus Variable-Number Tandem Repeat Analysis (MLVA) in Eight European Countries, 2012. *Euro Surveill.* **2013,** *18,* 20385.

152. Noller, A. C.; McEllistrem, M. C.; Pacheco, A. G.; Boxrud, D. J.; Harrison, L. H. Multilocus Variable-Number Tandem Repeat Analysis Distinguishes Outbreak and Sporadic *Escherichia coli* O157:H7 Isolates. *J. Clin. Microbiol.* **2003,** *41,* 5389–5397.

153. Keys, C.; Kemper, S.; Keim, P. Highly Diverse Variable Number Tandem Repeat Loci in the *E. coli* O157:H7 and O55:H7 Genomes for High-Resolution Molecular Typing. *J. Appl. Microbiol.* **2005,** *98,* 928–940.

154. Lindstedt, B. A.; Heir, E.; Gjernes, E.; Vardund, T.; Kapperud, G. DNA Fingerprinting of Shiga-Toxin Producing *Escherichia coli* O157 Based on Multiple-Locus Variable-Number Tandem-Repeats Analysis (MLVA). *Ann. Clin. Microbiol. Antimicrob.* **2003,** *2,* 12–18.

155. Lindstedt, B. A.; Brandal, L. T.; Aas, L.; Vardund, T.; Kapperud, G. Study of Polymorphic Variable-Number of Tandem Repeats Loci in the ECOR Collection and in a Set of Pathogenic *Escherichia coli* and Shigella Isolates for use in a Genotyping Assay. *J. Microbiol. Methods* **2007,** *69,* 197–205.

156. Hyytia-Trees, E.; Lafon, P.; Vauterin, P.; Ribot, E. M. Multilaboratory Validation Study of Standardized Multiple-Locus Variable-Number Tandem Repeat Analysis Protocol for Shiga Toxin-Producing *Escherichia coli* O157: A Novel Approach to Normalize Fragment Size Data Between Capillary Electrophoresis Platforms. Foodborne Pathog Dis **2010,** *7,* 129–136.

157. Hyytia-Trees, E.; Smole, S. C.; Fields, P. A.; Swaminathan, B.; Ribot, E. M. Second Generation Subtyping: A Proposed PulseNet Protocol for Multiple-Locus Variable Number Tandem Repeat Analysis of Shiga Toxin-producing *Escherichia coli* O157 (STEC O157). *Foodborne. Pathog. Dis.* **2006,** *3,* 118–131.

158. Franci, T.; Sanso, A. M.; Bustamante, A. V.; Lucchesi, P. M. A.; Parma, A. E. Genetic Characterization of Non-O157 Verocytotoxigenic *Escherichia coli* Isolated from Raw Beef Products Using Multiple-Locus Variable-Number Tandem Repeat Analysis. *Foodborne Pathog. Dis.* **2011,** *8,* 1019–1023.

159. Timmons, C.; Trees, E.; Ribot, E. M.; Gerner-Smidt, P.; LaFon, P.; Im, S.; Ma, L. M. (2016) Multiple-Locus Variable-Number Tandem Repeat Analysis for Strain Discrimination of Non-O157 Shiga Toxin-Producing *Escherichia coli. J. Microbiol. Methods* **2011,** *125,* 70–80.

160. Bustamante, A. V.; Sanso, A. M; Lucchesi, P. M. A.; Parma, A. E. Genetic Diversity of O157:H7 and Non-O157 Verocytotoxigenic *Escherichia coli* from Argentina Inferred from Multiple-Locus Variable-Number Tandem Repeat Analysis (MLVA). *Int. J. Med. Microbiol.* **2010,** *300,* 212–217.

161. Izumiya, H.; Pei, Y. Terajima, J.; Ohnishi, M.; Hayashi, T.; Iyoda, S.; Watanabe, H. (2010) New System for Multilocus Variable-Number Tandem-Repeat Analysis of the Enterohemorrhagic *Escherichia coli* Strains Belonging to Three Major Serogroups: O157, O26, and O111. *Microbiol. Immunol.* **2010,** *54,* 569–577.

162. Lobersli, I.; Haugum, K.; Lindstedt, B-A. Rapid and High Resolution Genotyping of all *Escherichia coli* Serotypes Using 10 Genomic Repeat-Containing Loci. *J. Microbiol. Methods* **2012,** *88,* 134–139.

163. Urdahl, A. M.; Strachan, N. J. C.; Wasteson, Y.; MacRae, M.; Ogden, I. D. (2008) Diversity of *Escherichia coli* O157 in a Longitudinal Farm Study Using Multiple-Locus Variable-Number Tandem-Repeat Analysis. *J. Appl. Microbiol.* **2012,** *105,* 1344–1353.

164. Dyet, K.H.; Robertson, I.; Turbitt, E.; Carter, P. E. (2011) Characterization of *Escherichia coli* O157:H7 in New Zealand Using Multiple-Locus Variable-Number Tandem-Repeat Analysis. *Epidemiol. Infect.* **2012**, *139,* 464–471.

165. Prendergast, D. M.; Lendrum, L.; Pearce, R.; Ball, C.; McLernon, J.; O'Grady, D.; Scott, L.; Fanning, S.; Egan, J.; Gutierrez, M. Verocytotoxigenic *Escherichia coli* O157 in Beef and Sheep Abattoirs in Ireland and Characterization of Isolates by Pulsed-Field Gel Electrophoresis and Multi-Locus Variable Number of Tandem Repeat Analysis. *Int. J. Food Microbiol.* **2011**, *144,* 519–527.

166. Naseer, U. Olsson-Liljequist, B. E.; Woodford, N.; Dhanji, H.; Canton, R.; Sundsfjord, A.; Lindstedt, B. A. Multi-Locus Variable Number of Tandem Repeat Analysis for Rapid and Accurate Typing of Virulent Multidrug Resistant *Escherichia coli* Clones. *PLoS ONE* **2012**, *7,* e41232.

167. Soderlund, R.; Hedenström, I. Nilsson, A.; Eriksson, E.; Aspán, A. (2012) Genetically Similar Strains of *Escherichia coli* O157:H7 Isolated from Sheep, Cattle and Human Patients. *BMC Vet. Res.* **2012**, *8,* 200.

168. Perry, N.; Cheasty, T.; Dallman, T.; Launders, N.; Willshaw, G. Application of Multilocus Variable Number Tandem Repeat Analysis to Monitor Verocytotoxin-Producing *Escherichia coli* O157 Phage Type 8 in England and Wales: Emergence of a Profile Associated with a National Outbreak. *J. Appl. Microbiol.* **2013**, *115,* 1052–1058.

169. Kumar, A.; Taneja, N.; Sharma, R. K.; Sharma, H.; Ramamurthy, T.; Sharma, M. Molecular Characterization of Shiga-Toxigenic *Escherichia coli* Isolated from Diverse Sources from India by Multi-Locus Variable Number Tandem Repeat Analysis (MLVA). *Epidemiol. Infect.* **2014**, *142,* 2572–2582.

170. Bai, L.; Guo, Y.; Lan, R.; Dong, Y.; Wang, W.; Hu, Y.; Gan, X.; Yan, S.; Fu, P.; Pei, X. Xu, J.; Liu, X.; Li, F. (2015) Genotypic Characterization of Shiga Toxin-Producing *Escherichia coli* O157:H7 Isolates in Food Products from China Between 2005 and 2010. *Food Control* **2014**, *50,* 209–214.

171. Rumore, J. L.; Tschetter, L.; Nadon, C. The Impact of Multilocus Variable-Number Tandem-Repeat Analysis on PulseNet Canada *Escherichia coli* O157:H7 Laboratory Surveillance and Outbreak Support, 2008–2012. *Foodborne Pathog. Dis.* **2016**, *13,* 255–261.

172. Feil, E. J. Small Change: Keeping Pace with Microevolution. *Nat. Rev. Microbiol.* **2004**, *2,* 483–495.

173. Maiden, M. C. Multilocus Sequence Typing of Bacteria. *Annu. Rev. Microbiol.* **2006**, *60,* 561–588.

174. Salcedo, C.; Arreaza, L.; Alcala, B.; de la Fuente, L.; Vazquez, J. A. Development of a Multilocus Sequence Typing Method for Analysis of *Listeria monocytogenes* Clones. *J. Clin. Microbiol.* **2003**, *41,* 757–762.

175. Revazishvili, T.; Kotetishvili, M.; Stine, O. C.; Kreger, A. S.; Morris, J. G. Jr.; Sulakvelidze, A. (2004) Comparative Analysis of Multilocus Sequence Typing and Pulsed-Field Gel Electrophoresis for Characterizing *Listeria monocytogenes* Strains Isolated from Environmental and Clinical Sources. *J. Clin. Microbiol.* **2004**, *42,* 276–285.

176. Ragon, M.; Wirth, T.; Hollandt, F.; Lavenir, R.; Lecuit, M.; Le Monnier, A.; Brisse, S. A New Perspective on *Listeria monocytogenes* Evolution. *PLoS Pathog.* **2008,** *4,* e1000146.

177. Parisi, A.; Latorre, L.; Normanno, G.; Miccolupo, A.; Fraccalvieri, R.; Lorusso, V.; Santagada, G. Amplified Fragment Length Polymorphism and Multi-Locus Sequence Typing for High-Resolution Genotyping of *Listeria monocytogenes* from Foods and the Environment. *Food Microbiol.* **2010,** *27,* 101–108.

178. Mammina, C.; Parisi, A.; Guaita, A.; Aleo, A.; Bonura, C.; Nastasi, A.; Pontello, M. Enhanced Surveillance of Invasive Listeriosis in the Lombardy Region, Italy, in the Years 2006-2010 Reveals Major Clones and an Increase in Serotype 1/2a. *BMC Infect. Dis.* **2013,** *13,* 152.

179. Chenal-Francisque, V.; Diancourt, L.; Cantinelli, T.; Passet, V.; Tran-Hykes, C.; Bracq-Dieye, H.; Leclercq, A.; Pourcel, C.; Lecuit, M.; Brissed, S. Optimized Multi-locus Variable-Number Tandem-Repeat Analysis Assay and its Complementarity with Pulsed-Field Gel Electrophoresis and Multilocus Sequence Typing for *Listeria monocytogenes* Clone Identification and Surveillance. *J. Clin. Microbiol.* **2013,** *51,* 1868–1880.

180. Martin, B.; Perich, A.; Gomez, D.; Yanguela, J.; Rodriguez, A.; Garriga, M.; Aymerich, T. Diversity and Distribution of *Listeria monocytogenes* in Meat Processing Plants. *Food Microbiol.* **2014,** *44,* 119–127.

181. Haase, J. K.; Didelot, X.; Lecuit, M.; Korkeala, H. The Ubiquitous Nature of *Listeria monocytogenes* Clones: A Large-Scale Multilocus Sequence Typing Study. *Environ. Microbiol.* **2014,** *16,* 405–416.

182. Zhang, W.; Jayarao, B. M.; Knabel, S. J. Multi-Virulence-Locus Sequence Typing of *Listeria monocytogenes*. *Appl. Environ. Microbiol.* **2004,** *70,* 913–920.

183. Kalpana, K.; Mitra, S.; Muriam, P. M. Subtyping of *Listeria monocytogenes* by Multi-locus Sequence Typing. Animal Science Research Report. Oklahoma State University 2004. Publication no. 1008. http://www.ansi.okstate.edu/research/2004/2004%20 Kalpana%20Research%20Report.pdf9

184. Chen, Y.; Zhang, W.; Knabel, S. J. Multi-Virulence-Locus Sequence Typing Clarifies Epidemiology of Recent Listeriosis Outbreaks in the United States. *J. Clin. Microbiol.* **2005,** *43,* 5291–5294.

185. Chen, Y.; Zhang, W.; Knabel, S. J. Multi-Virulence-Locus Sequence Typing Identifies Single Nucleotide Polymorphisms Which Differentiate Epidemic Clones and Outbreak Strains of *Listeria monocytogenes*. *J. Clin. Microbiol.* **2007,** *45,* 835–846.

186. Cantinelli, T.; Chenal-Francisque, V.; Diancourt, L.; Frezal, L.; Leclercq, A.; Wirth, T.; Lecuit, M.; Brisse, S. "Epidemic Clones" of *Listeria monocytogenes* are Widespread and Ancient Clonal Groups. *J. Clin. Microbiol.* **2013,** *51,* 3770–3779.

187. Doijad, S.; Lomonaco, S.; Poharkar, K.; Garg, S.; Knabel, S.; Barbuddhe, S.; Jayarao, B. Multi-Virulence-Locus Sequence Typing of 4b *Listeria monocytogenes* Isolates Obtained from Different Sources in India over a 10-Year Period. *Foodborne Pathog. Dis.* **2014,** *11,* 511–516.

188. Torpdahl, M.; Skov, M. N.; Sandvangb, D.; Baggesen, D. L. Genotypic Characterization of *Salmonella* by Multilocus Sequence Typing, Pulsed-Field Gel Electrophoresis

and Amplified Fragment Length Polymorphism. *J. Microbiol. Methods.* **2005,** *63,* 173–184.

189. Harbottle, H.; White, D. G.; McDermott, P. F.; Walker, R. D.; Zhao, S. Comparison of Multilocus Sequence Typing, Pulsed-Field Gel Electrophoresis, and Antimicrobial Susceptibility Typing for Characterization of *Salmonella enterica* Serotype Newport Isolates. *J. Clin. Microbiol.* **2006,** *44,* 2449–2457.

190. Almeida, F.; Pitondo-Silva, A.; Oliveira, M. A., Falcao, J. P. Molecular Epidemiology and Virulence Markers of Salmonella Infantis Isolated over 25 Years in Sao Paulo State, Brazil. *Infect. , Genet. Evol.* **2013,** *19,* 145–151.

191. Toboldt, A.; Tietze, E.; Helmuth, R.; Junker, E.; Fruth, A. Malorny, B. Molecular Epidemiology of *Salmonella enterica* Serovar Kottbus Isolated in Germany from Humans, Food and Animals. *Vet. Microbiol.* **2014,** *170,* 97–108.

192. Fakhr, M. K.; Nolan, L. K., Logue, C. M. Multilocus Sequence Typing Lacks the Discriminatory Ability of Pulsed-Field Gel Electrophoresis for Typing *Salmonella enterica* Serovar Typhimurium. *J. Clin. Microbiol.* **2005,** *43,* 2215–2219.

193. Tankouo-Sandjong, B.; Sessitsch, A.; Liebana, E.; Kornschober, C.; Allerberger, F.; Hachler, H.; Bodrossy L. MLST-v Multilocus Sequence Typing Based on Virulence Genes, for Molecular Typing of *Salmonella enterica* Subsp. enterica Serovars. *J. Microbiol. Methods* **2005,** *69,* 23–36.

194. Liu, F.; Barrangou, R.; Gerner-Smidt, P.; Ribot, E. M.; Knabel, S. J.; Dudley, E. G. Novel Virulence Gene and Clustered Regularly Interspaced Short Palindromic Repeat (CRISPR) Multilocus Sequence Typing Scheme for Subtyping of the Major Serovars of *Salmonella enterica* subsp. enterica. *Appl. Environ. Microbiol.* **2011,** *77,* 1946–1956.

195. Liu, F.; Kariyawasam, S.; Jayarao, B. M.; Barrangou, R.; Gerner-Smidt, P.; Ribot, E. M.; Knabel, S. J.; Dudley, E. G. Subtyping *Salmonella enterica* Serovar Enteritidis Isolates from Different Sources by Using Sequence Typing Based on Virulence Genes and Clustered Regularly Interspaced Short Palindromic Repeats (CRISPRs). *Appl. Environ. Microbiol.* **2011,** *77,* 4520–4526.

196. Singh, P.; Foley, S. L; Nayak, R.; Kwon, Y. M. Multilocus Sequence Typing of *Salmonella* Strains by High-Throughput Sequencing of Selectively Amplified Target Genes. *J. Microbiol. Methods* **2012,** *88,* 127–133.

197. Shariat, N.; Sandt, C. H.; Di Marzio, M. J.; Barrangou, R.; Dudley, E. G. CRISPR-MVLST Subtyping of *Salmonella enterica* Subsp. enterica Serovars Typhimurium and Heidelberg and Application in Identifying Outbreak Isolates. *BMC Microbiol.* **2013,** *13,* 254.

198. Shariat, N.; Kirchner, M. K.; Sandt, C. H.; Trees, E.; Barrangou, R.; Dudley, E. G. Subtyping of *Salmonella enterica* Serovar Newport Outbreak Isolates by CRISPR-MVLST and Determination of the Relationship Between CRISPR-MVLST and PFGE Results. *J. Clin. Microbiol.* **2013,** *51,* 2328–2336.

199. Shariat, N.; Di Marzio, M. J.; Yin, S.; Dettinger, L.; Sandt, C. H.; Lute, J. R.; Barrangou, R.; Dudley, E. G. The Combination of CRISPR-MVLST and PFGE Provides Increased Discriminatory Power for Differentiating Human Clinical Isolates of *Salmonella enterica* subsp. enterica serovar Enteritidis. *Food Microbiol.* **2013,** *34,* 164–173.

200. Reid, S. D.; Herbelin, C. J.; Bumbaugh, A. C.; Selander, R. K.; Whittam, T. S. Parallel Evolution of Virulence in Pathogenic *Escherichia coli. Nature* **2000**, *406*, 64–67.

201. Wirth, T.; Falush, D.; Lan, R.; Colles, F.; Mensa, P.; Wieler, L. H.; Karch, H.; Reeves. P. R.; Maiden, M. C.; Ochman, H.; Achtman, M. Sex and Virulence in *Escherichia coli*: An Evolutionary Perspective. *Mol. Microbiol.* **2006**, *60*, 1136–1151.

202. Jaureguy, F.; Landraud, L.; Passet, V.; Diancourt, L.; Frapy, E.; Guigon, G.; Carbonnelle, E.; Lortholary, O.; Clermont, O.; Denamur, E.; Picard, B.; Nassif, X.; Brisse, S. Phylogenetic and Genomic Diversity of Human Bacteremic *Escherichia coli* Strains. *BMC Genomics* **2008**, *9*, 560.

203. Weissman, S. J.; Johnson, J. R.; Tchesnokova, V.; Billig, M.; Dykhuizen, D.; Riddell, K.; Rogers, P.; Qin, X.; Butler-Wu, S.; Cookson, B. T.; Fang, F. C.; Scholes, D.; Chattopadhyay, S.; Sokurenko, E. High-Resolution Two-Locus Clonal Typing of Extraintestinal Pathogenic *Escherichia coli. Appl Environ Microbiol* **2012**, *78*, 1353–1360.

204. Jores, J.; Rumer, L.; Wieler, L. H. Impact of the Locus of Enterocyte Effacement Pathogenicity Island on the Evolution of Pathogenic *Escherichia coli. Int. J. Med. Microbiol.* **2004**, *294*, 103–113.

205. Jønsson, R.; Struve, C.; Boisen, N.; Mateiu, R. V.; Santiago, A. E.; Jenssen, H.; Nataro, J. P.; Krogfelt, K. A. Novel Aggregative Adherence Fimbria Variant of Enteroaggregative *Escherichia coli. Infect. Immun.* **2015**, *83*, 1396–1405.

206. Torres, A. G.; Zhou, X.; Kaper, J. B. Adherence of Diarrheagenic *Escherichia coli* Strains to Epithelial Cells. *Infect. Immun.* **2005**, *73*, 18–29.

207. Beutin, L.; Aleksic, S.; Zimmermann, S.; Gleier, K. Virulence Factors and Phenotypical Traits of Verotoxigenic Strains of *Escherichia coli* Isolated from Human Patients in Germany. *Med. Microbiol. Immunol.* **1994**, *183*, 13–21.

208. Burgos, Y.; Beutin, L. Common Origin of Plasmid Encoded Alpha-Hemolysin Genes in *Escherichia coli. BMC Microbiol.* **2010**, *10*, 193.

209. Hunt, S.; Green, J.; Artymiuk, P. J. Hemolysin E (HlyE, ClyA SheA) and related toxins. *Adv. Exp. Med. Biol.* **2010**, *677*, 116–126.

210. Nair, G. B.; Takeda, Y. Heat-Labile and heat-Stable Enterotoxins of *E. coli, Escherichia coli: mechanisms of virulence;* Sussman, M., Ed.; Cambridge University Press, Cambridge, UK: 1997; pp 237–256.

211. Klapproth, J-M. A.; Scaletsky, I. C.; McNamara, B. P.; Lai, L-C.; Malstrom, C.; James, S. P.; Donnenberg, M. S. A Large Toxin from Pathogenic *Escherichia coli* Strains that Inhibits Lymphocyte Activation. *Infect. Immun.* **2000**, *68*, 2148–2155.

212. Menard, L-P.; Debreuil, J. D. Enteroaggregative *Escherichia coli* Heat Stable Enterotoxin 1 (EAST1): A New Toxin with an Old Twist. *Crit. Rev. Microbiol.* **2002**, *28*, 43–60.

213. Leblanc, J. J. Implication of Virulence Factors in *Escherichia coli* O157:H7 Pathogenesis. *Crit. Rev. Microbiol.* **2003**, *29*, 277–296.

214. Cherla, R. P.; Lee, S-Y.; Tesh, V. L. Shiga Toxins and Apoptosis. *FEMS Microbiol. Lett.* **2003**, *228*, 159–166.

215. Huang, D. B.; Okhuysen, P. C.; Jiang, Z-D.; Dupont, H. L. Enteroaggregative *Escherichia coli*: An Emerging Enteric Pathogen. *Am. J. Gastroenterol.* **2004**, *99*, 383–389.

216. Chapman, P. A. Methods Available for the Detection of *Escherichia coli* O157 in Clinical, Food and Environmental Samples. *World J. Microbiol. Biotechnol.* **2000**, *16*, 733–740.

217. Ateba, C. N.; Bezuidenhout, C. C. Characterization of *Escherichia coli* O157 Strains from Humans, Cattle and Pigs in the North-West Province, South Africa. *Int. J. Food Microbiol.* **2008,** *128,* 181–188.
218. Sallam, K. I.; Mohammed, M. A.; Ahdy, A. M.; Tamura, T. Prevalence, Genetic Characterization and Virulence Genes of Sorbitol-Fermenting *Escherichia coli* O157:H- and *E. coli* O157:H7 Isolated from Retail Beef. *Int. J. Food Microbiol.* **2013,** *165,* 295–301.
219. Martins, F. H.; Guth, B. E. C.; Piazza, R. M.; Leao, S. C.; Ludovico, A.; Ludovico, M. S.; Dahbi, G.; Marzoa, J.; Mora, A.; Blanco, J.; Pelayo, J. S. Diversity of Shiga Toxin-Producing *Escherichia coli* in Sheep Flocks of Parana State, Southern Brazil. *Vet. Microbiol.* **2015,** *175,* 150–156.
220. Jiang, Y.; Yin, S.; Dudley, E. G.; Cutter, C. N. Diversity of CRISPR Loci and Virulence Genes in Pathogenic *Escherichia coli* Isolates from Various Sources. *Int. J. Food Microbiol.* **2015,** *204,* 41–46.
221. Douellou, T.; Delannoy, S.; Ganet, S.; Mariani-Kurkdjian, P.; Fach, P.; Loukiadis, E.; Montel, M.; Thevenot-Sergentet, D. Shiga Toxin-Producing *Escherichia coli* Strains Isolated From Dairy Products—Genetic Diversity and Virulence Gene Profiles. *Int. J. Food Microbiol.* **2016,** *232,* 52–62.
222. Al-Arfaj, A. A.; Ali, M. S.; Hessain, A. M.; Zakri, A. M.; Dawoud, T. M.; Al-Maary, K. S.; Moussa, I. M. Phenotypic and Genotypic Analysis of Pathogenic *Escherichia coli* Virulence Genes Recovered from Riyadh, Saudi Arabia. *Saudi J. Biol. Sci.* **2016,** *23,* 713–717.
223. Ranjbar, R.; Hosseini, S.; Zahraei-Salehi, T.; Kheiri, R.; Khamesipour, F. Investigation on Prevalence of *Escherichia coli* Strains Carrying Virulence Genes *ipaH, estA, eaeA* and *bfpA* Isolated from Different Water Sources. *Asian Pac. J. Trop. Dis.* **2016,** *6,* 278–283.
224. Burgos, M. J. G.; Marquez, M. L. F.; Pulido, R. P.; Galvez, A.; Lopez, R. L. Virulence Factors and Antimicrobial Resistance in *Escherichia coli* Strains Isolated from Hen Egg Shells. *Int. J. Food Microbiol.* **2016,** *238,* 89–95.
225. Khan, S. B.; Zou, G.; Cheng, Y-T.; Xiao, R.; Li, L. Wu, B.; Zhou, R. Phylogenetic Grouping and Distribution of Virulence Genes in *Escherichia coli* Along the Production and Supply Chain of Pork Around Hubei, China. *J. Microbiol., Immunol. Infect.* **2017,** *50,* 382–385. DOI: 10.1016/j.jmii.2016.03.006.
226. McClure, P. *Escherichia coli*: Virulence, Stress Response and Resistance. In *Understanding Pathogen Behavior. Virulence, Stress Response and Resistance;* Griffiths, M., Ed.; CRC Press, Boca Raton, FL, USA: 2005; pp 240–278.
227. Carey, C. M.; Kostrzynska, M.; Thompson, S. *Escherichia coli* O157:H7 Stress and Virulence Gene Expression on Romaine Lettuce Using Comparative Real-Time PCR. *J Microbiol Methods* **2009,** *77,* 235–242.
228. Garcia-Heredia, A.; Garcia, S.; Merino-Mascorro, J. A.; Feng, P.; Heredia, N. Natural Plant Products Inhibits Growth and Alters the Swarming Motility, Biofilm Formation, and Expression of Virulence Genes in Enteroaggregative and Enterohemorrhagic *Escherichia coli*. *Food Microbiol.* **2016,** *59,* 124–132.
229. Paramithiotis, S.; Hadjilouka, A.; Drosinos, E. H. *Listeria* Pathogenicity Island 1. Structure and Function. In *Listeria monocytogenes: Food Sources, Prevalence and Management Strategies*; Hambrick, C., Nova Publishers: 2014; pp 265–282.

230. Sabet, C.; Lecuit, M.; Cabanes, D.; Cossart, P.; Bierne, H. LPXTG Protein InlJ (2005) a Newly Identified Internalin Involved in *Listeria monocytogenes* Virulence. *Infect. Immun.* 2005, *73*, 6912–6922.

231. Roche, S. M.; Velge, P.; Liu, D. Virulence Determination. In: *Hanbook of Listeria monocytogenes*; Liu D., CRC Press, Boca Raton, FL, USA: 2008; pp 241–270.

232. Bierne, H.; Sabet, C.; Personnic, N.; Cossart, P. Internalins: A Complex Family of Leucine-Rich Repeat-Containing Proteins in *Listeria monocytogenes*. *Microbes Infect.* **2007**, *9*, 1156–1166.

233. Liu, D.; Lawrence, M. L.; Austin, F. W.; Ainsworth, A. J. A Multiplex PCR for Species-and Virulence-Specific Determination of *Listeria monocytogenes*. *J. Microbiol. Methods* **2007**, *71*, 133–140.

234. Orsi, R. H.; den Bakker, H. C.; Wiedmann, M. *Listeria monocytogenes* Lineages: Genomics, Evolution, Ecology, and Phenotypic Characteristics. *Int. J. Med. Microbiol.* **2011**, *301*, 79–96.

235. Jamali, H.; Radmehr, B. Frequency, Virulence Genes and Antimicrobial Resistance of *Listeria* Spp. Isolated from Bovine Clinical Mastitis. *Vet J.* **2013**, *198*, 541–542.

236. Moreno, L. Z.; Paixão, R.; de Gobbi, D. D. S.; Raimundo, D. C.; Ferreira, T. S. P.; Moreno, A. M.; Hofer, E.; dos Reis, C. M. F.; Matté, G. R.; Matté, M. H. Phenotypic and Genotypic Characterization of a Typical *Listeria monocytogenes* and *Listeria innocua* Isolated from Swine Slaughterhouses and Meat Markets. *BioMed Res. Int.* **2014**, *2014*, Article ID 742032.

237. Soni, D. K.; Singh, M.; Singh, D. V.; Dubey, S. K. Virulence and Genotypic Characterization of *Listeria monocytogenes* Isolated from Vegetable and Soil Samples. *BMC Microbiol.* **2014**, *14*, 241.

238. Jacquet, C.; Doumith, M.; Gordon, J. I.; Martin, P. M.; Cossart, P.; Lecuit, M. A Molecular Marker for Evaluating the Pathogenic Potential of Foodborne *Listeria monocytogenes*. *J. Infect. Dis.* **2004**, *189*, 2094–2100.

239. van Stelten, A.; Simpson, J. M.; Ward, T. J.; Nightingale, K. K. Revelation by Single Nucleotide Polymorphism Genotyping that Mutations Leading to a Premature Stop Codon in *inlA* are Common Among *Listeria monocytogenes* Isolates from Ready-to-eat Foods but not Human Listeriosis Cases. *Appl. Environ. Microbiol.* **2010**, *76*, 2783–2790.

240. Shen, J.; Rumpb, L.; Zhang, Y.; Chen, Y.; Wang, X.; Meng, J. Molecular Subtyping and Virulence Gene Analysis of *Listeria monocytogenes* Isolates from Food. *Food Microbiol.* **2013**, *35*, 58–64.

241. Lomonaco, S.; Patti, R.; Knabel, S. J.; Civera, T. Detection of Virulence-Associated Genes and Epidemic Clone Markers in *Listeria monocytogenes* Isolates from PDO Gorgonzola Cheese. *Int. J. Food Microbiol.* 2012, 160, 76–79.

242. Kaur, S.; Malik, S. V. S.; Vaidya, V. M.; Barbuddhe, S. B. *Listeria monocytogenes* in Spontaneous Abortions in Humans and its Detection by Multiplex PCR. *J. Appl. Microbiol.* **2007**, *103*, 1889–1896.

243. Behari, J.; Youngman, P. Regulation of Hly Expression in *Listeria monocytogenes* by Carbon Sources and pH Occurs Through Separate Mechanisms Mediated by PrfA. *Infect. Immun.* **1998**, *66*, 3635–3642.

244. Ripio, M. T.; Vazquez-Boland, J. A.; Vega, Y.; Nair, S.; Berche, P. Evidence for Expressional Crosstalk Between the Central Virulence Regulator PrfA and the Stress

Response Mediator ClpC in *Listeria monocytogenes*. *FEMS Microbiol. Lett.* **1998,** *158,* 45–50.

245. Hanawa, T.; Fukuda, M.; Kawakami, H.; Hirano, H.; Kamiya, S.; Yamamoto, T. The *Listeria monocytogenes* DnaK Chaperone is Required for Stress Tolerance and Efficient Phagocytosis with Macrophages. *Cell Stress Chaperones* **1999,** *4,* 118–128.

246. Hanawa, T.; Yamanishi, S.; Murayama, S.; Yamamoto, T.; Kamiya, S. Participation of DnaK in Expression of Genes Involved in Virulence of *Listeria monocytogenes*. *FEMS Microbiol. Lett.* **2002,** *214,* 69–75.

247. Gaillot, O.; Pellegrini, E.; Bregenholt, S.; Nair, S.; Berche, P. The ClpP Serine Protease is Essential for the Intracellular Parasitism and Virulence of *Listeria monocytogenes*. *Mol. Microbiol.* **2000,** *35,* 1286–1294.

248. van der Veen, S.; Hain, T.; Wouters, J. A.; Hossain, H.; de Vos, W. M., Abee, T.; Chakraborty, T.; Wells-Bennik, M. H. J. The Heat-Shock Response of *Listeria monocytogenes* Comprises Genes Involved in Heat Shock, Cell Division, Cell Wall Synthesis, and the SOS Response. *Microbiology* **2007,** *153,* 3593–3607.

249. Phan-Thanh, L.; Gormon, T. Analysis of Heat and Cold Shock Proteins in *Listeria* by Two Dimensional Electrophoresis. *Electrophoresis* **1995,** *16,* 444–450.

250. Nelson, K. E.; Fouts, D. E.; Mongodin, E. F.; Ravel, J.; DeBoy, R. T.; Kolonay, J. F.; Rasko, D. A.; Angiuoli, S. V.; Gill, S. R.; Paulsen, I. T.; Peterson, J.; White, O.; Nelson, W. C.; Nierman, W.; Beanan, M. J.; Brinkac, L. M.; Daugherty, S. C.; Dodson, R. J.; Durkin, A. S.; Madupu, R.; Haft, D. H.; Selengut, J.; Van Aken, S.; Khouri, H.; Fedorova, N.; Forberger, H.; Tran, B. Kathariou, S.; Wonderling, L. D.; Uhlich, G. A.; Bayles, D. O.; Luchansky, J. B.; Fraser, C. M. Whole Genome Comparisons of Serotype 4b and 1/2a Strains of the Food-Borne Pathogen *Listeria monocytogenes* Reveal New Insights into the Core Genome Components of this Species. *Nucleic Acids Res.* **2004,** *32,* 2386–2395.

251. Schmid, B.; Klumpp, J.; Raimann, E.; Loessner, M. J.; Stephan, R.; Tasara, T. Role of Cold Shock Proteins in Growth of *Listeria monocytogenes* Under Cold and Osmotic Stress Conditions. *Appl. Environ. Microbiol.* **2009,** *75,* 1621–1627.

252. Durack, J.; Ross, T.; Bowman, J. P. Characterisation of the Transcriptomes of Genetically Diverse *Listeria monocytogenes* Exposed to Hyperosmotic and Low Temperature Conditions Reveal Global Stress-Adaptation Mechanisms. *PLoS ONE* **2013,** *8,* e73603.

253. Cotter, P. D.; Gahan, C. G.; Hill, C. A Glutamate Decarboxylase System Protects *Listeria monocytogenes* in Gastric Fluid. *Mol. Microbiol.* **2001,** *40,* 465–475.

254. Cotter, P. D.; Gahan, C. G. M.; Hill, C. Analysis of the Role of the *Listeria monocytogenes* F0F1ATPase Operon in the Acid Tolerance Response. *Int. J. Food Microbiol.* **2000,** *60,* 137–146.

255. Milecka, D.; Samluk, A.; Wasiak, K.; Krawczyk Balska, A. An Essential Role of a Ferritin Like Protein in Acid Stress Tolerance of *Listeria monocytogenes*. *Arch. Microbiol.* **2015,** *197,* 347–351.

256. Sleator, R. D.; Gahan, C. G.; Abee, T.; Hill, C. Identification and Disruption of BetL, a Secondary Dlycine Betaine Transport System Linked to the Salt Tolerance of *Listeria monocytogenes* LO28. *Appl. Environ. Microbiol.* **1999,** *65,* 2078–2083.

257. Sleator, R. D.; Gahan, C. G. M.; Hill, C. Identification and Disruption of the proBA Locus in *Listeria monocytogenes*: Role of Proline Biosythesis in Salt Tolerance and Murine Infection. *Appl. Environ. Microbiol.* **2001,** *67,* 2571–2577.

258. Sleator, R. D.; Gahan, C. G. M.; Hill, C. Mutations in the Listerial proB Gene Leading to Proline Overproduction: Effects on Salt Tolerance and Murine Infection. *Appl. Environ. Microbiol.* **2001,** *67,* 4560–4565.

259. Sleator, R. D.; Hill, C. A Novel Role for the LisRK Two-Component Regulatory System in Listerial Osmotolerance. *Clin. Microbiol. Infect.* **2005,** *11,* 599–601.

260. Duche, O.; Tremoulet, F.; Glaser, P. Labadie, J. (2002) Salt Stress Proteins Induced in *Listeria monocytogenes. Appl. Environ. Microbiol.* **2005,** *68,* 1491–1498.

261. Brondsted, L.; Kallipolitis, B. H.; Ingmer, H.; Knochel, S. kdpE and a Putative RsbQ Homologue Contribute to Growth of *Listeria monocytogenes* at High Osmolarity and Low Temperature. *FEMS Microbiol. Lett.* **2003,** *219,* 233–239.

262. Renzoni, A.; Klarsfeld, A.; Dramsi, S.; Cossart P. Evidence that PrfA, the.Pleiotropic Activator of Virulence genes in *Listeria monocytogenes*, can be Present but Inactive. *Infect. Immun.* **1997,** *65,* 1515–1518.

263. Milenbachs, A-A.; Brown, D-P.; Moors, M.; Youngman, P. Carbon-Source Regulation of Virulence Gene Expression in *Listeria monocytogenes. Mol. Microbiol.* **1997,** *23,* 1075–1085.

264. Milenbachs Lukowiak, A.; Mueller, K-J.; Freitag, N-E.; Youngman, P. (2004) Deregulation of *Listeria monocytogenes* Virulence Gene Expression by Two Distinct and Semi-Independent Pathways. *Microbiology* **1997,** *150,* 321–333.

265. Bowman, J-P.; Bittencourt, C-R.; Ross, T. Differential Gene Expression of *Listeria monocytogenes* During High Hydrostatic Pressure Processing. *Microbiology* **2008,** *154,* 462–475.

266. Kastbjerg, V-G.; Larsen, M-H.; Gram, L.; Ingmer, H. Influence of Sublethal Concentrations of Common Disinfectants on Expression of Virulence Genes in *Listeria monocytogenes. Appl. Environ. Microbiol.* **2010,** *76,* 303–309.

267. Hadjilouka, A.; Mavrogiannis, G.; Mallouchos, A.; Paramithiotis, S.; Mataragas, M.; Drosinos, E. H. Effect of Lemongrass Essential Oil on *Listeria monocytogenes* Gene Expression. *LWT--Food Sci. Technol.* **2017,** *77,* 510–516.

268. Romanova, N. A.; Wolffs, P. F.; Brovko, L. Y.; Griffiths, M. W. Role of Efflux Pumps in Adaptation and Resistance of *Listeria monocytogenes* to Benzalkonium Chloride. *Appl. Environ. Microbiol.* **2006,** *72,* 3498–3503.

269. Elhanafi, D.; Dutta, V.; Kathariou, S. Genetic Characterization of Plasmid-Associated Benzalkonium Chloride Resistance Determinants in a *Listeria monocytogenes* Strain from the 1998–1999 Outbreak. *Appl. Environ. Microbiol.* **2010,** *76,* 8231–8238.

270. van der Veen, S.; Abee, T. Importance of SigB for *Listeria monocytogenes* Static and Continuous-Flow Biofilm Formation and Disinfectant Resistance. *Appl. Environ. Microbiol.* **2010,** *76,* 7854–7860.

271. Stasiewicz, M-J.; Wiedmann, M.; Bergholz, T-M. The Transcriptional Response of *Listeria monocytogenes* During Adaptation to Growth on Lactate and Diacetate Includes Synergistic Changes that Increase Fermentative Acetoin Production. *Appl. Environ. Microbiol.* **2011,** *77,* 5294–5306.

272. Dutta, V.; Elhanafi, D.; Kathariou, S. Conservation and Distribution of the Benzalkonium Chloride Resistance Cassette bcrABC in *Listeria monocytogenes. Appl. Environ. Microbiol. 2013, 79,* 6067–6074.

273. Shi, H.; Trinh, Q.; Xu, W.; Luo, Y.; Tian, W.; Huang, K. The Transcriptional Response of Virulence Genes in *Listeria monocytogenes* During Inactivation by Nisin. *Food Control* **2013**, *31*, 519–524.

274. Liu, X; Basu, U; Miller, P.; McMullen, L. M. Stress Response and Adaptation of *Listeria monocytogenes* 08–5923 Exposed to a Sublethal Dose of Carnocyclin. *Appl. Environ. Microbiol.* **2014**, *80*, 3835–3841.

275. Pleitner, A. M.; Trinetta, V.; Morgan, M. T.; Linton, R. L.; Oliver, H. F. Transcriptional and Phenotypic Responses of *Listeria monocytogenes* to Chlorine Dioxide. *Appl. Environ. Microbiol.* **2014**, *80*, 2951–2963.

276. Laursen, M. F.; Bahl, M. I.; Licht, T. R.; Gram, L.; Knudsen, G. M. A Single Exposure to a Sublethal Pediocin Concentration Initiates a Resistance-Associated Temporal Cell Envelope and General Stress Response in *Listeria monocytogenes*. *Environ. Microbiol.* **2015**, *17*, 1134–1151.

277. Hadjilouka, A.; Nikolidakis, K.; Paramithiotis, S.; Drosinos, E. H. Effect of Co-culture with Enterocinogenic *E. faecium* on *L. monocytogenes* Key Virulence Gene Expression. AIMS. *Microbiology* **2016**, *2*, 304–315.

278. NicAogáin, K.; O'Byrne, C. P. The Role of Stress and Stress Adaptations in Determining the Fate of the Bacterial Pathogen *Listeria monocytogenes* in the Food Chain. *Front. Microbiol.* **2016**, *7*, 1865.

279. Alessandria, V.; Rantsiou, K.; Dolci, P.; Zeppa, G.; Cocolin, L-A. Comparison of Gene Expression of *Listeria monocytogenes* in Vitro and in the Soft Cheese Crescenza. *Int. J. Dairy Technol.* **2013**, *66*, 83–89.

280. Olesen, I.; Thorsen, L.; Jespersen, L. Relative Transcription of *Listeria monocytogenes* Virulence Genes in Liver pâtés with Varying NaCl Content. *Int. J. Food Microbiol.* **2010**, *141*, S60–S68.

281. Rantsiou, K.; Greppi, A.; Garosi, M.; Acquadro, A.; Mataragas, M.; Cocolin, L. Strain Dependent Expression of Stress Response and Virulence Genes of *Listeria monocytogenes* in Meat Juices as Determined by Microarray. *Int. J. Food. Microbiol.* **2012**, *152*, 116–122.

282. Rantsiou, K.; Mataragas, M.; Alessandria, V.; Cocolin, L. Expression of Virulence Genes of *Listeria monocytogenes* in Food. *J. Food Saf.* **2012**, *32*, 161–168.

283. Duodu, S.; Holst-Jensen, A.; Skjerdal, T.; Cappelier, J-M.; Pilet, M-F.; Loncarevic, S. Influence of Storage Temperature on Gene Expression and Virulence Potential of *Listeria monocytogenes* Strains Grown in a Salmon Matrix. *Food Microbiol.* **2010**, *27*, 795–801.

284. Hadjilouka, A.; Molfeta, C.; Panagiotopoulou, O.; Paramithiotis, S.; Mataragas, M.; Drosinos, E. H. Expression *of Listeria monocytogenes* Key Virulence Genes During Growth in Liquid Medium, on Rocket and Melon at 4, 10 and 30ºC. *Food Microbiol.* **2016**, *55*, 7–15.

285. Williams, J-R.; Thayyullathil, C.; Freitag, N-E. Sequence Variations Within PrfA DNA Binding Sites and Effects on *Listeria monocytogenes* Virulence Gene Expression. *J. Bacteriol.* **2000**, *182*, 837–841.

286. Kazmierczak, M-J.; Wiedmann, M.; Boor, K-J. Contributions of *Listeria monocytogenes* sB and PrfA to Expression of Virulence and Stress Response Genes During Extra-and Intracellular Growth. *Microbiology* **2006**, *152*, 1827–1838.

287. Sabbagh, S. C.; Forest, C. G.; Lepage, C.; Leclerc, J-M.; Daigle, F. S. So Similar yet so Different: Uncovering Distinctive Features in the Genomes of *Salmonella enterica* Serovars Typhimurium and Typhi. *FEMS Microbiol. Lett.* **2010**, *305*, 1–13.

288. Garai, P.; Gnanadhas, D. P.; Chakravortty, D. *Salmonella enterica* Serovars Typhimurium and Typhi as Model Organisms: Revealing Paradigm of Host-Pathogen Interactions. *Virulence* **2012**, *3*, 377–388.

289. Rosselin, M.; Abed, N.; Virlogeux-Payant, I; Bottreau, E.; Sizaret, P. Y.; Velge, P.; Wiedemann, A. Heterogeneity of Type III Secretion System (T3SS)-1-Independent Entry Mechanisms Used by *Salmonella Enteritidis* to Invade Different Cell Types. *Microbiology* **2011**, *157*, 839–847.

290. Velge, P.; Wiedemann, A.; Rosselin, M.; Abed, N.; Boumart, Z.; Chausse, A. M.; Grepinet, O.; Namdari, F.; Roche, S. M.; Rossignol, A.; Virlogeux-Payant, I. Multiplicity of *Salmonella* Entry Mechanisms, a New Paradigm for *Salmonella* Pathogenesis. *MicrobiologyOpen* **2012**, *1*, 243–258.

291. Ramos-Morales, F. Impact of *Salmonella enterica* Type III Secretion System Effectors on the Eukaryotic Host Cell. *ISNR Cell Biol.* **2012**, *2012*, Article ID 787934.

292. Blondel, C. J.; Jimenez, J. C.; Contreras, I.; Santiviago, C. A. Comparative Genomic Analysis Uncovers 3 Novel Loci Encoding Type Six Secretion Systems Differentially Distributed in *Salmonella* serotypes. *BMC Genomics* **2009**, *10*, 354.

293. Fookes, M.; Schroeder, G. N.; Langridge, G. C.; Blondel, C. J.; Mammina, C.; Connor, T. R.; Seth-Smith, H.; Vernikos, G. S.; Robinson, K. S.; Sanders, M.; Petty, N. K.; Kingsley, R. A.; Bäumler, A. J.; Nuccio, S. P., Contreras, I.; Santiviago, C. A.; Maskell, D.; Barrow, P.; Humphrey, T.; Nastasi, A.; Roberts, M.; Frankel, G.; Parkhill, J.; Dougan, G.; Thomson, N. R. *Salmonella bongori* Provides Insights into the Evolution of the *Salmonellae*. *PLoS Pathog.* **2011**, *7*, e1002191.

294. Pezoa, D.; Blondel, C. J.; Silva, C. A.; Yang, H-J.; Andrews-Polymenis, H.; Santiviago, C. A.; Contreras, I. Only One of the Two Type VI Secretion Systems Encoded in the *Salmonella enterica* Serotype Dublin Genome is Involved in Colonization of the Avian and Murine Hosts. *Vet. Res.* **2014**, *45*, 2.

295. Blondel, C. J.; Jimenez, J. C.; Leiva, L. E.; Alvarez, S. A.; Pinto, B. I.; Contreras, F.; Pezoa, D.; Santiviago, C. A.; Contreras, I. The Type VI Secretion System Encoded in *Salmonella* Pathogenicity Island 19 is Required for *Salmonella enterica* Serotype Gallinarum Survival Within Infected Macrophages. *Infect. Immun.* **2013**, *81*, 1207–1220.

296. Blondel, C. J.; Yang, H. J.; Castro, B.; Chiang, S.; Toro, C. S.; Zaldivar, M.; Contreras, I.; Andrews-Polymenis, H. L.; Santiviago, C. A. Contribution of the Type VI Secretion System Encoded in SPI-19 to Chicken Colonization by *Salmonella enterica* Serotypes Gallinarum and Enteritidis. *PLoS One* **2010**, *5*, e11724.

297. Kuang, D.; Xu, X.; Meng, J.; Yang, X.; Jin, H.; Shi, W.; Pan, H.; Liao, M.; Su, X.; Shi, X.; Zhang, J. Antimicrobial Susceptibility, Virulence Gene Profiles and Molecular Subtypes of *Salmonella* Newport Isolated from Humans and Other Sources. *Infect. Genet. Evol.* **2015**, *36*, 294–299.

298. Zadernowska, A.; Chajecka-Wierzchowska, W. Prevalence, Biofilm Formation and Virulence Markers of *Salmonella* sp. and *Yersinia enterocolitica* in Food of Animal Origin in Poland. LWT--*Food Sci. Technol.* **2017**, *75*, 552–556.

299. Li, K.; Ye, S.; Alali, W. Q.; Wang, Y.; Wang, X.; Xia, X.; Yang, B. Antimicrobial Susceptibility, Virulence Gene and Pulsed-Field Gel Electrophoresis Profiles of *Salmonella enterica* Serovar Typhimurium recovered from retail raw chickens, China. *Food Control* **2017,** *72,* 36–42.

300. Suez, J.; Porwollik, S.; Dagan, A.; Marzel, A.; Schorr, Y. I.; Desai, P. T.; Agmon, V.; McClelland, M.; Rahav, G.; Gal-Mor, O.(Virulence Gene Profiling and Pathogenicity Characterization of Non-typhoidal *Salmonella* Accounted for Invasive Disease in Humans. *PLoS ONE* **2013,** *8,* e58449.

301. Spector, M. P.; Kenyon, W. J. Resistance and Survival Strategies of *Salmonella enterica* to Environmental Stresses. *Food Res. Int.* **2012,** *45,* 455–481.

302. Jakociune, D.; Herrero-Fresno, A.; Jelsbak, L.; Olsen, J. E. Highly Expressed Amino Acid Biosynthesis Genes Revealed by Global Gene Expression Analysis of *Salmonella enterica* Serovar Enteritidis During Growth in Whole Egg are Not Essential for this Growth. *Int. J. Food Microbiol.* **2016,** *224,* 40–46.

303. Doulgeraki, A. I.; Papaioannou, M.; Nychas, G-JE. Targeted Gene Expression Study of *Salmonella enterica* During Biofilm Formation on Rocket Leaves. *LWT—Food Sci. Technol.* **2016,** *65,* 254–260.

304. Lindstedt, B. A.; Heir, E.; Gjernes, E.; Vardund, T.; Kapperud, G. DNA Fingerprinting of Shiga-Toxin Producing *Escherichia coli* O157 Based on Multiple-Locus Variable-Number Tandem-Repeats Analysis (MLVA). *Ann. Clin. Microbiol. Antimicrob.* **2003,** *2,* 12.

CHAPTER 2

MICROBIAL BETA GLUCANASE IN AGRICULTURE

LEKSHMI K. EDISON, S. SHIBURAJ, and N. S. PRADEEP*

*Microbiology Division, Jawaharlal Nehru Tropical Botanic Garden and Research Institute, Palode, Trivandrum, Kerala 695562, India, *E-mail: drnspradeep@gmail.com*

CONTENTS

ABSTRACT

β-Glucanases are a group of glycosyl hydrolases, which hydrolyse β-glucan substrates and have been broadly characterised in microorganisms like bacteria and fungi. Extracellular β-glucanases act as exo and endo hydrolases and are further categorized into different types based on the substrate specificity. Comprehensive exhaustion of β-glucan substrates needs the synergic action of various β-glucanases. They have great ecological significance and extensive applications in various industries like brewing, textile, food and also agriculture. The agricultural applications, which include pathogen and disease control, seed germination, a supplement of farm animal nutrition, production of plant and fungal protoplasts, etc. are currently attaining increased scientific attention for averting present and future agricultural crises. This chapter discusses the various β-glucanases, their significance and applications in agriculture. Microbial enzyme practices decrease environmental degradation and, thus maintain sustainable agriculture by economically increasing productivity and food safety.

2.1   INTRODUCTION

Microorganisms are the simplest forms of life, generating great interest due to their precise internal chemistry. Active natural enzymes from microorganisms are capable of catalysing numerous chemical reactions, depending on the substrate binding domain. Employing enzymes for practical applications and industrial usage can be traced centuries back. Obtained from readily accessible microbial sources, they have been preferred because they are affordable, eco-friendly, efficient, have low energy consumption and improve product properties. The growing enzyme studies as well as the knowledge regarding their benefits surprisingly increase their commercial applications every year.[1] Microbial enzymes are involved insupporting the agricultural sciences, too. Their role in agricultural technique development maximizes eco-friendly agricultural production and reduces labour. The emerging necessity of higher food production and sustainability requires the enhanced use of biocontrol agents and fertilizers, which adversely affects the global biodiversity. The increasing use of microorganisms and their valuable products contribute to eco-friendly sustainable agricultural development.

Cereals, which are a major source of animals' nutrients, comprise different quantities of β-glucans.[2] β-glucans are the chief cell wall component in the form of non-starch polysaccharide and constitute more than 70% of plant cell wall as β-D-glucopyranose residue. They are mainly present in cereals like oats and barley and also in fungal and microbial cell walls.[3] In plant cells, they act as a vital source of stockpiled glucose for growing seedlings. Majority of β-glucans are water soluble, but some are covalently bound to cell wall proteins. Mixed glycosidic linkages such as β-1,4, β-1,3 and β-1,6 make them more soluble than major cellulose because they contain only long β-1,4-linked glucose units. β-glucans have both positive and negative properties. In human beings, their ingestion increases satiety, thus diminishing obesity. However, β-glucans have many adverse effects, too. In poultry, they intensify the energy metabolism, particularly in mono-gastric animals and are also responsible for deprived nutritive value.[4] In brewing industry, high molecular weight β-glucans create severe problems by increasing brewer mash viscosity and turbidity which in turn causes reduction in yields and leaves scums in beer.[5] Several pathogenic fungi that are detrimental in plants contain β-glucans in their cell wall.

Many complications related to β-glucan can be effectively solved by the sensible use of β-glucan degrading enzymes typically known as β-glucanases or β-D-glucan hydrolases, a glycosyl hydrolase, respon-sible for the hydrolysis of insoluble β-glucan substrates. The hydrolysis of β-glucan molecules by the action of β-glucanases leads to the produc-tion of D-glucose, thus serving as a carbon source. β-glucanase can act both as exo- as well as endo-hydrolase. Endo-β-glucanase randomly slices β-glucan chain internally while exo-β-glucanase acts on the nonreducing ends where they release glucose residues. On the basis of the type of glycosidic linkage cleaved, they are further classified as 1,3-β-glucanases, 1,4-β-glucanases and 1,6-β-glucanases.[6]

A wide range of β-glucanases such as bacterial or fungal β-glucanases are exclusively active on β-glucan substrates and cleave within the main chain of mixed linkage β-glucan with a different point of action. This is significant in choosing specific substrates for the assay of these groups of enzymes.[7] The known specificity and mechanism of the enzyme activity within the immense substrate range would allow us to realize that the β-glucanases are diverse from cellulases. These enzymes are believed to have appreciable ecological implication and have commercial usage in several industrial productions. β-glucanase application is well-established in many biotechnological indus-tries for processes such as preparation of protoplasts from fungi and bacteria; and also from plant cells, beer production, wine extract clarification, barley β-glucan degradation for animal feed enzyme industry, saccharification of agricultural and industrial wastes and coffee processing. It is also used as an additive in detergents and bio-control agent for disease protection in plants.

2.2   BETA GLUCANASE: DIFFERENT TYPES

β-glucanases are predominant O-Glycosyl hydrolases (EC:3.2.1.x) that cleave the glycosidic bond between two or more carbohydrates, or between a carbohydrate and a noncarbohydrate moiety via general acid catalysis necessitating a proton donor and a nucleophile base. Based on their action patterns, they are divided into two types exo- and endo-hydrolases. Exo-hydrolase catalyses the hydrolysis of the nonreducing ends of the β-glucan chain and sequentially releases glucose residues as the sole hydrolysis product. Endo-hydrolase cleaves β-glycosidic linkages at random sites laterally in the polysaccharide chains and release smaller oligosaccharides. According to the type of cleavages of β-glucoside linkages and their hydrolytic action patterns against specific substrates, they have been diversified into numerous types.

2.2.1   ENDO-1,3-β-GLUCANASE (EC 3.2.1.39)

Endo-1,3-β-Glucanase hydrolyse internal 1,3-β-D-glucosidic linkages in 1,3-β-D-glucans. Commonly used synonyms are glucan endo-1,3-beta-D-glucosidase; 3-beta-D-glucan glucanohydrolase; 1–3-β-glucan 3-glucanohydrolase; 1–3-β-glucan endohydrolase; 1,3-β-D-glucan 3-glucanohydrolase, β-1,3-glucanase; callase; endo-(1,3)-β-D-glucanase; endo-(1→3)-β-D-glucanase; endo-1,3-β-D-glucanase; endo-1,3-β-glucanase; endo-1,3-β-glucosidase; kitalase; laminaranase; laminarinase; oligo-1,3-glucosidase and 1,3-β-D-glucan glucanohydrolase (Fig. 2.1). In plants, it has a vital role in functional and developmental processes, as well as in microbial pathogen defence mechanisms, primarily on antifungal activity. It hydrolyses substrates like curdlan (Fig. 2.2), laminarin, paramylon and pachyman.[8]

FIGURE 2.1   Molecular model of endo-1,3-beta-glucanase from *Hordeum vulgare*. *Source:* http://www.ebi.ac.uk/thornton-srv/databases/cgi-bin/enzymes/GetPage. pl?ec_number=3.2.1.39

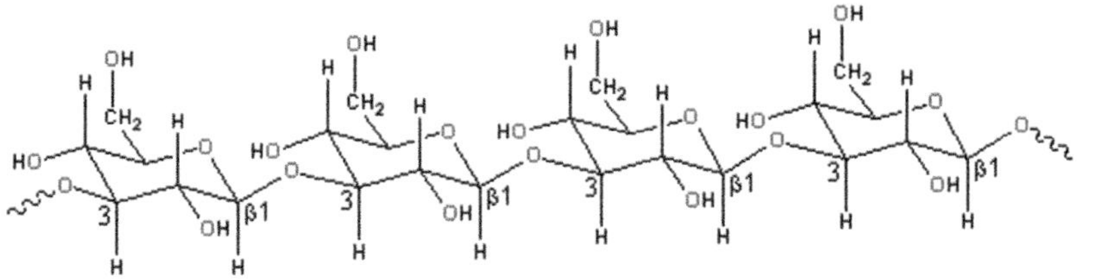

FIGURE 2.2 Chemical structure of curdlan, unbranched linear 1–3-β-D glucan with no side chains. *Source:* http://www1.lsbu.ac.uk/water/curdlan.html

2.2.2 ENDO-1,4-β-GLUCANASE (EC 3.2.1.4)

Endo-1,4-β-Glucanase brings about the endohydrolytic cleavage of (1,4)-β-glucosyl linkages in various substrates (Fig. 2.3). Synonyms include 9.5 cellulase; β-1,4-endoglucan hydrolase; β-1,4-glucanase; alkali cellulose; cellulase A; celludextrinase; cellulase A 3; cellulosin AP; endo-1,4-β-D-glucanase; endoglucanase D; pancellase SS; endo-1,4-β-glucanase; carboxymethyl cellulose; endoglucanase; endo-1,4-β-D-glucanohydrolase and 1,4-(1,3; 1,4)-β-D-glucan 4-glucanohydrolase (Fig. 2.4). It has been associated with cellulose or noncellulosic polysaccharide hydrolysis during cell wall breakdown and grain germination. There is also some evidence that it has important role in cellulose synthesis as a part of cell growth. An endo-(1,4)-β-glucanase, KORRIGAN, has significant role in cellulose synthesis.[9] Specific substrates for endo-1,4-β-glucanase are cellulose, avicel, lichenin and cereal β-D-glucans.

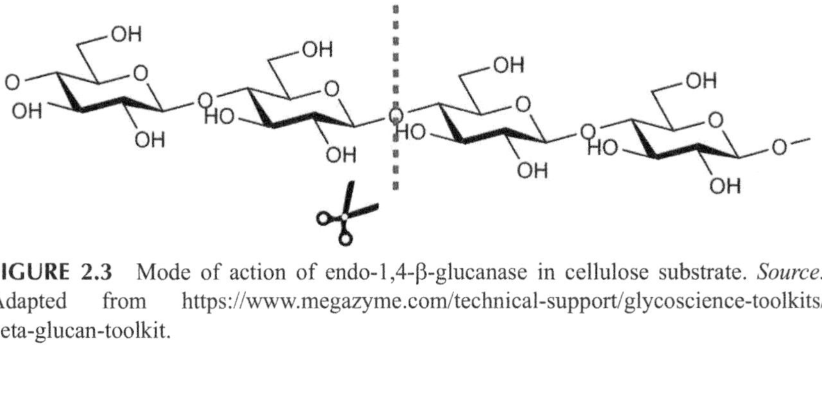

FIGURE 2.3 Mode of action of endo-1,4-β-glucanase in cellulose substrate. *Source:* Adapted from https://www.megazyme.com/technical-support/glycoscience-toolkits/beta-glucan-toolkit.

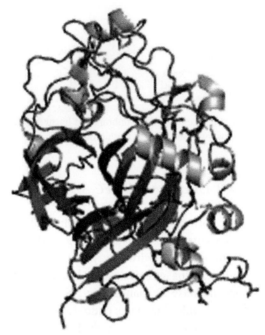

FIGURE 2.4   Molecular model of endo-1,4-beta-glucanase from *Humicola insolens.*

2.2.3   *ENDO-1,6-β-GLUCANASE (EC 3.2.1.75)*

Endo-1,6-β-Glucanase hydrolysis (1–6)-linkages randomly in (1–6)-beta-D-glucans. Its synonyms are β-1,6-glucan 6-glucanohydrolase; β-1,6-glucan hydrolase; β-1,6-glucanase-pustulanase; β-1→6-Glucan hydrolase; endo-(1→6)-beta-D-glucanase; endo-1,6-β-glucanase; endo-β-1,6-glucanase; endo-1,6-β-D-glucanase; β-1,6-glucanase; 1,6-β-D-glucan glucanohydrolase. The specific substrates include lutean, pustulan (Fig. 2.5) and 1,6-oligo-beta-D-glucosides. 1,6-β-glucans are commonly found in lichens and fungal cell walls.

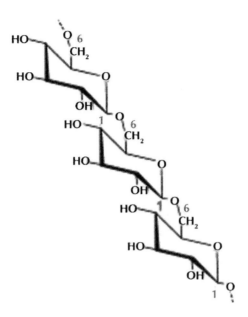

FIGURE 2.5 Schematic structure of Pustulan, a-1,6-β-glucan. Reprinted with permission from Martin K, McDougall BM, McIlroy S, Jayus, Jiezhong Chen J, Seviour RJ (2007) Biochemistry and molecular biology of exocellular fungal β-(1,3)- and β -(1,6)-glucanases. Microbiol Rev 31: 168–192. © 2007 Oxford University Press.

2.2.4 ENDO-1,3 (4)-β-GLUCANASE (EC 3.2.1.6)

It catalyses the endohydrolysis of (1→3) or (1→4)-linkages within the β-D-glucan chains when the reducing group of glucose residue is participated in the linkage to be hydrolysed and replaced at C-3. Other names are endo-1,3-β-D-glucanase; laminarinase; β-1,3-glucanase; β-1,3–1,4-glucanase; endo-1,3-β-glucanase; endo-β-1,3 (4)-glucanase; endo-β-1,3–1,4-glucanase; endo-β-(1→3)-D-glucanase; endo-1,3–1,4-β-D-glucanase; endo-β-(1–3)-D-glucanase; endo-β-1,3-glucanase IV; endo-1,3-β-D-glucanase and 1,3-(1,3; 1,4)-β-D-glucan 3 (4)-glucanohydrolase (Fig. 2.6). Specific substrates for endo-1,3 (4)-β-glucanase include lichenin, laminarin and cereal D-glucans.

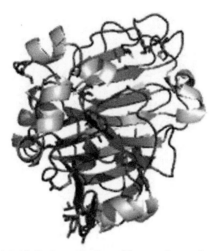

FIGURE 2.6   Endo-1,3 (4)-β-glucanase from *Phanerochaete chrysosporium.*

2.2.5   ENDO-β-1,3–1,4-GLUCANASE (EC 3.2.1.73)

It is certainly different from endo-1,3 (4)-β-glucanase since it acts only on cereal β-glucans and lichenin, not on β-glucans comprising individual 1,3-or 1,4-bonds. It hydrolyses (1→4)-β-D-glucosidic bonds in β-D-glucans comprising (1→3)-and (1→4)-bonds (Fig. 2.7). Synonyms are licheninase; 1,3; 1,4-β-glucan 4-glucanohydrolase; 1,3; 1,4-β-glucan endohydrolase; β-(1→3), (1→4)-D-glucan 4-glucanohydrolase; lichenase; endo-β-1,3–1,4 glucanase; mixed linkage β-glucanase; β-glucanase and 1,3–1,4-β-D-glucan 4-glucanohydrolase (Fig. 2.8).

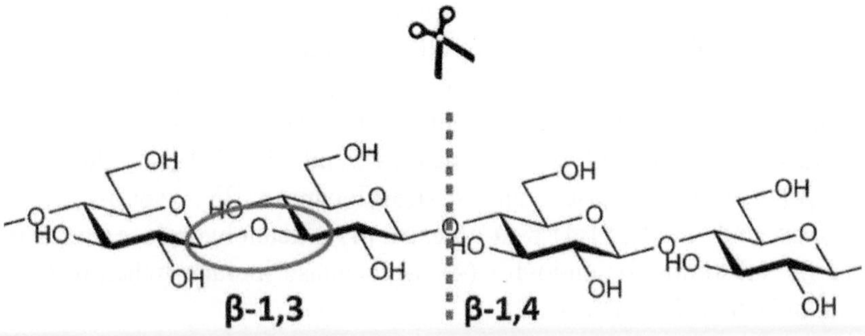

β-1,3   **β-1,4**

FIGURE 2.7   Mode of action of endo-1,3–1,4-β-glucanase. *Source:* Adapted from https://www.megazyme.com/technical-support/glycoscience-toolkits/beta-glucan-toolkit.

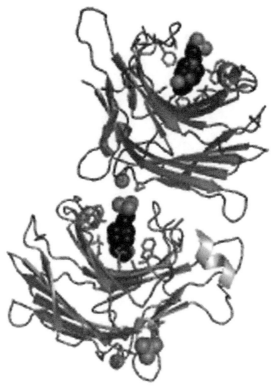

FIGURE 2.8 Endo-1,3–1,4-β-glucanase from *Paenibacillus macerans. Source:* http://www.ebi.ac.uk/thornton-srv/databases/cgi-bin/pdbsum/GetPage. pl?pdbcode=1ajk&template=main.html

2.2.6   EXO-1,3-β-GLUCANASE (EC 3.2.1.58)

Enzyme exo-1,3-β-glucanases acts on the nonreducing tails of (1→3)-β-D-glucans and releases α-glucose. Other names of exo-1,3-β-glucanases are 1,3-β-glucosidase; β-1,3-glucan exo-hydrolase; exo (1→3)-β-glucanase; exo-1,3-β-D-glucanase; exo-1,3-β-glucosidase; exo-β-(1→3)-D-glucanase; exo-β-(1→3)-glucanohydrolase; exo-β-1,3-D-glucanase; exo-β-1,3-glucanase and 1,3-β-glucan glucohydrolase (Fig. 2.9).

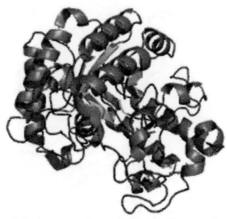

FIGURE 2.9 Exo-1,3-β-glucanase from *Candida albicans. Source:* http://www.ebi.
ac.uk/thornton-srv/databases/PDBsum/cz/1cz1//chainA.jpg

2.2.7 EXO-1,4-β-GLUCANASE (EC 3.2.1.91)

It hydrolyses (1→4)-β-D-glucosidic bonds in cellulose and cellote-
traose and liberating cellobiose from the nonreducing tails of the chains.
Alternative terms are cellulose 1,4-β-cellobiosidase (nonreducing end);
1,4-β-glucan cellobiosidase; β-1,4-glucan cellobiohydrolase; β-1,4-glucan
cellobiosylhydrolase; C1 cellulase; CBH 1; avicelase; cellobiohydrolase; cellobiohydrolase I; cellobiosidase; exo-β-1,4-glucan cellobiohydrolase; exo-cellobiohydrolase; exoglucanase; exocellobiohydrolase;
1,4-β-cellobiohydrolase and exo-1,4-β-D-glucanase (Fig. 2.10).

FIGURE 2.10 Three-dimensional structures of cellulose-binding domains of exo-1,4-
β-glucanase from *Trichoderma reesei. Source:* http://www.ebi.ac.uk/thornton-srv/
databases/PDBsum/az/1az6//chainA.jpg

2.3 BETA GLUCANASE AS A BIOCONTROL AGENT

One of the major issues in crop production is pest and disease control. Fungi are the most conspicuous organisms inducing a great deal of damages in plants. According to Fisher et al., 2012, diseases caused by fungi are highly lethal and cause severe host extinctions.[10] Cell walls of pathogenic fungi like *Sclerotium rolfsii, Rhizoctonia solani,* etc. are predominantly constituted by polysaccharides like β-1,3-glucan, β-1,6 glucan, chitin, β-1,3–1,4-glucan, galactomannan, etc. are crucial for disease transmission and pathogenesis. These fungal-specific polysaccharides are the main targets of antifungal agents. The second major cell wall polysaccharide in the fungal pathogen is Beta-glucan, which comes after chitin. Some evidence propose that β-glucan plays a prominent role in fungal pathogenesis in plants.[11] Enzyme β-Glucanase commonly known as pathogenesis-related protein (PR) collapse the pathogenic fungi by cleaving the β-glucan polysaccharide and leads to the leakage of cell contents. Synergistic action of different β-glucanases need complete degradation of ß-glucans cell wall. Microorganisms having the ability to produce β-Glucanase accomplish extensive crop protection and also promote plant growths.[12, 13]

Pathogenic fungus *Botrytis cinerea* causes widespread *Botrytis* blight or gray mold rot in many field crops all around the world.[14] Endo-β-1,3–1,4-glucanases produced by *Paenibacillus polymyxa* strain A21 has shown significant antagonistic activity against *Botrytis cinerea*. Antifungal activity was shown to improve by cloning and expression of glucanase gene from *P. polymyxa* strain A21. Tremendous inhibitory properties were observed in *B. cinerea* spore germination and mycelial growth.[15] Endo-β-1,3–1,4-glucanase from *P. polymyxa* also showed antifungal activity against *Pyricularia oryzae*.[16] The recombinant *P. polymyxa* β-1,3–1,4-glucanases also exhibited significant antagonism against fungi, such as *Colletotrichum musae* and *F. oxysporum*.[17] *Colletotrichum gloeosporioides* and *Curvularia affinis* are major fungal pathogens in oil palms and broadly attack *Elaeis guinensis* Jacq. leaves. The disease caused by these organisms is very harmful and also extends to lifespan breeding as well as immature plants. β-glucanase produced from *Bacillus clausii, B. subtilis, B. circulans, B. halodurans, B. licheniformis, Arthrobacter* sp. *and Pseudomonas cepacia* are good inhibitors of these fungal pathogens. Purified β-glucanase from *Bacillus subtilis* SAHA 32.6 strain exhibited better activity than crude enzyme.[18]

Lysobacter species have been documented as prospective biological control agents. They are most effective in wide range of plant pathogens like bacteria, oomycetes, fungi and nematodes by abolishing its growth and infection activity. *Lysobacter enzymogenes* is the most commonly reported biocontrol effective strain. Palumbo et al., (2003) found that three beta-1,3-glucanase genes responsible for the production of β-1, 3-glucanase enzymes in *L. enzymogenes* strains C3 and N4–7, enable to break down the cell walls.[19] The enzyme activity was confirmed by the mutagenesis of beta-1,3-glucanase genes in strain C3, showed reduced biological control.[19, 20]

Streptomyces species are also frequently used for producing lysing enzymes against a large spectrum of pathogenic fungi. An important factor of *Actinomycetes* fungus antagonism is the extracellular enzymes having β-glucanase activities. *Streptomyces* sp. EF-14 produced β-1,6 glucanase and β-1,3 glucanase have been recognised as the best effective fungal antagonist of *Phytophthora* spp. Soil supplemented with *Streptomyces nigellus* NRC 10 strain reduced the percentage of damping off diseases in tomato plants caused by *Pythium ultimum*. Studies indicated that the strains are excellent producers of the β-1,3 and β-1,4 glucanases.[21] In vitro produced β-1,4, β-1,3 and β-1,6 glucanases from *Actinoplanes philippinesis*, *Microbispora rosea* and *Micromonospora chalcea* caused hyphae lysis of *Pythium aphanidermatum* and also minimised cucumber damping-off disease under regulated glass house conditions.[22] Some *Streptomyces* strains (CAI-24, CAI-121, CAI-127, KAI-32 and KAI-90) exhibited significant biocontrol activity against *Fusarium oxysporum* f. sp. Cicero causes Fusarium wilt of chickpea. All these five Streptomyces strains produced β-1,3-glucanase.[13] Extracellular β-glucanase from *Trichoderma harzianum* is a good suppressor of plant pathogens like *Rhizoctonia solani*, *Sclerotium rolfsii* and *Pythium* spp. It degrades the cell wall β-glucans, thus destroying cell wall integrity.[23]

2.4   BETA GLUCANASE AS A FEED ADDICTIVE

All animals require enzymes for digestive processes. These can either be, produced by the animals themselves or by endogenous microorganisms in their digestive tract. Feed ingredients that contain starches with fibre-rich cell walls, minerals, proteins and many anti-nutritional factors that may not be accessible to endogenous enzymes present in animal's digestive tract, may cause poor digestive performances and digestive upsets. The supplementation of exogenous enzymes in animal feeds improves the diet utilization by increasing the efficiency of overall digestion. The commercial use of microbial enzymes as feed enzymes significantly improves the nutritional quality of barley-based poultry feeds. In fact, the first report on the prospective use of enzymes to enhance animal performance had appeared around 100 years ago. Nowadays, the practice of microbial enzymes as feed additives is well-established and most of the poultry and swine feeds contain feed enzymes. The widely used enzymes for the significant feed utilization are xylanases, β-glucanases, phytases, amylases, pectinases and proteases.

Diets with viscous cereals like wheat, barley, rye or triticale contain a relatively large proportion of soluble and insoluble fibres and non-starch polysaccharides (NSP). Monogastric animals such as pigs and poultry are unable to digest fibres due to the lack of digestive enzyme production. A most prominent soluble fibre is β–glucan. A major source of β-glucan in animal feeds is from barley grain. Oats also contain certain levels of β-glucan and these are infrequently served to chickens and pigs. Water-soluble, viscous β–glucans have lethal effect in livestock varieties particularly poultry.[24] They increase the viscosity of the fillings in the small intestine to significantly disturb the rate of movement of barley-based diets. This hinders the digestion of nutrients and their absorption, consequently reducing animal growth. The problem is also linked with certain digestive disorders such as non-specific colitis in swine and hock burns and sticky litter in poultry. Addition of β–glucan degrading enzyme β–glucanase can effectively solve these problems by improving the digestive performance of the feeds and also decrease the chance of digestive disorders.

The anti-nutritional effects of mixed linked β-glucans due to their viscosity-inducing properties have been known for several years.[25] Early studies indicate that the addition of supplemental enzymes likes β-glucanase have a positive impact on animal growth. The inclusion of β-glucanase in poultry feeds with barley and oats showed better performance. β-glucanase was normally added in poultry feeds in dry form and it got activated upon hydration inside the digestive tract. This widespread application largely adopted barley into poultry feeds. According to Campbell and Bedford (1992), many practical feeding trials have been conducted using a chick model because the substantial growth depression due to the use of high-viscosity hulless barley applied feeds in chicks is very high when compared to others.[26] Supplementation of the enzyme β-glucanase in chick feed gives good results and also increases its overall market value. However, the age is a major factor for enzyme addition. In adult cocks, the adequately developed gastrointestinal tracts negate the harmful effects of β-glucan. Also, supplementation of β-glucanase to barley feed of broiler chicks during their first 3 weeks, develops them better. There are even fewer reports in turkeys when compared to chickens. According to Mathlouthi et al., (2003), wheat and barley-based diets with β-glucanase supplementation have shown more significant responses in young turkeys than older birds.[27] However, the influence of enzymes in pig digestibility is variable. They differ physiologically from young chicks as the digesta has higher water content. So, the viscosity problem can essentially be eliminated by simple dilution. Earlier studies indicated that ileum of pigs is more efficient to digest β-glucans than starch indicating the presence of endogenous β-glucanase. But its original source which has not been recognized may be from bacterial or from feed origin.[28]

The action of β-glucanase in β-glucan containing foods is more eluci-dated. It improves the efficiency of utilisation of β-glucan containing food-stuff by hydrolysing specific β-1,3 and 1,4 bonds in barley, oats and other β-glucan containing feed ingredients that are not commonly hydrolysed by endogenous digestive enzymes. It degrades anti-nutritional factors that decrease nutrient digestion and may increase digesta viscosity. The enzyme disrupts the integrity of endosperm cell wall and releases cell wall bound or entrapped nutrients and also transits nutrient digestion to more capable digestion sites. It helps in reducing endogenous secretions and protein losses from the gut by lowering manure output and excreta mois-ture content and thus the maintenance requirements. β-glucanase action reduces intestinal tract weights, changes the intestinal morphology and microflora profiles in the large and small intestine and provides overall animal health. As the enzyme utilizes the substrate present in the gut, there is a direct impact on the microfloral populations. This helps uplifting the growth of poorly performing animals and keeps uniformity in the market.

2.5   GERMINATION ENHANCEMENT BY BETA GLUCANASE

Recently explored the new function of β-glucanase was endosperm weak-ening as a part of seed germination. Endosperm weakening or dissolution of the mechanical resistance of the micropylar endosperm seems to be necessary for the successful completion of seed sprouting. According to Ikuma and Thimann (1963), the ultimate phase in the germination regulating process is the production of an enzyme, which facilitates penetration of the tip of the radicle through the coat.[29] Leubner-Metzger (2003) proposed that class β-1,3-glucanase induced within the micropylar endosperm before its rupture is closely accompanying with altered endosperm rupture in the regulation of germination in response to light, gibberellins, abscisic acid (ABA) and ethylene.[30] These findings support that cell-wall-modifying proteins such as enzyme β-1,3-glucanase have significant contributions in the regulation of various plant physiological and developmental processes like microsporogenesis, pollen germination, pollen tube growth, fertilization, embryogenesis, seed development, seed germination, mobilization of endosperm, cereal grain storage reserves, dormancy release and after-ripening of dicot seeds.

In Poaceae family, the starchy endosperm walls consist of abundant β-1,3–1,4-D-glucans, which is, however, less abundant in other tissue walls. It represents a chief source of stored glucose for the developing seedlings. The water-soluble β-1,3–1,4-D-glucans comprise 1000 or more glucosyl residues linked with (1,3)-and (1,4)-β-glucosidic linkages at a ratio of 2:1 to 3:1. During normal development of a plant endo-and exo-hydrolases, that is endo-and exo-β-1,3–1,4-glucanase are used to degrade or transform cell wall 1,3–1,4-β-glucans. The primary function of endo-β-1,3–1,4-glucanase in germinating grain is to participate in cell wall mobilization. In the elongating coleoptile, the endo-β-1,3–1,4-glucanase loosen wall structure by partial hydrolysis of wall polysaccharides and facilitate turgor-driven cell elongation in developing tissues. Exo-β-1,4-glucanase also take part in trimming of wall β-1,3–1,4-D-glucans. Another important role of exo-β-1,4-glucanase during plant growth development is the release of active phytohormones from inactive hormone-glucoside conjugates. But further evidence about this function is not available.[31] Figure 2.11 represents the enzymatic hydrolysis of cell wall β-1,3–1,4-D-glucans by β-glucanases.

The molecular mechanisms behind the effects of β-glucanase on seed dormancy and germination are less explored. Leubner-Metzger (2003) proposed a hypothesis that β-glucanase degrades cell-wall material, causes endosperm weakening and radicle protrusion, that is endosperm ruptures.[30] The possibility is the enzyme digesting β-1,3-glucan callose, placed between the plasma membrane and the cell wall of tissues. It is largely present in the seed-covering layers of many dicot species. In legumes, it causes water impermeability in the seed coat. In muskmelon, the embryo enclosed in a single layer of endosperm cells covered by a thick deposit of callose and a thin waxy suberin-and outer lipid layer creates a semi-permeable apoplastic envelope. Artificial degradation of the callose layer in decoated muskmelon seeds with exo-β-1,3-glucanase induces semiper-meability loss. The elicitor-active β-glucan oligosaccharides released by the indirect action of β-glucanase from plant cell wall components act as signaling molecules that mediate the processes like endosperm weakening, cell death of seed layers, dormancy release or pathogen defence.

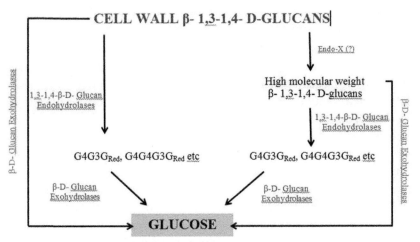

FIGURE 2.11   Enzymatic hydrolysis of cell wall β-1,3–1,4-D-glucans.

2.6   BETA GLUCANASE IN PROTOPLAST PREPARATION

Protoplasts, the naked plant cells, are the most commonly used research tool in plant cells for the parasexual modification, provide numerous benefits for crop improvement. Due to the high specificity and discriminating hydrolysis of cell wall polymers, β-glucanases are widely applied in the protoplast preparations from plant, bacterial and fungal cells. Protoplasts are valuable in numerous areas of research such as for isolating high molecular weight DNA from cells with the tough cell wall, transformation, propagating wild plant species using tissue culture etc. For successful protoplast preparation, it is important that adequate cell wall must be removed to allow fusion between protoplasts and sufficient cell wall must remain for the successful regeneration of the complete cell wall. Nowadays different β-glucanase formulations are commercially available. Some are scientifically proved as best for the preparation of protoplasts.

Zymolyase® is a commercially available enzyme preparation, purified form of *Arthrobacter luteus* culture fluid. It is very efficient for preparing protoplasts, and also not impeding the redevelopment of protoplasts into viable cells. It contains four enzymes with different cell wall hydrolysing properties. They are β-1,3 glucan laminaripentaohydrolase, β-1,3 glucanase, protease and mannase, all of which act in a combined form for efficient protoplast formation. The first enzyme β-1,3 glucan laminaripentaohydrolase hydrolyses cell wall glucans into pentamers. Then, the second key enzyme beta-1,3 glucanase fragments glucans to glucose, thus completely degrading the cell wall. The more dilute concentration of Zymolyase can be used for practical applications; this makes it more cost-effective.[32] Novozym 234, is a mixture of β-glucanase which has been used to prepare protoplasts. Yatalase, prepared from Corynebacterium culture supernatants, is used for preparing protoplasts from filamentous fungi. It consists mainly of combinations of chitinase, chitobiase and beta-1,3-glucanase.

2.7 CONCLUSION

Microbial β-glucanases are considered as environmental safe resolutions for many environmentally dangerouscomplications like accumulation of toxic compounds due to intense use of fungicides as a part of modern agricultural practices. Currently, they have been deliberated to be of great significance in agricultural practices and also have potential applications in some other fields like food and beverages industry. They may also be used as a sophisticated tool in β-glucan structural analysis. The most remarkable problem related to its commercial availability is the cost. A few kinds of pure β-glucan products available from some companies can be used as the substrate for β-glucanase are highly expensive, which in turn increases the overall production outlays of the enzymes. The increased cost of enzyme limits its applications in large-scale industries. As a part of the extensive production, some of the β-glucanase genes have been effectively cloned and expressed in heterologous host cells, this also facilitated a better understanding of the regulation of synthesis. In addition, genetic engineering approach is a powerful tool to construct recombinant expression cassette encoding different β-glucanase genes and production of enzymes with commercially reasonable scale. At present only a few organisms are exploited for screening β-glucanase enzyme systems. Further research for the purification and characterization of more enzymes will be necessary for better understanding of their properties. The use of cost-effective microbial enzymes is an efficient tool for maintaining a green environment.

KEYWORDS

- β-glucan
- barley
- pathogenesis
- curdlan
- endosperm
- crop protection
- microsporogenesis

REFERENCES

1. Duza, M. B.; Mastan, S. A. Microbial Enzymes and their Applications–A Review. Indo Am. *J. Pharm. Res.* **2013**, *3*, 6208–6219.
2. Genc, H.; Ozdemir, M.; Demirbas, A. Analysis of Mixed Linked (1–3), (1–4)-β-D-Glucans in Cereal Grains from Turkey. *Food Chem.* **2001**, *73*, 221–224.
3. Demirbas, A. β-Glucan and Mineral Nutrient Contents of Cereals Grown in Turkey. *Food Chem.* **2005**, *90*, 773–777.
4. Annison, G., Choct, M. Anti-Nutritive Activities of Cereal Non-Starch Polysaccharides in Boiler Diets and Strategies Minimizing their Effects. *World's Poult. Sci. J.* **1991**, *47*, 232–242.
5. Plans, A. Bacterial 1,3–1,4-β-Glucanases: Structure, Function and Protein Engineering. *Biochim. Biophys. Acta.* **2000**, *1543*, 361–382.
6. Dake, M. S.; Jhadav, J. P.; Patil, N. B. Induction and Properties of 1,3-β-D-Glucanases from *Aureobasidium pullulans*. *Indian J. Biotechnol.* **2004**, *3*, 58–64.

7. McCleary. Analysis of Feed Enzymes. In *Enzymes in Farm Animal Nutrition*. Wallingford, Oxfordshire, UK: CABI Pub. CABI, 2001; p.86.

8. Xu, X.; Feng, Y.; Fang, S.; Xu, J.; Wang, X.; Guo, W. Genome-Wide Characterization of the β-1,3-Glucanase Gene Family in Gossypium by Comparative Analysis. *Sci. Rep.* **2016**, *6*, 29044.

9. Buchanan, M.; Burton, R. A.; Dhugga, K. S.; Rafalski, A. J.; Tingey, S. V.; Shirley, N. J.; Fincher, G. B. Endo-(1,4)-β-Glucanase Gene Families in the Grasses: Temporal and Spatial Co-Transcription of Orthologous Genes. *BMC Plant Biol.* **2012**, *12*, 235. http://www. biomedcentral.com/1471–2229/12/235.

10. Fisher, M. C.; Henk, D. A.; Briggs, C. J.; Brownstein, J. S.; Madoff, L. C.; Mccraw, S. L.; Gurr, S. J. Emerging Fungal Threats to Animal, Plant and Ecosystem Health. *Nature* **2012**, *484*, 186–194.

11. Ruel, K.; Joseleau, J. Involvement of an Extracellular Glucan Sheath during Degradation of Populous Wood by *Phanerochaete chrysosporium*. *Appl. Environ. Microbiol.* **1991**, *57*, 374–384.

12. Mouyna, I.; Hartl, L.; Latgé, J. β-1,3-Glucan Modifying Enzymes in *Aspergillus fumigatus*. *Front. Microbiol.* **2013**, *4*, 1–9.

13. Gopalakrishnan, S.; Srinivas, V.; Sree Vidya, M.; Rathore, A. Plant Growth-Promoting Activities of Streptomyces spp. in Sorghum and Rice. *SpringerPlus* [Online] **2013**, *2*, 574. http://www. springerplus.com/content/2/1/574.

14. Alfonso, C.; Raposo, R.; Melgarejo, P. Genetic Diversity in *Botrytis Cinerea* Populations on Vegetable Crops in Greenhouses in South-Eastern Spain. *Plant Pathol.* **2000**, *49*, 243–51.

15. Li, J.; Liu, W.; Luo, L.; Dong, D.; Liu, T.; Zhang, T.; Lu, C.; Liu, D.; Zhang, D.; Wu, H. Expression of *Paenibacillus polymyxa* β-1,3–1,4-Glucanase in *Streptomyces lydicus* A01 improves its Biocontrol Effect against Botrytis cinerea. *Biol. Control.* **2015**, *90*, 141–147.

16. Yao, W. L.; Wang, Y. S.; Han, J. G.; Li, L. B.; Song, W. Purification and Cloning of an Antifungal Protein from the Rice Diseases Controlling Bacterial Strain *Paenibacillus polymyxa* WY110. *Acta Genet. Sin.* **2004**, *31*, 878–887.

17. Wen, F. Y.; Liao, F. P.; Lin, J. R.; Zhong, Y. S. Cloning, Expression and Application of β-1,3–1,4-Glucanase Gene from *Paenibaccillus polymyxa* CP7. *J. Integr. Agric.* **2010**, *43*, 4614–4623.

18. Dewi, R.; Mubarik, N.; Suhartono, M. Medium Optimization of Beta-Glucanase Production by *Bacillus subtilis* SAHA 32.6 used As Biological Control of oil Palm Pathogen. *Emir. J. Food Agri.* **2016**, *28*, 116–125.

19. Palumbo, J. D.; Sullivan, R. F.; Kobayashi, D. Y. Molecular Characterization and Expression in *Escherichia coli* of Three-1,3-Glucanase Genes from *Lysobacter enzymogenes* Strain N4–7. *J. Bacteriol.* **2003**, *185*, 4362–4370.

20. Palumbo, J. D.; Yuen, G. Y.; Jochum, C. C.; Tatum, K.; Kobayashi, D. Y. Mutagenesis of β-1,3-Glucanase Genes in *Lysobacter enzymogenes* Strain C3 Results in Reduced Biological Control Activity toward Bipolaris Leaf Spot of Tall Fescue and Pythium Damping-Off of Sugar Beet. *Phytopathology* **2005**, *95*, 701–707.

21. Helmy, S. M.; Zeinat, K.; Mahmoud, S. A.; Moataza, S.; Nagwa, M.; Amany, H. *Streptomyces nigellus* as a Biocontrol Agent of Tomato Damping-Off Disease Caused by *Pythium ultimum*. *Electron. J. Pol. Agric. Univ.* **2010**, *13*, 4.

22. El-Tarabily, K. A. Rhizosphere-Competent Isolates of Streptomycete and Non-Streptomycete Actinomycetes Capable of Producing Cell-Wall-Degrading Enzymes to Control *Pythium aphanidermatum* Damping-Off Disease of Cucumber. *Can. J. Bot.* **2006**, *84*, 211–222.

23. Lisboa, B. B.; Bochese, C. C.; Vargas, L. K.; Silveira, J. R.; Radin, B.; Oliveira, A. M. Efficiency of *Trichoderma harzianum* and *Gliocladium viride* in Decreasing the Incidence of *Botrytis cinerea* in Tomato Cultivated in Protected Environment. *Cienc. Rural.* **2007**, *37*, 1255–1260.

24. Walsh, G. A.; Murphy, R. A.; Killeen, G. F.; Headon, D. R.; Power, R. F. Technical Note: Detection and Quantification of Supplemental Fungal Beta-Glucanase Activity in Animal Feed. *J. Anim. Sci.* **1995**, *73*, 1074.

25. Aastrup, S. The Effect of Rain on β-Glucan Content in Barley Grains. *Carlsberg Res. Commun.* **1979**, *44*, 381–393.

26. Campbell, G. L., Bedford, M. R. Enzyme Applications for Monogastric Feeds: A Review. *Can. J. Anim. Sci.* **1992**, *72*, 449–466.

27. Mathlouthi, N.; Juin, H.; Larbier, M. Effect of Xylanase and β-Glucanase Supplementation of Wheat-or Wheat-and Barley-Based Diets on the Performance of Male Turkeys. *Br. Poult. Sci.* **2003**, *44*, 291–298.

28. Graham, H.; Fadel, J. G.; Newman, C. W.; Newman, R. K. Effect of Pelleting and Betaglucanase Supplementation on the Ileal and Fecal Digestibility of a Barley-Based Diet in the Pig. *J. Anim. Sci.* **1989**, *67*, 1293–1298.

29. Ikuma, H.; Thimann, K. V. The Role of the Seedcoats in Germination of Photosensitive Lettuce Seeds. *Plant Cell Physiol.* **1963**, *4*, 169–185.

30. Leubner-Metzger, G. Functions and Regulation of β-1,3-Glucanases during Seed Germination, Dormancy Release and After-Ripening. *Seed Sci. Res.* **2003**, *13*, 17–34.

31. Hrmova, M.; Fincher, G. B. Structure-Function Relationships of β-D-Glucan Endo- and Exo-hydrolases from Higher Plants. *Plant Mol. Biol.* **2001**, *47*, 73–91.

32. Comparison Zymolyase™ vs Lyticase & Glusulase http://www. amsbio.com/ brochures/Zymolyase_Comparison.pdf

CHAPTER 3

APPLICATION OF PLANT GROWTH PROMOTING RHIZOBACTERIA IN AGRICULTURE

SUSHANTO GOUDA[1], ROUT GEORGE KERRY[2], DIBYARANJAN SAMAL[2], GYANA PRAKASH MAHAPATRA[2], GITISHREE DAS[3], and JAYANTA KUMAR PATRA[3] *

[1]*Amity Institute of Wildlife Science, Noida, Uttar Pradesh 201303, India*

[2]*Department of Biotechnology, AMIT College, Khurda, Odisha 752057, India*

[3]*Research Institute of Biotechnology & Medical Converged Science, Dongguk University-Seoul, Ilsandong-gu, Gyeonggi-do 10326, Republic of Korea,*E-mail: jkpatra.cet@gmail.com*

CONTENTS

ABSTRACT

The development of life in all forms is dependent on agriculture since civilization. The coevolution between plant and microbes can be best explained by a group of unique microbes termed as plant growth promoting rhizobacteria (PGPR) that show antagonistic and synergistic interactions between microorganisms, soil and plants. Plant growth promoting rhizo-bacteria are an essential group of microorganisms that boost the growth rate of plants directly or indirectly. Different modes of action of PGPR in plant growth and development have been conceptualized but, the role of the PGPR as bio-fertilizer for sustainable agriculture is still procras-tinated. However, with the development of advance techniques such as nano-encapsulation and microencapsulation, the efficacy of PGPR has been greatly improved in agricultural sustainability. Hence, this chapter provides brief information on sustainable agriculture by PGPR using different advance approaches.

3.1   INTRODUCTION

Agriculture has been the largest source of wealth since civilization, earth being a planet with the population around 7.41 billion, inhabiting 6.38 billion hectares of land[1] and about 1.3 billion people directly dependent on agriculture for their survival. Degradation of soil is something that cannot be neglected.[2] Estimation by the Food and Agriculture Organization (FAO) Food Balance Sheet 2004 has shown that 99.7% of food for the total population comes from the terrestrial environment alone. With about 79 million people being added to the world population every year, there has been a continuous increase in the demand for food and scarcity in the supply.[3] In India, 60.6% of land is used for agricultural purposes by half of its population for growing several forms of cereals, vegetables, pulses and so forth. Agricultural productivity, water quality and climate change are greatly influenced by the effectual exchange of nutrients, energy and carbon between soil organic matters and plant system.[4] The nature of soil has been regulated by a number of aspects such as organic carbon content, moisture, nitrogen (N), phosphorus (P) and potassium (K) (NPK) content and several other factors.

Direct manifestation of soil through microbes present in soil and through leguminous plants as a holobiont relationship by the process of biomineralization and synergetic coevolution have been observed to have great potential in the improvement of soil quality and fertility.[5,6] The soil physiology and performance are critically affected by plant associated microorganisms, which suggest that there might be a holobiont link in the coevolution and ecology of plant and animals.[7] It has also been evident that accumulating coevolution of soil microbes with plants is vital in response to transmute or extreme abiotic environments resulting in the improvement of economic viability, soil fertility and environmental sustainability.[8,9] The coevolution between plant and microbes can be best explained in plant growth promoting rhizobacteria (PGPR) which show antagonistic and synergistic interactions with microorganisms and soil, either directly or indirectly boosting the plant growth rate.[10,11] The increase in plant growth is directly linked with crop improvement which in turn is unswervingly proportional to the development in agriculture. Plant growth promoting rhizobacteria have greatly influenced the soil characteristic and have a vital role to play in the conversion of barren and poor quality land into cultivable and fertile land that can be used for agricultural purposes.

Microbial revitalization by PGPR, which includes different groups of bacteria that can colonize the roots of plants and enhance their differentiation, has been an area that is currently exploited for agriculture productivity in many parts of the world.[12] It is generally achieved through direct or indirect approaches. In the direct approach, a bacterium directly provides the plant with essential compounds that can promote plant growth and can be achieved through techniques such as bio-fertilization, rhizo-remediation and plant stress control.[13] The ability of absorbing water, temperature and other nutrients are some of the environmental factors that strongly restrict terrestrial plant's growth. The introduction of PGPR improves the plant growth by increasing its water accessibility and uptake of nutrients from a limited nutrient pool in the soil through the process of bio-fertilization. It also involves rhizo-remediation, which is the interaction of soil and microbes with the plant to enhance plant growth. Plant stress neutralization is another important factor of PGPR where it basically neutralizes two types of stress, that is biotic and abiotic stress.[14] Therefore, the following chapter highlights some of the suitable approaches that can be adapted for the implementation of PGPR as tools to combat plant diseases and enhance agricultural productivity.

3.2   PLANT GROWTH PROMOTING RHIZOBACTERIA (PGPR)

Plants have always been in a symbiotic relationship with several forms of soil microbes for their growth, development and other requirements. The symbiotic microorganisms inhabiting the rhizosphere of many plant species have diverse beneficial effects on the host plant.[15] These beneficial free-living soil bacteria are usually referred to as PGPR. In the current era, rhizobacteria possess a conspicuous impact on plants which can be a significant tool to defend the health of plants in an eco-friendly manner.[16] Applications of PGPR associations with different plant species have been investigated in certain cases such as oat, canola, soy, potato, maize, peas, tomatoes, lentils, barley, wheat, radicchio and cucumber.[17] They have been involved in various biotic activities of the soil ecosystem to make it dynamic for alimental turn over and sustainable for crop production by enhancing its physiological properties.[18]

3.3   DIFFERENT CLASSES OF PGPR

Plant growth promoting rhizobacteria can be broadly classified into two main types, extracellular plant growth promoting rhizobacteria (ePGPR) and intracellular plant growth promoting rhizobacteria (iPGPR).[19] The ePGPRs basically inhabit the rhizosphere (on the rhizoplane) or in the spaces between the cells of root cortex, whereas iPGPRs mainly inhabit the specialized nodular structures of root cells. The bacterial genera that belong to ePGPR are *Azotobacter, Serratia, Azospirillum, Bacillus, Caulobacter, Chromobacterium, Agrobacterium, Erwinia, Flavobacterium, Arthrobacter, Micrococcous, Pseudomonas* and *Burkholderia*. The endophytic microbial families belonging to iPGPR are *Allorhizobium, Bradyrhizobium, Mesorhizobium, Rhizobium* and also *Frankia* species which can fix atmospheric nitrogen specifically for higher plants.[20]

The rhizosphere is the zone of the soil surrounding a plant root where the biological and the chemical features of the soil are influenced by the plant root. All of the microbial activity is carried out in this phase and is, thus, known as the store house of microbes. These bacteria may be symbiotic or nonsymbiotic, and are determined by the direct beneficiary action or by only deducing nutrients for their own survival.[21] The composition of these transudes depend upon the physiological status and species of plants and microorganisms.[22]

There are many applications of PGPR which are provided in Table 3.1. Plant growth promoting rhizobacteria enhance the plant growth in a well-known natural event due to certain specific traits.[18] Plant growth promoting rhizobacteria may enhance plant growth by direct and indirect mechanism. These approaches are involved in the enhancement of plant physiology and resistivity from different phytopathogens through various modes and actions.[23] The various mechanisms exhibited by PGPR as plant growth enhancers are as follows.

TABLE 3.1 Plant Growth Promoting Rhizobacteria (PGPR) and its Application in Plant Growth and Development.

Name of microbes	Source/host plant	Uses
Azospirillum lipoferum	*Triticum aestivum*	Promoted development of root system
Pseudomonas fluorescens	*Medicago sativa*	Increased metabolism, sequestration of cadmium from solution and degradation of trichloroethylene
Pseudomonas putida	*Arabidopsis thaliana*	Improves utilization of plant secondary metabolites
Azotobacter chroococcum	*Brassica juncia*	Stimulation of plant growth
Bacillus subtilis	*Brassica juncia*	Facilitated nickel accumulation
Azospirillum brasilense	*Saccharum officinarum*	Able to alter plant root architecture by increasing the formation of lateral and adventitious roots and root hairs
Pseudomonas aeruginosa	*Cicer arietinum*	Positively stimulated potassium and phosphorus uptake
Herbaspirillum seropedicae	*Oryza sativa*	Produces production of gibberellins
Pseudomonas cepacia	*Cucumis sativus*	Prevent from pathogens of *Pythium ultimum*
P. fluorescens	*Phaseolus vulgaris*	Prevents halo blight
P. cepacia	*P. vulgaris*	Prevents from *Sclerotium rolfsii*
P. fluorescens	*Gossypium hirsutum*	Helps to cease damping off of cotton
P. cepacia	*G. hirsutum*	Helps plant to stand against *Rhizoctonia solani* virus
Pseudomonas gladioli	*G. hirsutum*	Give resistance against *Helicoverpa armigera* virus
Bacillus subtilis	*G. hirsutum*	Prevents from *Meloidogyne incognita* and *Meloidogyne arenaria*

3.3.1   DIRECT MECHANISM

The mechanism by which PGPR directly facilitates nutrient uptake or increases nutrient availability comprises of nitrogen fixation, mineralize organic compounds, solubilization of mineral nutrients and production of phytohormones (Fig. 3.1).[20]

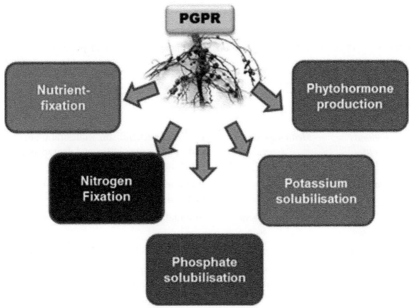

FIGURE 3.1   Different types of direct mechanism of plant growth promoting rhizobacteria (PGPR) for nutrient uptake.

3.3.1.1   NUTRIENT-FIXATION

Plant growth promoting rhizobacteria can act as a direct growth enhancer to plants as they have the tendency to increase the accessibility of nutrient concentration through the roots of plants by fixing or locking the supply of nutrients for plant growth and productivity.[24]

3.3.1.2   NITROGEN FIXATION

Biological nitrogen fixation is an astounding process which accounts for approximately two-thirds of the nitrogen fixed globally. This outstanding biological process is carried out either by symbiotic or nonsymbiotic interactions between microbes and plants.[25] Plant growth promoting rhizobacteria strains such as *Rhizobium* sp., *Azoarcus* sp., *Beijerinckia* sp., *Pantoea agglomerans* and *Klebsiella pneumoniae* are some symbiotic microbes which are most notifiable for fixing atmospheric N_2 in soil.[26]

3.3.1.3   PHOSPHATE SOLUBILIZATION

Phosphorus plays an important role for almost all major metabolic processes of a plant including energy transfer, signal transduction, respiration, biosynthesis of macromolecules and photosynthesis.[27] However, 95–99% phosphorus is presented mostly in insoluble, immobilized and precipitated form and is, therefore difficult for most plants to absorb. Plants can absorb phosphate only in two soluble forms that is, the monobasic (H_2PO_4) and the dibasic (HPO_4^{2-}) ions. Phosphate solubilizing PGPR such as *Arthrobacter, Bacillus, Burkholderia, Pseudomonas, Rhizobium, Flavobacterium, Rhodococcus* and *Serratia* have attracted the attention of agriculturists as soil inoculums to improve plant growth and yield in many plant species.[28]

3.3.1.4   POTASSIUM SOLUBILIZATION

In the present scenario, potassium deficiency has become one of the major constraints in crop production. The ability of PGPR to solubilize potassium rocks through the production and secretion of organic acids is being widely investigated. Potassium solubilizing PGPR such as *Acidothiobacillus, Bacillus edaphicus, Ferrooxidans, Bacillus mucilaginosus, Pseudomonas, Burkholderia* and *Paenibacillus* sp. are being reported to release potassium in accessible form from potassium bearing minerals in soils.[29]

3.3.1.5   PHYTOHORMONE PRODUCTION

Phytohormones or plant growth regulators are organic substances, which at low concentrations (<1mM) promote, inhibit or modify the augmentation and amplify plant products.[30] Ironically production of these phytohormones could also be induced by certain microbes like PGPR in plants. Common groups of phytohormones include gibberellins, cytokinins, abscisic acid, ethylene, brassinosteroids and auxins that can affect the cell proliferation in the root architecture by over production of lateral roots and root hairs and consecutive increase of nutrient and water uptake capacity.[31]

3.3.1.6   BIO-FERTILIZATION

The phenomenon where soil fertility is increased by means of biological entities can be regarded as bio-fertilization and the chemical supplements fixed from the surroundings by the biological entities are regarded as bio-fertilizers.[32] Increased fertility of the soil by biological means can be easily achieved by certain natural processes such as phosphate solubilization, siderophores production, exopolysaccharides production, bio-fixation of atmospheric nitrogen.[33]

3.3.1.7   PHOSPHATE SOLUBILIZATION

Microorganisms play an important role in mediating phosphorus available to plants by playing a role in the soil phosphorus succession.[34] Though the microbes can solubilize phosphorus which results in the increase in soil fertility, further studies regarding their use in the field as bio-fertilizer are still awaiting as limited research and literature are available on the microbes.

3.3.1.8   SIDEROPHORES PRODUCTION

Several microorganisms belonging to the group of PGPR have the capacity to utilize siderophores produced by diverse species of bacteria and fungi. The bacteria, *Pseudomonas putida* can utilize the heterologous siderophores produced by other rhizospheres to enhance the level of iron available to it in the natural habitat.[35]

3.3.2   INDIRECT MECHANISM

Indirect mechanism involves the process through which PGPR prevents or neutralizes the deleterious effects of phytopathogens on plants by the production of repressive substances which result in an increased natural resistance of the host.[36] The contribution of PGPR in such a mechanism includes the production of hydrolytic enzymes (chitinases, cellulases, proteases etc.), production of various antibiotics in response to plant pathogen or disease resistance, and so forth (Fig. 3.2).[37,38]

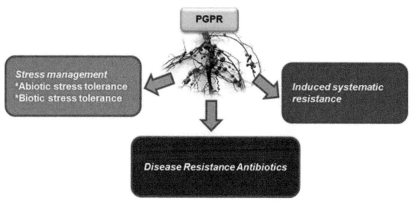

FIGURE 3.2   Different types of indirect mechanism of PGPR.

3.3.2.1   STRESS MANAGEMENT

Plants are frequently exposed to various environmental stress and have developed specific mechanisms to conflict this stress.[39] Over the past few decades the understanding of the molecular mechanisms implicated in abiotic and biotic stress tolerance has been achieved through various studies. Some of the active participation of PGPR and their role in stress management are discussed below.

3.3.2.2   ABIOTIC STRESS TOLERANCE

The abiotic stresses (high winds, extreme temperature, drought, salinity, flood etc.) bestow highly negative impact on the survival and biomass production. There are several other strategies that have been proposed to alleviate the degree of cellular damage caused by water stress and to improve tolerance in plants. Among them, exogenous application of PGPR in compatible osmolytes such as proline, glycinebetaine, trehalose and so forth, has gained considerable attention in mitigating the effect of stress. The use of PGPR in plant abiotic stress management has been studied comprehensively where it neutralizes the toxic effect of cadmium pollution on barley plants due to its ability of scavenging the cadmium ions present in soil.[40]

3.3.2.3   BIOTIC STRESS TOLERANCE

Biotic stress is another stress tolerance activity in plant that causes reduction in agricultural yield due to pathogenic impact on crops by different pathogens, such as bacteria.[41] Such a problem could be solved by the use of PGPR such as *Paenibacillus polymyxa* B2, B3, B4, *Bacillus amyloliquefaciens* strain HYD-B17, *Bacillus licheniformis* strain HYTAPB18, *Bacillus thuringiensis* strain HYDGRFB19, *Paenibacillus favisporus* strain BKB30 and *B. subtilis* strain RMPB44.[42]

3.3.2.4   DISEASE RESISTANCE ANTIBIOTICS

Use of microbial antagonists against plant pathogens in agricultural crops has been suggested as a substitute to certain chemical pesticides due to their unique biological properties. Basically microorganisms belonging to PGPRs such as *Bacillus* and *Pseudomonas* are known to play a major part in the inhibition of pathogenic microorganism by producing antibiotics.[43]

3.3.2.5   INDUCED SYSTEMATIC RESISTANCE

Induced systematic resistance can be defined as a physiological state in which improved defensive capacity is evoked in response to particular environmental stimuli. It is the state of increased resistance at particular sites of plants at which induction had occurred. PGPR has demonstrated 'induction of systemic resistance' in many plants against several environmental conditions.[44]

3.4 CONCLUSION

Plant growth-promoting rhizobacteria (PGPR) in addition to their role in enhancing plant growth also serve in the remediation and management of contaminated and degraded wastelands, eutrophication of water bodies, controlling pesticide pollution, nitrogen and phosphorous runoff. Studies have also indicated that such changes can alter plant-microbe interactions, through modification of microbial biology and biogeochemical cycles. Emerging of plant growth promoting rhizomes or microbes can serve as key players in sustainable agriculture by improving soil fertility, producing tolerant crop, productivity and maintaining a balanced nutrient cycling. Further studies of the selection of suitable rhizosphere microbes, productive microbial communities along with exploring multidisciplinary research combined with the application of biotechnology, nanotechnology, agrobiotechnology, chemical engineering and material science can also provide new formulations and opportunities with immense potential.

KEYWORDS

- **agriculture**
- **PGPR**
- **bio-fertilization**
- **nano-encapsulation**
- **revitalization**
- **sustainable development**

REFERENCES

1. World Population Data Sheet 2016. http://www.prb.org/Publications/Datasheets/2016/2016-world-population-data-sheet.aspx.
2. FAO. 2011. http://www.fao.org/3/a-bc595e.pdf.
3. Alexandratos, N.; Bruinsma, J.. Introduction and Overview. In *World Agriculture: Towards 2015/2030 an FAO Perspective*; Bruinsma, J., Ed.; Earthscan Publications, London, UK, 2003; pp 1–28.

4. Lehmann, J.; Kleber, M. The Contentious Nature of Soil Organic Matter. *Nature* **2015**, *528*, 60–68.

5. Paredes, S. H.; Lebeis, S. L. Giving Back to the Community: Microbial Mechanisms of Plant–Soil Interactions. *Funct. Eco.* **2016**, *30*(7), 1–10.

6. Rosenberg, E.; Rosenberg, I. Z. Microbes Drive Evolution of Animals and Plants: The Hologenome Concept. *Microbial. Bio.* **2016**, *7*(2), 1–8.

7. Agler, M. T.; Ruhe, J.; Kroll, S.; Morhenn, C. M.; Kim, S. T.; Weigel, D.; Kemen, E. M. Microbial Hub Taxa Link Host and Abiotic Factors to Plant Microbiome Variation. *PLoS Biol.* **2016**, *14*(1), 1–31.

8. Khan, Z.; Rho, H.; Firrincieli, A.; Hung, S. H.; Luna, V.; Masciarelli, O.; Kim, S. H.; Doty, S. L. Growth Enhancement and Drought Tolerance of Hybrid Poplar upon Inoculation with Endophyte Consortia. *Curr. Plant Biol.* **2016**, *6*, 1–10.

9. Compant, S.; Saikkonen, K.; Mitter, B.; Campisano, A.; Blanco, J. M. Soil, Plants and Endophytes. *Plant Soil.* **2016**, *405*, 1–11.

10. Rout, M. E.; Callaway, R. M. Interactions between Exotic Invasive Plants and Soil Microbes in the Rhizosphere Suggest that Everything is Not Everywhere. *Ann. Bot.* **2012**, *110*, 213–222.

11. Bhardwaj, D.; Ansari, M. W.; Sahoo, R. K.; Tuteja, N. Biofertilizers Function as Key Player in Sustainable Agriculture by Improving Soil Fertility, Plant Tolerance and Crop Productivity. *Microb. Cell Fact.* **2014**, *13*(66), 1–10.

12. Gabriela, F.; Casanovas, E. M.; Quillehauquy, V.; Yommi, A. K.; Goni, M. G.; Roura, S. I.; Barassi, C. A. Azospirillum Inoculation Effects on Growth, Product Quality and Storage Life of Lettuce Plants Grown under Salt Stress. *Sci. Hortic.* **2015**, *195*, 154–162.

13. Goswami, D.; Thakker, J. N.; Dhandhukia, P. C. Portraying Mechanics of Plant Growth Promoting Rhizobacteria (PGPR): A Review. *Cogent Food Agric.* **2016**, *2*, 1–19.

14. Gonzalez, A. J.; Larraburu, E. E.; Llorente, B. E. *Azospirillum brasilense* Increased Salt Tolerance of Jojoba during In Vitro Rooting. *Ind. Crops Prod.* **2015**, *76*, 41–48.

15. Raza, W.; Ling, N.; Yang, L.; Huang, Q.; Shen, Q. Response of Tomato Wilt Pathogen *Ralstonia solanacearum* to the Volatile Organic Compounds Produced by a Biocontrol Strain *Bacillus amyloliquefaciens* SQR-9. *Sci. Rep.* **2016**, *6*, 24856.

16. Akhtar, N.; Qureshi, M. A.; Iqbal, A.; Ahmad, M. J.; Khan, K. H. Influence of Azotobacter and IAA on Symbiotic Performance of Rhizobium and Yield Parameters of Lentil. *J. Agric. Res.* **2012**, *50*, 361–372.

17. Gray, E. J.; Smith, D. L. Intracellular and Extracellular PGPR: Commonalities and Distinctions in the Plant-Bacterium Signaling Processes. *Soil Biol. Biochem.* **2005**, *37*, 395–412.

18. Gupta, G.; Parihar, S. S.; Ahirwar, N. K.; Snehi, S. K.; Singh, V. Plant Growth Promoting Rhizobacteria (PGPR): Current and Future Prospects for Development of Sustainable Agriculture. *J. Microb. Biochem. Technol.* **2015**, *7*, 96–102.

19. Viveros, O. M.; Jorquera, M. A.; Crowley, D. E.; Gajardo, G.; Mora, M. L. Mechanisms and Practical Considerations Involved in Plant Growth Promotion by Rhizobacteria. *J. Soil Sci. Plant Nutr.* **2010**, *10*, 293–319.

20. Bhattacharyya, P. N.; Jha, D. K. Plant Growth-Promoting Rhizobacteria (PGPR): Emergence in Agriculture. *World J. Microbiol. Biotechnol.* **2012**, *28*, 1327–1350.

21. Kundan, R.; Pant, G.; Jado, N.; Agrawal, P. K. Plant Growth Promoting Rhizobacteria: Mechanism and Current Prospective. *J. Fertil. Pestic.* **2015**, *6*, 2.

22. Kang, B. G.; Kim, W. T.; Yun, H. S.; Chang, S. C. Use of Plant Growth-Promoting Rhizobacteria to Control Stress Responses of Plant Roots. *Plant Biotechnol. Rep.* **2010**, *4*, 179–183.

23. Zakry, F. A. A.; Shamsuddin, Z. H.; Khairuddin, A. R.; Zakaria, Z. Z.; Anuar, A. R. Inoculation of *Bacillus sphaericus* UPMB-10 to Young Oil Palm and Measurement of its Uptake of Fixed Nitrogen using the N Isotope Dilution Technique. *Microbes Environ.* **2012**, *27*, 257–262.

24. Kumar, A. Phosphate Solubilizing Bacteria in Agriculture Biotechnology: Diversity, Mechanism and Their Role in Plant Growth and Crop Yield. *Int. J. Adv. Res.* **2016**, *4*(4), 116–124.

25. Shridhar, B. S. Review: Nitrogen Fixing Microorganisms. *Int. J. Microbiol. Res.* **2012**, *3*(1), 46–52.

26. Ahemad, M.; Kibret, M. Mechanisms and Applications of Plant Growth Promoting Rhizobacteria: Current Perspective. *J. King Saud Uni. Sci.* **2014**, *26*, 1–20.

27. Anand, K.; Kumari, B.; Mallick, M. A. Phosphate Solubilizing Microbes: An Effective and Alternative Approach as Biofertilizers. *Int. J. Pharm. Pharm. Sci.* **2016**, *8*(2), 37–40.

28. Oteino, N.; Lally, R. D.; Kiwanuka, S.; Lloyd, A.; Ryan, D.; Germaine, K. J.; Dowling, D. N. Plant Growth Promotion Induced by Phosphate Solubilizing Endophytic Pseudomonas Isolates. *Front. Microbiol.* **2015**, *6*, 745

29. Liu, D.; Lian, B.; Dong, H. Isolation of Paenibacillus Sp. and Assessment of Its Potential for Enhancing Mineral Weathering. *Geomicrobiol. J.* **2012**, *29*, 413–421

30. Damam, M.; Kaloori, K.; Gaddam, B.; Kausar, R. Plant Growth Promoting Substances (Phytohormones) Produced by Rhizobacterial Strains Isolated from the Rhizosphere of Medicinal Plants. *Int J Pharm Sci Rev Res.* **2016**, *37*(1), 130–136.

31. Sureshbabu, K.; Amaresan, N.; Kumar, K. Amazing Multiple Function Properties of Plant Growth Promoting Rhizobacteria in the Rhizosphere Soil. *Int. J. Curr. Microbiol. Appl. Sci.* **2016**, *5*(2), 661–683.

32. Suhag, M. Potential of Biofertilizers to Replace Chemical Fertilizers. *Int. Adv. Res. J. Sci., Eng. Technol.* **2016**, *3*(5), 163–167.

33. Noumavo, P. A.; Agbodjato, N. A.; Baba-Moussa, F.; Adjanohoun, A.; Baba-Moussa, L. Plant Growth Promoting Rhizobacteria: Beneficial Effects for Healthy and Sustainable Agriculture. *Afr. J. Biotechnol.* **2016**, *15*(27), 1452–1463.

34. Kaur, H.; Kaur, J.; Gera, R. Plant Growth Promoting Rhizobacteria: A Boon to Agriculture. *Int. J. Cell Sci. Biotechnol.* **2016**, *5*, 317–322.

35. Rathore, P. A Review on Approaches to Develop Plant Growth Promoting Rhizobacteria. *Int. J. Recent. Sci. Res.* **2015**, *5*(2), 403–407.

36. Singh, R. P.; Jha, P. N. Molecular Identification and Characterization of Rhizospheric Bacteria for Plant Growth Promoting Ability. *Int. J. Curr. Biotechnol.* **2015**, *3*, 12–18.

37. Nivya, R. M. A Study on Plant Growth Promoting Activity of the Endophytic Bacteria Isolated from the Root Nodules of *Mimosa pudica* Plant. *Int. J. Innovative. Res. Sci. Eng. Technol.* **2015**, *4*, 6959–6968.

38. Gupta, S.; Meena, M. K.; Datta, S. Isolation, Characterization of Plant Growth Promoting Bacteria from the Plant *Chlorophytum borivilianum* and In-Vitro Screening for Activity of Nitrogen Fixation, Phospthate Solubilization and IAA Production. *Int. J. Curr. Microbiol. Appl. Sci.* **2014**, *3*, 1082–1090.

39. Ramegowda, V.; Senthil-Kumar, M. The Interactive Effects of Simultaneous Biotic and Abiotic Stresses on Plants: Mechanistic Understanding from Drought and Pathogen Combination. *J. Plant Physiol.* **2015**, *176*, 47–54.

40. Vejan, P.; Abdullah, R.; Khadiran, T.; Ismail, S.; Boyce, A. N. Role of Plant Growth Promoting Rhizobacteria in Agricultural Sustainability—A Review. *Molecules* **2016**, *21*, 573. DOI: 10.3390/molecules21050573.

41. Haggag, W. M.; Abouziena, H. F.; Abd-El-Kreem, F.; Habbasha, E. S. Agriculture Biotechnology for Management of Multiple Biotic and Abiotic Environmental Stress in Crops. *J. Chem. Pharm. Res.* **2015**, *7*(10), 882–889.

42. Ngumbi, E.; Kloepper, J. Bacterial-Mediated Drought Tolerance: Current and Future Prospects. *Appl. Soil. Ecol.* **2016**, *105*, 109–125.

43. Ulloa-Ogaz, A. L.; Munoz-Castellanos, L. N.; Nevarez-Moorillon, G. V. Biocontrol of Phytopathogens: Antibiotic Production as Mechanism of Control. In *The Battle Against Microbial Pathogens. Basic Science, Technological Advance and Educational Programs;* Mendez Vilas, A., Ed.; 2015, 1, pp 305–309.

44. Prathap, M.; Ranjitha, K. B. D. A Critical Review on Plant Growth Promoting Rhizobacteria. *J. Plant Pathol. Microbiol.* **2015**, *6*(4), 1–4.

CHAPTER 4

MICROBIAL PESTICIDES—POTENTIAL ALTERNATIVES

BIGHNESWAR BALIYARSINGH[1], ASEEM MISHRA[2], and SAKTIKANTA RATH[1,*]

[1]*School of Life science, Ravenshaw University, Cuttack, Odisha, India*

[2]*P.G. Department of Biotechnology, Utkal University, Bhubaneswar, Odisha, India, *E-mail: saktirath@gmail.com*

CONTENTS

ABSTRACT

Global reduction in agricultural produce with burgeoning human population is beginning to unleash an era of food shortage where any loss of food during storage may lead to heavy socio-economic burden on any country. Though natural calamities that affect food crop production and their loss during storage are beyond human control, conservative estimates of loss exclusively by pests amount to almost 30% in present times. This is of utmost relevance especially when we are discovering that the use of traditional chemo-pesticides has led to resistance amongst target pests. Any more increase in the dose of the presently used pesticides can be associated with dramatic ecological derangements and off-site actions. This necessitates the development of eco-friendly alternatives in the present times. Biopesticides represent an interesting class of interventions wherein living organisms or their products (phytochemicals, microbial toxins) that are biocidal towards the target insect are used. This concept is being extensively applied in the bio-intensive integrated pest management systems (IPM) for protection of crops. A subcategory, microbial pesticide, is particularly interesting due to the fundamental reason that invertebrate pathogens cause epizootic on a specific pest population being noninfective to humans and hence potentially non-damaging to the ecosystem. The typical examples of live microbes used as biopesticides are biofungicides (*Trichoderma*), bioherbicides (*Phytophthora*) and bioinsecticides (*Bacillus thuringiensis [Bt]*) apart from bacteriophage, bacteria, yeast and fungi that are being studied worldwide. The most popular bacterial insecticides are derived from different species of *Bacillus*, most notably the *Bt*. Bt-based microbial insecticides are the formulations of either spores or/ and metabolites produced during fermentation culture—known as crystal (Cry) toxins. Cry toxicity results in damage to the insect midgut, followed by bacterial septicemia and death. The limitations such as special formulations, storage, toxicity to only a specific species and end-user availability do pose some challenges but do not limit the acceptance of microbial insecticides. Fungi such as *Beauveria bassiana, Nomuraea rileyi, Verticillium lecanii* and *Lagenidium giganteum* are also used as broad-spectrum pesticides. Virus-based biopesticides are mostly composed of nuclear polyhedrosis viruses (NPVs), a package of numerous virus particles or granulosis viruses (GVs), in which one or two virus particles are embedded in a protein capsule. A truly efficacious product could compete effectively in the current scenario if improvements to efficacy can be achieved through better formulation and application methods and reductions in the production expenditure. In order to improve efficacy while retaining the desired host range, there is also a need to investigate additional strains and species of pathogens.

4.1 INTRODUCTION

The biggest challenge for agriculture worldwide is to supply food to the demand of burgeoning human population. According to Food and Agriculture Organization (FAO), every year, one-third of the food produced for consumption, that is roughly 1.3 billion tones, is wasted which costs about $ 750 billion (more than Rs. 47 lakh crore). The food loss happens from production level to consumption level. In industrialized countries, the wastage and loss of food is approximately US$ 680 billion while in the developing countries it is about US$ 310 billion. Especially in developing countries, the scenario is more griming where 40% of food losses happen during storage and postharvest processing levels. This is evident from the fact that about 21 million tons of wheat is wasted in India. Globally, 50% of produced food meets the similar fate of wastage and never reaches the needy consumer.

The sustaining of the global economy and socio-environmental system depend on the resources of the forest and agriculture. Various pests hamper the growth of agro-economy by reducing the yield in terms of quality and quantity. Commonly, pests are referred as unwanted organisms that cause harm to humans or livestock or have the ability to destroy agro-products. Since the time immemorial, various pests', such as microbes, insects, mice and weeds, are the leading causes of radical decrease in crop yields. About 35% of worldwide crops and foods are destroyed during production by the pests. Postharvest damage by microorganisms, insects, rodents and birds add 10–20% more loss, amounting to a total of 40–50%. Apart from traditional pests, each area is threatened by the addition of new invasive non-native pest species by global trade and cultural mix. The global food production faces paramount challenges in controlling established as well as new invasive pests.

In general, any compound or mixture used or formulated to kill, repel or mitigate any kind of pest is considered as pesticide. These pesticides facilitate protection of land and crops of people and control vectors from spreading disease. Traditionally various physical and chemical methods have been in use for pest control. The foremost objective of the pest control is to reduce the population of insects, plant pathogens or weeds to an acceptable range. The concept of pesticide was widely popularized by the introduction of 1,1,1-trichloro-di-(4-chlorophenyl) ethane, popularized as dichlorodiphenyltrichloroethane (DDT), just after the Second

World war. Though, it was first isolated in Germany in 1874 but was recognized in 1939 by Swiss Nobel Prize-winner chemist Paul Müller as a neurotoxic to insects.[1] Initially, DDT was able to eliminate disease-causing and crop-feeding insects and sometimes doubling the yields of crops. DDT being the first efficient synthetic pesticide had all the good properties of insecticide, any person at that time could imagine. However, in 1990s, people began to visualize and started grouping together various negative properties and effects of chemosynthetic pesticides, including the DDT, which are hazardous to a diverse group of animals and aquatic life. The terminologies like *pollutants, environmental contaminants, pesticides, DDT, mercury* and so forth were depicted as hazardous, as they are predisposed to bioaccumulate. Even though, most of the modern farming operations involve huge amounts of chemical pesticides as these chemical means of plant protection are cheaper and convenient way of managing and preventing pests and diseases of crops.

Moreover, during the period of 2011–2016, the requirements of agro-active compounds have increased at a rate of about 1.4% per year with the valuation of US$ 4.8 billion. At the beginning of 2016, it is estimated that the demand for production of formulated pesticide will grow at 2.6% per annum in the United States alone. Growth in production of pesticides is even rapid in agro-economy-based developing countries of Latin America (especially Brazil and Argentina) and Asia. By the end of 2016, the gross revenue generated by three countries India, China and Japan is expected to be US$ 16.2 billion with annual compound annual growth rate (CAGR) of 6.8%.[2, 3] The projected global market value of pesticide will reach US$ 81.13 billion by 2019 at an expected CAGR of 6.9%. It is forecasted that sophisticated farming techniques and demand for industrial feedstock push Brazil to become the largest global producer of pesticide, surpassing the United States.

There has been a dilemma in the usage of conventional chemosynthetic insecticides; although these are widely used in controlling pests, various factors such as development of insecticide resistance, environmental toxicity and deregistration of insecticides due to effect on human health, are limiting the effectiveness and continued use of these compounds. Besides, the growing global populations in the coming years demand an increase of crop production by 30% above the present level[4] and also an underlying concern over the safety of synthetic pesticides runs high among the public. Therefore, researchers are focusing on the eco-friendly alternatives, in order to address the current and future challenges of controlling crop-harming insects, mites and fungi. Biopesticides, being biodegradable and harmless to habitat, prove to be effective and successful alternatives of chemosynthetic pesticides.

4.2   OVERVIEW OF MICROBIAL INSECTICIDES

Interest in microbial metabolites being potent pesticides, arises from attempts to solve the above dilemma. Microbes are the treasure box of various pharmaco-active drugs and antibiotics. Metabolites of microbes are expected to counter the pesticide resistance and pollution likely caused by synthetic pesticides. Almost 30 years ago, compounds of microbial origin, such as kasugamycin, blasticidin S, polyoxin, validamycin and mildiomycin as fungicides and tetranactin as a miticide, were introduced into the fields. Some recent examples of microbial pesticides include avermectin, milbemycin and bialaphos. The excellent activity of these microbial products suggests the possible discovery of new improved agro-active compounds from microbial metabolites.[5]

Natural materials of microbes, animals and plants are the source of biological pesticides (biopesticides). Living organisms acting as natural enemies or extracted products (phytochemicals of plants, microbial proteins) or secreted semiochemicals can be grouped under biopesticides. They are less harmful to soil ecosystem or human health. The specificity of biopesticides against target pest provides an effective and ecologically amicable solution to the pest problems as they show minimal toxicity towards environment or human health. Because of distinctive modes of action, they reduce the risk associated with pesticides. In general, biopesticides are naturally derived compounds that control undesired organisms via non-toxic mechanism. Typically, biopesticides are grouped into three major groups of (i) biochemical pesticides, (ii) plant-incorporated protectants (PIPs) and (iii) microbial pesticides.[6]

4.2.1   BIOCHEMICAL PESTICIDES

Biochemical pesticides are synthetically formulated equivalents of the natural compounds. A diverse group of products ranging from natural

repellents, semiochemicals (pheromones and kairomones), pest regulator compounds from plants, resistance inducer in crops and biological
enzymes comprises biochemical pesticides. Suffocating agents such as
soybean oil, acetic acid as desiccants and diatomaceous earth as abrasives
are also included in biochemical pesticides. Nontoxic mode of action to
the target pest(s), minimal toxicity to the environment, rapid degradation
and nonpersistance in the environment are some of its advocating points.
Despite all these advantageous points, most of them are non-specific and
have a broad spectrum of target pests.

4.2.2   PLANT-INCORPORATED PROTECTANTS (PIPs)

PIPs are proteins or metabolites produced by genetically modified (GM)
crops/plants that provide resistance against harmful agents. For example,
in the United States, maize crops are engineered to gain pesticidal property
by inserting gene from the bacterium *Bacillus thuringiensis (Bt)* which
produces toxins targeted against the western corn rootworm *(Diabrotica
virgifera virgifera)*. Environmental Protection Agency along with Department of Agriculture of states regulates these proteins (e.g. *Bt* crystal [*Cry*]
proteins) and its genetic material (e.g. Cry genes), baring the plants and
its own genome. However, there is concern over the increasing resistance
property of the pests towards the PIPs.

4.2.3   MICROBIAL PESTICIDES

Microorganisms that exhibit antagonistic effect against any pest are
grouped under microbial pesticides. The mode of action of these microbial
pesticides often includes competition for food and space by the organism,
release of inhibitory or toxic substance to its niche or target pests being
the growth substance. Though different kinds of pests can be controlled by
the microbial pesticides, each microbial active compound is more specific
against its target pest(s). For example, different strains of *Bt* produces
a mixture of proteins but specifically lethal to one or few related insect
species. However, most of the marketed microbial pesticides are strains
or subspecies of *Bt*. Microorganisms like viruses, fungi and protozoa also
show antagonistic effects against crop-damaging pest or disease-causing

organism which needs thorough exploration and investigation. Broadly, these microbial pesticides can be grouped into but not limited to (i) prokaryotic microorganisms, (ii) viruses and (iii) eukaryotic microorganisms which include fungi, protozoa and algae.

The discovery and development of pesticides face multifaceted challenges. The screening and evaluation of pesticidal properties of microbes or its metabolites are of strategic importance in these development biopesticides. Advance in research on the chemical basis of plant–pathogen interaction, compounds likely to be involved in these interactions and their pharmacological activities will provide insight in the production and development of microbial metabolites into biocides. The development and screening methods can be summarized as:[5]

i) Exploitation of novel microbial groups as sources for the production of active compounds. Basidiomycetes, blue-green algae and myxobacteria were affluent producers of agro-active metabolites.

ii) Design of specific fermentation techniques and use in screening methods.

iii) Devising new strategies for the identification and detection of agro-activity, viz. introduction of unconventional test organisms to find new compounds.

iv) Application of recombinant DNA techniques for breeding mutants that produce an altered spectrum of metabolites.

v) Advances in synthetic chemistry and *in-silico* studies are helpful in designing new screening methods for the formulation and production of pesticides.

Many new microbial metabolites discovered by improved screening and evaluation methods are focused towards the possible uses in agrochemical areas of pest control for which synthetic pesticides are insufficiently effective. Avermectin, an antiparasitic product, initially intended to use in animal husbandry for the control of endoparasitic nematodes and ectoparasitic arthropods prove subsequently its applicability as a pesticide. Phthoxazolin, isolated from *Streptomyces* sp. intended to use against *Phytophthora parasitica* as an inhibitor of cellulose biosynthesis[7] has undergone many chemical modifications to be used as an herbicide. Similarly, microbial compounds like phthoramycin and rustmicin are used as growth inhibitors of *Phytophthora* and *Puccinia*, respectively.[8]

4.3   MICROBES: THE SOURCE OF PESTICIDES/INSECTICIDES

Most likely microbes are potential sources of various agro-active compounds that may act against certain microbes, insects or weeds (herbicides). On the basis of their target organism, biocides are broadly categorized into biofungicides (e.g. against *Trichoderma*), bioherbicides (e.g. against *Phytophthora*) and bioinsecticides. *Bt* is one of the popular examples of microbial biopesticides, applied to control infestations in potato, cabbage or economic crops like cotton. A list of different microbial pesticide products with their target pests is given in Table 4.1.

4.3.1   BACTERIA

Biopesticides derived from bacteria are widely favoured among users because of their variety and cheapness. Dynamic soil ecosystem provides large number of bacterial pathogens active against various insects which are spore-forming, rod-shaped bacteria of *Bacillus* genus. The effectiveness of single type of bacterial strains ranges from controlling of entire order of insects to one or few species of target pests. For example, *Bt* var. *kurstaki* is lethal against caterpillar stages of various butterflies and moths. While *Bacillus popilliae* (milky spore disease) is effective against larvae of Japanese beetle, it has no toxicity against the more related annual white grubs (masked chafers, *Cyclocephala* genus). More than one hundred *Bt*-based bioinsecticides have been developed that are mostly active against genus *Lepidopteran, Dipteran* and *Coleopteran* larvae.[9]

Pesticidal formulations of *Bt* provide an effective alternative in controlling insects. The efficiency of *Bt* and its Cry proteins is due to their target specificity and minimal toxicity towards humans.[9,10] Various molecular techniques were employed to produce *Bt*-toxin-based pesticides which are encoded by several *Cry* genes.[11] Apart from commonly available pesticides of *Bt* endotoxin derivatives, some *Bt* strains produce another exotoxin called *thuringiensin,* in significant amount. Being produced from living organisms and effective against wide variety of animals, *thuringiensin* is pitched as natural bioactive compound. Moreover, strains of *Bt* like *Bt* var. *san diego* and *Bt* var. *tenebrinos* are in use to control beetles. The product of *Bt* var. *san diego* (trade name of M-One®) is effective against larvae of the Colorado potato beetle as well as larvae and adults

TABLE 4.1 Comprehensive List of Microbial Pesticide Products and Their Target Pests and Applied Crops/Plants or Areas.

S. no.	Microorganism	Registered products	Important target pests	Applied crops/ plants/ areas
	Bacteria			
1.	*Agrobacterium radiobacter*	Galltrol-A, Dygall, Norbac 84C, Nogall	Controls crown galls	Ornamental nursery stock. Soil treatment
2.	*Bacillus thuringiensis (Bt)* subsp. *Aizawai*	Certan, Agree WP, Solbit #8, Solbit Chae, Turex, Xentari AS, Xentari WDG	Fruitworm, looper, leafrollers, caterpillars, cotton bollworm, soybean looper, cabbage worm, european corn borer, twig borer, gypsy moth, wax moth, lepidopteran larvae, *pluella xylostella* and *pieris* spp., Control of wax moth infestations in honeybee hives	Mostly small fruit plants and leafy vegetables. Berry plantings, cranberry, grape, citrus, corn, orange, tangerine, nut trees, almond, walnut, tomato, onion, potato, radish, alfalfa, bean, tobacco, celery, cotton, cucumber, melon, cauliflower, cabbage
3.	*Bacillus thuringiensi* subsp. *israelensis (Bti)*	Bactimos, Mosquito Attack, Skeetal, Teknar, Vectobac,	Effective against larvae of *Aedes* and *Psorophora* mosquitoes, black flies, and fungus gnats	Applications generally made over wide areas by mosquito and blackfly abatement districts
4.	*Bacillus thuringiensis* subsp. *israelensis* H-14	Bactimos Granules/ Pellets, Gnatrol DG, VectoBac 1200L	Mosquitos, mushroom fly (sciaridae), black fly (streams only), fungus gnat	Rice fields, pastures, ditches, catch basins, polluted water, salt marshes, tidal water, vegetable plants, ornamental plantings

TABLE 4.1 *(Continued)*

S. no.	Microorganism	Registered products	Important target pests	Applied crops/ plants/ areas
5.	*Bacillus thuringiensis* subsp. *Kurstaki* (*btk*)	Bioprotec 3p,bactur, bactospeine, dipel, caterpillar killer, futura, javelin, thuricide, topside, tribactur, worm attack, batik,	Effective against foliage-feeding caterpillars. Leafrollers, loopers and moths. Rice borer (*spodoptera frugiperda*), european corn borer, fruitree leafroller, fruitworms, winter moth (apples), diamondback moth, cabbage looper, cotton bollworm, alfalfa caterpillar, redhumped caterpillar, lepidopterous spp.	Rice, cotton, forestry, rangeland, forage crops, fruiting vegetables, legume vegetables, root and tuber vegetables, tomato,potato, cucumber, corn, cabbage,tobacco, bean (all types), sugar beet, apple, pear, apricot, peach, cherry, plum, lettuce, cranberry, stone fruit, pome fruit, berries fruit, grape, melon (all types), nuts
		Bioprotec hp,costar Bioprotec xhp,	Eastern hemlock, forest tent caterpillar, fall spanworm, gypsy moth, cankerworm,	Widely applied on forests, rangelands
6.	*Bacillus thuringiensis* subsp. *Kurstaki* sa-11	Bactec bt 32, biobit hp wp, dipel 150 dust, britz bt dust, delfin deliver, dipel dust	Omnivorous looper, caterpillar (various), head moth, webworm, banana skipper, saddleback caterpillar, tussock moth, indian meal moth, citrus cutworm, european corn borer, pine butterfly, cabbage looper,	Blackberry, raspberry, blueberry, grape, strawberry, citrus, tropical fruits, banana, pomegranate, cabbage, vegetable crops, garlic, leek, corn, alfalfa, stored grain, small grains, tobacco, almond, chestnut, pome fruit, apple
		Bonide Bt	Controlling Larvae Of Moths.	Fruit trees, leafy vegetables, ornamental trees
7.	*Bt kurstaki* abts351	Dipel-df	Looper, grape leafroller, webworm, green cloverworm,tent caterpillar, fall webworm,	Hay and other forage crops; berries and small fruit such as grapes, berries, bulbs

TABLE 4.1 (Continued)

S. no.	Microorganism	Registered products	Important target pests	Applied crops/ plants/ areas
8.	*Bacillus thuringiensis* subsp. *Kurstaki* bmp-123	Bmp 123 10g, bmp 123 2x wp	Pests include target pests of *btk* sa-11. Bud moths, achemon sphinx, mimosa webworm	Corn, tobacco, small fruit, root crop vegetables, leafy vegetables, ornamental plants
9.	*Bacillus thuringiensis* subsp. *Kurstaki* eg2348	Bactec bt16 , condor wdg, condor wp	Leaf roller, caterpillar, webworm, looper, variegated cutworm, pine butterfly, indian meal moth, tufted apple bud moth	Berry plantings, nut trees, pear, tropical fruit, melon, cucumber, cabbage, stored grain, small grains, sugar beet, sunflower, forest trees
10.	*Bacillus thuringiensis* subsp. *Kurstaki* eg7673 toxin, eg7826 toxin	Raven, lepinox g, lepinox wdg, crymax, (gene. Engineered strains)	Colorado potato beetle, cottonwood leaf beetle, elm leaf beetle, imported willow leaf beetle, mexican bean beetle, corn earworm, fall armyworm, european corn borer	All beans, green bean, lima bean, snap bean, eggplant, irish/white potato, soybean, tomato, forest land, nursery trees, shade trees, corn (pop, sweet, field)
11.	*Bacillus thuringiensis* subsp. *Kurstaki* hd-1	Dipel 2x df, dipel es, dipel es-nt, dipel wp, foray 48ba, foray 76b	Target pests include leafrollers, loopers, webworm, caterpillars, moths, fall webworm, tomato pinworm, spanworm, banana skipper, essex/european skipper, tent caterpillar, douglas-fir tussock moth,	Small fruit, blackberry, cranberry, grape,citrus, nut trees, mustard greens, parsley, stem vegetables, root crop vegetables, seed and pod vegetables, sugar beet, sunflower, soybean, pastures, stored seeds, forage crops, rapeseed
12.	*Bacillus thuringiensis* subsp. *San-diego*	M-one	Larvae of colorado potato beetle, elm leaf beetles, mosquitos	Pools in woodlots, ditches, natural marshes, estuarine areas, catch basins, sewage, lagoons

TABLE 4.1 *(Continued)*

S. no.	Microorganism	Registered products	Important target pests	Applied crops/ plants/ areas
13.	*Bacillus thuringiensis* subsp. *Tenebrinos*	Trident,foil, m-one, m-track,novodor	Larvae of colorado potato beetle, elm leaf beetles	Eggplant, tomato, potato, ornamental plants and shade trees. Similar to m-one.
14.	*Bacillus lentimorbus*	Doom, Japidemic	Grubs of japanese beetle except illinois lawn	Corn, soybean, small grains and forage
15.	*Bacillus popilliae*	Milkyspore	Japanese beetle	Ornamental plants, domestic dwellings
16.	*Bacillus sphaericus*	(in development)	Larvae of *culex psorophora,* and *culiseta*	Applied in stagnant or turbid water
Viruses				
1.	Gypsy moth nuclear polyhedrosis virus (npv)	Gypchek	Gypsy moth caterpillars	Forest, woodland, ornamental
2.	Pine sawfly npv	Neochek-S	Pine sawfly larvae	Eastern white pine,red and jack pines
3.	Douglas-fir tussock moth NPV	Virtuss WP, TM Biocontrol-1	Douglas-fir tussock moth (*orgyia pseudotsugata*)	Forest, woodland, ornamental,
4.	Celery looper NPV	CLV-LC	Primary target celery looper (*anagrapha falcifera*), cotton bollworm, tobacco budworm, beet armyworm, cabbage looper, diamond back moth	Beets, clover, corn, lettuce, cucumber, pepper, tomato, broccoli, cabbage, cauliflower, onion, potato, sweet potato, corn, wheat, tobacco, peanut, pea, sugar beet, sunflower, soybean

TABLE 4.1 (Continued)

S. no.	Microorganism	Registered products	Important target pests	Applied crops/ plants/ areas
5.	Occlusion bodies of beet armyworm npv	Spod-x lc, gemstar-lc	Beet armyworm (*spodoptera exigua*), tobacco budworm, cotton bollworm, corn earworm, tomato fruit worm	Cucumber, tomato, broccoli, cabbage, lettuce, onion, potato, sweet potato, corn, wheat, tobacco, saintpaulia, rose, bouvardia
6.	Codling moth granulosis virus (gv)	Madex-3,granupom, cyd-x, virosoft, carpovirusine	Codling moth (*cydia pomonella*)	Apple, pear, quince, plum, prune, walnut trees, apricot, peach, almond, kakis, medlar, orange. (Current use have been limited)
7.	Indian meal moth gv	Nutguard-v or fruitguard-v	Indian meal moth (*plodia interpunctella*)	Dried fruit (raisin, currant, prune, date, fig), shelled nuts (almond, walnut, peanut)
8.	Summer fruit tortrix gv	Capex-2	Summer fruit tortrix (*adoxophyes orana*)	Apple, pear, quince, plum, prune
	Fungi			
1.	*Beauveria bassiana*	Botanigard, Mycotrol, Naturalis, Ostrinil, Trichobass-L, Trichobass-P	Tagets mainly soil-dwelling insects, aphids, fungus gnats, mealy bugs, mites, thrips, whiteflies. Homopterous spp., Coleopterous spp., Phyllophaga spp., Leptophobia spp.	Maize (corn), cotton, potato, banana, tropical pineapple, sugarcane, flowers, ornamentals, strawberry, lettuce, tobacco, cabbage
2.	*Beauveria bassiana* gha	Botanigard ES, Botanigard WP, Mycotrol 22WP, Mycotrol-ES and O	Targets like aphids, whiteflies, thrips, weevil and mealybug such as, citrus black fly/ whitefly, bean aphid, melon/ cotton aphid, pear thrip, potato/onion thrip, root weevil	Blasam pear, bamboo shoots, bean, beet, blackeyed pea, cabbage, cantaloupe, carrot, melon, garlic, ginger, potato, pumpkin, radish, mustard greens, coffee, papaya, peach

TABLE 4.1 *(Continued)*

S. no.	Microorganism	Registered products	Important target pests	Applied crops/ plants/ areas
3.	*Beauveria Brongniartii*	Betel,engerlingspilz	Sugar cane beetle, cockchafer	Sugar cane, meadow, turf
4.	*Verticillium lecanii*	Mycotal, vertalec	Whitefly, thrip, aphid (except the chrysanthemum aphid)	Tomato, cucumber, bean, aubergine, lettuce, pepper, ornamental plants
5.	*Lagenidium giganteum*	Laginex	Watermold that paralyzes larvae of mosquito species (infective in the dry periods)	*Chaoborus astictopus*, the clear lake gnat
6.	*Metarhizium anisopliae strain fi-1045*	Bio-cane granules	Greyback canegrub (*dermolepida albohirtum*)	New plant cane crops
7.	*M. Anisopliae strainf001*	Bio-green granules	Redheaded cockchafer(*Adoryphorus couloni*)	Pasture, turf, lawns
	Protozoa			
1.	*Nosema locustae*	Nolo bait, grasshopper attack, semaspore bait	Grasshoppers and mormon crickets Black field cricket (useful for rangeland grasshopper control.)	All field crops (not recommended for use on a small scale, such as backyard gardens)
	Entomogenous nematodes			
1.	*Heterorhabditis bacteriophora*	Heteromask, larvanem, b-green, grubstake-hb	Larvae of curculionids, strawberry & black vine weevil, garden chafer, japanese beetle (*popillia japonica*), billbugs, dung beetle	Crops, leaves and roots of ornamentals, strawberry liter and plants, turf,

TABLE 4.1 (Continued)

S. no.	Microorganism	Registered products	Important target pests	Applied crops/ plants/ areas
2.	*Heterorhabditis heliothidis*	-	Larvae of soil-dwelling and boring insects	Products are not widely popular
3.	*Heterorhabditis marilatus*	Grubstake-hm	Citrus/ strawberry root weevil, sugar-cane stalk borer, japanese beetle, banana weevil	Citrus, strawberry potato, sugarcane, cucumber, carrot, banana
4.	*Heterorhabditis megidis*	Larvanem-m, dickmaulrussler-	Curculionids (*otiorhynchus sulcatus*), black vine weevil (*otiorhynchus sulcatus*)	Unspecified crops, nursery, strawberry, ornamental plants, young vine
5.	*Steinernema carpocapsae*	Biovector 25, ecomask	Mint root borer, mint flea beetle, cranberry girdler, strawberry root weevil, white grub	Cranberry, peppermint, spearmint, black vine
6.	*Steinernema feltiae*	Biosafe, ecomask, scanmask, vector	Larvae of a wide variety of soil-dwelling and boring insects	Plant growing in wet soils because of moisture requirement for its effectiveness.
7.	*Steinernema riobravis*	Biovector 355, biovector 355 1	West indian sugarcane borer, blue-green citrus weevil, citrus root weevil,	Sugarcane, citrus
8.	*Steinernema scapterisci*	Nematac-s	Late nymph and adult stages of mole crickets	

of elm-leaf beetle. In addition to this, some strains of genus *Bacillus* like *Bacillus sphaericus* are lethal to the larvae of *Culex, Psorophora* and *Culiseta* genera.[12] However, it exhibits a range of effectiveness against the larvae of *Aedes* species; *Aedes vexans* being very susceptible whereas *Aedes aegypti* (yellow fever vector) and *Aedes albopictus* are resistant. The field results of *B. sphaericus* against *Anopheles* are not promising even if the laboratory tests are successful. It is due to the fact that bacteria have to be present for longer time on the surface of water in order to get fed by *Anopheles* mosquitoes. However, it still remains a valuable pesticide as it is effective in turbid or stagnant water and naturally maintains and cycles itself in the water to infect subsequent generations of mosquito larvae.

Bacterial insecticides marketed under the trade names Doom®, Iapidemic® and Grub Attack® contain the bacteria *B. popilliae* and *Bacillus lentimorbus*. These bacteria grow and proliferate on the insect larvae. As an example, pesticides containing these bacteria are applied to turf of soil which then watered into the sub-surface of soil to kill larval (grub) stage of the Japanese beetle. When susceptible grubs feed on the spores of these bacteria, the bacteria proliferate within the insect and the grub's internal organs are liquefied, turning into milky white coloration (hence named, milky spore disease). Studies have reported limited effectiveness of these *Bacillus* strains towards larvae of beetles of genera *Cyclocephala*. However, *B. popilliae* and *B. lentimorbus*, unlike *Bt*, can cycle and survive as spores for 15–20 years under the undisturbed soil.

Apart from entomopathogenic species of *Bacillus* genera, *Clostridium bifermentans* serovar *Malaysia* are useful in controlling mosquitoes and blackflies.[13] The identified insecticidal proteins of these serovars (Cbm71) show similarity with *Bt* δ-endotoxins. Similarly from *Pseudomonas entomophila*, a ubiquitous bacterium, different insecticidal toxin complexes have been identified.[14] The members of the genera *Xenorhabdus* and *Photorhabdus* are effective against nematodes of the genus *Steinernema* and *Heterorhabditis*, respectively. Even rare actinomycetes like *Kitasatospora* (order *Actinomycetales*) produces 16-membered macrolide insecticidal protein termed as setamycin. In addition, members of this genus produce phosalacine, an herbicidal tripeptide;[15] cystargin, an antifungal peptide[16] and variety of antibiotics.

Techniques of biotechnology are not only helpful in widening the prospects for development of new improved pesticide but also enable to recombine the genes of toxins with the genome of crops or plant-dwelling bacteria. With the growth and multiplication of these plant-dwelling bacteria, the bacterial toxin is produced within the plant that provides natural defence against pests. Such strategies are now being tested on corn in controlling the European corn borer by help of *Bt*. Although it seems very effective, the season-long of high-level toxin may pose risk of developing insect resistance towards the *Bt* toxin. Hence, during design and formulation of pesticide, the possible development of resistance in insects must be addressed.

4.3.1.1 TYPICAL EXAMPLE OF MICROBIAL INSECTICIDES-BT

The most widely used biological controls of insects in the world are the formulations of the entomopathogenic bacterium, *Bt*. Because of the user safety, narrow host range and compatibility with other pesticides in pesticide management programmes have resulted in more public acceptance towards the *Bt*-derived microbial insecticides in controlling pests of agriculture and forest. Since the cloning of the insecticidal Cry proteins of *Bt* in 1981, the number of genes encoding proteins with pesticidal property is increasing continuously.[17] New insecticidal genes of *Bt* are now being expressed in vitro or inserted into crops to express formulations of *Bt* insecticides.[18] Over 190 *Bt*-derived insecticides were registered and marketed to control various pests of genera *Homoptera, Hymenoptera, Orthoptera, Lepidoptera, Diptera* and *Coleoptera*.

Bt-based microbial insecticides are formulations of spores and materials produced during fermentation that include crystalline insecticidal proteins, known as Cry toxins. More than 700 *Cry* genes have been identified that code for Cry proteins.[11,19] A list of *Bt* toxins along with their organisms is given in Table 4.2. The insecticidal activity of the *Bt* depends on the Cry toxins that are ingested by the host. Cry toxicity results in damage to the insect midgut, followed by bacterial septicemia and death. Apart from Cry toxin, various *Bt* strains synthesize cytolytic (Cyt) toxins (also known as δ-endotoxins) during sporulation stage and parasporal crystalline inclusions during stationary phase of growth have also insecticidal property. Also during vegetative stage, *Bt* synthesizes another insecticidal protein and secretes outside of cell. These are termed as vegetative insecticidal proteins (Vip) and the secreted insecticidal protein (Sip).[20,21]

Advances in Microbial Biotechnology

TABLE 4.2 Bt-toxins with Their Target Organism and Receptors.

Target pest	Order	Designated *Bt Toxin*	Target receptors
Butterflies and moths	Lepidoptera	Cry1Aa-e, Cry1Ba, Cry1Ca-b, Cry1D-J, Cry2Aa-c, Cry9Aa, Cry9Ba, Cry9Ca,	Cadherin-like protein (CADR), (GPI)-anchored amino peptidase-N (APN), BTR-270, P252
Two-winged or true flies	Diptera	Cry1Ca, Cry2Aa, Cry4Aa, Cry4Ba, Cry10Aa, Cry11Aa-b, Cry44A, Cry47A, Cry48A and Cyt1A-B, Cyt2A-B	GPI-anchored alkaline phosphatase (ALP), cadherin-like protein (CADR)
Beetles and weevils	Coleoptera	Cry3Aa, Cry3Ba, Cry3Bb, Cry3Ca Cry7Aa, Cry7Ab, Cry8Aa, Cry8Ab, Cry8Ca, Cry14Aa, Cry1Ba, Cry1Ia, Cry1Ib,	Cadherin-like receptor (CADR)-binding Cry3Aa
Nematode	Rhabditida	Cry5Aa, Cry5Ab, Cry12Aa, Cry13Aa, Cry21A, Cry55A	GPI-anchored receptor, bre-1 and bre-5 (*Bt*-toxin resistant gene '*bre*')
Aphids, leafhoppers	Hemiptera	Cry2A, Cry3A, Cry11A and Vip1Ae, Vip2Ae	ALP receptor
Bees, wasps and sawflies	Hymenoptera	Cry3A, Cry5A, Cry22A,	GPI-anchored receptor
Snails, slugs	Gastropoda	Cry1Ab	

Structural analyses of Cry-toxins display a high degree of similarity with a three-domain organization, suggesting a similar mode of action on the target organism by damaging their mid gut lining and binding to similar receptors.

The Cry toxin name is derived from the structural representations with parasporal crystals. Based on structure, Cry toxins cannot be grouped into single homologous family instead they are subdivided into groups of different lineages. Most of the Cry toxins belong to three-domain Cry toxin family that shares remarkable conserved three-domain structure (hence the family name), even though the exact amino acid sequence is different.[22–24] The perforating domain or Domain I, an N-terminal seven α-helix cluster, is responsible for penetration and pore formation on cuticle.[11,25,26] The toxin–receptor interactions are carried out by the central or middle domain (Domain II) which is composed of three antiparallel β-sheets.[27] Domain III or galactose-binding domain, two antiparallel β-sheet sandwiches, is involved in pore formation by binding to cell receptors.[26] Although the active toxin is three-domain folded structure, the unprocessed pro-toxin has four additional domains made up of two alpha helix (domains IV and VI) and two beta roll. The 16 cysteine residues are involved in stabilizing the *Bt* crystals by disulphide bonds. Such disulphide cross-links are prone to break under alkaline and acidic conditions which explain the solubilization property of *Bt* toxins inside the gut of insect. Most of the Cry toxins possess up to five conserved blocks inside the core of active toxin and three conserved blocks outside the core that is c-terminal of pro-toxin. However, several shorter pro-toxins (70 kDa) are also synthesized, such as Cry3 and Cry11, which lack the extended C-terminal region.

The *cyt* genes-encoded Cyt proteins (cytotoxic) are small and distinct group of Cry proteins of the *Bt*-produced insecticidal proteins.[28] According to *Bt* Toxin Nomenclature Committee, the Cyt proteins were grouped into three different families named as Cyt1, Cyt2 and Cyt3 which show toxicity mostly against larvae of mosquitoes and black flies (*Dipteran* sp.).[25,29,30] The three-dimensional structures of the Cyt1Aa and Cyt2Ba are single domain of three-layered alpha-beta proteins.[31,32] The Cyt1Ca protein isolated from *Bt* subsp. *israelensis* has another domain at the c-terminal end which has homology with carbohydrate-binding domain of ricin; although no significant larvicidal or haemolytic activity was reported.[33] The Cyt2C protein toxicity is observed against nematodes (*Rhabditida*) and to an extent towards cancer cells.[19] The synergistic mode of action of Cyt proteins can be observed by induction of the insecticidal activity in other Cry or Vip3 toxins and thereby reducing the development of resistance to Cry toxins alone.[30, 34] For example, Cyt1Aa protein, toxic against *Chrysomela scripta* prevents the development of resistance to Cry3Aa.[35] Synergic activity is observed between Cry11Aa to Cyt1Aa during oligomerization of Cry11Aa toxin and pore formation. Even Cyt1Aa also show synergism Mtx1 toxin family against *Culex quinquefasciatus*.[36] On the other hand, in vitro studies have suggested cytolytic (haemolytic) mode of action for Cyt proteins toxicity. Two different models have been suggested for the Cyt proteins mechanism of action: a conventional pore-formation model, while other suggests a less specific detergent action mechanism.[29,30,23]

Apart from Cry or Cyt toxins, the overall diversity of *Bt* strains activity against different insects is augmented with the discovery of secretory extracellular proteins, produced during vegetative growth phase.[37,38] These proteins are grouped into two, that is Vips[20] and Sips.[39] On the basis of amino acid sequence homology, the Vips are categorized into four subfamilies of Vip1, Vip2, Vip3 and Vip4. The Vip1 (~100 kDa) and Vip2 (~50 kDa) proteins are binary toxins translated from *vip1* and *vip2* genes present in single operon.[23,40] The binding of proteolyticly activated Vip1 with the monomeric cell surface receptors and formation of homo-heptamers facilitates the transfer of cytotoxic Vip2 into the cell by acid endosomes. The cytotoxic action of Vip2 is likely by ADP-ribosylation of actin at Arg-177 which sterically inhibits actin polymerization leading to disarrangement of cytoskeleton and cell death.[41,42] Some coleopteran pests and the sap-sucking pest *Aphis gossypii* (*Hemiptera*) are affected by the Vip1 and Vip2 binary toxin. While *Lepidopteran* species viz. *Agrotis ipsilon, Spodoptera exigua* and *Sodoptera frugiperda* show sensitivity towards single-chain Vip3 proteins. As detailed mode of action of Vip3A is yet to be discovered, the suggested mechanism involves attachment of Vip3A to midgut epithelial cells followed by gut paralysis and larval death. The Sip is the only protein among secreted proteins of *Bt* that possesses insecticidal activity against coleopteran larvae. The designated secretary gene (*Sip1Aa1*) is 1104 bp long which encodes a peptide of 367 amino acids and molecular weight of ~41-kDa. Apart from consensus secretion signal of 30 amino acids, it shows noteworthy similarity with 36-kDa Mtx3 mosquitocidal toxin of *B. sphaericus* SSII-1. Sip1Aa1 is the only secretory protein reported to date to exhibit strong inhibitory effect against *Leptinotarsa decemlineata, Diabrotica undecimpunctata howardi* and *D. virgifera virgifera* species.

4.3.2 VIRUSES

Many viruses cause epidemic among insect population. In 1975, Sandoz Inc. first introduced a virus-based insecticide called Elcar™.[43] The lethality of the viruses is very specific, affecting whole insect genus or only single type of species. In general, nuclear polyhedrosis viruses (NPVs) and granu-losis viruses (GVs) have been exploited by as potential insecticides. NPVs are the crystalline package of numerous viruses within insect cell nuclei while GVs are granular or capsule-like protein Cry of one or two virus particles. During 1980s, the insecticide Elcar™ helpful in controlling cotton bollworm was a *Heliothis zea* NPV whose production and usage was later limited. Still, large numbers of nucleo-polyhedrosis virus-based pesticides are registered or marketed worldwide. Viral-based pesticides are commonly used in controlling forest pests. Viral pathogens are effective against forest pests is due to cycling of the pathogens, protection from ultraviolet radiation and persistence in forest environment. Most of the viral pathogens or products are specific to single species or group of related forest pests like gypsy moth, Douglas-fir tussock moth, spruce budworm and pine sawfly. Recent studies on development of viral insecticides focus against alfalfa looper, soybean looper, armyworms, cabbage looper and cabbage worm.

During World War II, the baculovirus was first introduced into the environment to control pest. At that time, an NPV-specific against spruce sawfly was introduced in Canada which is so successful that no other pesticides were required to control this hymenopteran species. More than 50 formulations of *baculovirus-based pesticides* induce toxic epizootics which have catastrophic effect on the host insect population. One of the typical examples is *Heliothis zea* NPV preparations, marked as Elcar™, show lethality towards genera *Helicoverpa* and *Heliothis*. The resistance to chemical pesticides like pyrethroids can be avoided by using *Heliothis zea* single NPV (HzSNPV)-based pesticides. Natural control of pests viz. *Helicoverpa armigera,*[44] cotton bollworm and genera attacking tomato, maize, sorghum and soybean is also achieved by HzSNPV. This is the reason being, countries like India and china are introducing these microbial pesticides in various pest management program where cotton, pepper tomato, and maize are grown in large areas.

Another *baculovirus*, Helicoverpa armigera single NPV (HaSNPV), similar to HzSNPV, used in large scale in china on cotton crops[45] has broad spectrum of host.[46,47] In Brazil, the bioregulation of velvet-been caterpillar in soybean by baculovirus-based pesticide, *Anticarsia gemmatalis* nucleopolyhedrovirus was an inspiring example.[48] Caterpillars of moths belonging to *Spodoptera* genus is controlled by *Spodoptera*-specific NPV. Commercial preparations like SPOD-X™ containing *Spodoptera exigua* NPV was used control caterpillars on vegetables while Spodopterin™ containing *Spodoptera littolaris* NPV has striking result in protecting cotton, tomatoes and corn.[48,49] On limited scale, *Autographa californica* and *Anagrapha falcifera* NPVs show broad spectrum of protection against *Spodoptera* and *Helicoverpa* infestation in sugarcane, legume and rice.

Granulovirus, another group of viruses, were effective in controlling *Cydia pomonella*, coddling moth of fruits. In Latin American countries, the cassava plantations are protected from *Erinnyis ello* (cassava hornworm) by GVs.[50] *Amsacta moorei,* an entomopoxvirus, was active against lepidopteran pests such as *Estigmene acrea* and *Lymantria dispar.*[51] In Japan tea tortricidae, *Homona magnanima* and *Adoxophyes honmai* are controlled by target-specific GVs. However, favourable weather, slow action and microbial contamination, damage of viral pesticides due to direct sunlight and expensive production cost have curtailed their production. Some viral-based pesticides, codling moth GV (Decyde®) and the Heliothis NPV (Elcar®) are not successful enough in US or Europe, like it was. Nonetheless, researchers address such constraints by manipulating insecticidal genes in the viruses.

4.3.3   FUNGI

Fungi, like viruses and bacteria, have also the potential to biocontrol various agricultural pests. Entomopathogenic fungi are effective in controlling wide variety of pests/insects which spread by asexual spores called conidia. Unlike the bacterial spores, fungal infection is initiated with attachment of single-celled fungus, called as conidia or blastospores, to the insect cuticle. Then the fungal spores secrete various hydrolytic enzymes like proteases, chitinases and lipases, which facilitates the germination and penetration through cuticle by specialized structures, penetration pegs and/or appressoria.[52,53] By reaching the haemocoel, they utilize nutrients to grow in numbers as well as escape from insect's immune system.[54] However, the activity of fungal pesticides varies not only towards different life stages of insects' that is eggs, immature pupas or adults but also towards wide variety of species. One of the major advantages of fungal-based pesticides is that it does not have to be ingested by the host to cause infections. In most instances, as fungal infections progress, infected insects are killed by fungal toxins, not by the chronic effects of parasitism.

Beauveria bassiana, a soil fungus, has adverse effect on immature grubs and adults of many species of beetles. The synergistic action of *Beauveria bassiana* with sub-lethal dose of other pesticides show greater mortality rate in controlling Colorado potato beetle (*Leptinotarsa decemlineata*). Another strain of *B. bassiana* (Balsamo) var. *vuillemin* along with plant agro-active compounds of neem are successful in controlling *Bemisia tabaci* (Gennadius) (Hemiptera order), sweet potato whitefly, on eggplants and tomato.[55] The *B. bassiana*-based mycoinsecticide 'Boverin' coupled with low doses of trichlorophon have been effective in suppressing the second-generation outbreaks of *Cydia pomonella* L.[56] Studies in Great Britain reported that the fungi *Verticillium lecanii* (marketed as Vertelec®) are successful in controlling aphids and whiteflies in greenhouses farming as well as foliage-feeding caterpillars of soybeans plants. The *Neozygites tanajoae* (entomophthorales order) was identified as one of the important natural enemies of cassava green mite, *Mononychellus tanajoa* (Bondar), discovered in north-eastern Brazil.[57] Laboratory tests have suggested that *Neozygites floridana* (natural fungal pathogen) active against *Tetranychus evansi,* red spider mite of tomato and adults of *Ceratitis capitata.*[58] Similarly, the insect-pathogenic fungus *Metarhizium anisopliae* has significantly reduces the longevity of adults of *Aedes aegypti* and *Aedes albopictus* mosquitos.[59] Fungal pathogens also show significant ovicidal activity. Bioassay studies have proved that fungal biocontrol agents like *Beauveria bassiana, Metarhizium anisopliae* and *Paecilomyces fumosoroseus* show significant lethality on the eggs of carmine spider mite, *Tetranychus cinnabarinus.*[60] *Lagenidium giganteum,* an aquatic fungus, is effective in controlling mosquitoes in aquatic environment by cycling itself even it low mosquito density that is spores of affected larvae infect larvae of subsequent generations.

On the other hand, blue-green algae, myxobacteria, and red algae and bacidiomycete possess many bioactive antifungal, antibacterial and cytocidal compounds. Blue-green algae, *Lyngbya majuscula,* produces antifungal compound called malyngolide.[61] Myxobacterial strains produce agricultural antifungal compounds like phenoxan, ambruticin and pyrrolnitrin[62] preventing fungal diseases. The typical herbicidal agent, pereniporin, is produced from bacidiomycete strain of *Perenniporia madullaenpanis.*

4.3.4 PROTOZOA

Apart from lethality, the protozoans show chronic and debilitating effects in the insect host which reduces production of offspring from infected insects. The insecticidal potential has been evaluated for the species of *Nosema* and *Vairimorpha* genera. These protozoans are active against larvae of *Lepidopteran* and adults of order *Orthoptera* (the grasshoppers). The biopesticides against grasshoppers like NOLO Bait® and Grasshopper Attack® are the microsporidian formulations of *Nosema locustae.* Adequate control over hoppers in large areas like rangelands is achieved by *Nosema locustae*-based protozoan insecticides. Although, not all grasshoppers in the treated area are wiped out by *Nosema locustae*, the consumption of foliage by hoppers and laying of vival eggs by the infect females is reduced. However, the early nymphal stages (immature grasshoppers) are mostly affected by protozoan pesticides.[63,64]

4.4 BASIC MECHANISM OF ACTION OF MICROBIAL INSECTICIDES

Even though various biological researches are being carried out and knowledge is growing but they rarely tell us where and how pesticides interfere with the normal processes. Many toxicants or pesticides differ from one another in mode of action. Since the discovery in 1901, *Bt* has been widely used to control insects or pests in agronomy and forestry whose detailed mode of action still remains elusive. The basic mode of *Bt* depends on the synthesis of a crystalline inclusion of Cry proteins.

The conventional explanation for the toxicity of the three-domain Cry protein is the formation of trans-membrane pores or ion channels resulting in cell lysis by osmosis.[9] Three different models that is (i) classical model, (ii) sequential binding model and (iii) signalling pathway model have been proposed for the Cry toxin mechanism of action on host cells. According to 'classical' model, the lyses of epithelial cells of insect gut begin with ingestion of Cry toxins and dissolution at midgut lumen in alkaline condition. Followed by proteolytic cleavage of pro-toxin into smaller peptides of protease-resistant and attachment of peptide fragments to specific receptors of epithelial cells for the formation of pores permitting inorganic ions, amino acids and sugars into cell.[65,66] The series of action breaks arrangement of cell gut wall, leading to insect death. According to sequential binding model, the three-domain Cry toxin is activated by proteolytic cleavage of α-1 helix from domain I and binds to epithelial cell receptor of intestine, similar to the cadherin transmembrane glycoprotein receptors. The interaction of toxin with receptor induces the conformational change to form an oligomeric pre-pore structure.[22,24] On the other hand, the signalling-pathway model proposes that the toxin binds to the specific cadherin receptors which leads to Mg^{2+}-dependent and adenylyl cyclase/protein kinase-A-mediated necrosis of cell.[36] Although recent studies support both the sequential and signalling-pathway models, but the 'classical' model, emphasizing on Cry toxins-mediated pore formation, is still being widely accepted.

The infection in insects/pests begins after ingestion of viral and bacterial pesticides. Viral-based pesticides are more of symptomatic type of toxicity towards pests affecting various organs. The infection by viral pesticides is more prominent in caterpillars as viral particles pass through the gut wall and spread to other body tissue or organs. Some of the typical symptoms of infection progress are the liquefaction of internal organs and decolourization of cuticle leading to death of the host. Caterpillars killed by virus infection appear limp and soggy. However, even similar viral-based pesticides show different mode of action towards same or different pests. For example, in sawfly larvae, the viral infections are confided to the gut wall and poor symptomatic disease progression compared to distinguished disease progression in caterpillars. The effectiveness of viral pathogens is higher in rainy season as these pathogens can spread easily throughout insect/pest population by producing new virus particles from infected and killed pests.

4.5 ADVANTAGES AND DISADVANTAGES OF MICROBIAL INSECTICIDES

Though individual pesticide products are different in various ways, but in general, the beneficial characteristics of insecticides of microbial origin can be summarized as follows:

1. Most of the microbial insecticides contain organisms that are non-pathogenic to humans or to commercially important animals.
2. In contrast to synthetic products, microbial insecticides demonstrate high target (pest) specificity while affecting least to other beneficial insects including predators or parasites in the treated areas.
3. Unlike the chemo-pesticides, microbial pesticides show low residual effect after their application. So, they can be applied even during or just before pest attack. They show minimal off-site effects due to absence of run-off water contamination and soil percolation.
4. The emerging pest resistance can be alleviated by rotating conventional products with biopesticides.

5. They do not add to environmental pollution as they often decompose easily and quickly. Even they are effective in small quantities which reduces the environmental load. These improve management of pesticides and their residues.

6. As pathogenic microorganisms can establish in a pest population and cycle to next generations, the inherent bio-amplification of the microbial population reduces the need for repeated application of microbial pesticides.

7. The microbial products are not deactivated or damaged by other pesticide or even residues of conventional insecticides applied in combination during various IPM programmes. Thus, the use of conventional pesticides can be decreased significantly.

8. It increases the yield by reducing unwanted risk as it provides flexibility to farmers in application of pesticides.

The use of biopesticides is also met with a few limitations that are listed below. Understanding these limitations will help users to choose effective products and take necessary steps to achieve successful results.

1. The specificity of microbial insecticide allows it to control only a specific pest in a field. However, in case of multiple infestations with various pests, protection is not effective and pests cannot be eradicated. This can contribute to increased damage by altering the population dynamics of the pests.

2. Associated with the above limitation, it is obvious that a single microbial pesticide is also not effective against all pests even within the same genera or species of insects. For example, *Bt* var. *san diego* products are not pathogenic to key beetles like rootworms of corn arising due to differences in receptors protein of gut wall of the insects.

3. Susceptibility to microbial pesticides varies significantly at different life stages of pests. In addition to this, treatments must be directed to the plant parts or other material that are fed by target pest. These are some observations in *Bt* isolates.

4. Although protozoan pathogens show significant effect on insect populations, not all infected hoppers are killed and kills the grasshopper slowly (effects visible after three to six weeks). Sometimes the mobility of larger pests is likely to reduce the effect the

pathogenesis of insecticides. The infective spores of fungal insecticides need to be carefully produced and stored.

4.6   FUTURE PROSPECTS

The need for a continuous supply of food has led conventional agriculture to be strongly dependent on chemicals, the outcome of the first Green Revolution. The increasing concern of consumers and government on food safety has led cultivators and agriculturists to explore new environmentally friendly methods to replace, or at least supplement current chemical-based practices. In this scenario, the use of biopesticides has emerged as promising alternative to chemical pesticides. Their role appears to be promising in agriculture and forestry in the modern world. Problems in acceptance among cultivators have begun to reduce leading to a steady increase in their demand in all parts of the world. When incorporated within the IPM systems, biopesticides show equal or better efficacy than purely chemical pesticide use, especially for crops like fruits, vegetables, nuts and flowers.

Further discovery of new microbes and their formulation strategies will increase their adoption in farming practices. While bioinsecticides account for only a small fraction of the world pesticide market presently, there are optimistic predictions for an expansion of the whole market for bioinsecticides[67] despite entrenched pesticides market, low per capita usage of pesticides in several regions, shrinking farm lands and burgeoning research and development costs. Criteria like technical efficacy, regulatory compliance and commercial viability will be key determinants in their incorporation into pest management strategies. Governments in several countries have come up with initiatives aimed at promoting the development and use of biopesticides because of their low toxicity, greater safety, and higher effectiveness in controlling pests. A truly efficacious product could compete effectively in some current markets even at a greater cost than conventional chemicals because there is strong willingness to move away from synthetic insecticides and a perceptual shift in appreciation for a 'greener' product.

KEYWORDS

- **biopesticides**
- *Bacillus, Bt*
- NPVs
- cry toxins

REFERENCES

1. Pimentel, D. *CRC Handbook of Pest Management in Agriculture;* CRC Press: New York, NY, USA, 1990; Vol. 1.
2. Olson, S.; Ranade, A.; Kurkjy, N.; Pang, K.; Hazekamp, C. *Green Dreams or Growth Opportunities: Assessing the Market Potential for "Greener" Agricultural Technologies;* Lux Research Inc: Boston, MA, USA, 2013.
3. Thakore, Y. The Biopesticide Market for Global Agricultural Use. *Ind. Biotechnol.* **2006,** *2*(3), 194–208.
4. Eue, L. World Challenges in Weed Science. *Weed Sci.* **1986,** *34,* 155–160.
5. Tanaka, Y.; Omura, S. Agroactive Compounds of Microbial Origin. *Annu. Rev. Microbiol.* **1993,** *47*(1), 57–87.
6. Gupta, S.; Dikshit, A. K. Biopesticides: An Ecofriendly Approach for Pest Control. *J. Biopestic.* **2009,** *3*(1), 186–188.
7. Omura, S.; Tanaka, Y.; Kanaya, I.; Shinose, M.; Takahashi, Y. Phthoxazolin, a Specific Inhibitor of Cellulose Biosynthesis, Produced by a Strain of *Streptomyces* sp. *J. Antibiot.* **1990,** *43,* 1034–1036.
8. Nakayama, H.; Takatsu, T.; Abe, Y.; Shimazu, A.; Furihata, K., et al. Rustmicin, a New Macrolide Antibiotic Active Against Wheat Stem-Rust Fungus. *Agric. Biol. Chem.* **1987,** *51,* 853–859.
9. Usta, C. Microorganisms in Biological Pest Control—A Review (Bacterial Toxin Application and Effect of Environmental Factors). In *Biochemistry, Genetics and Molecular Biology Current Progress in Biological Research;* Silva-Opps, M., Ed.; 201, Intech. 2013.
10. Roh, J. Y.; Choi, J. Y.; Li, M. S.; Jin, B. R.; Je, Y. *Bacillus thuringiensis* as a Specific, Safe, and Effective Tool for Insect Pest Control. *J. Microbiol. Biotechnol.* **2007,** *17*(4), 547–559.
11. Schnepf, E.; Crickmore, N.; Rie, J. V.; Lereclus, D.; Baum, J.; Feitelson, J.; Dean, D. H. *Bacillus thuringiensis* and its Pesticidal Crystal Proteins. *Microbiol. Mol. Biol Rev.* **1998,** *62*(3), 775–806.
12. Scholte, E. J.; Knols, B. G. J.; Samson, R. A.; Takken, W. Entomopathogenic Fungi for Mosquito Control: A Review. *J. Insect Sci.* **2004,** *4,* 19.

13. Nicolas, L.; Hamon, S.; Frachon, E.; Sebald, M.; Barjac, H. de. Partial Inactivation of the Mosquitocidal Activity of *Clostridium bifermentans* Serovar *Malaysia* by Extracellular Proteinases. *Appl. Microbiol. Biotechnol.* **1990**, *34*, 36–41.

14. Vodovar, N.; Vinals, M.; Liehl, P.; Basset, A.; Degrouard, J.; Spellman, P.; Boccard, F.; Lemaitre, B. Drosophila Host Defense After Oral Infection by an Entomopathogenic *Pseudomonas* Species. *Proc. Natl. Acad. Sci. U. S. A.* **2005**, *102*, 11414–11419.

15. Omura, S. Trends in the Search for Bioactive Microbial Metabolites. *J. Ind. Microbiol.* **1992**, *10*(3–4), 135–156.

16. Uramoto, M.; Itoh, Y.; Sekiguchi, R.; Shin-Ya, K.; Kusakabe, H.; Isono, K. A New Antifungal Antibiotic, Cystargin: Fermentation, Isolation, and Characterization. *J. Antibiot.* **1988**, *41*(12), 1763–1768.

17. Schnepf, H. E.; Whiteley, H. R. Cloning and Expression of the *Bacillus thuringiensis* Crystal Protein Gene in *Escherichia coli*. *Proc. Natl. Acad. Sci. U. S. A.* **1981**, *78*(5), 2893–2897.

18. Sanchis, V. From Microbial Sprays to Insect-Resistant Transgenic Plants: History of the Biospesticide *Bacillus thuringiensis*. A Review. *Agron. Sustainable Dev.* **2011**, *31*(1), 217–231.

19. Van Frankenhuyzen, K. Insecticidal Activity of *Bacillus thuringiensis* Crystal Proteins. *J. Invertebr. Pathol.* **2009**, *101*(1), 1–16.

20. Estruch, J. J.; Warren, G. W.; Mullins, M. A.; Nye, G. J.; Craig, J. A.; Koziel, M. G. Vip3A, a Novel *Bacillus thuringiensis* Vegetative Insecticidal Protein with a Wide Spectrum of Activities Against Lepidopteran Insects. *Proc. Natl. Acad. Sci. U. S. A.* **1996**, *93*(11), 5389–5394.

21. Palma, L.; Muñoz, D.; Berry, C.; Murillo, J.; Caballero, P. *Bacillus thuringiensis* Toxins: An Overview of Their Biocidal Activity. *Toxins* **2014**, *6*(12), 3296–3325.

22. Bravo, A.; Gill, S. S.; Soberón, M. Mode of Action of *Bacillus thuringiensis* Cry and Cyt Toxins and Their Potential for Insect Control. *Toxicon: Off. J. Int. Soc. Toxinol.* **2007**, *49*(4), 423–435.

23. de Maagd, R. A.; Bravo, A.; Berry, C.; Crickmore, N.; Schnepf, H. E. Structure, Diversity, and Evolution of Protein Toxins from Spore-Forming Entomopathogenic Bacteria. *Annu. Rev. Genet.* **2003**, *37*, 409–433.

24. Pardo-López, L.; Soberón, M.; Bravo, A. *Bacillus thuringiensis* Insecticidal Three-Domain Cry Toxins: Mode of Action, Insect Resistance and Consequences for Crop Protection. *FEMS Microbiol. Rev.* **2013**, *37*(1), 3–22.

25. Ben-Dov, E. *Bacillus thuringiensis* subsp. *israelensis* and its Dipteran-Specific Toxins. *Toxins* **2014**, *6*(4), 1222–1243.

26. Xu, C.; Wang, B.-C.; Yu, Z.; Sun, M. Structural Insights into *Bacillus thuringiensis* Cry, Cyt and Parasporin Toxins. *Toxins* **2014**, *6*(9), 2732–2770.

27. Jenkins, J. L.; Dean, D. H. Exploring the Mechanism of Action of Insecticidal Proteins by Genetic Engineering Methods. *Genet. Eng.* **2000**, *22*, 33–54.

28. Guerchicoff, A.; Delécluse, A.; Rubinstein, C. P. The *Bacillus thuringiensis* Cyt Genes for Hemolytic Endotoxins Constitute a Gene Family. *Appl. Environ. Microbiol.* **2001**, *67*(3), 1090–1096.

29. Butko, P. Cytolytic Toxin Cyt1A and its Mechanism of Membrane Damage: Data and Hypotheses. *Appl. Environ. Microbiol.* **2003**, *69*(5), 2415–2422.

30. Soberón, M.; López-Díaz, J. A.; Bravo, A. Cyt Toxins Produced by *Bacillus thuringiensis*: A Protein Fold Conserved in Several Pathogenic Microorganisms. *Peptides* **2013**, *41*, 87–93.

31. Cohen, S.; Dym, O.; Albeck, S.; Ben-Dov, E.; Cahan, R.; Firer, M.; Zaritsky, A. High-Resolution Crystal Structure of Activated Cyt2Ba Monomer from *Bacillus thuringiensis* subsp. *israelensis*. *J. Mol. Biol.* **2008**, *380*(5), 820–827.

32. Cohen, S.; Albeck, S.; Ben-Dov, E.; Cahan, R.; Firer, M.; Zaritsky, A.; Dym, O. Cyt1Aa Toxin: Crystal Structure Reveals Implications for its Membrane-Perforating Function. *J. Mol. Biol.* **2011**, *413*(4), 804–814.

33. Kelker, M. S.; Berry, C.; Evans, S. L.; Pai, R.; McCaskill, D. G.; Wang, N. X., et al. Structural and Biophysical Characterization of *Bacillus thuringiensis* Insecticidal Proteins Cry34Ab1 and Cry35Ab1. *PloS One* **2014**, *9*(11), e112555.

34. Yu, X.; Liu, T.; Sun, Z.; Guan, P.; Zhu, J.; Wang, S.; Li, P. Co-expression and Synergism Analysis of Vip3Aa29 and Cyt2Aa3 Insecticidal Proteins from *Bacillus thuringiensis*. *Curr. Microbiol.* **2012**, *64*(4), 326–331.

35. Federici, B. A.; Bauer, L. S. Cyt1Aa Protein of *Bacillus thuringiensis* is Toxic to the Cottonwood Leaf Beetle, *Chrysomela scripta* and Suppresses High Levels of Resistance to Cry3Aa. *Appl. Environ. Microbiol.* **1998**, *64*(11), 4368–4371.

36. Zhang, B. H.; Liu, M.; Yang, Y. K.; Yuan, Z. M. Cytolytic Toxin Cyt1Aa of *Bacillus thuringiensis* Synergizes the Mosquitocidal Toxin Mtx1 of *Bacillus sphaericus*. *Biosci. Biotechnol. Biochem.* **2006**, *70*, 2199–2204.

37. Donovan, W. P.; Donovan, J. C.; Engleman, J. T. Gene Knockout Demonstrates that Vip3A Contributes to the Pathogenesis of *Bacillus thuringiensis* Toward *Agrotis ipsilon* and *Spodoptera exigua*. *J. Invertebr. Pathol.* **2001**, *78*(1), 45–51.

38. Milne, R.; Liu, Y.; Gauthier, D.; van Frankenhuyzen, K. Purification of Vip3Aa from *Bacillus thuringiensis* HD-1 and its Contribution to Toxicity of HD-1 to Spruce Budworm (*Choristoneura fumiferana*) and Gypsy Moth (*Lymantria dispar*) (*Lepidoptera*). *J. Invertebr. Pathol.* **2008**, *99*(2), 166–172.

39. Donovan, W. P.; Engleman, J. T.; Donovan, J. C.; Baum, J. A.; Bunkers, G. J.; Chi, D. J.; Walters, M. R. Discovery and Characterization of Sip1A: A Novel Secreted Protein from *Bacillus thuringiensis* with Activity Against Coleopteran Larvae. *Appl. Microbiol. Biotechnol.* **2006**, *72*(4), 713–719.

40. Gatehouse, J. A. Biotechnological Prospects for Engineering Insect-Resistant Plants. *Plant Physiol.* **2008**, *146*(3), 881–887.

41. Shi, Y.; Xu, W.; Yuan, M.; Tang, M.; Chen, J.; Pang, Y. Expression of *vip1/vip2* Genes in *Escherichia coli* and *Bacillus thuringiensis* and the Analysis of Their Signal Peptides. *J. Appl. Microbiol.* **2004**, *97*(4), 757–765.

42. Shi, Y.; Ma, W.; Yuan, M.; Sun, F.; Pang, Y. Cloning of *vip1/vip2* Genes and Expression of Vip1Ca/Vip2Ac Proteins in *Bacillus thuringiensis*. *Res. Gate* **2007**, *23*(4), 501–507.

43. Ignoffo, C. M.; Couch, T. L. *The Nucleopolyhedrosis Virus of Heliothis Species as a Microbial Pesticide. Microbial Control of Pests and Plant Diseases;* Academic Press: London, 1981; pp 329–362.

44. Mettenmeyer, A. Viral Insecticides Hold Promise for Biocontrol. *Farming Ahead* **2002**, *124*, 50–51.

45. Zhang, G.; Sun, X.; Zhang, Z. Production and Effectiveness on the New Formulation of Heliothis Virus (NPV) Pesticide-Emulsifiable Suspension. *Virol. Sin.* **1995,** *10*(3), 242–247.

46. Srinivasa, M.; Jagadeesh Babu C. S.; Anitha, C. N.; Girish, G. Laboratory Evaluation of Available Commercial Formulations of HaNPV Against *Helicoverpa armigera* (Hub.). *J. Biopestic.* **2008,** *1,* 138–139.

47. Grzywacz, D.; Richards, A.; Rabindra, R. J.; Saxena, H.; Rupela, O. P. *Efficacy of Biopesticides and Natural Plant Products for Heliothis/Helicoverpa Control. Heliothis/Helicoverpa Management-Emerging Trends and Strategies for Future Research;* Oxford and IBH Publishing Co. Pvt. Ltd.: New Delhi, 2005; pp 371–389.

48. Moscardi, F. Assessment of the Application of Baculoviruses for Control of Lepidoptera. *Annu. Rev. Entomol.* **1999,** *44*(1), 257–289.

49. Kumari, V.; Singh, N. P. *Spodoptera litura* Nuclear Polyhedrosis Virus (NPV-S) as a Component in Integrated Pest Management (IPM) of *Spodoptera litura* (Fab.) on Cabbage. *J. Biopestic. India* **2009,** *2,* 84–86.

50. Bellotti, A. C. Recent Advances in Cassava Pest Management. *Annu. Rev. Entomol.* **1999,** *44,* 345–370.

51. Muratoglu, H.; Nalcacioglu, R.; Demibag Z. Transcriptional and Structural Analyses of *Amsacta moorei* Entomopoxvirus Protein Kinase Gene (AMV197, pk). *Ann. Microbiol.* **2010,** *60,* 523–530.

52. Xiao, G.; Ying, S. H.; Zheng, P.; Wang, Z. L.; Zhang, S.; Xie, X. Q.; Feng, M. G. Genomic Perspectives on the Evolution of Fungal Entomopathogenicity in *Beauveria bassiana. Sci. Rep.* **2012,** *2,* 483.

53. Zheng, P.; Xia, Y.; Xiao, G.; Xiong, C.; Hu, X.; Zhang, S.; Wang, C. Genome Sequence of the Insect Pathogenic Fungus *Cordyceps militaris,* a Valued Traditional Chinese Medicine. *Genome Biol.* **2011,** *12*(11), R116.

54. Hajek, A. E.; Leger, R. J. S. Interactions Between Fungal Pathogens and Insect Hosts. *Annu. Rev. Entomol.* **1994,** *39*(1), 293–322.

55. Islam, M. T.; Castle, S. J.; Ren, S. Compatibility of the Insect Pathogenic Fungus *Beauveria bassiana* with Neem Against Sweetpotato Whitefly, *Bemisia tabaci,* on Eggplant. *Entomol. Exp. Appl.* **2010,** *134*(1), 28–34.

56. Mazid, S.; Kalita, J. C.; Rajkhowa, R. C. A Review on the use of Biopesticides in Insect Pest Management. *Int. J. Sci. Adv. Technol.* **2011,** *1*(7), 169–178.

57. Delalibera, I.; Gomez, D. R. S.; De Moraes, G. J.; De Alencar, J. A.; Araujo, W. F. Infection of *Mononychellus tanajoa* (Acari: Tetranychidae) by the Fungus Neozygites sp. (Entomophthorales) in Northeastern Brazil. *Fla. Entomol.* **1992,** *75*(1)145–147.

58. Castillo, M.-A.; Moya, P.; Hernández, E.; Primo-Yúfera, E. Susceptibility of *Ceratitis capitata* Wiedemann (Diptera: Tephritidae) to Entomopathogenic Fungi and Their Extracts. *Biol. Control.* **2000,** *19*(3), 274–282.

59. Scholte, E. J.; Takken, W.; Knols, B. G. J. Infection of Adult *Aedes aegypti* and Ae. Albopictus Mosquitoes with the Entomopathogenic Fungus Metarhizium Anisopliae. *Acta Trop.* **2007,** *102*(3), 151–158.

60. Shi, W. B.; Feng, M. G. Lethal Effect of *Beauveria bassiana, Metarhizium anisopliae,* and *Paecilomyces fumosoroseus* on the Eggs of *Tetranychus cinnabarinus* (Acari: Tetranychidae) with a Description of a Mite Egg Bioassay System. *Biol. Control.* **2004,** *30*(2), 165–173.

61. Ichimoto, I.; Machiya, K.; Kirihata, M.; Ueda, H. Stereoselective Synthesis of Marine Antibiotic-Malyngolide and its Stereoisomers. *Agric. Biol. Chem.* **1990,** *54*(3), 657–662.

62. Kunze, B.; Jansen, R.; Pridzun, L.; Jurkiewicz, E.; Hunsmann, G.; Höfle, G.; Reichenbach, H. Phenoxan, a New Oxazole-Pyrone from Myxobacteria: Production, Antimicrobial Activity and its Inhibition of the Electron Transport in Complex I of the Respiratory Chain. *J. Antibiot.* **1992,** *45*(9), 1549–1552.

63. Patrick, C. D. Grasshoppers and Their Control. Texas FARMER Collection, 2004.

64. Shapiro-Ilan, D. I.; Gouge, D. H.; Koppenhöfer, A. M. 16 Factors Affecting Commercial Success: Case Studies in Cotton, Turf and Citrus. In *Entomopathogenic Nematology;* Gaugler, R. ed. CABI; 2002, 333.

65. Vachon, V.; Laprade, R.; Schwartz, J. L. Current Models of the Mode of Action of *Bacillus thuringiensis* Insecticidal Crystal Proteins: A Critical Review. *J. Invertebr. Pathol.* **2012,** *111*(1), 1–12.

66. Kirouac, M.; Vachon, V.; Noël, J.-F.; Girard, F.; Schwartz, J.-L.; Laprade, R. Amino Acid and Divalent Ion Permeability of the Pores Formed by the *Bacillus thuringiensis* Toxins Cry1Aa and Cry1Ac in Insect Midgut Brush Border Membrane Vesicles. *Biochim. Biophys. Acta—Biomembr.* **2002,** *1561*(2), 171–179.

67. Olson, S. An Analysis of the Biopesticide Market Now and Where it is Going. *Outlooks Pest Manage.* **2015,** *26*, 203–206.

CHAPTER 5

MICROALGAE: A POTENTIAL SOURCE OF BIOFUEL

DEVENDRA KUMAR[1,*], PRIYANKA NEHRA[1], ANUJ KUMAR[2,3], and NEERAJ KUMAR[4]

[1]Centre for Conservation and Utilization of Blue Green Algae, Indian Agricultural Research Institute Pusa New Delhi 110012, India, *E-mail: devendra2228@gmail.com

[2]School of Chemical Engineering, Yeungnam University, 280 Daehak-Ro, Gyeongsan 38541, South Korea

[3]Department of Nano, Medical and Polymer Materials, College of Engineering, Yeungnam University, 280 Daehak-Ro, Gyeongsan 38541, South Korea

[4]Department of Microbiology, Kurukshetra University, Kurukshetra, Haryana, India

CONTENTS

ABSTRACT

Microalgae are also known for third generation biofuel as well as alternative feedstock because some species contain high percentage of lipid content, which could be extracted from their cells, processed and esterified into transportation fuels by using currently developed technology; they have high biomass, permit the use of nonarable land and unpotable water, use far less water and do not displace food crop cultures; the production of this organism is not depending on the season and it can be easily harvested. Bioprospecting of microalgal strains encompasses searching, collection and preservation of unique strains from different aquatic environments for exploiting the potential applications of nutraceuticals, value-added products such as polyunsaturated fatty acids (PUFA). High lipid contents were extracted from the biomass through different solvent extraction methods and the composition of individual fatty acids had been found out through gas chromatography–mass spectrometry (GC-MS) analysis based on their retention time and molecular weight. These results make a better understanding of rapid metabolic responses in the percentage of saturated fatty acids (SFA), PUFA and mono unsaturated fatty acids (MUFA).

5.1 INTRODUCTION

Developing a large number of supplies of clean energy for the near future is one of the society's most daunting challenges and is intimately linked with economic prosperity, global stability and quality of life. Fuels represent around 73% of the total global energy requirements, particularly in manufacturing, transportation and domestic heating. Biodiesel is the only important source promising renewable transportation fuels that have achieved remarkable success worldwide. According to the World Bank report (2008), 6.8 billion L of biodiesel was produced worldwide in 2006, 78% of which by the European Union and 14% by the USA. The recent contribution of biodiesel to global transportation fuel consumption is, however, only 0.17% and the favourable policies of major countries in the world are expected to increase this contribution by 5 times, by 2020. It is therefore predictable that massive global demand for renewable energy will continue to drive the rapid growth of biodiesel production on an unprecedented scale. Nevertheless, there is an increase in fuel prices worldwide that had brought about public awareness and concerns regarding the competition for agricultural resources and the energy sectors. For that, we need to develop sustainable and cost-effective alternatives to the traditional agricultural and forestry crops is therefore of the urgent need for sustainable biofuel production. In recent years, biodiesel fuel has received considerable attention, as it is made from renewable resources, biodegradable and non-toxic, and provides environmental benefits, since its use leads to a decrease in the harmful emissions of many primary pollutants like carbon monoxide, hydrocarbons with a consequent decrease in the greenhouse effect, according to the Kyoto Protocol agreement (1992) at Japan. Biodiesel usage will allow for a balance to be achieved between economic development, agriculture and the ecosphere.[1]

Extensive studies have been conducted on using vegetable oils, lingo cellulose from plants as biodiesel fuel. The focus has mainly been on oils, like rapeseed, soybean, safflower and sunflower oils which are mostly edible in nature.[2] Most recently, research efforts have been aimed to provide high-energy outputs to replace conventional fossil fuels by identifying suitable biomass species.[3] However, few attempts have been made to produce biodiesel from non-edible sources, like used frying oil, greases, mahua oils, lard and jatropha. Due to the high feed cost of vegetable oil, the cost of biodiesel production is still a major obstacle for large-scale commercial exploitation.[4]

Microalgae have been suggested as a potential feedstock for fuel production because of the number of advantages, including higher photosynthetic efficiency and higher biomass production as compared to other energy sources.[5] Microalgae represent an exceptionally diverse, according to cell composition, but highly specialized group of microorganisms because of adaptations of various ecological habitats. Many microalgae have the ability to produce huge amounts (e.g., 25–60% dry biomass) of triacylglycerol (TAG) as a storage biofuel under photooxidative stress or other adverse environmental conditions. Microalgae with high oil productivities, which means have high lipid content, are desired for producing biodiesel. Depending on the nature of species, microalgae produce many different kinds of hydrocarbons, lipids and complex oils.[6,7]

As a matter of fact, average biodiesel production yield from microalgae can be 20–30 times greater than the yield obtained from oleaginous vegetable oils and/or seeds.[8] Some microalgal species have high oil content and can be induced to produce a higher concentration of lipids (e.g. low nitrogen media, Fe^{3+} concentration and light intensity). [9] The main characteristic of algae is to fix CO_2, can also be an interesting method of removing gases from industries, and thus can be used to reduce greenhouse gases with a higher production of microalgal biomass and consequently higher biodiesel yield.[10] Algal biomass production technologies can be easily adapted to various levels of operational skills; some microalgae have also a good fatty acid profile, an unsaponifiable fraction of this allowing a biodiesel production with high oxidation stability.[11] The physical, chemical and fuel properties of biodiesel from algal strains oil in general (e.g. acid value, density, heating value and viscosity, etc.) are comparable to those of fuel diesel.[12,13]

In recent years, the main focus of research in the field of biofuels from natural resources like microalgae has been centred on downstream aspects such as biomass and lipid production from microalgae, biomass harvesting techniques, bioreactor designs and the chemistry of biofuel production because of unicellular nature of green algae. Microalgal bioprospecting encompasses isolation, identification and collection of unique microalgal strains from different aquatic environments and habitats for exploiting the potential applications of value-added products such as polyunsaturated fatty acids (PUFAs).[14]

The need to the replacement of fossil fuels with fuels derived from renewable biomass is currently focused on biodiesel from oleaginous plant seeds and ethanol from sugarcane/corn, but environmental and social concerns are shifting the attention towards the development of third generation algal biofuels.[15] The third generation is represented by biofuels from micro- and macroalgae but in the case of terrestrial plant sources the production of second-generation biofuels relies on the conversion of the highly abundant and widespread nonedible lignocellulosic fraction. Benefits rising from the utilization of aquatic over terrestrial biomass include (i) utilization of marginal areas (e.g. desert and coastal regions), (ii) higher light use efficiency (about 5% vs. 1.8%),[16] (iii) minor dependence on climatic conditions, (iv) possible coupling with other activities (e.g. waste-water treatment, CO_2 sequestration), (v) easier genetic manipulation to modify chemical composition (e.g. lipid content) and (vi) availability of a larger number of species.[17] However, the industrial development of fuels from microalgae is still hampered by higher overall costs with respect to fossil fuel counterpart: bioreactors and operating raceways are expensive and the major problem is the harvesting of algal biomass, it is costly and time-consuming. For this reason, the net energy balance from microalgae cultivation is still in conversation.[18,19]

Besides the cost for cultivating microalgae, downstream processes are to be taken into account to evaluate the overall productivity. Lipids, which include hydrocarbons and acylglycerols, represent the most valuable fraction of microalgal biomass as their high lipid content per mass unit is similar to conventional fuels. Several oleaginous microalgae (with lipid content exceeding 25% of their dry weight) have been exploited to this purpose and the biodiesel obtained has been claimed to be more conventional than convenient biodiesel from lignocelluloses plants.[20] Recovery of lipids from algal biomass by means of mechanical (cold press) and chemical processes (extraction with solvents) or both[21] the latter being more effective; mainly for analytical purposes on a laboratory scale, the extraction from freeze-dried samples of algae is usually performed with a chloroform/methanol/water mixture and then the oil is fractionated through column chromatography to obtain hydrocarbons, and polar and nonpolar lipids. Nonpolar organic solvents such as n-hexane are good alternatives for hydrocarbons extraction.[22]

Microalgae are capable of producing many useful products for use as biofuels, including methane and hydrogen, but the most versatile and applicable to the current transport infrastructure is algal lipids. By trans-esterification processes, algal lipids can be converted into biodiesel, which means conversion into ester form, chemically. When triglycerides within the algal lipids are broken down with methanol, it converts into glycerol and methyl-esters (biodiesel). The reaction can be catalysed by many different alkalis, acids or lipase enzymes, each providing their own disadvantages and advantages to the overall process.[23]

Typically algal lipids classify into two main categories, the neutral lipids (those without large charges) and polar lipids (hydrophobic and hydrophilic, different charges). Neutral triglycerides (or TAGs) are the most important type of lipid for biodiesel production. Because of their low levels of molecular cross-linkages, these lipids are of most use for the process. When it is stored for prolonged periods or used in combustion engines, these bonds can cause oxidation issues in biodiesel. Other algal lipids can also be used for the biofuel process, but they require further downstream processing (such as catalytic hydrogenation) to comply with the standards of fuels (Table 5.1).

TABLE 5.1   Oil Content of Some Microalgae. (*Source:* Adapted from Chisti, 2007; Gouveia and Oliveira, 2009.)

Microalgae	Oil content (Per cent dry weight)
Botrycoccus braunii	25–80
Chlorella protothecoides	23–30
Chlorella vulgaris	14–40
Crypthecodinium cohnii	20
Cylindrotheca sp.	16–37
Dunaliella salina	14–20
Neochloris oleoabundans	35–65
Nitzschia sp.	45–47
Phaeodactylum tricornutum	20–30
Schizochytrium sp.	50–77
Spirulina maxima	4–9
Tetraselmis suecia	15–23

5.2   SOURCES OF BIODIESEL

5.2.1   EDIBLE OILS

It was studied that a blend of 25% vegetable oil and 85% diesel fuel was found to be successful. It has been proved that the use of 100% vegetable oil was also possible with some minor modifications in the fuel system.[24] A diesel fleet was powered with a blend of 92% filtered used cooking oil and 5% diesel in 1982, and also studied many nonedible oils as alternative feedstock for biodiesel. Therefore, it is necessary to develop the techniques for reduction of the price of biodiesel by decreasing process steps in downstreaming process.[25]

5.2.2   NONEDIBLE PLANT OILS

Some report on nonedible oils like used frying oil, grease, tallow and lard are a major source of biodiesel.[26] There are many nonedible tree-based oilseeds available in India with an estimated annual production of more than 25 MT. These oilseeds have a greater potential of being transesterified for making biodiesel and the prospects of fatty acid methyl esters (FAMEs) analyses of some 27 nontraditional plant seed oils including Jatropha, potential biodiesel source in India. Among them, *Calophyllum inophyllum*, *Azadirachta indica*, *Jatropha curcas* and *Pongamia pinnata* were found more suitable for use as biodiesel and they meet the major specification of biodiesel for use in diesel engines. Moreover, they reported that 73 oil-bearing plants contain 36% or more oil in their seed, fruit or nut. Reddy and Ramesh[27] reported the comparison of properties of Jatropha oil, diesel and biodiesel from Jatropha. Biodiesel produced from Jatropha oil has similar characteristics as that of petroleum diesel which shows that Jatropha oil is a strong alternative source of diesel. Subramanian et al.[28] reported that there are over 326 different species of tree of which seeds are having a large amount of oil. Thus, there is a significant potential for nonedible oil sources from different plants for biodiesel production as an alternative to petrodiesel. Pedro et al.[29] reported soapnut seeds as a biodiesel source and named it as 'waste-to-energy' scheme. Furthermore, planting soapnut trees in community forestry and in barren lands provides feedstock for biodiesel production as well as a sink for carbon sequestration.

5.2.3   AGRICULTURAL WASTES AND BYPRODUCTS

There is an ability of every organism to metabolize a large variety of sugars suggested that other agricultural and food processing wastes might support the growth and oil production. Some researchers conducted preliminary experiments which indicated ripened bananas as a good source of carbon and supported good growth and lipid production by *Apiotrichum curvatum*. [30]

Tao et al.[31] reported exploitation of carbohydrate-based microbial oil, because it consists of similar fatty acids to that of plant oil. Corn straw can produce biodiesel at 25.5 kg and glycerol at 25.7 kg per ton (fatty acid calculates as stearic acid). Lignocellulosic materials such as corn stalk, *Saraca indica* leaves and rice straw provide abundant and renewable energy sources. Lignocellulosics contain sugars that are polymerized to cellulose and hemicellulose, and they can be liberated by hydrolysis and subsequently converted to biofuel.[32]

5.2.4 MICROBIAL SOURCES

Oleaginous yeasts and moulds accumulate TAGs rich in PUFAs.[33,34] *Yarrowia lipolytica* strain has been reported to produce a significant amount of lipids (up to 43% w/w of dry weight), rich in oleic and linoleic acid, when grown on industrial glycerol.[35] The major lipid component of oleaginous yeasts and fungi is triglycerides, composed of C16 and C18 series long-chain fatty acids, which are quite similar to those of vegetable oils, such as rapeseed oil and soybean oil. Therefore, microbial lipids can potentially be used as biodiesel.

It was also studied that in *Rhodotorula spp., Lipomyces starkeyi* and *Cryptococcus curvatus* (also known as *Apiotricum curvatum*) can accumulate between 40 and 70% of their biomass as lipid. Direct methanolysis of oleaginous microbial biomass for biodiesel production was recently studied.[36] Oleaginous yeast was isolated from the flower samples of *Rosa centifolia* and was studied for biodiesel production. According to some researchers, few prokaryotes that accumulate TAG belong to Gram-positive genera such as *Streptomyces* or *Rhodococcus*. Indeed, in *Streptomyces coelicolor*, TAG can constitute 80% of the total cell volume during the stationary phase.

Rainer et al.[37] established biosynthesis of biodiesel adequate fatty acid ethyl esters, referred to as Microdiesel, in metabolically engineered *Escherichia coli*. This was achieved by heterologous expression of unspecific acyltransferase from *Acinetobacter baylyi* strain ADP1 and in *E. coli* of the *Zymomonas mobilis* pyruvate decarboxylase and alcohol dehydrogenase.

Fungal lipids investigation as biodiesel fuel was studied less extensively. A recent report on oleaginous fungal (*Mortierella isabellina*) biomass was directly methanolised for biodiesel production.[38]

5.2.5 ALGAL SOURCES

Microalgae are unicellular and able to accumulate up to 87% of their total cell dry weight as neutral lipid, particularly in response to environmental stresses such as nitrogen limitation, high CO_2, salinity or high tempera-ture.[39] Unfortunately, biodiesel from animal and plant sources cannot realistically satisfy even a small fraction of the existing demand for transport fuels. At present, microalgae appear to be the major source of renewable biodiesel that is capable of meeting the global demand for transport fuel.

Oil from microalgae was studied extensively for the source of methyl ester diesel fuel (Nagle and Lemke, 1990). Moreover, microalgae exhibit properties that make them well suited for use in a commercial-scale biodiesel production facility. Many microalgae exhibit high productivity and rapid growth, and many species can be induced to accumulate substan-tial amounts of lipids, often greater than 65% of their biomass.[40]

Microalgae like higher plants produce storage lipids in the form of TAGs. Although TAGs could be used to produce a large variety of chemi-cals, this FAMEs which can be used as a substitute for fossil-derived diesel fuel is the main focus for the researchers.

5.3 ALGAE: AN OVERVIEW

Algae are cosmopolitan; we see it flourish in both saline and fresh water, in cold mountain streams and hot inland swamps and ponds. Algae grow in almost any aquatic environment and use light and carbon dioxide (CO_2) for their sustainable biomass. Algae range in size from a few micrometres to over 32 m in length.

5.3.1 *MICROALGAE VERSUS MACROALGAE*

Phycologists classified algae into two main categories: macroalgae and microalgae. Macroalgae are the multi-cellular, larger (in the size of inches and greater) algae often seen growing in lakes and ponds. These larger algae can grow in a variety of ways. The largest multi-cellular algae are called seaweed; an example of this is the giant kelp plant, which could grow well over 27 m in length. On the other hand, Microalgae are tiny (in the size of micrometres) unicellular algae that normally grow in suspen-sion within aquatic conditions.[41] Microalgae cells can double every few hours during their exponential growth period.[42] The fact that they grow so quickly makes them a promising crop for human use. Microalgae are often responsible for the appearance of cloudiness within a pond or even an aquarium. Both types of algae grow extremely quickly. The largest seaweed, giant kelp, is known to grow as fast as 52 cm/day, and can reach a length up to 82 m.[43] Microalgae are known to contain large amounts of lipids by dry weight within their cell structure, so they are increasingly becoming an interest in a biodiesel feedstock.

5.3.2 MAJOR COMPOSITION OF MICROALGAL BIOMASS

Microalgal cells are a type of eukaryotic cells, and as such contain the same internal organelles such as chloroplasts, a nucleus and so forth. The biomass of microalgae contains a number of compounds such as proteins and lipids, which make up the organelles. The composition of the biomass is useful for characterizing how the microalgae species is best used. For example, with the knowledge that biodiesel is made from oils, if microalga does not have a sufficient amount of lipid and protein content, it would not be used as a biofuel feedstock.

Three main components of algal biomass are lipids/natural oil, carbo-hydrates and protein. Because the bulk of the natural oil made by micro-algae is present in the form of triacylglycerides (TAGs) (Fig. 5.1), which is the right kind of oil for producing biodiesel, microalgae are the exclu-sive focus in the algae-to-biofuel arena.[44] Oil contents of microalgae are usually between 25% and 55% (dry weight), while some strains can reach as high as 85%. Compared to terrestrial crops, microalgae grow very quickly because of their high growth rate. Their doubling time or gener-ation time is near about 24 h. At their log phase, some microalgae can double every 4 h.

$$CH_2O-OCHCH_2CH_2.........CH_2CH_3$$
$$|$$
$$CHO-OCHCH_2CH_2.........CH_2CH_3$$
$$|$$
$$CH_2O-OCHCH_2CH_2.........CH_2CH_3$$

FIGURE 5.1   Molecular structure of TAG.

The fatty acids attached to the TAG within the algal cells can be both short- and long-chain hydrocarbons. The shorter chain length acids are ideal for the creation of biodiesel, and some of the longer ones can have other benefits.

5.3.3 TAXONOMIC CHARACTERISTICS OF BIODIESEL-PRODUCING MICROALGAE

Dunaliella: Domain: Eukaryota, Kingdom: Viridiplantae, Phylum: Chlorophyta, Class: Chlorophyceae, Order: Chlamydomonadales, Family: Dunaliellaceae, Genus: *Dunaliella*, Species: *D. acidophila, D. bardawil, D. bioculata, D. lateralis, D. maritima* and *D. tertiolecta*. The latter is a marine green flagellate species and the cell size is 12–14 μm. The oil yield of that strain about 40% (organic basis). *D. tertiolecta* is a very fast growing strain having doubling time 5 h (Fig. 5.2A).

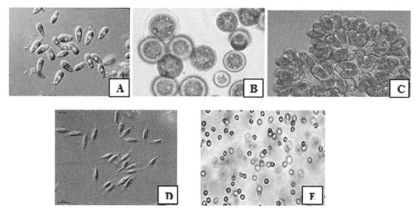

FIGURE 5.2   Photo micrographs of algal strains: (A) dunaliella; (B) nannochloropsis; (C) botryococcus braunii; (D) isochrysis and (E) phaeodactylum.

Nannochloropsis: Domain: Eukaryota, Kingdom: Chromalveolata, Phylum: Heterokontophyta, Class: Eustigmatophyceae, Genus: *Nannochloropsis, Nannochloropsis* is a genus of alga comprising approximately six species, *N. gaditana, N. granulate, N. limnetica, N. oceanic, N. oculata, N. salina*. The characterization is mostly done by 18S rDNA and *rbcL* gene sequence analysis. Chlorophyll b and c are not found in this strain like others. The species have mostly found in the marine environment but sometimes occur in brackish and fresh water. All species are nonmotile spheres and small which do not express any distinct morphological features, and cannot be distinguished by either electron or light microscopy. *Nannochloropsis* is able to build up a high concentration of a range of pigments such as astaxanthin, zeaxanthin and canthaxanthin. The algae have a very simple ultrastructure, compared to neighbouring taxa. They have a diameter of about 3 μm. It is considered to be a promising alga for industrial applications because it has a high level of lipid contents. It is also used as an energy-rich food source for rotifers and fish larvae (Fig. 5.2B).

Botryococcus braunii: Kingdom: Plantae, Division: Chlorophyta, Class: Chlorophyceae, Order: Chlorococcales, Family: Dictyosphaeriaceae, Genus: *Botryococcus*, Species: *Botryococcus braunii* (Bb) is a pyramid-shaped and green planktonic microalga of the order Chlorococcales (class Chlorophyceae) that is of potentially very important in the field of applied biotechnology. It is notable for its ability to produce high amounts of hydrocarbons, especially oils in the form of triterpenes, which are typically around 35%–45% of their dry weight. It has a relatively thick cell wall compared to other algal strains that are accumulated from previous cellular divisions; making extraction of cytoplasmic components is rather difficult. The useful hydrocarbon oil is present outside of the cell colonies held together by a lipid biofilm matrix and can be found in tropical or temperate oligotrophic lakes and estuaries and will bloom in the presence of elevated levels of dissolved inorganic phosphorus (Fig. 5.2C).

Dunaliella salina: Kingdom: Plantae, Phylum: Chlorophyta, Class: Chlorophyceae, Order: Volvocales, Family: Dunaliellaceae, Genus: *Dunaliella*, Species: *Dunaliella salina, Dunaliella salina* strain is mostly of halophile pink microalgae especially found in sea salt fields. Few organisms can survive in such highly saline conditions as salt evaporation ponds. It has a large amount of carotenoids, so it has high antioxidant activity. To survive, these organisms have high concentrations of β-carotene to protect themselves against the high concentrations of glycerol and intense light and to provide protection against osmotic pressure. This offers an opportunity for commercial biological production of these substances.

Isochrysis: Kingdom: Chromalveolata, Phylum: Haptophyta, Class: Prymnesiophyceae, Order: Isochrysidales, Family: Isochrysidaceae, Genus: *Isochrysis*, Species: *I. galbana. Isochrysis galbana* is a microalga. It is a type of species of the genus *Isochrysis*. It is desirable in the bivalve aquaculture industry because it is an outstanding food for various bivalve larvae (Fig. 5.2D).

Phaeodactylum: Kingdom: Chromalveolata, Phylum: Heterokontophyta, Class: Bacillariophyceae, Order: Naviculales, Family: Phaeodactylaceae, Genus: *Phaeodactylum*, Species: *P. tricornutum*. It is a diatom. It is the only species in the genus *Phaeodactylum*. Unlike other diatoms, *P. tricornutum* can exist in different morphotypes (fusiform, oval and triradiate), and changes in environmental conditions can change cell shape. Molecular basis of cell shape control and morphogenesis can be used to explore by this features. Furthermore, *P. tricornutum* can grow in the absence of silicon, and the biogenesis of silicified frustules is facultative, thereby providing opportunities for experimental exploration of silicon-based nanofabrication in diatoms (Fig. 5.2E).

5.3.4   SAMPLING OF MICROALGAE STRAINS

For the sampling of microalgal strains, we consider some aspects like environmental factors (abiotic and biotic stress), type of aquatic system, parameters measured onsite and sampling equipment. For successful biofuel production using microalgae, the crucial step is isolation, identification of that strains which have high lipid content. According to Borowitzka,[45] it is crucial to evaluate harvesting costs at the time of choosing the species. Damaged or dead cells may lead to failure of the experiments, so we keep that thing in mind during harvesting. Therefore, the temporal and spatial collection strategy should be adopted to cater for any succession that can occur at the sampling site.[46,47]

5.3.4.1   ENVIRONMENTAL SAMPLING

Mainly microalgal strains are found in diverse environmental conditions and habitats such as brackish, lacustrine, hypersaline, wastewater maturation ponds, fresh water, rivers, marine, coastal areas and dams. Due to the changing of environmental conditions and selection pressure, there is a diverse range of microalgal species worldwide found in extreme environments.[48]

Brackish aquatic systems constitute a mixture of fresh water and seawater and this usually occurs at the river mouth on the coastline. Because of the presence of the high nutritional composition of the warmer temperatures and aquatic system, microalgal strains prefer brackish conditions.[49] Most of the microalgae found in brackish conditions are found in suspension as a consequence of the rapid water movements.

In any habitat, microalgae have been shown to have successional tendencies due to inclement weather, seasonal variations and variable nutrient availability. In bioprospecting, it is important to isolate microalgal samples temporally and spatially so as to determine if there are any successional tendencies in the habitat. According to Brahamsha,[50] microalgal biomass has shown clear spatial and temporal patterns during the heterogeneous conditions of the closed and open phases in estuaries. The population dynamics of the microalgae in any habitat is complex and are found as a mixed consortium. Different types of habitats are found in different environmental conditions, where microalgae can be collected. Table 5.1 depicts lipid content of some microalgal strains collected from different aquatic environments.

5.3.4.2   MEASURING PARAMETERS

Isolating pure microbial cultures and cultivating them in the laboratory on defined media is used to more fully characterize the metabolism and physiology of organisms. However, identifying an appropriate growth medium for a novel isolate remains a challenging task. The conditions include nutrient concentration (nitrates and phosphates), abiotic factors such as light (quality and quantity), dissolved CO_2, dissolved O_2, water temperature, pH and salinity. In laboratories, it is impractical to measure biotic factors on site such as pathogens (fungi, bacteria and viruses) and any other competitors in the habitat. After collecting the water samples, these can be analysed by microbiological methods in the laboratory and also through microscopically. However, if resources and time are available, it is desirable to measure the operational factors such as shear produced by dilution rates, mixing, depth and water velocity that is if the water is flowing. The global positioning system (GPS), coordinates of the sampling location, must be recorded for future reference and resampling.[51]

5.3.4.3   SAMPLING EQUIPMENT AND PROCEDURE

There are no simple and cheap methods of collecting microalgal samples in any documented literature which researchers can follow. Basically, samples can be collected from the natural substrata by scrapping, chipping and brushing from rock surfaces and bottom sediments. The brushing method was reported to be an effective and reproducible method of collecting microalgal cells and also that it does not damage them.[52]

The microalgal selection and sampling process are well established although they require specialised equipment and may be time consuming. The equipment required for microalgal sampling includes a mesh net (0.81 m mesh), vials for collecting samples, knife, scooping jar, scalpels, O_2 analyser with a data logger, dissolved CO_2, GPS, salinity meter, multiprobe system (measuring pH, light meter, temperature, turbidity, light intensity and conductivity simultaneously).

5.4   ISOLATION AND PURIFICATION

5.4.1   CULTURE MEDIA

Many artificial culture media have been developed for isolation and cultivation of microalgae. On the basis of nutrient requirement of the organism, some of them are formulated and modified after detailed research. In case of marine algae, though artificial sea water media is common, good growth of algae can be achieved by adding small quantities of non-polluted natural seawater (less than 1–7%) to the artificial seawater. The main problem is bacterial contamination in microalgal purification but by treatment with a detergent and phenol it can be prevented. By carefully manipulating major nutrients, a wide range of microalgae can be formulated[53] (Table 5.2).

TABLE 5.2   Common Media Used for Microalgal Strains from Different Aquatic Environments.

Media	Suitable for	Reference(s)
AF-6 medium	Xanthophyceae, Dinoflagellate	Watanabe et al. (2000)
AK medium	Marine Algae	Barsanti and Gualtieri (2000)
Beijerinck medium	Fresh water algae	Barsanti and Gualtieri (2000)
BG-11	Chlorophyceae	Barsanti and Gualtieri (2000)
Bold Basal Medium	Broad spectrum medium for fresh water algae	Barsanti and Gualtieri (2000)
Combo Medium	Green algae and Diatoms	Kilham et al. (1998)
Diatom Medium	Fresh water diatoms	Cohn et al. (2003)

5.4.2 SINGLE-CELL ISOLATION

All the microalgal strains are unicellular, so it is a difficult task to isolate them. By using micropipette or by making Pasteur capillary tube, we can easily isolate the unicellular strains, although automation is more advantageous. Isolation through micropipette is usually performed with a glass capillary or Pasteur pipette having a straight, bent or curved tip. Rogerson et al.[54] employed repeated introduction and ejection of cells, suspended in a 2% crude papain solution, into and from a micropipette to generate ca. 11% naked cells of *Nannochloropsis*. These microscopic cells were re-isolated via micropipette into fresh medium. The traditional method of micropipette isolation can be successful in identifying the single cells. Because the tiny cells cannot be easily distinguished from particles, especially when working with seawater, ultraclean droplets for rinsing are necessary.

5.4.3 BY USING AGAR PLATES

It is a very old method for isolation of any microorganisms. If we have coccoid algae and soil algae, it is the best isolation method not only for ease of use but also because axenic cultures can often be directly established without further treatment. For successful isolation onto agar plates, the alga must be able to grow on agar as pour plates or streak.[55] Some algae do not grow on the surface of agar plates, but they do grow embedded in agar and by further purification they were isolated.

5.4.4 BY CELL SPRAY

An atomized or fine spray of algal cells can be used to inoculate agar plates. The technique can vary, but in general a liquid cell suspension is atomized with forced sterile air; by this, all the cells are scattered on to the agar plate. For best results, we must keep in mind that the spray is totally sterilized. The agar plates are incubated in incubators, and when colonies are formed then these selected cells are removed and inoculated for further processes.

5.4.5   GRAVIMETRIC SEPARATION

This method totally depends on sedimentation coefficient of cells. Usually, the goal is to separate the heavier and larger cells from smaller bacteria and algae. The concept of gravity is perhaps more frequently used to concentrate on the target organisms rather than to establish unialgal cultures. If we do centrifugation for a short duration, it brings large diatoms and dinoflagellates to a loose pellet, and the smaller cells can be decanted. Centrifugation with use of density gradients (e.g. Percoll, Silica sol) has been used to separate mixed laboratory cultures where each species was separated into a sharp band.[56] Settling is more effective for non-swimming large or heavy cells. Both centrifugation and settling techniques are effective for concentrating larger cells, but single-cell isolation is nearly impossible and, hence, combined with other technique.

5.4.6   BY SERIAL DILUTION

This is the best method for isolation of single pure cell of any organism; the main purpose of serial dilution is to deposit only one cell into flask, a test tube or well of a multiwell plate, thereby establishing a single-cell isolate. The technique can be used in several ways like dilution with distilled water, seawater, filtered water from the sample site, culture medium or some combination of these. Also, selenium, ammonium or another element can be added to some isolation tubes or cell wells to specifically select for needed species that require these nutrients.[57]

5.5   ADVANCED METHODS

Mostly algal cultures are unicellular, which means they contain only one kind of algal species, usually a clonal population (but which may contain fungi, bacteria or protozoa), or cultures may be axenic meaning that they do not contain any type of contamination. Spraying and streaking are useful conventional techniques for colonial, single-celled and filamentous algae growing on the agar surface. Many for isolation and cell separation with high specificity are limiting because cell populations are frequently heterogeneous and the cells are suspended in a solution of different biomolecules, chemicals and cells. Micro-fabricated devices have small working volume and subsequently reduced throughput if we compared it with traditional separation methods.

5.5.1 MICROMANIPULATION

From liquid culture for selection of single cells, we can use micromanipulator. This would mean that from an enriched environmental sample we isolate a single cell and grown on plates as well as liquid monoculture. This will afford a time saving of ~60% over the current serial plating technique. The successful application of micromanipulation techniques requires the expertise and experience. It requires the handling of a stereo microscope or inverted microscope with magnification up to 200X; Dark field, phase contrast or optics is an advantage. Hematocrit tubes or capillary tubes of approximately 1 mm diameter × 100 mm long are used for picking individual.[58] This equipment is the ideal tool for algal screening and isolation and also stipulated in literature.[59]

5.5.2 FLOW CYTOMETRY

To perform a quantitative and rapid version of the experiments that could otherwise be performed by fluorescent microscopy analysis permits the investigator by flow cytometry. One very important step further allows the process to be taken as fluorescence-activated cell sorting (FACS). Electronic cell sorting for culture and isolation of marine eukaryotic phytoplankton and dinoflagellates was compared to the traditional method of manually picking cells using a micropipette.[60] Trauma to electronically sorted cells was not a limiting factor, as fragile dinoflagellates, such as *Karenia brevis* (Dinophyceae), survived electronic cell sorting to yield viable cells. The rate of successful isolation of large-scale (>42) cultures was higher for manual picking than for electronic cell sorting (3% vs. 0.7%, respectively). However, manual picking of cells is more time consuming and labour intensive. Direct electronic single-cell sorting was more successful than utilizing a pre-enrichment sort followed by electronic single-cell sorting. Seventy per cent of isolated cells were recovered in a new medium, which was optimized for axenic dinoflagellate cultures.

The appropriate recovery medium may enhance the rate of successful isolations. The greatest limiting factor to the throughput of electronic cell sorting is the need for manual post-sort culture maintenance. However, when combined with newly developed automated methods for electronic single-cell sorting growth and screening it has the potential to accelerate the discovery of new algal strains.

5.5.3   OTHER METHODS

On the basis of immunological and non-immunological methods, we can also separate cells of our interest. Conventional cell separation can be carried out by immunoreactions of a membrane protein with the capturing antibodies because the type of integrated proteins is specific for their function. The immunological technique is a mainstay of commercialized cell separation methods such as the magnetic-activated cell sorting,[61] FACS[62] and affinity-based cell sorting;[63] one of the advantages of this method is selectivity and high specificity. The main disadvantage is that the immunologically isolated cells may suffer from damages and overall separation system involves complicated processes such as immunoreactions and elution of cells from the capturing antibodies and high cost.

On the other hand, non-immunological methods are relatively simple and fast techniques. They include techniques such as hydrodynamic separation;[64] ultrasound separation,[65] aqueous two-phase system[66] and dielectrophoresis;[67] these exploit an interactive physical property of cell with the surrounding media. However, a disadvantage is its low specificity for cell separation, as cells do not show remarkable differences between each cell type with the exception of immunological properties. In this method, the type of cells is determined and separated according to their shape, cell size and other physical properties.

5.5.4   SCREENING OF MICROALGAE

The main focus of bioprospecting of algal strains for biofuel production is targeting organisms with tolerance to environmental parameters and rapid growth rate. The most conventional method routinely used for knowing the lipid content involves gravimetric determination and solvent extraction. A major disadvantage of the conventional method is that it is labour-intensive and time-consuming; it has a low throughput screening rate. Moreover, approximately 15–20 mg wet weight of cells must be cultured for the extraction and derivatization. Consequently, there is greater interest on a rapid in situ measurement of the lipid content.[68]

Lipid soluble fluorescent dye like Nile red (NR) (9-diethylamino-5H-benzo-[a]-phenoxazine-5-one) has been commonly used to evaluate the lipid content of microorganisms and animal cells and especially microalgae. NR possesses several characteristics advantageous to in situ screening. It is relatively intensely fluorescent, photostable in hydrophobic environments and organic solvents. As the polarity of the medium decreases, the maximum emission of NR is blue-shifted, which allows one to differentiate between polar and neutral lipids at the excitation and emission wavelengths. Elsey et al.[69] showed the technical emission spectra for NR in various solvents. In hexane, the peak emission intensity of NR is near 576 nm, when excited at 486 nm. The ethanol and chloroform peak excitation was recorded at 632 and 600 nm, respectively. In acetone, NR is excited at 488–525 nm and the fluorescent emission measured at 570–600 nm using various instruments.[70] Measurement of neutral lipids using the NR application requires the instrument to be calibrated using the stain dissolved in an organic solvent and account for the nonlinear intensity emission with respect to time. Per unit cell measurements of lipid requires a calibration curve that correlates fluorescence to lipid content, whether determined gravimetrically or by the use of lipid standards. Thick cell walls of microalgae inhibit the permeation of NR and may indicate the absence of oil, even though gravimetric analysis shows high yields of neutral lipids. It has been noted that the permeation of NR dye is also variable among algal species, requiring the use of high levels of DMSO (20–30% v/v) and elevated temperatures (40°C).

Alternatively, recently boron-dipyrromethene (BODIPY) 505/515 (4,4-difluoro-1,3,5,7-tetramethyl-4-bora-3a, 4adiaza-s-indacene) the lipophilic fluorescent dye has been used as a vital stain to monitor algal oil storage within viable cells. Lipid bodies are stained green and chloroplasts red and visualized in live oleaginous (oil-containing) algal cells. High lipid yielding cells may be easily identified and isolated microscopically using a micromanipulator system, FACS or flow cytometry. Subsequently, pure cultures may be propagated from the isolated viable cells. Nile red has broader emission spectrum than BODIPY 505/515, making it potentially more useful for confocal imaging, where fluorescence contrast enhancement of lipid bodies is important for image resolution.[71] BODIPY 505/515 has the advantage that it does not bind to cytoplasmic compartments other than lipid bodies and chloroplasts.

A recent study demonstrated the use of Fourier transform infrared micro-spectroscopy (FTIR) to determine carbohydrate and lipid content of fresh water microalgae. Fourier transform infrared micro-spectroscopy was shown to be an efficient and rapid tool for monitoring lipid accumulation of microalgae. This study had reported highly significant correlations between the NR-based lipid measurements and FTIR. For the purposes of bioprospecting for high lipid yielding microalgae, a rapid throughput of sample processing is required. The semi-quantification of neutral lipids using BODIPY or NR and fluorescence microscopy allows for an initial visualization and rapid screening of lipid globules. FTIR spectroscopy may be used thereafter to quantify the yield of lipids. Once the high lipid-producing microalgae have been isolated, purified and identified; a further step in the screening would be done to determine photosynthetic efficiency of the culture.

Subsequent to understanding the physiology and screening of the algal isolate is imperative. Pulse amplitude modulated (PAM) chlorophyll—fluorescence measurements are widely used as a rapid, simple and noninvasive method to assess the physiological state of microalgae.[72] It is also a valuable tool to assess the optimum growth conditions required to maximize the biomass yield to quantify extreme environmental stresses (potassium adsorption ratio [PAR], pH salinity and temperature) and the effect of nutrient on the algal culture. Neutral lipid synthesis is stimulated under limited conditions or nutrient depleted. Many microalgae have the ability to produce up to almost 80% dry cell weight of TAG as a storage lipid under environmental stress or nutrient. The maximum quantum efficiency of Photosystem II [FV/Fm], PAM fluorometer parameters (electron transport rate and non-photochemical quenching) may be used as indicators of nutrient stress and consequently the possibility of neutral lipid synthesis and can be a valuable instrument in the screening process. Neutral lipid synthesis is likely to occur during the stationary phase of growth due to nutrient limitation.

Kinetics of growth and tolerance is also considered for the screening process of microalgae for higher lipid-producing potential. The success of downstream processing is dependent on reliable physiological and biochemical screening tools such as FTIR spectroscopy, PAM, fluorometry and BODIPY lipid stain.

5.6 IMPROVEMENT OF MICROALGAE THROUGH GENETIC ENGINEERING

Throughout the past few decades, obtaining large quantities of algal biomass has been achieved, but obtaining large amounts of lipids has not been an easy task. The key to unlock the maximum oil-producing capacity of algae may be through genetic and metabolic engineering and it has gained considerable importance in recent decades.[73] Although the application of genetic engineering to improve energy production phenotypes in eukaryotic microalgae is in its infancy, significant advances in the development of genetic manipulation tools have recently been achieved with microalgal model systems and are being used to manipulate central carbon metabolism in these organisms. Some of the microalgal genomes have been sequenced such as the diatoms *Phaeodactylum tricornutum* and *Thalassiosira pseudonana* and these aid in understanding the functional genes and the genome structure.

Molecular studies focusing on the enzymatic change in the cell when exposed to chemical inducers will help to optimize growth conditions and, hence maximize oil production by steering the synthesis to produce preferred substrate. *Chlamydomonas* has been studied extensively regarding the biochemical pathways and the recent sequencing has shown to be an interest for many research groups.[74]

Genetic and metabolic engineering involve the altering of genetic material to change its structure and characteristics. This is achieved by many processes such as the insertion of a transposon linked to various promoter and regulatory regions, insertion of site-directed mutagenesis or recombinant DNA. In algae, transformation can occur at the chloroplast and nuclear level. Stable recombination of the foreign DNA has proven to be problematic, and by knowing the exact gene sequence precise insertion of the recombinant DNA will be possible. This known insertion may help to eliminate the problems associated with genetic transformation such as little or no gene and protein expression and even gene silencing (nuclear transformation) of the other genes due to unfavourable insertion of the transgene.

To date, gene silencing has been successful in knocking out or silencing certain genes in diatoms. RNA interference (RNAi) or antisense RNA expression knocks out the target mRNA and the respective gene, therefore altering the phenotype. The light-harvesting antenna protein complex has been downregulated via gene silencing.[75]

The progress in identifying relevant bioenergy genes and pathways in microalgae, and powerful genetic techniques have been developed to engineer some strains via the targeted disruption of transgene expression and/or endogenous genes in these areas is rapidly advancing our ability to genetically optimize the production of targeted biofuels.

5.7   ALGAE MASS CULTIVATION SYSTEMS

Producing micro algal biomass is generally more expensive and difficult task than growing crops. Photosynthetic growth requires water, inorganic salts, light and carbon dioxide. Temperature must remain generally within the range of 20–30°C. To minimize the expense, biodiesel production must rely on freely available sunlight, despite daily and seasonal variations in light levels. Growth medium must provide the inorganic elements that constitute the algal cell. Essential elements include phosphorus (P), nitrogen (N), iron and, in some cases, silicon. Microalgal biomass contains approximately 55% carbon by dry weight.[76]

Microalgae require several things to grow. Since they are photosynthetic, they need carbon dioxide, a light source, water and inorganic salts. The temperature should be between 20 and 30°C (~60–80°F) for optimal growth. Some microalgae may be grown without light, using an organic carbon source rather than CO_2; this is known as heterotrophic growth. However, in order to minimize costs, algae is often grown using sunlight as a free source of light, even though it lowers productivity due to daily and seasonal variations in the amount of available light. The growth medium must contain essential nutrients such as phosphorus, nitrogen, iron and sometimes silicon for algal growth. For large-scale production of microalgae, algal cells are continuously mixed to prevent the algal biomass from settling and nutrients are provided during daylight hours, when the algae are reproducing. However, up to one-quarter or the algal biomass produced during the day can be lost due to respiration during the night.[77]

5.7.1   OPEN RACEWAYS

Open ponds are the simplest and oldest systems for mass cultivation of microalgae. It is standardized system like the shallow pond is usually with about 1 ft deep; algae are cultured under conditions identical to the natural environment. The pond is usually designed in a 'track' or 'raceway' configuration, in which a paddlewheel provides circulation and mixing of the nutrients and algal cells. The raceways are typically made from poured concrete, or they are simply dug into the earth and lined with a plastic liner to prevent the ground from soaking up the liquid. To minimize space and loss, we used baffles in the channel guide. In front of the paddlewheel, we added medium and algal broth is harvested behind the paddle wheel, after it has circulated through the loop. Open ponds are relatively easy to maintain since they have large open access to clean off the biofilm that builds up on the surfaces. Open ponds lose water by evaporation at a rate similar to land crops and are also susceptible to contamination by unwanted species, this is the main disadvantage of open systems. A new open pond is typically inoculated with the desired algal culture with the aim of initiating growth and dominating the pond flora. From the aquatic species program collection of over 3,500 photosynthetic organisms, none were found to be able to continually dominate an open pond and have desirable biofuel properties, that is high lipid contents (TAG)[78] (Fig. 5.3A).

FIGURE 5.3   Mass production of strains in (A) photobioreactor and (B) raceways.

5.7.2   CLOSED PHOTOBIOREACTORS

Besides energy, chemicals and saving water, closed bioreactors have many other advantages, which are increasingly making them the reactor of choice for biofuel production, as their costs are reduced. Photobioreactors is the best method to cultivate microalgae. This process eliminates many of the problems experienced with raceway systems and open pond; the advantages seem obvious. The main advantage of photobioreactors can overcome the problems of evaporation encountered and contamination in open ponds.

On a yield basis, the most important among these aspects is that they support up to fivefold higher productivity with respect to reactor volume and consequently have a smaller 'footprint'. The latter point is the deciding factor, as the goal is to collect as much solar energy as possible from a given piece of land. Higher bioreactor costs are, therefore compensated for higher productivity as recent studies show. The biomass productivity of traditional raceway pond can be 13 times lesser than that of a photo-bioreactor on average. Harvest of biomass from photobioreactors is less expensive than that from a raceway pond, since the typical algal biomass is about 25 times as concentrated as the biomass found in raceways[79] (Fig. 5.3B).

5.8 DOWNSTREAM PROCESSING

5.8.1 LIPID EXTRACTION

Lipids are storage products of microalgae with high nutritional value, their accumulation and synthesis by microalgae is a principle source of bioenergy as they supply essential PUFAs. Furthermore, microalgal lipids have been suggested as a potential diesel fuel substitute with an emphasis on the neutral lipids due to their lower degree of unsaturation and their accumulation in microalgal cells at end of growth stage for biodiesel. Extraction of lipids from microalgae is one of the costliest and debated processes involved in biodiesel production.

A wide range of lipid extraction methods are available and the choice of each method is based on efficiency, accuracy, cost effectiveness, easy to carry out, robustness, high throughput capability and most importantly reproducibility and precision. Widely reported methods for the extraction of lipids from microalgal cells include the following: Folch method, gravimetric method and Bligh and Dyer method. Two conventional methods that are frequently used by many lipid analysts involve solvent extraction and

gravimetric determination. The main drawback of the conventional gravimetric method for lipid determination is that it involves several complicated steps such as separation and concentration, biomass harvesting and
lipid extraction, which can result in loss of lipids content.[80] Hence, in
order to get around this drawback, increasing attention is focused on in
situ measurements of the lipid contents.

Time-domain nuclear magnetic resonance and NR staining methods
have been investigated for the quantification of lipid content in microalgae since they are rapid, simple and feasible as compared to gravimetric methods. One major drawback of the NR staining method is that
a number of green algae, which produce lipids cannot be detected by this
method because of walls that prevent the dye from penetrating the cell
wall.[81]

Lipids can also be removed from algal biomass by means of physical
(cold press) and chemical processes (extraction with solvents) or both,
the latter being more effective; on a laboratory scale, mainly for analytical purposes, the extraction from freeze-dried samples of algae is usually
performed with a chloroform/methanol/water mixture and then the oil is
fractionated by column chromatography to obtain hydrocarbons, nonpolar
and polar lipids. Nonpolar organic solvents such as n-hexane are good
alternatives for hydrocarbons extraction.

Some researchers have proposed a new procedure to extract hydrocarbons from water and dried suspended samples of the microalga *Botryococcus braunii* by using switchable polarity solvents based on alcohol
and 1,8-diazabicyclo-[5.4.0]-undec-7-ene (DBU) content. DBU/octanol
exhibited the highest yields of extracted hydrocarbons from both liquid
and freeze-dried algal samples (9.2 and 16%, respectively against 7.8 and
5.6% with *n*-hexane).

5.8.2   THIN-LAYER CHROMATOGRAPHY (TLC) FOR SCREENING HIGH LIPID CONTENT

The microassay of lipids by fluorometric scanning of thin-layer chromatograms has several analytical advantages. A number of general procedures
have been described for the fluorometric quantitation of lipids separated by thin-layer chromatography (TLC). The technique combines the
convenience, speed and versatility of TLC with the great sensitivity of

fluorometry. These include treating chromatograms with fluorescent dyes such as 2,7-dichlorofluorescein, 1-anilino-8-naphthalene,[82] rhodamine, 6-*p*-toluidino-2-naphthalenesulfonic acid or primuline. Other methods involve formation of fluorescent reaction products by heating lipids in the presence of strong acids, ammonium hydrogen salts or silicon tetrachloride. These latter procedures have the same disadvantages as acid charring for densitometry, that is difficulty in controlling the reaction process and destruction of the lipid samples.

On the other hand, only rhodamine has been popular but most of the fluorescent dyes proposed fade rapidly. The use of rhodamine for the fluorometric TLC microassay of lipids, however, requires a preliminary hydrogenation of samples with platinum oxide catalyst to saturate fatty acid moieties before quantitative analysis and chromatographic separation is carried out. The fluorescent dye, NR can be used as a sensitive stain for the detection of other hydrophobic compounds such as drugs and lipids separated by TLC.[83]

5.8.3   QUALITATIVE AND QUANTITATIVE ANALYSIS OF LIPID BY GC-MS

Hydrocarbon extract was purified by column chromatography on silica gel. The hydrocarbon sample was analysed on gas chromatography–mass spectrometry (GC/GC-MS) (PerkinElmer Instruments, USA) using Elite-5 (cross bond 5% di-phenyl-95% dimethyl polysiloxane) capillary column (30 m × 0.25 mm; ID × 0.25 μm film thickness, PerkinElmer Instruments, USA) with temperature gradient from 130°C (5 min)/8 to 200°C (2 min)/5 to 280°C (15 min) at a flow rate of 1 ml/min carrier gas. The detector and injector port temperatures were maintained at 250 and 240°C, respectively.

The mass spectra were recorded under electron impact ionization at 70 eV electron energy with a mass range from 40 to 600 at a rate of one scans. Hydrocarbons were identified by their $M^{+\bullet}$ ions and comparison of the mass spectra with those of the standards as reported in literature .Comparison of the botryococcenes major daughter ions and their relative abundances afforded a high degree of (>96%) match. Further, relative retention data of these compounds conformed with respect to squalene with those reported in the literature.[84]

5.8.4 TRANSESTERIFICATION OF FATTY ACIDS

The viscosities of microalgal oils and vegetable oils are usually higher than that of diesel oils. Hence, it is clear that it cannot be applied to engines as such. The original viscosity of microalgal oils will greatly reduce by transesterification and we can increase the fluidity, too. Many reports on the production of biodiesel from microalgal oils are available.

The effects of the molar ratio of glycerides to alcohol; reaction temperature, catalysts and time, the contents of water in oils, free fatty acids (FFAs) and fats have been reported in some recent reports. It was reported that biodiesel could be produced by acidic transesterification with 57:2 M ratio of methanol to microalgal oil at 33°C temperature and 100% catalyst quantity (based on oil weight) was achieved. The most abundant composition of microalgal oil transesterified with methanol is $C_{19}H_{36}O_2$, which is suggested to accord with the standard of biodiesel. In recent years, there have been many reports about the applications of transesterified technologies for biodiesel by Stavarache et al.

The transesterification process is affected by various factors depending upon the reaction condition used. The free fatty acid and moisture content, catalyst type and concentration, molar ration of alcohol to oil and type of alcohol, reaction time and temperature, mixing intensity and organic co-solvents are key parameters for determining the viability of the oil transesterification process. As per the reported literature, most of the transesterification studies have been done on edible oils by using methanol and NaOH/KOH as catalyst. There are very few studies reported on oils which are produced by microalgae. This transesterification process can be further improved to get good quality of biodiesel from the lipids of microalgae.

Transesterification of *Nannochloropsis oculata* microalga strains for lipid to biodiesel on AI_2O_3 supported MgO and CaO catalysts is described by Umdu et al. The activities of AI_2O_3 supported MgO and CaO catalysts in the transesterification of lipid of yellow green microalgae, *Nannochloropsis oculata*, as a function of methanol amount and the CaO and MgO loadings at 50°C. It was found that pure CaO and MgO were not active and CaO/AI_2O_3 catalyst among all the mixed oxide catalysts showed the highest activity. Not only the basic strength but also basic site density is important to increase the yield of biofuels. When methanol/lipid molar ratio was 33%, biodiesel yield was over 82 wt.%; CaO/AI_2O_3 catalyst increased the rate to 97.6% from 25%.

Recently, enzymatic approaches for biofuel production have received much attention since these have many advantages over chemical methods such as lower alcohol to oil ratio, moderate reaction conditions, environmental friendly and easier product recovery.[85]

5.9 CONCLUSION

There are so many problems like rising energy imports; high energy prices concerns about petroleum supplies and greater recognition of the environmental consequences of fossil fuels have driven interest in the more production of renewable transportation biofuels. To assuage the depleting fossil reserves, environmental concerns and the rising cost of fuels in the world market, there is a spurring demand to look for sustainable, greener fuels that are economically cost effective with sustainable environmental benefits. As the emphasis switched to production of natural oils for biodiesel, microalgal strains as biofuel feedstock has attracted increasing attention, because of high lipid concentration by dry weight. There is a need to develop the strains with high lipid content through genetic engineering and deploy microalgae technology as it provides an exciting option for the recycling of fossil fuel emissions. Algal biofuel opens up a promising avenue for producing several 'quads' of biofuel, because microalgae generally produce more of the right kind of high density lipid content for the production of biofuels by esterification. Though, a nascent field today, microalgae seem to present the only bio-solution to replace fossil fuels completely. Hence, the present study was aimed to study the algal growth rate, biomass production, lipid content and productivity of microalgal cultures isolated from fresh water and marine ecosystems.

KEYWORDS

- microalgae
- biofuels
- esterification
- PUFA

REFERENCES

1. Meher, L. C.; Sagar, D. V.; Naik, S. N. Technical Aspects of Biodiesel Production by Transesterification—a Review. *Renew. Sustain. Energy Rev.* **2006,** *10,* 248–268.

2. Lang, X.; Dalai, A. K.; Bakhshi, N. N.; Reaney, M. J.; Hertz, P. B. Preparation and Characterization of Bio-Diesels from Various Bio-Oils. *Bioresour. Technol.* **2001,** *8,* 53–62.

3. Miao, X.; Wu, Q. Biodiesel Production from Heterotrophic Microalgal Oil. *Bioresour. Technol.* **2006,** *97*(6), 841–846.

4. Montefri, M. J.; Tai, X. W.; Obbard, J. P. Recovery and Pre-Treatment of Fat, Oil and Grease from Grease Interceptors for Biodiesel Production. *Appl. Energy* **2010,** *87,* 3155–3161.

5. GuanHua, H.; Chen, F.; Wei, D.; Zhang, W.; Chen, G. Biodiesel Production by Microalgal Biotechnology. *Appl. Energy* **2010,** *87,* 38–46.

6. Lin, L.; Cunshan, Z.; Vittayapadung, S.; Xiangqian, S.; Mingdong, D. Opportunities and Challenges for Biodiesel Fuel. *Appl. Energy* **2011,** *88,* 1020–1031.

7. Metzger P.; Templier, J.; Largeau, C.; Casadevall, E. An n-Alkatriene and Some Alkadienes from the a Race of the Green Algae *Botryococcus braunii*. *Phytochem* **1986,** *25:* 1869–1872.

8. Chisti, Y. Biodiesel from Microalgae. *Biotechnol. Adv.* **2007,** *25,* 294–306.

9. Illman, A. M.; Scragg, A. H.; Shales, S. W. Increase in *Chlorella* Strains Calorific Values when Grown in Low Nitrogen Medium. *Enzyme Microb. Technol.* **2000,** *27,* 631–635.

10. Maeda, K.; Owada, M.; Kimura, N.; Omata, K.; Karube, J. CO_2 Fixation from the Flue Gas on Coal-Red Thermal Power Plant by Microalgae. *Energy Convers. Manage.* **1995,** *36,* 717–720.

11. Dote, Y.; Sawayama, S.; Inoue, S.; Minowa, T.; Yokoyama, S. Recovery of Liquid Fuel from Hydrocarbon Rich Microalgae by Thermochemical Liquefaction. *Fuel* **1994,** *73,* 1855–1857.

12. Miao, X.; Wu, Q. Biodiesel Production from Heterotrophic Microalgal Oil. *Bioresour. Technol.* **2006,** *97*(6), 841–846.

13. Rana, R.; Spada, V. Biodiesel Production from Ocean Biomass. In *Proceedings of the 15th European Conference and Exhibition,* Berlin, 2007.

14. Olaizola, M. Commercial Development of Microalgal Biotechnology: From the Test Tube to the Market Place. *Biomol. Eng.* **2003,** *20,* 459–466.

15. Williams, P. B. Biofuel: Microalgae at the Social and Ecological Costs. *Nature* **2007,** *450,* 478.

16. Posten, C.; Schaub, G. Microalgae and Terrestrial Biomass as Source for Fuels – A Process View. *J. Biotechnol.* **2009,** *142*(1), 64–69.

17. An, J. Y.; Sim, S. J.; Lee, J. S.; Kim, B. W. Hydrocarbon Production from Secondarily Treated Piggery Wastewater by the Green Alga *Botryococcus braunii*. *J. Appl. Phycol.* **2003,** *15*(2–3), 185–191.

18. Benemann, J. R. CO_2 Mitigation with Microalgae Systems. *J. Energy Convers. Manage.* **1997,** *38,* 475–479.

19. Molina, G. E.; Belarbi, E. H.; Acien, F. G.; Medina, M. A.; Chisti, Y. Recovery of Microalgal Biomass and Metabolites: Process Options and Economics. *Biotechnol. Adv.* **2003,** *20,* 491–515.

20. Hu, Q.; Sommerfeld, M.; Jarvis, E.; Ghirardi, M.; Posewitz, M.; Seibert, M.; Darzins, A. Microalgal Triacylglycerols as Feed-Stocks for Biofuel Production: Perspectives and Advances. *Plant J.* **2008**, *54*, 621–639.
21. Jae-Yon, L.; Chan, Y.; So-Young, J.; Chi-Yong, A.; Hee-Mock, O. Comparison of Several Methods for Effective Lipid Extraction from Microalgae. *Bioresour. Technol.* **2010**, *101*, 575–577.
22. Bligh, E. G.; Dyer, W. J. A Rapid Method of Total Lipid Extraction and Purification. *Can. J. Biochem. Physiol.* **1959**, *37*, 911–917.
23. Gouveia, L.; Marques, A. E.; de Silva, T. L.; Reis, A. *Neochloris oleabundans* UTEX #1185: A Suitable Renewable Lipid Source for Biofuel Production. *Ind. Microbiol. Biotechnol.* **2009**, *36*, 821–826.
24. Ramadhas, A. S.; Jayaraj, S.; Muraleedharan, C. Use of Vegetable Oils as I.C. Engine Fuels. *Renewable Energy* **2003**, *10*, 1–16.
25. Azam, M. M.; Waris, A.; Nahar, N. M. Prospects and Potential of Fatty Acid Methyl Esters of Some Non-Traditional Seed Oils for Use as Biodiesel in India. *Biomass Bioenergy* **2005**, *29*, 293–302.
26. Alcantara, R.; Amores, J.; Canoira, L.; Fidalgo, E.; Franco, M. J.; Navarro, A. Catalytic Production of Biodiesel from Soy-Bean Oil, Used Frying Oil and Tallow. *Biomass Bioenergy* **2000**, *18*, 515–527.
27. Reddy, J. N.; Ramesh, A. Parametric Studies for Improving the Performance of a Jatropha Oil Fuelled Compression Ignition Engine. *Renewable Energy* **2005**, *31*, 1994–2016.
28. Subramanian, A. K.; Singal, S. K.; Saxena, M.; Singhal, S. Utilization of Liquid Biofuels in Automotive Diesel Engines: An Indian Perspective. *Biomass Bioenergy.* **2005**, *9*, 65–72.
29. Pedro, B.; Agudelo, J.; Agudelo, A. Basic Properties of Palm Oil Biodiesel-Diesel Blends. *Fuel* **2008**, *87*, 2069–2075.
30. Glatz, B. A.; Floetenmeyer, M. D.; Hammond, E. G. Fermentation of Bananas and Other Food Wastes to Produce Microbial Lipid. *J. Food Prot.* **1985**, *48:* 574–577.
31. Tao, J.; Dai, C. C.; Dai, Q. The Conversion Efficiency and Economic Feasibility of Microbial Energy. *Chinese J. Microbiol.* **2006**, *26*, 48–54.
32. Dai, C. J.; Lengxie, Y. T.; Mozhao, R. Biodiesel Generation from Oleaginous Yeast *Rhodotorula glutinis* with Xylose Assimilating Capacity. *Afr. J. Biotechnol.* **2007**, *6*(18), 2130–2134.
33. Matsunaga, T.; Takeyama, H.; Nakao, T.; Yamazawa, A. Screening of Marine Microalgae for Bioremediation of Cadmium-Polluted Seawater. *J. Biotechnol.* **1999**, *70*, 33–38.
34. Ratledge, C.; Wynn, J. P. The Biochemistry and Molecular Biology of Lipid Accumulation in Oleaginous Microorganism. *Adv. Appl. Microbiol.* **2002**, *51*, 1–51.
35. Papanikolaou, S.; Muniglia, L.; Chevalot, I.; Aggeles, G.; Marc, I. *Yarrowia lipolytica* as a Potential Producer of Citric Acid from Raw Glycerol. *J. Appl. Microbiol.* **2002**, *92*, 737–744.
36. Liu, B.; Zongbao, Z. Biodiesel Production by Direct Methanolysis of Oleaginous Microbial Biomass. *J. Chem. Technol. Biotechnol.* **2007**, *82*, 775–780.
37. Rainer, K.; Torsten, S.; Alexander, S. C. Microdiesel: *Escherichia coli* Engineered for Fuel Production. *Microbiology* **2006**, *152*, 2529–2536.

38. Li, M.; Hu, C. W. Outdoor Mass Culture of the Marine Microalga *Pavlova viridis* Prymnesiophyceae for Production of Eicosapentaenoic Acid (EPA). *Cryptogam. Algol.* **2007,** *28*(4) 397–410.

39. Murphy, J. The Biogenesis and Functions of Lipid Bodies in Animals, Plants and Microorganisms. *Prog. Lipid Res.* **2001,** *40,* 325–438.

40. Michele A.; Dibenedetto, A.; Carone, M.; Colonna, T.; Carlo, T. Fragile Production of Biodiesel from Macroalgae by Supercritical CO_2 Extraction and Thermochemical Liquefaction. *Environ. Chem. Lett.* **2005,** *3*(3), 136–139.

41. Chang, H. Marine Biodiversity and Systematics. Research Programmes. April 14, 2009, 2007.

42. Metting, F. B. Biodiversity and Application of Microalgae. *J. Ind. Microbiol. Biotechnol.* **1996,** *17*(5–6): 477–489.

43. Thomas, D. Seaweeds. London, 2002.

44. Danielo, O. An Algae-Based Fuel. *Biofutur* **2005,** 255.

45. Borowitzka, M. A. Microalgae for Aquaculture: Opportunities and Constraints. *J. Appl. Phycol.* **1997,** *9,* 393–401.

46. Anandraj, A.; Perissinotto, R.; Nozais, C. A Comparative Study of Microalgal Production in a Marine Versus a River-Dominated Temporarily Open/Closed Estuary, South Africa. *Estuarine, Coastal Shelf Sci.* **2007,** *73,* 768–780.

47. Bernal, C. B.; Vázquez, G.; Quintal, I. B.; Bussy, A. L. Microalgal Dynamics in Batch Reactors for Municipal Wastewater Treatment Containing Dairy Sewage Water. *Water Air Soil Pollut.* **2008,** *190,* 259–270.

48. Rodolfi, L.; Chini, Z. G.; Bassi, N.; Padovani, G.; Biondi, N.; Bonini, G.; Tredici, M. R. Microalgae for Oil: Strain Selection, Induction of Lipid Synthesis and Outdoor Mass Cultivation in a Low-Cost Photobioreactor. *Biotechnol. Bioeng.* **2009,** *102,* 100–112.

49. Woelfel, J.; Schumann, R.; Adler, S.; Hubener, T.; Karsten, U. Diatoms Inhabiting a Wind Flat of the Baltic Sea: Species Diversity and Seasonal Variations. *Estuarine, Coastal Shelf Sci.* **2007,** *75,* 296–307.

50. Brahamsha, B. A Genetic Manipulation System for Oceanic Cyanobacteria of the Genus *Synechococcus. Appl. Environ. Microb.* **1996,** *62,* 1747–1751.

51. Umdu, E. S.; Tuncer, M.; Seker, E. Transesterification of *Nanochloropsis oculata* Microalga's Lipid to Biodiesel on Al_2O_3 Supported CaO and MgO Catalysts. *Bioresour. Technol.* **2009,** *100,* 2828–2831.

52. MacLulich, J. H. Experimental Evaluation of Methods for Sampling and Assaying Intertidal Epilithic Microalgae. *Marine Ecol.—Prog. Ser.* **1986,** *34,* 275–280.

53. Scott, S. A.; Davey, M. P.; Dennis, J. S.; Horst, I.; Howe, C. J.; Lea-Smith, D. J.; Smith, A. Biodiesel from Algae: Challenges and Prospects. *Curr. Opin. Biotechnol.* **2010,** *21,* 1–10.

54. Rogerson, A.; DeFreitas, A. S. W.; McInnes, A. C. Observations on Wall Morphogenesis in *Coscinodiscus asteromphalus* (Bacillariophyceae). *Trans. Am. Microsci. Soc.* **1986,** *105:* 59–67.

55. Beer, L. L.; Boyd, E. S.; Peters, J. W.; Posewitz, M. C. Engineering Algae for Biohydrogen and Biofuel Production. *Curr. Opin. Biotechnol.* **2009,** *20,* 264–271.

56. Andersen, R. A.; Kawachi, M. Traditional Microalgae Isolation Techniques. In *Algal Culturing Techniques;* Andersen R. A., Eds.; Elsevier, Academic Press, London, UK. 2005; pp 83–100.

57. Stavarache, C.; Vinatoru, M.; Maeda, Y.; Bandow, H. Ultrasonically Driven Continuous Process for Vegetable Oil Transesterification. *Ultrason. Sonochem.* **2007**, *14*, 413–417.

58. Godhe, A.; Anderson, D. M.; Rehnstam-Holm, A. S. PCR Amplification of Microalgal DNA for Sequencing and Species Identification: Studies on Fixatives and Algal Growth Stages. *Harmful Algae* **2002**, *27*, 1–8.

59. Kacka, A.; Donmez, G. Isolation of *Dunaliella spp.* from a Hypersaline Lake and Their Ability to Accumulate Glycerol. *Bioresour. Technol.* **2008**, *99*, 8348–8352.

60. Sinigalliano, C. D.; Winshell, J.; Guerrero, M. A.; Scorzetti, G.; Fell, J. W.; Eaton, R. W.; Brand, L.; Rein, K. S. Viable Cell Sorting of Dinoflagellates by Multiparametric Flow Cytometry. *Phycologia* **2009**, *48*, 249–257.

61. Han, K. H.; Frazier, A. B. A Microfluidic System for Continuous Magnetophoretic Separation of Suspended Cells Using Their Native Magnetic Properties. *Proc. Nanotech.* **2005**, *1*, 187–190.

62. Takahashi, K.; Hattori, A.; Suzuki, I.; Ichiki, T.; Yasuda, K. Non-Destructive on Chip Cell Sorting System with Real-Time Microscopic Image Processing. *J. Nanobiotechnol.* **2004**, *2*, 5.

63. Cooksey, K. E.; Guckert, J. B.; Williams, S.; Callis, P. R. Fluorometric Determination of the Neutral Lipid Content of Microalgal Cells Using Nile Red. *J. Microbiol. Meth.* **1987**, *6*, 333–345.

64. Shevkoplyas, S. S.; Yoshida, T.; Munn, L. L.; Bitensky, M. W. Biomimetic Autoseparation of Leukocytes from Whole Blood in a Microfluidic Device. *Anal. Chem.* **2005**, *77*, 933–937.

65. Petersson, F.; Nilsson, A.; Holm, C.; Jönsson, H.; Laurell, T. Separation of Lipids from Blood Utilizing Ultrasonic Standing Waves in Microfluidic Channels. *Analyst* **2004**, *129*, 938–943.

66. Yamada, M.; Kasim, V.; Nakashima, M.; Edahiro, J.; Seki, M. Continuous Cell Partitioning Using an Aqueous Two-Phase Flow System in Microfluidic Devices. *Biotechnol. Bioeng.* **2004**, *88*, 489–494.

67. Doh, I.; Cho, Y. H. A Continuous Cell Separation Chip Using Hydrodynamic Dielectrophoresis (DEP) Process. *Sensor Actuat. A-Phys.* **2005**, *12*, 59–65.

68. Genicot, G.; Leroy, J. L.; Soom, A. V.; Donnay I. The Use of a Fluorescent Dye, Nile Red, to Evaluate the Lipid Content of Single Mammalian Oocytes. *Theriogenology* **2005**, *63*, 1181–1194.

69. Elsey, D.; Jameson, D.; Raleigh, B.; Cooney, M. J. Fluorescent Measurement of Microalgal Neutral Lipids. *J. Microbiol. Meth.* **2007**, *68*, 639–642.

70. Elsey, D.; Jameson, D.; Raleigh, B.; Cooney, M. J. Fluorescent Measurement of Microalgal Neutral Lipids. *J. Microbiol. Meth.* **2007**, *68*, 639–642.

71. Cooper, M. S.; Hardin, W. R.; Petersen, T. W.; Cattolico, R. A. Visualizing "Green Oil" in Live Algal Cells. *J. Biosci. Bioeng.* **2010**, *102*(2), 198–201.

72. Schreiber, U.; Bilger, W.; Neubauer, C. Chlorophyll Fluorescence As a Nonintrusive Indicator for Rapid Assessment of in vivo Photosynthesis. In *Ecophysiology*

of Photosynthesis; Schulze, E. D., Caldwell, M. M., Eds.; Springer: Berlin, 1994; pp 49–70.

73. Brown, M. R.; Dunstan, G. A.; Norwood, S. J.; Miller, K. A. Effects of Harvest Stage and Light on the Biochemical Composition of the Diatom *Thalassiosira pseudonana*. *J. Phycol.* **1996,** *32,* 64–73.

74. Hu, Q.; Sommerfeld, M.; Jarvis, E.; Ghirardi, M.; Posewitz, M.; Seibert, M.; Darzins, A. Microalgal Triacylglycerols As Feed-Stocks for Biofuel Production: Perspectives and Advances. *Plant J.* **2008,** *54,* 621–639.

75. Mussgnug, J. H.; Thomas-Hall, S.; Rupprecht, J.; Foo, A.; Klassen, V.; McDowall, A.; Schenk, P. M.; Kruse, O.; Hankamer, B. Engineering Photosynthetic Light Capture: Impacts on Improved Solar Energy to Biomass Conversion. *Plant Biotechnol. J.* **2007,** *5,* 802–814.

76. Sánchez, J. A.; Rodríguez, E. M.; Casas, J. L.; Fernández, J. M.; Chisti, Y. Shear Rate in Stirred Tank and Bubble Column Bioreactors. *Chem. Eng. J.* **2006,** *124,* 1–5.

77. Molina, G. E.; Fernandez, G. A.; Garcia, C. F.; Chisti, Y. Photobioreactors: Light Regime, Mass Transfer, and Scaleup. *J. Biotechnol.* **1999,** *70,* 231–247.

78. Sheehan, J.; Dunahay, T.; Benemann, J.; Roessler, P. *A Look Back at the US Department of Energy's Aquatic Apecies Program—Biodiesel from Algae;* Report- NREL/ TP-580-24190. National Renewable Energy Laboratory: Golden, CO., 1998.

79. Barbosa, B.; Jansen, M.; Ham, N. Microalgae Cultivation in Air-Lift Reactors: Modeling Biomass Yield and Growth Rate As a Function of Mixing Frequency. *Biotechnol. Bioeng.* **2003,** *82,* 170–179.

80. Miao, X.; Wu, Q. High Yield Bio-Oil Production from Fast Pyrolysis by Metabolic Controlling of *Chlorella prototothecoides*. *J. Biotechnol.* **2004,** *110*(1), 85–93.

81. Samori, C.; Torri, C.; Samori, G.; Fabbri, D.; Galletti, P.; Guerrini, F.; Pistocchi, R.; Tagliavini, E. Extraction of Hydrocarbons from Microalga *Botryococcus braunii* with Switchable Solvents. *Bioresour. Technol.* **2010,** *101,* 3274–3279.

82. Heyneman, R. A.; Bernard, D. M.; Vercauteren, R. E. Direct Fluorometric Micro Determination of Phospholipids on Thin-Layer Chromatograms. *J. Chromatogr.* **1972,** *68,* 285–288.

83. Fowler, S. D.; Brown, W. J.; Warfel, J.; Greenspan, P. Use of Nile Red for the Rapid in situ Quantitation of Lipids on Thin-Layer Chromatograms. *J. Lipid Res.* **1987,** *28,* 1225–1232.

84. Dayananda, C.; Sarada, R.; Kumar, V.; Aswathanarayana, G.; Ravishankar, G. A. Isolation and Characterization of Hydrocarbon Producing Green Alga *Botryococcus braunii* from Indian Freshwater Bodies. *J. Biotechnol.* **2007,** *10,* 78–91.

85. Meher, L. C.; Sagar, D. V.; Naik, S. N. Technical Aspects of Biodiesel Production by Transesterification—A Review. *Renew. Sustain. Energy Rev.* **2006,** *10,* 248–268.

CHAPTER 6

PROSPECTS AND APPLICATION OF *AZOSPIRILLUM* SPP. AS A NATURAL AGRICULTURAL BIOFERTILIZER

D. MOHAPATRA, N. R. SINGH, and S. K. RATH*

*School of Life Sciences, Ravenshaw University, Cuttack 753003, India, *E-mail: saktirath@gmail.com*

CONTENTS

ABSTRACT

Nutrient is the basic need of all living organisms for their existence on this planet. The main sources of nutrient for human being are plants and their products. To comply with these needs, a human completely depends on agriculture. Modern science and technology provides various means to improve the yield and quality of the agricultural products that ranges from application of synthetic fertilizers to chemical pesticides. Application of such synthetic chemicals increases the agricultural production to a large extent but such an increase was marred with long-term side-effects of those chemicals to both the flora and fauna. It is evident that the deposition of unused chemicals in the soil leads to the destruction of soil microflora, loss in productivity and many health hazards due the entry of such chemicals into the food chain of human being. These problems provoked biologists to look for sustainable alternative new product that does not have such drawbacks. One such sustainable alternative is to look for fertilizers that have biological origin. Biofertilizer, obtained from natural organisms, has almost no side-effects towards the environment and the living organisms including human. Microbes capable to fix nitrogen can play pivotal role in the development of biofertilizers as they can make nitrogen available to plants that is essentially required for their growth as macronutrient. One such free-living nitrogen-fixing bacteria, which can be used as biofertilizer, is *Azospirillum*. *Azospirillum* can be a potential agent for the increase in production of different agricultural product. The overall focus of this chapter is on the production, utilization and efficiency of *Azospirillum* and how it is helpful in improving the agricultural yield without side-effects.

6.1 INTRODUCTION

Agriculture is the most important field for the growth and development of any country as it not only provides the economic base for the growth of the country revenue but also is very much important for the sustainability and the very existence of the people in the country. With diversification of the sources of income in developing country, the contribution of agriculture sector to the national economy is steadily decreasing during last one decade. The main cause of such depletion is rapid industrialization, depletion of agricultural land, hike in the cost of farming but not much hike in the price of the products, drastic climate change leading to insufficient rainfall or too much of rain and the most important is the lack of interest of the young generation to take up farming as a source of income but rather prefer to official jobs. Nevertheless, agricultural sector is still the most important sector of the nation. However, due to rapid industrialization and urban encroachments, area under cultivation is depleting and it would be a great challenge to meet the requirements in terms of food grain and to minimize crop losses in the coming decades. This very important sector of country is under severe stress and needs revival soon if we want to exist in this surface of earth. The very solution to this problem is to increase the production of our food products in less land, which will leads to increase in the profit margin of farmers and motivate them for agriculture. This has been achieved in the past by using fertilizer and various pesticides to enhance productivity and to decrease the loss of the products by insects. This use of chemical fertilizers for the enhancement of agricultural products was seen as a blessing for the agricultural sector but with passing year this blessing soon became a curse in disguise. By exclusive use of these chemicals, there was seen net loss of productivity of the field which has now become a concern for the sector. When investigations were done regarding the cause of such effects, it was found out the use of these chemicals have degraded the quality of the soil due to their persistence nature. The good microflora of the soil is lost which play a big role in the mineralization process of the soil due to which essential minerals are depleting in the soil which has made plants prone to many deficiency diseases and led to decrease in productivity. These chemicals have also led to increase in population of new variety of microbes which have thrived on these new chemicals that have led to depletion of other essential nutrient in the soil. Extreme uses of inorganic fertilizer and pesticides have created various environmental issues ranging from greenhouse effect, degradation of water quality, ozone layer depletion and so forth. The best solution to these environmental problems is to use natural products derived from nature like biofertilizers and biopesticides as these are eco-friendly and have no such side-effects that are caused by the use of artificial chemicals. The advantages of biofertilizer are they not only provide nutrient to agricultural plant but also help to control soilborne disease to maintain the mineralization in soil and structure of the soil. These natural products act as the backbone for sustainable agriculture. Most of the terrestrial plants form symbiotic association with *Arbuscular mycorrhizal* fungi (AMF) which are one of the important microbes of the soil; these also are found beneficial when they are inoculated at early seedling stage. Mycorrhization of micropropagated plantlets at early stage obtained from plant tissue culture helps in overcoming of 'transplant shock'. The biggest limitation of the use of AMF is their limitation of not be cultured axenically, which makes them unsuitable for rapid and mass cultivation in the in vitro condition. The other groups of microbes used worldwide as growth promoters for plants are nitrogen-fixing bacteria and hold a promising future. With the passage of time, the use of Rhizobium is increasing in legume crops to increase agricultural productivity. The increase in the market demand of these products has made them a common market product in the fertilizer sector and is now available in different inoculation form. Another advantage of these products over other groups is that axenic culture of these products can be produced in mass culture in laboratories and be stored for future use. The other group of organisms which are very much important are the phosphate-solubilizing bacteria that convert insoluble phosphate to soluble form so that they can be absorbed and utilized by the plant. Two important bacteria namely *Azotobacter* and *Azospirillum* are efficient organisms responsible for the increase in crop yield. But there are other organisms like blue-green algae that also contribute to a greater extent to the nitrogen need for sustainable agriculture. The effectiveness of these products to increase the production in crops, vegetables and trees depends on the triple relationship between legume and rhizobium and mycorrhizae. History shows that blue-green algae are the highest producer of plant-utilized nitrogen. Therefore, the importance of these organisms to enhance the production should not be taken lightly. The mass culture of these organisms will be of great use to us as they can be easily marketed in various forms in the open market.

The last four decades have seen many changes in the population and food habit of the human being. In these last few decades, the human population has grown enormously and has become just double of what it was before which has bound us to increase the food production and we have achieved it.[1] The most important role in the increase of production is played by plant nutrition. We have achieved this by the implementation of inorganic fertilizer. To increase the productivity we have increased the use of these chemicals like nitrogen (N) by nine folds and chemical phosphorous (P) by almost four folds.[1] This huge increase in the use of these chemical fertilizers of N and P in agricultural practices has bound us to produce low cost fertilizer rather than high quality products. We have emphasized on the quantity of the product rather than quality and the after-effect of the use of these products.[1,2] The enhancement in the production of the agricultural product by the use of these chemicals has also created various issues starting from degradation of soil quality, deterioration of water quality of surface and groundwater, reduced biodiversity, air pollution and suppressed ecosystem function.[1,2,3] The pollution of the ecosystem due to greater availability of the nutrient is of two types: direct and indirect. The direct way of pollution includes poor management of the inorganic fertilizer by using excessive amount than required for the growth of the plant and misuse of the fertilizer, which leads to various problems like volatilization, leaching, denitrification and acidification. The indirect harmful effect of these chemicals is mainly due to the process by which they are produced in huge amount in industries which includes the use of fossil fuel in Haber-Bosch process and in the transport of chemicals from the factories to the place of implementation. This results in the airborne pollution by increasing the amount of CO_2 and N_2 in air, which is deposited in the terrestrial ecosystem. The more complete and accurate view of the N cycle, how they are deposited globally and various ways in which they are influenced by various physiochemical parameters of the nature, is given in the referred manuscript.[3] A pathway of nitrogen cycle in the nature is given in Fig. 6.1.

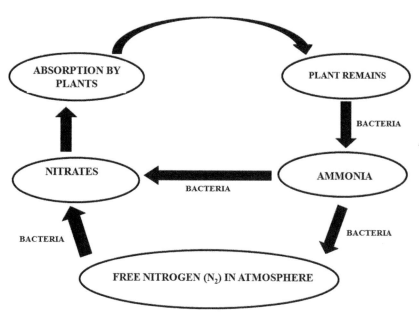

FIGURE 6.1   Pathway of nitrogen fixation in nature.

The next best alternative which can be used as replacement for the inorganic nitrogen fertilizer is household waste and sewage sludge, as they are readily available and costless. The biggest problem with this source is their high content of heavy metals in them. So, if used as source of plant nutrient, these will have many harmful effects not only on crop growth but also on consumers, rhizosphere and microorganisms in the soil.[4,5] This very problem makes these very good sources of inexpensive plant nutrient unsuitable to be used as fertilizer for growing crops for the consumption of human being, because the heavy metal content in these products cannot be reduced to consumable level where they can be utilized as fertilizer. The two most essential elements for plant growth are N and P.[6,7] Both these elements are available in huge amount in the soil but the problem is they are not present in suitable form, so they are not used by the plants for their growth. Maximum N in the soil is associated with the soil organic matter due to which even after addition of fertilizer, plant compete with soil micro-organisms to obtain the soluble nitrates. There is a very different problem with P, which varies from soil to soil. For an instance, if the soil is acidic then when huge amount of P fertilizer is added to the soil it precipitates with iron (Fe) or aluminium (Al) and in the case of basic soil it precipitates as calcium phosphate.[7] So, the amount of P in soil cannot be compensated only by addition of fertilizer but also the pH of the soil has to be taken into account and accordingly addition of the fertilizer should take place.

The exclusive use of P and N sub-surface runoff of these chemicals and these soluble nutrients easily end up in ground water and surface-water bodies like lake, reservoirs and so forth. One such example is that the loss of P from the agricultural soil ends up in lake and estuaries in the developing countries which leads to eutrophication and hypoxia in the water bodies.[1] But the end result of this is that these countries have achieved the increase in the production of food by such agricultural practices and they have not faced that big problem in terms of deterioration of the environment.[1] The problem in the developing countries is different; the problem here is the lack of adequate agricultural practices and fertilizer which has not allowed increase in crop production, which has led to undernourishment of large population of these counties. All these scenarios state that the heed of the hour is a sustainable agricultural practice at global level. As in the developing countries, there is a need in reduction of environmental cost and energy which will lead to efficient and cost-effective production of adequate nutrition for the growing population and also lead to sustainable development of the agricultural sector without much harmful effect on the future generation. The main aim would be to suppress the environmental problem caused due to the loss of plant nutrient which will lead to increase in the production of the agricultural product. But all these have to be achieved without the use of costly inorganic fertilizer and the implementation of microorganisms that will help in more efficient nutrient uptake and also help in increasing the nutrient availability for the plant which will provide long-term solution for the future agricultural practices and help to improve the current scenario of the agricultural practices. In this chapter, we try to give a brief overview of these types of organisms that hold good to be used in the future for the enhancement of agricultural products without causing much damage to the ecosystem and leading to a cleaner and greener world.

Nitrogen is one of the most essential elements in the growth of the plant as nitrogen is an important part of amino acid which combines by peptide bond to form the proteins. As it is known that proteins are the building blocks of an organism and also play major role in various biochemical reactions by acting as biocatalysts, thus are very much important in the development of plants. There are various types of nitrogenous fertilizer used in the field of agriculture: calcium nitrate 15.5% N, ammonium nitrate 33.5% N, ammonium 21% N, calcium ammonium nitrate 27% N, ammonium sulphate nitrate 26% N, urea 46% N, sodium nitrate (natural chile salpeter) 16% N, low biuret urea and binary compound fertilizers nitrogen (N), phosphorus (P) and potassium (K) (NPK) which are exclusively used to increase the production. However, our chapter focuses on the application and production of biofertilizer. There are two types of nitrogen fixing bacteria found in the nature that are symbiotic and free-living. The most exclusively used N_2-fixing bacteria are of the family *Rhizobiaceae* (Rhizobia) which include the following genera: *Sinorhizobium, Rhizobium, Azorhizobium, Bradyrhizobium, Allorhizobium* and *Mesorhizobium*. [5,8]

These bacteria are distributed worldwide ranging from high latitudes in Europe and North America to the equator, to tropics in Australia and South America and infect the leguminous plants in these areas. These are very much important in tropical and equatorial areas as legumes are very much important in the areas because they are utilized in silvopastoral and agroforestry systems.[9] The nitrogen fixing capacity of these organisms vary greatly up to 450 Kg·N ha^{-1} depending upon the type of host plant and the strain of bacteria infecting the host.[10,11,12] So, for the best biofertilizer to be obtained from these organisms, both the strains and the relationship between the strain and the host need to be considered properly to develop a suitable biofertilizer for the field use. Other additional features to be observed for the effective development of a biofertilizer are characters like high nitrogen-fixing ability in natural environment, ability to compete with the indigenous strain already present in the soil (but with the maximized infection in the targeted plant).[13] The real fact is the inoculums should have a high survivor rate and could be easily produced and must show substantial effect on the plant on whose seed they are applied.[14] However, the problem with this type of nitrogen fixation is its limitation to only few plant varieties to which it is associated symbiotically. This leads to the search for free-living nitrogen-fixing organisms that can be applied for all the plant groups irrespective of their genus or family. A figure of conversion of nitrogen into soluble form is given below (Fig. 6.2).

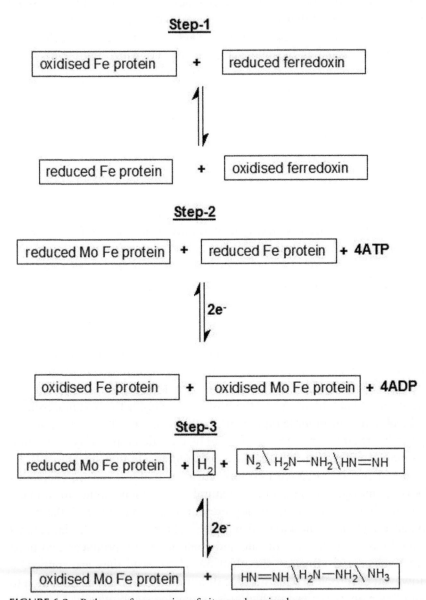

FIGURE 6.2   Pathway of conversion of nitrogen by microbes.

The above examples were of the symbiotic nitrogen-fixing bacteria that live in the root of the host plant and help in the nitrogen fixation, but there is an another group of bacteria which is not symbiotic but is still associated with nitrogen-fixing ability and includes a very huge number in nature and can be of great help to increase the production power of the crops. Such few bacteria are *Acetobacter diazotrophicus* and *Herbaspirillum* spp. that are associated with commercial crops like sugarcane, sorghum and maize;[15,16,17] *Azoarcus* spp. is associated with fodder like kallar grass (*Leptochloa fusca*);[18] and *Herbaspirillum, Alcaligenes, Klebsiella, Bacillus, Enterobacter, Rhizobium, Pseudomonas* and *Azospirillum* are associated with food crops like rice and maize.[19] One such bacteria is *Azospirillum* spp. which are omnipresent throughout the world and found associated with various standing crops grown in different variety of soil and also found in different climatic conditions like tropical, sub-tropical and temperate climatic conditions. This organism has nitrogen-fixing ability and has been isolated from any varieties of tropical and sub-tropical nonleguminous plant.[20] This species is found to colonise with many annual and perennial plants which have never been associated with symbiotic nitrogen fixing bacteria. The biggest beauty of this organism is it is a general root colonizer and has no such host specific limitations.[21]

Azospirillum is a common bacterium of the soil which was isolated and named as *Spirillum lipoferum* in the beginning. This organism was later isolated from dried seaweed in Indonesia and also from the soil,[22] it was also isolated from tropical plants as a phyllosphere bacterium[23,24] and was classified into the genus *Spirillum* in the order Pseudomononadales. Later, some researches by scientist stated that *S. lipoferum* should belong to the genus *Azospirillum*.[25] This organism accounts for about 1–10% of the total rhizosphere population.[26] The soil of the rhizosphere contains 100 times of *Azospirillum* as compared to that of the non-rhizosphere soil.[27] This organism has been isolated from different types of plants from different diverse conditions on earth like from extreme conditions (saline soil, desert soil and water flooded soil) and also from normal soil like tropical and temperate soil.[28] This organism is also found in the rhizosphere of many crop plants like rice, maze, wheat, pearl millet and sorghum in various locations worldwide.[29] The species of *Azospirillum* are one of the most effective nitrogen fixers when they are provided with the optimum conditions for the biological nitrogen fixation. Further studies show that *Azospirillum* inoculation with various crops play a major role in enhancing the growth of the plant,[30, 31] they also stated that in crop like sugarcane 50% of the content of the nitrogen was obtained by this source. Different wild verities of *Azospirillum* were found to be efficient nitrogen fixers which may be found either as free-living or in close association with the plants.[31] Further investigation reveals that certain mutant strains of *Azospirillum brasilense* have the ability to produce high amount of Iodole-3 acetic acid (IAA) in laboratory condition ranging from 16 $\mu g \cdot ml^{-1}$. When plants are inoculated with such mutants, there was seen an increase in root length, root number and root hair in wheat.[32] More investigations on this path disclose that there are many different plant growth-promoting compounds produced by the two strains of *Azospirillum* which consist of phytohormones such as IAA, Gibberellic acid, Abscisic acid and some growth regulators such as putrescine, spermine, spermidine and cadaverine were also found in culture supernatant of both strains.[33] All these properties make *Azospirillum* a suitable biofertilizer and it is being applied in various crops. The application of *Azospirillum* in combination with different living organisms, organic fertilizer, vermicompost and inorganic fertilizer is given below. A list of various organisms and their association with their host plants is given in the Table 6.1.

TABLE 6.1   List of Different Nitrogen Fixing Organism and their Relation with Host.

Name of organism	Mode of interaction with host
Anabaena	Free-living
Nostoc	Free-living
Azotobacter	Free-living
Beijerinckia	Free-living
Clostridium	Free-living
Trichodesmium	Free-living
Cyanothece	Free-living
Rhizobia	Free-living
Desulfovibrio	Free-living
Purple sulphur bacteria	Free-living
Purple non-sulphur bacteria	Free-living
Green sulphur bacteria	Free-living
Klebsiella	Free-living
Methylomonas	Free-living

TABLE 6.1 *(Continued)*

Name of organism	Mode of interaction with host
Pseudomonas	Free-living
Acaligenes	Free-living
Methanococcus	Free-living
Methanosarcina	Free-living
Thiobacillus	Free-living
Chlorobium	Free-living
Desulfovibro	Free-living
Chromatium	Free-living
Rhizobium meliloti	Symbiotic relation with Alfalfa
Rhizobium trifolii	Symbiotic relation with Clove
Rhizobium leguminosarum	Symbiotic relation with Pea, broad bean
Rhizobium phaseoli	Symbiotic relation with Beans
Rhizobium lupine	Symbiotic relation with Lupines
Rhizobium japonicum	Symbiotic relation with Soybean
Rhizobium fredii	Symbiotic relation with Soybean
Rhizobium loti	Symbiotic relation with Lotus
Rhizobium elti	Symbiotic relation with Mung bean, kidney bean
Rhizobium ciceri	Symbiotic relation with Chickpea
Rhizobium elkanii	Symbiotic relation with Soybean
Azospirillum	Symbiotic and free-living
Frankia	Symbiotic and free-living

6.2 APPLICATION OF *AZOSPIRILLUM* AS BIOFERTILIZER IN VARIOUS WAYS OF AGRICULTURE

6.2.1 AZOSPIRILLUM *ALONG WITH THE OTHER BACTERIA AND INORGANIC FERTILIZER*

Studies show that biofertilizers based on *Azospirillum brasilense* and *Glomus intraradices* increase the overall plant growth and yield of cherry tomato as compared to only chemical fertilizer.[34] *Azospirilla* along with the other bacteria and inorganic fertilizer have shown enhancement in the yield of many fruit productions like banana, strawberry, poovan, apple and citrus.[35] *Azospirillum brasilense* strongly enhanced root and shoot growth, germination value and vigour of tomato when inoculated by soaking; and also increased the level of endogenous plant IAA of cucumber and lettuce which helped in the enhancement of their growth.[36] Treatment with *Azotobacter* with *Azospirillum* shows significant increase in the yield of tomato as compared with single inoculations and control plant.[37] Another study on *Foeniculum vulgare* Mill shows that half dose of N, P and K fertilizer amended with different strains biofertilizer increases number of branches, plant height, plant fresh weight, plant dry weight and fruit yield as compared to the use of only 50 and 100% of inorganic fertilizer. The best result in terms of yield was obtained in a mixture of 50% NPK and mixture of biofertilizer obtained from mixture of strains of *Azotobacter chroococcum*, *Azospirillum lipoferum*, and *Bacillus megatherium*.[38] Research on the use of suitable carrier for biofertilizer shows that seed germination, plant height, plant biomass and yield of wheat was significantly increased in all the treatments compared to control. The maximum per cent seed germination, plant height, plant biomass were observed from the combination of *Azospirillum* and *Azotobacter* with 50% inorganic nitrogen. It shows the use of fly ash-based biofertilizer as a means for the reduction of the use of inorganic fertilizer without hampering the yield of the crop.[39] Another set of study shows that there was significant reduction in the use of inorganic nitrogen (urea) when farming of the bread wheat was done using the combination of nitrogen fertilizer along with biofertilizer produces from *Azotobacter* and *Azospirillum* showing a reduction in the use of inorganic fertilizer by 25–50% from the recommended value of 230 kg·N ha^{-1}. The best result was obtained using 75% inorganic fertilizer and 25% biofertilizer. But the combination of 50% inorganic fertilizer and 50% biofertilizer also showed high yield as compared to the use of only inorganic fertilizer. This shows the effectiveness of the biofertilizer in reducing the use of organic inorganic fertilizer.[40] A study was undertaken to determine the effect of combination of inorganic fertilizer and biofertilizer on the production of wheat in Upper Egypt. The use of the combination of biofertilizer derived from *Azotobacter* and *Azospirillum* with inorganic fertilizer showed that the best combination of mixture was 25% biofertilizer or 50% biofertilizer with the rest of inorganic fertilizer shows the best result. The result obtained also depicts that this combination of biofertilizer and inorganic fertilizer not only increases the yield of the plant but also helps in reducing the cost of agriculture which makes it more profitable for the farmer and also available to poor farmers.[41] A research at International Rice Research Institute, Philippines to study the effect of three different biofertilizers derived from *Azospirillum*, *Trichoderma* and unidentified rhizobacteria was conducted in four cropping seasons in between 2009 and 2011. For the experiment, different concentration of recommended rate (RR) of fertilizer like 100%, 50%, and 25% RR were used as control to study its effect in fully irrigated conditions in lowland rice. The result showed a considerable increase in the production of rice in all experiments with only exception during 2008/2009 dry season. An increase in production of around 5–18% was observed by the biofertilizer treatment and the best results were seen from *Azospirillum*-based biofertilizer. However, various factors such as type of biofertilizer, season and inorganic fertilizer treatment are associated with the increase in yield . The absolute grain yield was found to be 0.5 t·ha^{-1} but the effect of yield was not affected by inorganic fertilizer rate, but the best results were obtained at low and medium fertilizer rate. There was enhancement in the grain yield up to 5 t.·ha^{-1} by the use of biofertilizer which proves that biofertilizer will be most useful in rainfed environment. This increase in the rate of production may not be very much significant, but is helpful in increasing the profit level of the farmer by increasing the yield and decreasing the use of costly inorganic fertilizer. So, it has a future if it can also give good result under stress conditions like salinity, acidity and low fertile soil.[42] Another research at Rice Research Station of Tonekabon, Iran, showed that there was increase in grain and head yields of rice with increasing rate of the use of inorganic fertilizer, but biofertilizer helps only to increase the grain yield in the plant. There was a significant increase in the production seen until the increase was up to 75 kg·ha^{-1} but not that significant when the use increased to 150 kg·ha^{-1}. This shows that lower rate of inorganic fertilizer along with biofertilizer will be helpful in increasing the yield of the plant.[43] A research at the Tea Research Foundation Valparai, India, on the application of three different biofertilizers obtained from organisms like VAM fungi, phosphobacteria and *Azospirillum* was conducted. The finding showed that bio-inoculants help in increasing the efficiency of the root and shoot system of the plant by providing essential nitrogen and phosphorus to the plant. The biofertilizer also helped in the production of high quality seedling. This makes it an important method which can be applied in tea-growing areas to enhance productivity and improve the growth rate of the plant. It will also be helpful in reducing the cost of production by reducing the use of high cost chemical fertilizer and help in reduction of the cost of seedling production.[44]

6.2.2 AZOSPIRILLUM *ALONG WITH INORGANIC FERTILIZER*

Research shows *Azospirillum* inoculation prepared using fructose in single cell suspension culture when applies to maze there was a significant increase in the root and foliage dry weight, increase in root surface area of the maze seedling as compared to the plants which are grown without this treatment and also more effective than plant grown in *Azospirillum* which were grown in malate. Further, it was seen that the maze grown with the formulation showed increase in the dry weight of the seed by 59% which is similar to the result obtained by using 60 Kg urea·N ha^{-1}.[45] Study on *Carthamus tinctorius L* shows that inoculation of *Azospirillum* along with 50% inorganic nitrogen fertilizer shows good result and increase in the productivity of the plant.[46] Study shows that *Azospirillum brasilense* when inoculated with maze has a beneficial effect on the plant by increasing the vigour of the plant and also increase the production power of the plant in the climatic and soil condition of Wielkopolska region[47] and help in increasing the rate of germination in maize.[48] Combination of *Azospirillum* along with inorganic fertilizer shows enhancement in the height, flowering and the yield of *Tagetes erecta L*.[49] Study on *Coriandrum sativum L.* depicts that inoculation of *Azospirillum* with inorganic fertilizer increases the yield of the crop drastically as compared to only inorganic fertilizer.[50] Similar result are obtained with Malaysian Sweet Corn.[51]

6.2.3 AZOSPIRILLUM *ALONG WITH OTHER BACTERIA*

Studies show a significant increase in grain yield that is seen in *Zea mays* by the use of combined inoculation of *Azospirillum* and *Pseudomonas* strains[52] and cotton.[53] In crops like pearl millet and sorghum, there was observed an increase in the production of these crops by using the combined inoculation of *Azotobacter* and *Azospirillum*, which are reviewed by,[54] there was also seen increase in the yield of dry land crops by 11–12% with the use of this combination of inoculation. But maize, wheat and rice which get better conditions than pearl millet and sorghum show a much greater increase in the yield that is up to 15–20% as compared to dry land crops with the same inoculation. The result obtained from this study shows that incorporation of *Azotobacter* in combination with *Azospirillum* shows a better yield, enhancement of growth and increase in quality of onion. The storage capacity of the crop also increases when combination of *Azotobacter* and phosphate-soluble bacteria is effective.[55] The findings suggested that the combined application of *Azotobacter* and *Azospirillum* increased the seed oil yield of *Brassica napus L.* as compared to the use of only inorganic nitrogen fertilizer,[56] the same result was also seen in *Helianthus annuus L.*[57] A study in the hydroponic situation showed that inoculation of *Azospirillum* spp. and *Azotobacter* spp. along with nitrogen enhance the growth and yield of strawberry.[58]

6.2.4 AZOSPIRILLUM *ALONG WITH ORGANIC FERTILIZER*

A combination of *Azospirillum brasilense* along with organic fertilizer showed improved growth and yield of sweet potato in addition to production cost reduction and conservation of natural resources.[59] "Findings show that there is significant effect of *Azotobacter*, *Azospirillum* and *Pseudomonas* on sunflower grain yield and phytohormones content.[60] Also the yield is effective along with the application of animal manure.[60] Better grain yield and seed quality of sunflower was observed with farmyard manure.[61]

6.2.5 AZOSPIRILLUM *ALONG WITH VERMICOMPOST*

Azospirillum is also very much effective in increasing the yield of tomato along with vermicompost. In a research conducted at Research and Development Centre, Bharathiar University, Coimbatore, Tamil Nadu, on the application of vermicompost and biofertilizer on tomato. The result obtained shows significant increase in whole plant height (cm), number of branches, number of leaves per plant, length of the root, number of fruits per plant, and also increase in the harvest index. Total examination of the experiment shows that vermicompost with biofertilizer inoculation enhances plant mineral concentration through nitrogen fixation which alters fruit production in tomato plants.[62]

6.2.6 AZOSPIRILLUM *WITHOUT ANY COMBINATION*

Another study on 10 native *Azospirillum* strains (TNM01, TVM02, TKV03, TNM04, TVR05, TKC06, TMG07, TKR08, TTP09 and TMP10) on the growth and seedling of rice shows beneficiary effect over the untreated paddy seedling. Out of the ten strains taken for experiment, the treatment with TMP10 and TNM5 strains showed the best response among all the other strains in terms of increase in the seedling growth.[63]

Table of various ways of application of biofertilizer and their important crops associated with it are given in two tables (Tables 6.2 and 6.3), respectively:

TABLE 6.2   Various Mode of Action of *Azospirillum* as Biofertilizer

Organism	Mode of action	Reference
Azospirillum brasilense	With *Glomus intraradices* and inorganic fertilizer	[34]
Azospirilla	With other bacteria and inorganic fertilizer	[35]
Azospirillum brasilense	With other bacteria and inorganic fertilizer	[36]
Azospirillum	With *Azotobacter* and inorganic fertilizer	[37,39–41,43]
Azospirillum lipoferum	With *Azotobacter chroococcum* and *Bacillus megatherium* and inorganic fertilizer	[38]
Azospirillum	With Trichoderma, unidentified rhizobacteria and inorganic fertilizer	[42]
Azospirillum	With phosphobacteria and inorganic fertilizer	[44]
Azospirillum	Only with inorganic fertilizer	[45–51]
Azospirillum	Only with *Pseudomonas*	[52,53]
Azospirillum	Only with *Azotobacter*	[54,56–58]
Azospirillum	Only with *Azotobacter* and Pospahte soluble bacteria	[55]
Azospirillum brasilense	With organic fertilizer	[59]
Azospirillum	With organic fertilizer	[60,61]
Azospirillum	With vermicompost	[62]
Azospirillum	Directly on seedling	[63]

TABLE 6.3 Biofertilizer Applications in Various Plants.

Name of the organism	Applied plant	Reference
Acetobacter diazotrophicus	sugarcane, sorghum and maize	[15–17]
Herbaspirillum spp.	sugarcane, sorghum, rice and maize	[15–17]
Azoarcus spp.	kallar grass	[18]
Alcaligenes	rice and maze	[19]
Azospirillum	strawberry, tomato, tea, wheat, safflower, Mexican marigold, coriander, sweet corn, rice, cotton, pearl millet, onion, rapeseeds, sunflower and maze	[19, 35, 37, 39, 44, 46, 49–51, 53–57]
Klebsiella	rice and maze	[19]
Bacillus	rice and maze	[19]
Enterobacter	rice and maze	[19]
Rhizobium	rice and maze	19
Pseudomonas	sunflower, rice and maze	[19, 60]
Azospirillum brasilense	cherry tomato, tomato, maze, sweet potato	[34, 36, 47, 59]
Glomus intraradices	cherry tomato	34
Azotobacter	Tomato, wheat, pearl millet, onion, rapeseeds, sunflower, strawberry	[37, 39, 54–58]
Azotobacter chroococcum	fennel	[38]
Azospirillum lipoferum	fennel	[38]
Bacillus megatherium	fennel	[38]
Trichoderma	rice	[42]
phosphobacteria	tea	[44]
Pseudomonas strains	Maze, cotton	[52, 53]

6.3 CONCLUSION

Several other research shows that *Azospirillum* has growth enhancement effect on various food crops like carrot, sunflower, sugar beet, oak, eggplant, tomato, cotton and pepper and also has very prominent impact on the major food crops of the world like wheat and rice which is the staple food of the whole world.[64,21] From the data collected from the field experiments of the last two decades, it is clear that application of *Azospirillum* in the agriculture field shows considerable increase in crop yield in 60–70% of the case.[65] The increase in yield is considerable up to 30% but the general increase in yield ranges from 5 to 30%. Data collected throughout the world in the last 20 years from the use of *Azospirillum* inoculation in the field shows that the bacteria has ability to promote increase in the production of agriculturally important crops in different climate and different types of soil.[65] These increase in the production is not more due to nitrogen fixation but is rather due to the production of growth promoting substances like phytohormones along with nitrogen which increase the yield in these plants.[66] Another fact also comes into light that these organisms cannot bring about the enhancement all alone but are always used in association with other organism or along with inorganic fertilizer, organic fertilizer or vermicompost. The research on the use of *Azospirillum* as a potential biofertilizer is still at its infancy and a lot of research is required before any such formulation can be designed. But the potential of this organism should not be doubted as it has all the features required by any organism to become a potential nitrogen biofertilizer in future and for its application worldwide. The main area of research should be the isolation of indigenous strains of this organism which grow well in association with other bacteria or, in short, a search for a bacterial consortium which has promising result. Last but not the least suitable carriers for the preservation, marketing and field use of this formulation should be produced so that they can be made available to common man readily and in less cost. If all these criteria are fulfilled then in near future we will be having a biofertilizer, which can replace the inorganic fertilizer by combination with organic fertilizer, or will reduce the use of inorganic fertilizer largely for a sustainable and eco-friendly environment.

KEYWORDS

- agriculture
- sustainable development
- microbes
- biofertilizer
- *Azospirillum*

REFERENCES

1. Vance, C. P. Symbiotic Nitrogen Fixation and Phosphorus Acquisition. Plant Nutrition in a World of Declining Renewable Resources. *Plant Physiol.* **2001,** *127*(2), 390–397.

2. Schultz, R. C.; Joe, P. C.; Richard, R. F. Agroforestry Opportunities for the United States of America. *Agroforestry Syst. (United States of America)* **1995,** *31*(2), 117–132.

3. Socolow, R. H. Nitrogen Management and the Future of Food: Lessons from the Management of Energy and Carbon. *Proc. Natl. Acad. Sci.* **1999,** *96*(11), 6001–6008.

4. Giller, K. E.; Ernst, W.; Steve, P. M. Toxicity of Heavy Metals to Microorganisms and Microbial Processes in Agricultural Soils: A Review. *Soil Biol. Biochem.* **1998,** *30*(10), 1389–1414.

5. Graham, P. H.; Vance, C. P. Nitrogen Fixation in Perspective: An Overview of Research and Extension Needs. *Field Crops Res.* **2000,** *65*(2), 93–106.

6. Schachtman, D. P.; Robert, J. R.; Sarah, M. A. Phosphorus Uptake by Plants: From Soil to Cell. *Plant Physiol.* **1998,** *116*(2), 447–453.

7. Hinsinger, P. Bioavailability of Soil Inorganic P in the Rhizosphere As Affected by Root-Induced Chemical Changes: A Review. *Plant Soil* **2001,** *237*(2), 173–195.

8. Vance, C. P. *Legume Symbiotic Nitrogen Fixation: Agronomic Aspects. The Rhizobiaceae;* Springer: Netherlands, 1998; pp 509–530.

9. Dommergues, Y. R.; Subba, R. N. S. Introduction of N_2-Fixing Trees in Non-N_2-Fixing Tropical Plantations. *Microb. Interact. Agric. For.* **2000,** *2,* 131–154.

10. Stamford, N. P.; Ortega, A. D.; Temprano, F.; Santos, D. R. Effects of Phosphorus Fertilization and Inoculation of *Bradyrhizobium* and mycorrhizal fungi on growth of *Mimosa caesalpiniaefolia* in an Acid Soil. *Soil Biol. Biochem.* **1997,** *29*(5), 959–964.

11. Unkovich, M. J.; John, S. P.; Paul, S. Nitrogen Fixation by Annual Legumes in Australian Mediterranean Agriculture. *Aust. J. Agric. Res.* **1997,** *48*(3), 267–293.

12. Unkovich, M. J.; John, S. P. An Appraisal of Recent Field Measurements of Symbiotic N_2 Fixation by Annual Legumes. *Field Crops Res.* **2000,** *65*(2), 211–228.

13. Stephens, J. H. G.; Rask, H. M. Inoculant Production and Formulation. *Field Crops Res.* **2000,** *65*(2), 249–258.

14. Date, R. A. Inoculated Legumes in Cropping Systems of the Tropics. *Field Crops Res.* **2000,** *65*(2), 123–136.

15. Triplett, E. W. *Diazotrophic endophytes*: Progress and Prospects for Nitrogen Fixation in Monocots. *Plant Soil* **1996,** *186*(1), 29–38.

16. James, E. K.; Olivares, F. L.; Baldani, J. I.; Döbereiner J. *Herbaspirillum*, an endophytic diazotroph Colonizing Vascular Tissue 3Sorghum bicolor L. Moench. *J. Exp. Bot.* **1997,** *48*(3), 785–798.

17. Boddey, R. M.; Da Silva, L. G.; Reis, V.; Alves, B. J. R.; Urquiaga, S.; Triplett, E. W. Assessment of bacterial nitrogen fixation in grass species. Prokaryotic nitrogen fixation: a model system for the analysis of a biological process. 2000, 705–726.

18. Malik, K. A.; Bilal, R.; Mehnaz, S.; Rasul, G.; Mirza, M. S.; Ali, S. Association of Nitrogen-Fixing, Plant-Growth-Promoting Rhizobacteria (PGPR) with Kallar Grass and Rice. *Plant Soil* **1997,** *194*(1–2), 37–44.

19. James, E. K. Nitrogen Fixation in Endophytic and Associative Symbiosis. *Field Crops Res.* **2000,** *65*(2), 197–209.
20. Neyra, C. A.; Döbereiner, J. Nitrogen Fixation in Grasses. *Adv. Agron.* **1977,** *29,* 1–38.
21. Bashan, Y.; Holguin, G. Azospirillum-Plant Relationships: Environmental and Physiological Advances (1990–1996). *Can. J. Microbiol.* **1997,** *43*(2), 103–121.
22. Schroder, M. Die assimilation des Lufstick stoffs durch einige Bacterian Lentrall. Bakteriol. Parasitendkd. *Infekt. Skr. Hyg. Abt.* **1932,** *2*(85), 177–212.
23. Becking, J. H. Pleomorphism in Azospirillum. In Azospirillum III. Springer Berlin Heidelberg, 1985, 243–262.
24. Breed, R. S.; Murrary, E. G. D.; Smith, N. R. *Bergey's Manual of Determinative Bacteriology;* Williams and Wilkins: Baltimore, Maryland, 1957; pp 254–257.
25. Krieg, N. R. The Taxonomy of the *Chemoheterotrophic spirilla. Annu. Rev. Microbiol.* **1976,** *30,* 303–325.
26. Okon, Y. Azospirillum As a Potential Inoculant for Agriculture. *Trends Biotechnol.* **1985,** *3*(9), 223–228.
27. De Coninck, K.; Horemans, S.; Randombage, S.; Vlassak, K. Occurrence and Survival of *Azospirillum spp.* in Temperate Regions. *Plant Soil* **1988,** *110*(2), 213–218.
28. Michiels, K.; Vanderleyden, J.; Van Gool, A. Azospirillum—Plant Root Associations: A Review. *Biol. Fertil. Soils* **1989,** *8*(4), 356–368.
29. Sumner, M. E. Crop Responses to Azospirillum Inoculation. In *Advances in Soil Science 12;* Springer: New York, 1990; 53–123.
30. Rennie, R. J.; DeFreitas, J. R.; Ruschel, A. P.; Vose, P. V. 15N Isotope Dilution to Quantify Dinitrogen (N_2) Fixation Associated with Canadian and Brazilian Wheat. *Can. J. Bot.* **1983,** *61*(6), 1667–1671.
31. Lima, E.; Boddey, R. M.; Döbereiner, J. Quantification of Biological Nitrogen Fixation Associated with Sugar Cane Using a 15N Aided Nitrogen Balance. *Soil Biol. Biochem.* **1987,** *19*(2), 165–170.
32. Hartmann, A.; Fusseder, A.; Klingmuller, W. Mutants of an Azospirillum Affected in Nitrogen Fixation and Auxin Production. *Exper. Suppl.* **1983,** *48,* 78–88.
33. Perrig, D.; Boiero, M. L.; Masciarelli, O. A.; Penna, C.; Ruiz, O. A.; Cassan, F. D.; Luna, M. V. Plant-Growth Promoting Compounds Produced by Two Agronomically Important Strains of *Azospirillum brasilense*, and Implications for Inoculant Formulation. *Appl. Microbiol. Biotechnol.* **2007,** *75,* 1143–1150.
34. Lira-Saldivar, R. H.; Hernandez, A.; Valdez, L. A.; Cárdenas, A.; Ibarra, L.; Hernández, M.; Ruiz, N. Azospirillum brasilense and Glomus intraradices Co-Inoculation Stimulates Growth and Yield of Cherry Tomato Under Shadehouse Conditions. *Phyton (Buenos Aires).* **2014,** *83,* 133–138.
35. Hazarika, B. N.; Ansari, S. Biofertilizers in Fruit Crops–A Review. *Agric. Rev.* **2007,** *28*(1), 69–74.
36. Mangmang, J. S.; Deaker, R.; Rogers, G. Early Seedling Growth Response of Lettuce, Tomato and Cucumber to *Azospirillum brasilense* Inoculated by Soaking and Drenching. *Hortic. Sci.* (Prague). **2015,** *42,* 37–46.
37. Ramakrishnan, K.; Selvakumar, G. Effect of Biofertilizers on Enhancement of Growth and Yield on Tomato (*Lycopersicum esculentum* Mill.). *Int. J. Res. Bot.* **2012,** *2*(4), 20–23.

38. Mahfouz, S. A.; Sharaf-Eldin, M. A. Effect of Mineral vs. Biofertilizer on Growth, Yield, and Essential Oil Content of Fennel (*Foeniculum vulgare* Mill.). *Int. Agro. Phys.* **2007,** *21*(4), 361.

39. Kumar, V.; Chandra, A.; Singh, G. Efficacy of Fly-Ash Based Bio-Fertilizers vs Perfected Chemical Fertilizers in Wheat (*Triticum aestivum*). *Int. J. Eng. Sci. Technol.* **2010,** *2*(7).

40. El-Lattief, A. *Improving Bread Wheat Productivity and Reduce Use of Mineral Nitrogen by Inoculation with Azotobacter and Azospirillum Under Arid Environment in Upper Egypt.* INTECH Open Access Publisher. 2012.

41. El-Lattief, E. A. Impact of Integrated Use of Bio and Mineral Nitrogen Fertilizer on Productivity and Profitability of Wheat (*Triticum aestivum* L.) Under Upper Egypt Conditions. *Int. J. Agron. Agric. Res.* **2013,** *3,* 67–73.

42. Banayo, N. P. M.; Cruz, P. C.; Aguilar, E. A.; Badayos, R. B.; Haefele, S. M. Evaluation of Biofertilizers in Irrigated Rice: Effects on Grain Yield at Different Fertilizer Rates. *Agriculture* **2012,** *2*(1), 73–86.

43. FIROUZI, S. Grain, Milling, and Head Rice Yields As Affected by Nitrogen Rate and Bio-Fertilizer Application. *Acta Agric. Slov.* **2015,** *105*(2), 241–248.

44. Nepolean, P.; Jayanthi, R.; Pallavi, R. V.; Balamurugan, A.; Kuberan, T.; Beulah, T.; Premkumar, R. Role of Biofertilizers in Increasing Tea Productivity. *Asian Pac. J. Trop. Biomed.* **2012,** *2*(3), S1443–S1445.

45. Fulchieri, M.; Frioni, L. Azospirillum Inoculation on Maize (Zea mays): Effect on Yield in a Field Experiment in Central Argentina. *Soil Biology and Biochemistry.* **1994,** *26*(7), 921–923.

46. Sudhakar, C.; Rani, C. S.; Knights, S. E.; Potter, T. D. Effect of Inclusion of Biofertilizers As Part of INM on Yield and Economics of Safflower (*Carthamus tinctorius* L). In *Safflower: Unexploited Potential and World Adaptability,* 7th International Safflower Conference, Wagga Wagga, New South Wales, Australia, 3–6 November, 2008. Agri-MC Marketing and Communication.2008, 1–5.

47. Swędrzyńska, D.; Sawicka, A. Effect of Inoculation with *Azospirillum brasilense* on Development and Yielding of Maize (*Zea mays* ssp. *saccharata* L.) Under Different Cultivation Conditions. *Pol. J. Environ. Stud.* **2000,** *9*(6), 505–509.

48. Pathak, A.; Chakraborti, S. K. Impact of Bio-Fertilizer Seed Treatment on Seed and Seedling Parameters of Maize (*Zea mays* L.). *The Bioscan* **2014,** *9*(1), 133–135.

49. Thumar, B. V.; Barad, A. V.; Neelima, P.; Bhosale, N. Effect of Integrated System of Plant Nutrition Management on Growth, Yield and Flower Quality of African Marigold (*Tagetes erecta* L.) cv. PUSA. *Asian J. Hortic.* **2013,** *8*(2), 466–469.

50. Singh, S. P. Effect of Bio-Fertilizer Azospirillum on Growth and Yield Parameters of Coriander (*Coriandrum sativum* L.) cv. Pant Haritima. *Veg. Sci.* **2014,** *40*(1), 77–79.

51. Faruq, G.; Shamsuddin, Z.; Nezhadahmadi, A.; Prodhan, Z. H.; Rahman, M. Potentials of Azospirillum Species for Improving Shoot and Root of a Malaysian Sweet Corn Variety (J 58) Under In Vitro Condition. *Int. J. Agric. Biol.* **2015,** *17,* 395–398.

52. Prabakaran, J.; Ravi, K. B. Interaction Effect of A. Brasilense and *Pseudomonas species.* A Phosphate Solubilizer on the Growth of *Zea mays.* In Microbiology Abstracts, XXXI Annual Conference of the Association of Microbiologists of India, TNAU, Coimbatore. 1991, Jan 23–26.

53. Radhakrishnan, K. C. Role of Biofertilizers in Cotton Productivity. In National Seminar Biofertilizer Production Problem and Constraints, TNAU, Coimbatore. 1996, Jan, 24–25.

54. Wani, S. P. Inoculation with Associative Nitrogen Fixing Bacteria: Role in Cereal Grain Production Improvement. *Indian J. Microbiol.* **1990**, *30*(4), 363–393.

55. Ghanti, S.; Sharangi, A. B. Effect of Bio-Fertilizers on Growth, Yield and Quality of Onion cv. Sukhsagar. *J. Crop Weed* **2009**, *5*(1), 120–123.

56. Naderifar, M.; Daneshian, J. Effect of Seed Inoculation with Azotobacter and Azospirillum and Different Nitrogen Levels on Yield and Yield Components of Canola (*Brassica napus* L.). *Iran. J. Plant Physiol.* **2012**, *3*(1), 619–626.

57. Javahery, M.; Rokhzadi, A. Effects of Biofertilizer Application on Phenology and Growth of Sunflower (*Helianthus annuus* L.) cultivars. *J. Basic Appl. Sci. Res.* **2011**, *1*, 2336–2338.

58. Rueda, D.; Valencia, G.; Soria, N.; Rueda, B. B.; Manjunatha, B.; Kundapur, R. R.; Selvanayagam, M. Effect of *Azospirillum spp.* and *Azotobacter spp.* on The Growth and Yield of Strawberry (*Fragaria vesca*) in Hydroponic System Under Different Nitrogen Levels. *J. Appl. Pharm. Sci.* **2016**, *6*(01), 048–054.

59. Yassin, M. A.; El Hassan, G. A.; Alsamowal, M. M.; Hadad, M. A. Effect of Organic Fertilizer Application Rate and *Azospirillum brasilense* Inoculum Level on Growth and Growth Components of Sweet Potato (*Ipomoea batatas* L). *J. Agric. Res.* **2016**, *2*(4), 52–62.

60. Maziyar, M.; Reza, A. M.; Hamid, M.; Mohammad, Z.; Mohsen, A.; Saeed, M. Response of Sunflower Yield and Phytohormonal Changes to Azotobacter, Azospirillum, Pseudomonas and Animal Manure in a Chemical Free Agroecosystem. *Ann. Biol. Res.* **2011**, *2*(6), 425–430.

61. Akbari, P.; Ghalavand, A.; Modarres, S. A.; Alikhani, M. A. The Effect of Biofertilizers, Nitrogen Fertilizer and Farmyard Manure on Grain Yield and Seed Quality of Sunflower (*Helianthus annus* L.). *J. Agric. Sci. Technol.* **2011**, *7*(1), 173–184.

62. Gopinathan, R.; Prakash, M. Effect of Vermicompost Enriched with Bio-Fertilizers on the Productivity of Tomato (*Lycopersicum esculentum mill.*). *Int. J. Curr. Microbiol. Appl. Sci.* **2014**, *3*(9), 1238–1245.

63. Senthil, K. R. Panneerselvam, A. Studies on the Effect of Different Native Strains of Azospirillum on Paddy (*Oryza sativa* L.). *J. Microbiol. Biotechnol. Res.* **2012**, *2*(6), 888–893.

64. Bashan, Y.; Ream, Y.; Levanony, H.; Sade, A. Nonspecific Responses in Plant Growth, Yield, and Root Colonization of Noncereal Crop Plants to Inoculation with *Azospirillum brasilense* Cd. *Can. J. Bot.* **1989**, *67*(5), 1317–1324.

65. Okon, Y.; Labandera-Gonzalez, C. A. Agronomic Applications of Azospirillum: An Evaluation of 20 Years Worldwide Field Inoculation. *Soil Biol. Biochem.* **1994**, *26*(12), 1591–1601.

66. Okon, Y. Azospirillum As a Potential Inoculant for Agriculture. *Trends Biotechnol.* **1985**, *3*(9), 223–228.

CHAPTER 7

YEAST: A MULTIFACETED EUKARYOTIC MICROBE AND ITS BIOTECHNOLOGICAL APPLICATIONS

NAYAN MONI DEORI[$], RACHAYEETA DEB[$], RIDDHI BANERJEE[$], and SHIRISHA NAGOTU*

*Organelle Biology and Cellular Ageing Lab, Department of Biosciences and Bioengineering, Indian Institute of Technology Guwahati, Guwahati, Assam 781039, India, Tel: +91-361-258-3209, Fax: +91-361-258-2249, *E-mail: snagotu@iitg.ernet.in*

CONTENTS

ABSTRACT

Microorganisms are valuable tools for biotechnology applications. Yeast, a eukaryotic microbe is an immensely attractive and versatile tool for such applications. The unique physiological and biological properties of various

[$] Equal contribution of all authors

yeast model systems are tapped for several biotechnological applications. In this chapter, we aim to provide an overview of our current knowledge on such diverse applications of various yeast models in avenues such as fundamental research, agriculture, environment and industry.

7.1   INTRODUCTION

Yeast is a eukaryotic, unicellular, saprophytic organism that lives in different habitats with great species diversity. The size of a yeast cell varies depending on the type of species and environment, typically measuring about 3–4 μm in width and 5–15 μm in length. Yeast exhibits varied and remarkable physiological properties that have facilitated its use in the biotechnology industry. Bacteria, mainly *Escherichia coli*, has long been regarded as the preferred host for several biotechnological applications such as production of recombinant proteins, amino acids, metabolites, and so forth.[70,107,124,154] Gradually over the years, the need for eukaryotic microbial host systems became evident. Yeast is the most desirable choice, as it offers the advantages of a unicellular organism (ease of cultivation and genetic manipulation) along with the protein processing ability of eukaryotes (protein folding and other post-translational modifications). Partitioning of proteins and metabolites into membrane-bound organelles is another important eukaryotic characteristic exhibited by yeast. Mitochondria and vacuoles increase the production rates of compounds like isobutanol and methyl halide respectively.[12,15] This increase in production levels occurs due to higher concentration of enzymes and increased availability of intermediates involved in the biosynthetic pathways.[12,15] Peroxisome is another important organelle greatly explored for heterologous protein production. The absence of protein-modifying enzymes in peroxisomes protects the heterologous proteins from undesirable modifications.[165] Proteins sensitive to proteolysis also remain safe from degradation in the peroxisomal matrix.[121] We aim to summarize the various applications of yeast in this chapter.

7.2   YEAST MODELS

Present day biotechnological applications hugely rely on various yeast models. *Saccharomyces cerevisiae* is the most extensively studied and commonly used yeast model. It is one of the earliest used microorganisms in brewing and baking industries.[72,137] Apart from this, it is also used in biofuel and recombinant protein production.[24,109] It is considered as a desirable host for several reasons such as non-pathogenicity, ease of cultivation, rapid growth rate, well-defined haploid and diploid states in its life cycle.[22,115] *S. cerevisiae* consists of 16 chromosomes (2n = 16) and its genome is completely sequenced, which allows easy genetic isolation, manipulation and mapping.[115] Furthermore, *S. cerevisiae* can be transformed with very stable 2-μm plasmid, which is used for the construction of yeast episomal vectors (YEps). These vectors are extensively used for the expression of recombinant proteins, as they have a high copy number per cell and high transformation frequency.[59] However, the use of *S. cerevisiae* also has some limitations; for example, it is not resistant to stress caused due to pH, temperature and so forth and is thus not suitable for large-scale production purposes. These major drawbacks led to the search for alternate host systems, which are referred to as nonconventional yeast models. Some of the important nonconventional yeast models include *Pichia pastoris, Hansenula polymorpha, Yarrowia lipolytica, Kluyveromyces lactis, Candida boidinii, Schizosaccharomyces pombe, Zygosaccharomyces rouxii*.

P. *pastoris* is a methylotrophic yeast which uses methanol as sole source of carbon and energy. It has advantages over *S. cerevisiae* as a host for heterologous protein expression in two different aspects. First, P. *pastoris* can be easily cultured at high cell densities because it does not produce ethanol extracellulary during fermentation and thus does not build up toxicity in the medium, unlike *S. cerevisiae*, leading to large-scale production of recombinant proteins.[115,166] Second, during protein production in this yeast, several important parameters such as pH, aeration, carbon source, feed rate can be controlled. In addition, the alcohol oxidase 1 (AOX1) promoter in P. *pastoris* is more efficient and tightly regulated, as compared to the most often used galactosidase 1 (GAL1) promoter in *S. cerevisiae*. AOX1 promoter is highly repressed in the absence of methanol, unlike GAL1 promoter, which helps in regulated overexpression of desired recombinant proteins.[28,181] Another methylotrophic yeast *H. polymorpha,* is excellent for large-scale production of mammalian proteins and has certain advantages over P. *pastoris*. This yeast is thermotolerant with relatively high optimal growth temperature of 37–43°C as compared to the growth temperature for P. *pastoris* which is 30°C. It also has higher growth rates on simple defined media and is the only methylotrophic yeast with nitrate assimilation pathway.[60,93,186]

Y. lipolytica is a nonconventional dimorphic yeast which is unicellular in glucose medium and exists as mycelia in defined media containing olive oil and casein and also as a mixture of both the forms in complex medium.[121] It is preferred over the conventional model system *S. cerevisiae,* with respect to production of human therapeutic proteins. Unlike the post-translational translocation in *S. cerevisiae,* the newly synthesized proteins in *Y. lipolytica* are co-translationally translocated into the endoplasmic reticulum, similar to that in higher eukaryotes.[121] Müller and co-workers conducted a study in different yeast models, such as *S. cerevisiae, Y. lipolytica, H. polymorpha, K. lactis* and *S. pombe,* comprising various criteria such as transformation efficiency, growth rate of recombinant host, degree of glycosylation and the amount of recombinant proteins produced by the host.[117] In this study, *Y. lipolytica* emerged as the most desirable host for the expression and production of heterologous proteins.[117] *Y. lipolytica* has an inherent capability of secreting various extracellular enzymes such as lipases, proteases, ribonucleases, phosphatases and esterases naturally, which can be used for hydrophobic substrate utilization in detergent, environmental, food and pharmaceutical industries.[13,36] In addition, this yeast model can also be used in several other industrial processes such as single-cell protein (SCP) and secondary metabolite production, conversion of fatty acids to alcohols, and so forth. All these properties highlight *Y. lipolytica* as an interesting alternate host system for expression studies.

K. lactis is another widely used yeast model which can be utilized as a potential host for recombinant protein production, especially low-value products, due to its ability to grow on cheap substrates like lactose.[187] It has been largely used for lactase (β-galactosidase) production in the food industry and as an expression system for manufacturing milk-clotting enzyme bovine chymosin.[21,77,185]

There are some other yeast models, apart from the above, which can survive in extreme environments and thus are interesting candidates to be explored in order to discover novel applications in industry. Some of these extremophiles include psychrophilic yeast species of *Pseudozyma* and *Cryptococcus,* halotolerant and osmotolerant marine yeast *Debaromyces hansenii,* and so forth. Psychrophilic yeast are mostly found in cold environments like polar regions and marine deep waters. They grow at temperatures below 5°C and exhibit no growth above 20°C. Their optimum and maximum growth temperatures are ≤15 and ≤20°C, respectively.[67]

These psychrophilic yeast produce cold-adaptive enzymes such as lipase, amylase, protease, and so forth, which are structurally stable and have high specific activities at low temperatures.[67,163] *D. hansenii* can survive in hyper-saline environments and can tolerate salinity up to the levels of 4 M NaCl, unlike *S. cerevisiae* whose growth is inhibited at 1.7 M NaCl.[5] Such unique characteristics of these extremophiles can be of immense economic importance in the food and dairy industries.[5,158]

7.3   APPLICATIONS

7.3.1   FUNDAMENTAL RESEARCH

The yeast *S. cerevisiae* is considered as an ideal eukaryotic model organism for several biological studies. *S. cerevisiae* genome is usually employed as a reference in deciphering conserved functions of various proteins across species.[156] The stability in the haploid and diploid state of *S. cerevisiae* strains provides the ease of convenient isolation of recessive mutations in haploid strains and enabling complementation tests in diploid strains. Homologous recombination in yeast has led to the development of techniques for introducing mutations in normal wild type genes.[71] Furthermore, *S. cerevisiae* has made a great contribution in understanding the fundamental concepts of cell biology in higher eukaryotes like humans. A number of engineered yeast models have emerged as a powerful technique to understand the molecular mechanism of complex human diseases.[82] Basic cellular processes such as heterochromatin formation and maintenance, gene silencing and homologous recombination can be studied by *S. cerevisiae* mating-type system.[44] *S. cerevisiae* has served as an important model organism in understanding eukaryotic centromere biology due to its simple and conserved functions.[110] The genomic information present in *Saccharomyces* Genome Database (SGD) provides a comprehensive integrated database for the discovery of functional relationship between fungi and higher eukaryotes.[44] Yeast species such as *S. pombe* and *P. pastoris* have contributed in unraveling the etiology and pathogenesis of several neurodegenerative diseases such as Parkinson's, Alzheimer's and Huntington's disease.[128] These yeast models have laid down the platform to study proteins involved in neurodegeneration.[128] and also to understand the events leading to protein aggregation and molecular mechanisms that result in cellular toxicity.[57] This enables researchers to successfully perform genetic and compound screening to identify novel candidate therapeutic targets and drugs.[114]

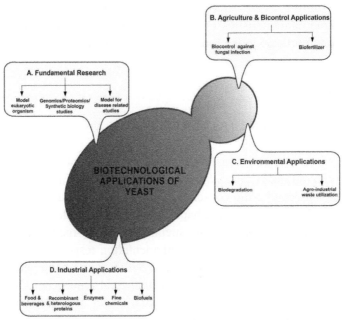

FIGURE 7.1 Schematic illustration of diverse biotechnological applications of yeast. (A) In fundamental research, yeast serves as a model organism to study various conserved biological mechanisms in eukaryotes. It is a valuable tool for genomic and proteomic studies, has also helped unravel several genetic features involving neurological disorders and ageing related diseases. (B) In agriculture and biocontrol applications, yeast is extensively utilised as a biofertilizer for increasing soil fertility and crop productivity. As a biological agent it also renders protection to crops from pests and plant diseases. (C) In environmental applications, yeast helps to maintain environmental sustainability by degrading harmful chemicals present in industrial and agricultural wastes. Other applications also include synthesis of lipids from agro-industrial wastes. (D) Yeast also plays a prominent role in various industrial applications, which include production of fermented food and beverages, vitamins, food additives in food industry, recombinant/ heterologous proteins for pharmaceutical field, several enzymes, fine chemicals and metabolites, biofuels, etc.

Yeast is also considered as an indispensable organism to study the events related to cancer and cell cycle progression. This is due to similarity between the genes involved in controlling cell cycle in yeast and those involved in the formation of tumour cells. It is worth mentioning that at the onset of characterization of the molecular events related to cancer, many of the mammalian cell cycle genes were first identified in yeast.[133] The DNA repair system in *S. cerevisiae* is also similar to the one in humans, which provides a platform to discover novel targets for cancer diagnosis.[65] The conserved features of autophagy events in many eukayotic organisms also render yeast, an important model system to study these events. Yoshinori Ohsumi, a yeast researcher was awarded the Nobel Prize for his contributions in the field of autophagy. Yeast, therefore, serves as an important model to understand the diseases related to autophagic dysfunctions such as diabetes, liver and heart diseases, cancer, neurodegeneration, and so forth.[38]

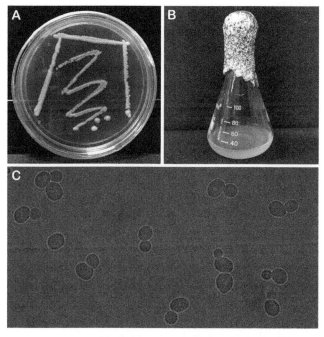

FIGURE 7.2   Various forms of *S. cerevisiae* as observed in laboratory conditions. (A) Wild type (BY4742) *S. cerevisiae* streaked into a freshly prepared YPD-agar plate. Distinct colonies are observed upon incubation of the streaked plate overnight at 30°C. (B) Liquid culture of wild type (BY4742) *S. cerevisiae* in YND media (selective). Turbidity in the medium represents cell growth resulting from overnight incubation maintained at 30°C at 200 rpm. (C) Budding yeast (*S. cerevisiae*) cells observed under a microscope using a 100X objective. Cells were cultured in liquid medium until they reached exponential growth phase and were visualized using a microscope.

7.3.2   AGRICULTURE AND BIOCONTROL APPLICATIONS

One of the most important contribution of yeast in agriculture is that it has antagonistic activities against undesirable bacteria and fungi which are a potential threat to crop yields.[69] They render their antagonism for other microorganisms by competing for nutrients, producing high concentrations of ethanol and by secretion of antibacterial compounds. They also release antimicrobial compounds such as the killer toxins or mycocins.[64,174] Yeast in combination with other microorganisms was found to enhance the yield and growth of plants.

7.3.2.1   AS BIOFERTILIZER TO ENHANCE CROP PRODUCTIVITY AND GROWTH

Biofertilizers are low-cost and effective substances, composed of microorganisms which enhance the fertility and provide nutrients to the soil. These are nontoxic and biodegradable and do not pollute the environment and are therefore preferred over chemical fertilizers in agricultural practices. These fertilizers are generally applied to seeds/seedlings or inoculated in soil. A specific strain of microorganisms such as *Rhizobium, Azotobacter, Azospirillum*, and so forth is used while applying the biofertilizer to seeds, as these strains are capable of growing in association with plant roots. Formulation of a biofertilizer involves the direct introduction of microbes to the soil, where it has to compete with the already existing microbial population.[30]

Yeast is largely used as biofertilizer in agriculture because of its safety to environment and human usage.[2] Yeast that resides in the rhizosphere of soil is directly or indirectly responsible for enhancement of plant root growth.[120] Yeasts such as *Candida, Geotrichum, Rhodotorula, Saccharomyces* and *Williopsis* are capable of nitrification of ammonium to nitrate.[7] Among the above, *S. cerevisiae* is considered to be the potential yeast for enhancement of crop growth. It was reported that *Saccharomyces* sp. can oxidize elemental sulphur to produce phosphate and sulphate thereby enhancing the amount of inorganic nutrients required for the plant growth.[46] A study conducted on sugar beet by Agami and colleagues reported that the application of *Kluyveromyces walti, Pachytrichospora transvaalensis* and *Saccharomyces catagenesis* considerably enhanced the concentration of photosynthetic pigments, total soluble proteins, sugars and sucrose of sugar beet plant.[2] This leads to a considerable increase in the growth and productivity of the plant.

7.3.2.2   AS BIOCONTROL AGAINST PLANT INFECTION

Biological control in agriculture is the use of natural or genetically modified microorganisms, genes or gene products to reduce the effect of pests and diseases on plant growth. Biocontrol agents can also be used in combination with biofertilizers, thereby augmenting the soil microflora to promote plant growth and root development. This leads to reduction in usage of other harmful pesticides and toxic chemicals in agriculture. Unlike viruses and bacteria, fungi need not to be ingested by the host organism to sustain their effective biocontrol activity. Rather, the invasion of fungi through the cuticles of insects can propel its antagonism towards its host and is therefore preferred over all other biocontrol agents.[84]

Yeast has a natural ability to grow quickly on leaf, fruit and flower surfaces where the sugar content is relatively high. They dominate these regions and compete with other microorganisms for space and nutrients, thereby acting as a biocontrol against other microorganisms.[145] To justify the biocontrol ability of yeast, yeast strains such as *Torulaspora globosa and Candida intermedia* isolated from sugar cane and different plant parts of maize such as rhizosphere, leaves and stalks were found to inhibit mold growth responsible for the anthracnose disease in these plants.[145] The role of *S. cerevisiae* as a biocontrol agent was demonstrated against the *Fusarium* infection of sugar beet plants.[160] The damping-off symptoms caused by *Fusarium oxysporum* F4 in sugar beet was inhibited by *S. cerevisiae* resulting in improved survival of the plant.[160] Another very important antagonistic characteristic of yeast against other microorganisms includes the secretion of antimicrobial compounds like 'mycocins' or killer toxin. Mycocins are extracellular proteins or glycoproteins which have antimicrobial activity against yeast strains that are closely related to the producer strain. The antimicrobial property of mycocins does not affect the producer strain, as they consist of a protective factor against the mycocin.[64] Mycocins disrupt the cell membrane and prevent the formation of β-1,3-glycan which is a cell wall component. This results in the formation of channels that lead to leakage of ions from the cytoplasmic membrane.[155] The killer toxin blocks DNA synthesis thereby interrupting cell division.[79] The antagonism of yeast against bacteria was first reported when an amine extract of yeast was found to inhibit the growth of *Staphylococci and E. coli*.[190] Growth of bacteria that causes spoilage of beer was reported to be inhibited by the yeast *Kluyveromyces thermotolerans*.[19] Extracellular glycolipids produced by *Candida bombicola* were reported to exhibit antibacterial activity against *Staphylococcus aureus*.[26]

7.3.3   ENVIRONMENTAL APPLICATIONS

Removal and treatment of wastes from industrial, domestic, municipal or agricultural sources[101] is a major component of environmental biotechnology. The major focus is on the degradation of xenobiotic compounds released by the industrial wastewater streams. This degradation of industrial waste effluents by aerobic and anaerobic treatment involves microorganisms. Yeast has potential application in the field of bioremediation. The yeast *Y. lipolytica* is known to degrade hydrocarbons such as alkanes, fatty acids, fats and oils.[203] *S. cerevisiae* also has the potential to degrade toxic pollutants like arsenic from industrial wastes.[169] Apart from this, yeast also has the ability to metabolize sugars into ethanol. The efficient production of ethanol from inexpensive feedstocks by yeast may serve as an alternative to vegetable oils used in the biodiesel industry.[99]

7.3.3.1   BIODEGRADATION OF HARMFUL CHEMICALS

A blended consortium of microorganisms such as bacteria, yeast and fungi mainly causes the degradation of harmful chemicals that contaminate soil and water.[173] Although bacteria is considered efficient in biodegradation of crude oil, there are reports that yeast can also be utilized in its biodegradation.[91] The yeast *Candida tropicalis* was reported in clearing oil spills at tropical soils.[78] *Y. lipolytica,* which is naturally found in oil-polluted environments, is used in the bioremediation of environments polluted with organic pollutants such as aliphatic and aromatic compounds. Since *Y. lipolytica* has a natural tendency to degrade hydrophobic substrates,[180] different strains of this fungus are widely used in the bioremediation of soils contaminated with diesel oil[105] and also for the biodegradation of hydrocarbons.[106] *Y. lipolytica* when used in combination with *Bacillus subtilis* or *Pseudomonas* sp. displays a considerable positive effect on the crude oil degradation process.[73,86] Biosurfactants produced by bacteria, yeast or fungi can remove hydrophobic organic compounds as well as heavy metals from wastewater effluents or polluted soils.[118] Higher biodegradability and low toxicity make biosurfactants a better choice as compared to synthetic surfactants. It was reported that the biosurfactant produced by the yeast *Candida lipolytica* can reduce the permeability of municipal solid waste leading to the recovery of heavy metals and organic hydrocarbons.[146]

7.3.3.2   AGRO-INDUSTRIAL WASTE UTILIZATION

Agricultural wastes are extensively used as raw materials for microbial lipid production. Oleaginous yeast are capable of accumulating up to 70% of their cellular dry weight as lipids.[4] Agricultural residues such as sugarcane bagasse, corn stover, corn stacks, and rice and wheat straw are the most common feedstocks with high lignocellulosic biomass content required for lipid production by microorganisms.[41,40] The cellulose unit present in the lignocellulosic biomasses when subjected to enzymatic treatment generates a hydrolysate that contains sugars in the form of mono and oligosaccharides. Oleaginous yeast can utilize these hydrolysates as a source of organic carbon to synthesize lipids in their cellular compartment.[9,129] Yeasts belonging to the genus *Rhodosporidium, Rhodotorula, Yarrowia, Cryptococcus, Candida, Lipomyces* and *Trichospora* are oleaginous. It was reported that *Rhodosporidium* produces the highest amount of lipid.[100] The lipid produced by oleaginous yeast is similar to vegetable oil in fatty acid composition. The oleaginous yeast is, therefore, considered as potential feedstock for the production of biodiesel and is hence economically and environmentally viable tool for biodiesel production.[139,140]

Waste generated by industrial activities can also be treated by oleaginous microorganisms.[116] Industrial wastes from pulp and paper, palm oil and natural rubber contains considerable amount of organic materials as contaminants.[3] Biological treatment of such contaminants utilizes microbes like bacteria and yeast. The ease of harvesting and accumulation of high lipid content make oleaginous fungi preferred microbes for such processes.[196] This microbial treatment process can be ultimately integrated in the production of biodiesel by oleaginous yeast utilizing such organic materials from the industrial wastewater effluents.[113] The municipal wastewater containing organic carbon can be utilized for yeast cultivation.[33] It was reported that when *Lipomyces starkeyi* was inoculated in fishmeal water supplemented with sewage water and glucose, higher cell growth and oil yield were obtained.[10]

7.3.4   INDUSTRIAL APPLICATIONS

Yeast has been widely studied as a model organism for industrial biotechnology since the past few decades due to its attractive physiological attributes such as production and extraction of intracellular as well as extracellular enzymes, recombinant proteins and fermentation products which are of great commercial significance.

7.3.4.1   FERMENTED FOOD AND BEVERAGE INDUSTRY

Fermentation is an essential step in the food and beverage industry. It involves the conversion of different carbohydrates into alcohols and organic acids under anaerobic conditions by microorganisms such as bacteria, yeast and algae. Food fermentation helps in enrichment of the diet by imparting diverse flavours and textures to raw food, preservation of food, fortification of food substrates and elimination of specific spoilage compounds. *S. cerevisiae* is known widely for the production of carbon dioxide required for leavening doughs in baking industry and generation of alcohols useful for the brewing industry.[98] Other strains belonging to non-Saccharomyces yeast such as *K. lactis, Candida, Zygosaccharomyces* are also used in the food industry.[17]

7.3.4.1.1   Fermented Meat, Milk and Cereal Products

Flavour development, curing, fermentation and enhancing the shelf life of meat products can be achieved by different yeast species.[123] *D. hansenii* and *Candida famata* are extensively used in the production of fermented sausages and cured hams whereas various other species belonging to the genera *Candida, Cryptococcus, Pichia, Rhodotorula, Trichosporon, Metschnikowia* and *Yarrowia* are also used in meat processing.[138] Zhang and colleagues reported that diet for broiler chickens containing yeast cells increased the meat tenderness and quality.[198] Enzymes from psychrophilic yeast such as *Rhodosporidium* and *Rhodotorula* are used for the tenderization of meat which provides the desired softness, fat content and enhanced flavours of stored meat.[25]

In the dairy industry, processing of animal milk followed by fermentation and generation of milk products such as buttermilk, yogurt, cheese, generally involves microorganisms like lactic acid bacilli. However, yeast is also involved in the production of distinctive flavours during ripening and aging of cheese.[54]

Some well-known examples of yeast are *D. hansenii* and *Kluyveromyces marxianus* in cheese flavouring;[11,96,157] *Y. lipolytica and D. hansenii* as co-starters in cheese making;[48] *Y. lipolytica, D. hansenii, K. marxianus, Saccharomyces exiguus, Candida milleri, Candida humilis, Candida krusei, Pichia anomala, Torulopsis* sp. in bread flavouring;[56,131,132] and *K. thermotolerans* and *Torulaspora delbrueckii* in the preparation of frozen dough breads.[8,81] During commercial bread production, *S. cerevisiae* transforms the fermentable carbohydrates present in the flour to ethanol and carbon dioxide. Enzymes present in *S. cerevisiae* help in protein hydrolysis leading to formation of free amino acids that further aid in the production of various aromatic compounds. Yeast along with lactic acid bacilli aids in sourdough fermentation, which gives a special sour taste to fermented breads. Traditional wheat-based, corn-based and sorghum-based fermented foods also involve the use of different yeast species.[20]

7.3.4.1.2 *Beverages: Alcoholic and Non-alcoholic*

Alcoholic beverages such as wine, beer or distilled spirits are fermented liquors that contain ethyl alcohol. Various microbes especially yeast, which can ferment sugars and withstand high ethanol concentrations are immensely useful in the production of alcoholic beverages. *S. cerevisiae* and *Saccharomyces bayanus* largely improve the sensory properties of wine during fermentation.[55,134] They maintain wine quality due to the production of profuse amounts of aromatic secondary metabolites.[176] Several yeasts such as *Candida tropicalis, Candida glabrata, Geotrichum penicillatum, Geotrichum candidum, Candida stellata, Schizosaccharomyces pombe, Torulaspora delbrueckii and Zygosaccharomyces bailii* along with lactic acid bacilli induce fermentation and impart significant taste to sweet, carbonated as well as special non-alcoholic beverages such as kombucha, boza, kvass, and so forth. [14,85,178]

7.3.4.1.3   Additional Applications in Food Processing

Some other important food applications where yeasts of various genera are utilized extensively are: in the production of different indigenous Asian fermented foods such as *naan*, *kulcha*, *dhokla* and pancakes such as *idli*, *dosa*, *appam*; fermentation of cocoa beans and soy sauce; development of flavours in chocolate production and additives in yoghurt fermentation.[6,168,199]

7.3.4.1.4   Food Grade Yeast

Food grade yeasts (*Saccharomyces*, *Candida*, *Kluyveromyces*, *Pichia* and *Torulopsis*) which produce SCPs are considered safe for human consumption worldwide. They are used as sources of high nutritional value proteins, enzymes and vitamins, with applications in the health and food industry.[141] For an overview of such yeast model systems, refer to Table 7.1.

TABLE 7.1   List of Various Food Grade Yeasts and Their Applications.

Name	Species	Potential use	References
Baker's yeast	*Saccharomyces cerevisiae*	Active dry yeast, instant dry yeast products	[17]
Brewer's yeast and nutritional brewer's yeast	*Saccharomyces uvarum, S. cerevisae* (Grown on cane sugar/beet molasses)	Beer bottom (lager) and top (ale) fermenters	[49,63, 189]
	S. cerevisiae (grown on malted barley), *Saccharomyces pastorianus*	Source of proteins containing essential amino acids, vitamin B3, B6, B9, phenolic compounds like gallic acid, catechin, etc.	
	Whole yeast/dried *S. cerevisiae*	Commercial food substitute for fish, fishmeal formulations, nutrient source for fastidious microbes like *Lactobacillus* sp.	

TABLE 7.1   *(Continued)*

Name	Species	Potential use	References
Wine yeast	*S. cerevisiae, Saccharomyces bayanus, S. uvarum, Saccharomyces oviformis, Saccharomyces carlsbergensis, Saccharomyces diastaticus*	Fermentation of sparkling wine and champagne	[17, 55, 95]
Distiller's yeast	*S. cerevisiae, Pichia galeiformis, Candida lactiscondensi, Hanseniaspora osmophila*	Production of brandy, whisky, rum and tequila	[183]
Probiotic yeast	*S. cerevisiae, Saccharomyces boulardii, Candida pintolopesii, Candida saitoana, Issatchenkia orientalis*	Probiotic approaches to treat diarrhoea, animal feed as probiotic additives	[35, 56, 88, 102]
Torula yeast	*Candida utilis*	Dietary food supplements, infant food, meat substitute and meat product, spices, sauces	[94, 192, 144]
Whey yeast	*Kluyveromyces lactis, Kluyveromyces marxianus*	Production of yeast biomass, Kefir, baby food and single-cell protein	[103]

The table gives a list of different food-grade yeasts and the various yeast models that belong to various categories along with their manifold uses in fermentation of distilled spirits, as leavening agents in baking, dietary supplements, nutrient sources and whey-based food products, and so forth.

7.3.4.2   *VITAMINS AND ADDITIVES IN FOOD PROCESSING*

In the modern food system, various naturally occurring as well as synthetic generally recognized as safe (GRAS) chemicals are used as food preservatives, thickeners, stabilizers, emulsifiers, acidity regulators and anti-caking agents. These food additives are an integral part of food processing. Yeast-derived nucleotides such as monosodium glutamate, 5′ guanosine monophosphate and 5′ inosine monophosphate are well-established food processing agents.[29] β-glucans, a group of β-D glucose polysaccharides are present in the yeast cell walls. These β-glucans extracted from brewer's yeast cells are often used as thickening agents because of their unique water-retaining, emulsifying and stabilizing characteristics.[29,66,135,179]

Vitamins are important organic biocompounds that are needed in substantial amounts by the human body as cofactors for enzyme biocatalysis, precursor molecules, antioxidants, regulators of cell and tissue growth and maintenance of mineral metabolism. Yeast is currently being explored to produce B-complex vitamins. Production of pyridoxine, that is vitamin B6.[191] and riboflavin from xylose and ribitol sources by *Pichia* sp. were studied by Sibirny and Boretsky.[164] Yeast belonging to genera *Rhodotorula, Sporobolomyces, Rhodosporidium* and *Sporidiobolus* produce carotenoids (beta carotene, gamma carotene, torulene and torularhodin) which have wide industrial applications as vitamin A precursors, singlet oxygen species scavenger, or potent antioxidant molecules. Carotenoids are also useful as bio-colouring agents, additives in food, beverage, meat products, poultry and fish biotechnological industries.[104]

Compounds like p-coumaroylcoA (used in the production of aromatic ketones like chalcones) and resveratrol (for lowering cholesterol levels) which have potent antioxidant properties are produced from metabolically engineered *S. cerevisiae*.[16]

7.3.4.3   RECOMBINANT/HETEROLOGOUS PROTEIN PRODUCTION

Bioengineered yeast is utilized as an efficient factory for the generation of several recombinant proteins of potent industrial importance (Table 7.2). Although more than 40 different proteins are produced by the major heterologous host *S. cerevisae*, other yeasts such *P. pastoris, H. polymorpha, Y. lipolytica, K. lactis, S. pombe* and *Arxula adeninivorans* are also used for their potential to produce such proteins.[27, 39, 43, 50, 74, 108, 130] Yeast is considered a better expression house over Chinese hamster ovarian cells and/ or bacteria because of their high growth rate and yield of the recombinant proteins. Efficient secretion, stability, durability, better post-translational modification strategies of the expressed proteins and low cost are the added advantages.[39] Yeast species such as *S. cerevisiae, P. pastoris, H. polymorpha, S. pombe, K. lactis* and *Y. lipolytica* are utilized for the production of antibodies.[52,58] Gidijala and colleagues reported the production of biologically active penicillin when complete penicillin biosynthetic pathway was introduced into the yeast *H. polymorpha*.[62]

TABLE 7.2   Various Important Recombinant Proteins Produced in Yeast.

Host	Product	Importance in	Reference
S. cerevisiae	Insulin	Diabetes	[182]
	Hepatitis A vaccine	Hepatittis A	
	Hepatitis B vaccine	Hepatitis B	
	Human growth hormone	Dwarfism	
	Erythropoietin	Renal disorder	
	Glucagon	Hypoglycaemia	
	Hirudin	Anticoagulant	
	Platelet-derived growth factor	Anaemia	
K. lactis	Interferon-α	Cancer therapy, treatment of acute and chronic hepatitis B and chronic hepatitis C	[31, 47, 53, 75, 92, 119, 143, 151, 171, 175]
	Human interleukin 1-β		
	Macrophage colony stimulating factor (M-CSF)		
	Insulin precursors, human serum albumin, antibodies	Autoimmune disorders, hematopoietic system Pharmaceutical applications	
P. pastoris	Recombinant human serum albumin	Inhibition of bacterial growth	[89, 112]
	Mytichitin-A (antimicrobial peptide)	Physiological application	

Recombinant proteins with numerous applications in therapeutics, basic research and biotechnology can be produced in yeast. The table highlights some of these proteins with relevant industrial applications.

7.3.4.4 YEAST-DERIVED ENZYMES, FINE CHEMICALS, METABOLITES AND ORGANIC COMPOUNDS

Enzymes are important biocatalysts that have diverse industrial applications. Recombinant DNA techniques have allowed the use of microorganisms especially yeast (Table 7.3) as enzyme-yielding factories. Different commercial enzymes can be grouped into four classes based on their applications, namely—(i) technical enzymes (detergent, paper, pulp, leather industry); (ii) food enzymes (dairy, wine, juice, baking industries); (iii) feed enzymes (animal feed); (iv) diagnostic enzymes (medical and drug industries).[1,182]

TABLE 7.3   Industrial Enzymes from Yeast.

Yeast	Enzyme	Applications	Reference
Kluyveromyces sp.	Chymosin, inulinase, lactase, β-galactosidase, phospholipase, chitinase, xylanase	Food and feed industry, cheese-making, production of single-cell proteins, fuel, wood pulp processing	[171]
Saccharomyces sp.	Chymosin	Milk clotting activity	[34 111, 167]
	α-galactosidase	Enzyme replacement therapy for Fabry disease	
	Invertase	Confectionaries, food industries and pharmaceuticals	
Candida sp.	Lipase	Food processing flavours	[37, 170, 182]
	Acid lipase	Wastewater degreasing	
	Inulinase	Bioremediation therapeutic	
	Invertase	Detergent	
	Phenylalanine dehydrogenase	Oleochemical and food industries	
		Food, pharmaceuticals	
		Confectionery industry	
		Pharmaceuticals	
Yarrowia sp.	Lipase,	Detergent	[36]
	Protease,	Food, pharmaceutical, and environmental industries	
	RNAase,	Production of high fructose syrups and fuel ethanol	
	Phosphatases		
	Inulinase	Food industry, treatment of selected types of haemopoietic diseases	
	Asparaginase		
	Laccase		
		Biorefineries	
Zygosaccharomyces sp.	L-glutaminase	Food industry	[87]

TABLE 7.3 *(Continued)*

Yeast	Enzyme	Applications	Reference
Psychrophilic yeast: *Cryptococcus cylindricus, Mrakia frigida* *Candida antartica* *Kluyveromyces* sp *C. lipolytica* *Cystofilobasidium capitatum*	Psychrophilic enzymes: Pectinase, amylase, protease, lipase, lactase	Biocatalysts for low-temperature reactions in industry, Fermentations	[67]

Yeast can be bioengineered for the production of a diverse group of enzymes. The table gives a comprehensive list of various yeast models used for the industrial production of enzymes and the various enzymes produced by these organisms and their applications.

Fine chemicals and metabolites are active ingredients and building blocks of many advanced chemical compounds. These include antibiotics, amino acids and nucleotides that are often synthesized by bioconversion procedures using microorganisms, as they are difficult to produce by chemical syntheses. Many pure, small compounds and metabolite-based products relevant to the pharmaceutical, biopharmaceutical, agrochemical, biotechnological and food industry are produced in different strains of recombinant yeast cells (Table 7.4).[83,122,130,188,193] Sakai and colleagues showed that methylotrophic yeasts (e.g. *C. boidinii*) are efficient producers of formaldehyde, an important organic compound from methanol.[150]

TABLE 7.4 Chemicals and Organic Acids Produced by Yeast.

Yeast	Chemicals	Potential use	Reference
Yarrowia lipolytica	Succinic acid, citric acid, Isocitric acid, α-ketoglutaric acid	Building block chemicals, health products, food and beverage industry, acidulant	[18, 51, 148, 153, 197, 182]
Candida glabrata	Pyruvic acid	Chemical, drug and agrochemical industries	[200]

TABLE 7.4   *(Continued)*

Yeast	Chemicals	Potential use	Reference
S. cerevisiae	Flavonoids, stilbenoids, phenylpropanoid-derived natural product, genistein, p-coumarin, circumin, resveratrol, naringenin, methylated flavonoids, etc.	Anti-infective agents, anti-cancerous properties, phytoestrogen activity	[45, 76]
Candida magnolia	Erythritol	Artificial sweetener, food additives	[61]
Candida glycerinogenes	Glycerol	Sweetener, humectants, pharmaceutical formulation	[202]
Xanthophyllomyces dendrorhous	Astaxanthin	Food colourant	[195]
Zygosaccharomyces rouxii	D-arabitol	Food industry	[149]

Yeast models are extensively used for the industrial production of a number of heterogeneous chemicals and organic acids. Table 7.4 lists various such useful chemicals that have a role in food, pharmaceutical, agrochemical and other industries.

7.3.4.5   BIOFUELS

Biofuels are liquid fuels derived from biomass materials such as plant and animal waste matter and are commonly used as transportation fuels. Medium-chain and long-chain n-alkanes are the major constituents of petroleum produced by a variety of algae, bacteria and marine yeast. The relatively easy conversion of fermentable sugars like glucose, sucrose, and so forth, into ethanol by *S.cerevisiae* and several other yeasts acting on sugarcane, molasses or other raw materials helps in the production of bioethanol and biodiesel. *S. cerevisiae* finds its application in the biofuel industry as a robust strain due to several factors such as high osmotolerance, acidic pH tolerance and permissive insertion of recombinant DNA through homologous recombination.[24] *S. cerevisiae* is bioengineered to produce several advanced biofuels such as 1-Butanol,[172] isobutanol[23,32,90,97] which are gasoline substitutes and better than ethanol. Fatty acid ethyl esters (FAEEs) and biodiesel[162] are useful as diesel substitutes along with the sesquiterpenoids like farnesene,[142,184] bisabolane.[127] Amorphadiene and other isoprenoids are proposed as alternative jet fuels.[126,127,147,194] Biodiesel can be used together with fossil diesel fuel or as diesel engine fuel in pure form. Nontraditional yeasts such as *Candida shehatae, Pachysolen tannophilus, Pichia stipitis* and a few marine yeasts also produce bioethanol.[152,159] *K. marxianus* can ferment inulin (naturally occurring polysaccharides produced by many types of plants, chemically found in dietary fibres as fructans) to ethanol from plant feedstock and cheese whey to bioethanol.[125] The ability of *Y. lypolytica* to accumulate lipid is harnessed for biofuel production.[177] Zhou and colleagues reported that compartmentalization of fatty acid biosynthesis pathways within yeast peroxisomes can maximize biofuel production.[201] Certain nonconventional yeast species such as *Z. rouxii, Dekkera bruxellensis, Pichia kudriavzevii* and *Z. bailii* can also be used for bioethanol production.[136]

7.4   CONCLUSION

In this chapter, we have highlighted some of the most important and recent applications of yeast in biotechnology. However, with numerous new applications emerging, we think the potential of yeast for biotechnological applications is still untapped largely. Advanced molecular biology and genetic engineering tools together with identification of new strains will enable to utilize this eukaryotic microbe to a greater scale. CRISPR-Cas system is one such example, which has influenced yeast genetic engineering tremendously and will be helpful in unfolding numerous other applications of this organism in research and industry.[42,80] Further studies related to compartmentalization of the products, enhancement of a broad range of substrate utilization and product synthesis, creation of robust and stress tolerant strains, and so forth, will undoubtedly help in unraveling valuable tools for numerous biotechnological applications.

7.5   ACKNOWLEDGEMENT

Our sincere apologies to all our colleagues for any publication not being cited which is due to space limitation. This work was supported by Science and Engineering Research Board (SERB), Department of Science and Technology, Government of India [Early Career Research Award; ECR/2015/000205] and IIT Guwahati Institutional Start-Up Grant.

KEYWORDS

- yeast
- nonconventional
- agriculture
- environment
- industry

REFERENCES

1. Aehle, W. *Enzymes in Industry: Production and Applications*. John Wiley & Sons; Hoboken, New Jersey, 2007.
2. Agamy, R.; Hashem, M.; Alamri, S. Effect of Soil Amendment with Yeasts as Biofertilizers on the Growth and Productivity of Sugar Beet. *Afr. J. Agric. Res.* **2013,** *8*(1), 46–56.
3. Agbor, V. B.; Cicek, N.; Sparling, R.; Berlin, A.; Levin, D. B. Biomass Pretreatment: Fundamentals Toward Application. *Biotechnol. Adv.* **2011,** *29*(6), 675–685.
4. Ageitos, J. M.; Vallejo, J. A.; Veiga-Crespo, P; Villa, T. G. Oily Yeasts as Oleaginous Cell Factories. *Appl. Microbiol. Biotechnol.* **2011,** *90*(4), 1219–1227.
5. Aggarwal, M.; Mondal, A. K. *Debaryomyces hansenii*: An Osmotolerant and Halotolerant Yeast. In *Yeast Biotechnology: Diversity and Applications;* Satyanarayana, T., Kunze, G. Eds.; Springer Netherlands: Dordrecht, 2009; pp 65–84.
6. Aidoo, K. E.; Nout, M. R.; Sarkar, P. K. Occurrence and Function of Yeasts in Asian Indigenous Fermented Foods. *FEMS Yeast Res.* **2006,** *6*(1), 30–39.
7. Al-Falih, A. M. Nitrogen Transformation in vitro by Some Soil Yeasts. *Saudi J. Biol. Sci.* **2006,** *13*, 135–140.
8. Alves-Araújo, C.; Almeida, M. J.; Sousa, M. J.; Leão, C. Freeze Tolerance of the Yeast *Torulaspora delbrueckii*: Cellular and Biochemical Basis. *FEMS Microbiol. Lett.* **2004,** *240*(1), 7–14.

9. Alvira, P.; Tomas-Pejo, E.; Ballesteros, M.; Negro, M. J. Pretreatment Technologies for an Efficient Bioethanol Production Process based on Enzymatic Hydrolysis: A Review. *Bioresour. Technol.* **2010,** *101*(13), 4851–4861.

10. Angerbauer, C.; Siebenhofer, M.; Mittelbach, M.; Guebitz, G. M. Conversion of Sewage Sludge into Lipids by *Lipomyces starkeyi* for Biodiesel Production. *Bioresour. Technol.* **2008,** *99*(8), 3051–3056.

11. Arfi, K.; Spinnler, H.; Tache, R.; Bonnarme, P. Production of Volatile Compounds by Cheese-Ripening Yeasts: Requirement for a Methanethiol Donor for S-methyl Thioacetate Synthesis by *Kluyveromyces lactis. Appl. Microbiol. Biotechnol.* **2002,** *58*(4), 503–510.

12. Avalos, J. L.; Fink, G. R.; Stephanopoulos, G. Compartmentalization of Metabolic Pathways in Yeast Mitochondria Improves Production of Branched Chain Alcohols. *Nat. Biotechnol.* **2013,** *31*(4), 335–341.

13. Bankar, A. V.; Kumar, A. R.; Zinjarde, S. S. Environmental and Industrial Applications of *Yarrowia lipolytica. Appl. Microbiol. Biotechnol.* **2009,** *84*(5), 847–865.

14. Basinskiene, L.; Juodeikiene, G.; Vidmantiene, D.; Tenkanen, M.; Makaravicius, T.; Bartkiene, E. Non-Alcoholic Beverages from Fermented Cereals with Increased Oligosaccharide Content. *Food Technol. Biotechnol.* **2016,** *54*(1), 36–44.

15. Bayer, T. S.; Widmaier, D. M.; Temme, K.; Mirsky, E. A.; Santi, D. V.; Voigt, C. A. Synthesis of Methyl Halides from Biomass Using Engineered Microbes. *J. Am. Chem. Soc.* **2009,** *131*(18), 6508–6515.

16. Becker, J. V.; Armstrong, G. O.; van der Merwe, M. J.; Lambrechts, M. G.; Vivier, M. A.; Pretorius, I. S. Metabolic Engineering of *Saccharomyces cerevisiae* for the Synthesis of the Wine-related Antioxidant Resveratrol. *FEMS Yeast Res.* **2003,** *4*(1), 79–85.

17. Bekatorou, A.; Psarianos, C.; Koutinas, A. A.; Production of Food Grade Yeasts. *Food Technol. Biotechnol.* **2006,** *44*(3), 407–415.

18. Berovic, M.; Legisa, M. Citric Acid Production. *Biotechnol. Annu. Rev.* **2007,** *13*, 303–343.

19. Bilinski, C. A.; Casey, G. P. Developments in Sporulation and Breeding of Brewer's Yeast. *Yeast* **1989,** *5*(6), 429–438.

20. Blandino, A.; Al-Aseeri, M.; Pandiella, S.; Cantero, D.; Webb, C.; Cereal-based Fermented Foods and Beverages. *Food Res. Int.* **2003,** *36*(6), 527–543.

21. Bonekamp, F. J.; Oosterom, J. On the Safety of *Kluyveromyces lactis*—A Review. *Appl. Microbiol. Biotechnol.* **1994,** 41(1), 1–3.

22. Branduardi, P.; Smeraldi, C.; Porro, D. Metabolically Engineered Yeasts: 'Potential' Industrial Applications. *J. Mol. Microbiol. Biotechnol.* **2008,** *15*(1), 31–40.

23. Brat, D.; Weber, C.; Lorenzen, W.; Bode, H. B.; Boles, E. Cytosolic Re-localization and Optimization of Valine Synthesis and Catabolism Enables Increased Isobutanol Production with the Yeast *Saccharomyces cerevisiae. Biotechnol. Biofuels* **2012,** *5*(1), 65.

24. Buijs, N. A.; Siewers, V.; Nielsen, J. Advanced Biofuel Production by the Yeast *Saccharomyces cerevisiae. Curr. Opin. Chem. Biol.* **2013,** *17*(3), 480–488.

25. Buzzini P, Margesin R, *Cold-adapted Yeasts: Biodiversity, Adaptation Strategies and Biotechnological Significance*; Springer Science & Business Media; Berlin, Germany, 2013.

26. Cavalero, D. A.; Cooper, D. G. The Effect of Medium Composition on the Structure and Physical State of Sophorolipids Produced by *Candida bombicola* ATCC 22214. *J. Biotechnol.* **2003**, *103*(1), 31–41.

27. Çelik, E.; Çalık, P. Production of Recombinant Proteins by Yeast Cells. *Biotechnol. Adv.* **2012**, *30*(5), 1108–1118.

28. Cereghino, J. L.; Cregg, J. M. Heterologous Protein Expression in the Methylotrophic Yeast *Pichia pastoris*. *FEMS Microbiol. Rev.* **2000**, *24*(1), 45–66.

29. Chae, H. J.; Joo, H.; In, M. J. Utilization of Brewer's Yeast Cells for the Production of Food-Grade Yeast Extract. Part 1: Effects of Different Enzymatic Treatments on Solid and Protein Recovery and Flavor Characteristics. *Bioresour. Technol.* **2001**, *76*(3), 253–258.

30. Chen, J. H. *The Combined Use of Chemical and Organic Fertilizers and/or Biofertilizer for Crop Growth and Soil Fertility*. Food and Fertilizer Technology Center: Taipei, 2006.

31. Chen, X.; Gao, B.; Shi, W.; Li, Y.; Expression and Secretion of Human Interferon Alpha A in Yeast *Kluyveromyces lactis*. *Yi chuan xue bao (Acta genetica Sinica)* **1991**, *19*(3), 284–288.

32. Chen, X.; Nielsen, K. F.; Borodina, I.; Kielland-Brandt, M. C.; Karhumaa, K. Increased Isobutanol Production in *Saccharomyces cerevisiae* by Overexpression of Genes in Valine Metabolism. *Biotechnol. Biofuels* **2011**, *4*(1), 21.

33. Chi, Z.; Zheng, Y.; Jiang, A.; Chen, S. Lipid Production by Culturing Oleaginous Yeast and Algae with Food Waste and Municipal Wastewater in an Integrated Process. *Appl. Biochem. Biotechnol.* **2011**, *165*(2), 442–453.

34. Chiba, Y.; Sakuraba, H.; Kotani, M.; Kase, R.; Kobayashi, K.; Takeuchi, M.; Ogasawara, S.; Maruyama, Y.; Nakajima, T.; Takaoka, Y.; Production in Yeast of α-Galactosidase A, a Lysosomal Enzyme Applicable to Enzyme Replacement Therapy for Fabry Disease. *Glycobiology* **2002**, *12*(12), 821–828.

35. Costalos, C.; Skouteri, V.; Gounaris, A.; Sevastiadou, S.; Triandafilidou, A.; Ekonomidou, C.; Kontaxaki, F.; Petrochilou, V. Enteral Feeding of Premature Infants with *Saccharomyces boulardii*. *Early Hum. Dev.* **2003**, *74*(2), 89–96.

36. Darvishi Harzevili, F. Yarrowia lipolytica *in Biotechnological Applications. Biotechnological Applications of the Yeast* Yarrowia lipolytica; Springer International Publishing: Cham, 2014; pp 17–74.

37. de Almeida, A. F.; Tauk-Tornisielo, S. M.; Carmona, E. C.; Acid Lipase from *Candida viswanathii*: Production, Biochemical Properties, and Potential Application. *BioMed Res. Int.* **2013**, *2013*, 435818.

38. Delorme-Axford, E.; Guimaraes, R. S.; Reggiori, F.; Klionsky, D. J. The Yeast *Saccharomyces cerevisiae*: An Overview of Methods to Study Autophagy Progression. *Methods* **2015**, *75*, 3–12.

39. Demain, A. L.; Vaishnav, P. Production of Recombinant Proteins by Microbes and Higher Organisms. *Biotechnol. Adv.* **2009**, *27*(3), 297–306.

40. Demirbaş, A. Bioethanol from Cellulosic Materials: A Renewable Motor Fuel from Biomass. *Energy Sources* **2005**, *27*(4), 327–337.

41. Demirbas, A. Progress and Recent Trends in Biodiesel Fuels. *Energy Convers. Manage.* **2009**, *50*(1), 14–34.

42. DiCarlo, J. E.; Norville, J. E.; Mali, P.; Rios, X.; Aach, J.; Church, G. M.; Genome Engineering in *Saccharomyces cerevisiae* Using CRISPR-Cas Systems. *Nucleic Acids Res.* **2013**, *41*(7), 4336–4343.

43. Dragosits, M.; Frascotti, G.; Bernard-Granger, L.; Vázquez, F.; Giuliani, M.; Baumann, K.; Rodríguez-Carmona, E.; Tokkanen, J.; Parrilli, E.; Wiebe, M. G. Influence of Growth Temperature on the Production of Antibody Fab Fragments in Different Microbes: A Host Comparative Analysis. *Biotechnol Prog.* **2011**, *27*(1), 38–46.

44. Duina, A. A.; Miller, M. E.; Keeney, J. B. Budding Yeast for Budding Geneticists: A Primer on the *Saccharomyces cerevisiae* Model System. *Genetics* **2014**, *197*(1), 33–48.

45. Falcone Ferreyra, M. L.; Rius, S.; Casati, P. Flavonoids: Biosynthesis, Biological Functions, and Biotechnological Applications. *Front. Plant Sci.* **2012**, *3*, 222.

46. Falih, A. M.; Wainwright, M. Nitrification, S-Oxidation and P-Solubilization by the Soil Yeast Williopsis *californica* and by *Saccharomyces cerevisiae*. *Mycol. Res.* **1995**, *99*(2), 200–204.

47. Feng, Y.; Zhang, B.; Zhang, Y.; Hiroshi, F. Secretory Expression of Porcine Insulin Precursor in *Kluyveromyces lactis* and its Conversion into Human Insulin. *Sheng wu hua xue yu sheng wu wu li xue bao Acta Biochim Biophys Sin (Shanghai)* **1996**, *29*(2), 129–134.

48. Ferreira, A.; Viljoen, B. Yeasts as Adjunct Starters in Matured Cheddar Cheese. *Int. J. Food Microbiol.* **2003**, *86*(1), 131–140.

49. Ferreira, I.; Pinho, O.; Vieira, E.; Tavarela, J. Brewer's *Saccharomyces* Yeast Biomass: Characteristics and Potential Applications. *Trends Food Sci. Technol.* **2010**, *21*(2), 77–84.

50. Ferrer-Miralles, N.; Domingo-Espín, J.; Corchero, J. L.; Vázquez, E.; Villaverde, A.Microbial Factories for Recombinant Pharmaceuticals. *Microb. Cell Fact.* **2009**, *8*(1), 17.

51. Finogenova, T.; Morgunov, I.; Kamzolova, S.; Chernyavskaya, O. Organic Acid Production by the Yeast *Yarrowia lipolytica*: A Review of Prospects. *Appl. Biochem. Microbiol.* **2005**, *41*(5), 418–425.

52. Fleer, R. Engineering Yeast for High Level Expression. *Curr. Opin. Biotechnol.* **1992**, *3*(5), 486–496.

53. Fleer, R.; Chen, X. J.; Amellal, N.; Yeh, P.; Fournier, A.; Guinet, F.; Gault, N.; Faucher, D.; Folliard, F.; Hiroshi, F. High-Level Secretion of Correctly Processed Recombinant Human Interleukin-1β in *Kluyveromyces lactis*. *Gene* **1991**, *107*(2), 285–295.

54. Fleet, G. Yeasts in Dairy Products. *J. Appl. Bacteriol.* **1990**, *68*(3), 199–211.

55. Fleet, G. The Microbiology of Alcoholic Beverages. *Microbiology of Fermented Foods*; Springer; Boston, MA: 1998; pp 217–262.

56. Fleet, G. H. Yeasts in Foods and Beverages: Impact on Product Quality and Safety. *Curr. Opin. Biotechnol.* **2007**, *18*(2), 170–175.

57. Franssens, V.; Bynens, T.; Van den Brande, J.; Vandermeeren, K.; Verduyckt, M.; Winderickx, J. The Benefits of Humanized Yeast Models to Study Parkinson's Disease. *Oxid. Med. Cell. Longevity* **2013**, *2013*, 760629.

58. Frenzel, A.; Hust, M.; Schirrmann, T. Expression of Recombinant Antibodies. *Front. Immunol.* **2013,** *4*:217.

59. Futcher, A. B.; Cox, B. S. Copy Number and the Stability of 2-Micron Circle-Based Artificial Plasmids of *Saccharomyces cerevisiae. J. Bacteriol.* **1984,** *157*(1), 283–290.

60. Gellissen, G. Heterologous Protein Production in Methylotrophic Yeasts. *Appl. Microbiol. Biotechnol.* **2000,** *54*(6), 741–750.

61. Ghezelbash, G.; Nahvi, I.; Malekpour, A. Erythritol Production with Minimum By-product Using *Candida magnoliae* Mutant. *Appl. Biochem. Microbiol.* **2014,** *50*(3), 324–328.

62. Gidijala, L.; Kiel, J. A.; Douma, R. D.; Seifar, R. M.; Van Gulik, W. M.; Bovenberg, R. A.; Veenhuis, M.; Van Der Klei, I. J.; An Engineered Yeast Efficiently Secreting Penicillin. *PLoS One* **2009,** *4*(12), e8317.

63. Goldammer, T. *The Brewers' Handbook*; KVP Publishers: Clifton, Virginia, 1999.

64. Golubev, W. *Antagonistic Interactions Among Yeasts. Biodiversity and Ecophysiology of Yeasts*; Springer; Berlin, Heidelberg, Germany, 2006; pp 197–219.

65. Guaragnella, N.; Palermo, V.; Galli, A.; Moro, L.; Mazzoni, C.; Giannattasio, S. The Expanding Role of Yeast in Cancer Research and Diagnosis: Insights into the Function of the Oncosuppressors p53 and BRCA1/2. *FEMS Yeast Res.* **2014,** *14*(1), 2–16.

66. Halász, A.; Lásztity, R. *Use of Yeast Biomass in Food Production*; CRC Press, Boca Raton, Florida, United States, 1990.

67. Hamid, B.; Rana, R. S.; Chauhan, D.; Singh, P.; Mohiddin, F. A.; Sahay, S.; Abidi, I. Psychrophilic Yeasts and Their Biotechnological Applications—A Review. *Afr. J. Biotechnol.* **2014,** *13*(22), 2188–2197.

68. Harzevili, F. D. *Biotechnological Applications of the Yeast* Yarrowia lipolytica; Springer; Berlin, Germany, 2014.

69. Hatoum, R; Labrie, S; Fliss, I. Antimicrobial and Probiotic Properties of Yeasts: From Fundamental to Novel Applications. *Front. Microbiol.* **2012,** *3*, 421.

70. Hermann, T. Industrial Production of Amino Acids by Coryneform Bacteria. *J. Biotechnol.* **2003,** *104*(1–3), 155–172.

71. Heslot, H.; Gaillardin, C.; *Molecular Biology and Genetic Engineering of Yeasts*; CRC Press; Florida, United States, 1991.

72. Hinchliffe, E. The Genetic Improvement of Brewing Yeast. *Biochem. Soc. Trans.* **1988,** *16*(6), 1077–1079.

73. Horáková, D. M.; Nemec, M.; RC1 Consortium for Soil Decontamination: Its Preparation and Use. *Environmental Science and Pollution Control Series*; 2000; pp 357–372.

74. Hou, J.; Tyo, K. E.; Liu, Z.; Petranovic, D.; Nielsen, J. Metabolic Engineering of Recombinant Protein Secretion by *Saccharomyces cerevisiae. FEMS Yeast Res.* **2012,** *12*(5), 491–510.

75. Hua, Z.; Liang, X.; Zhu, D. Expression and Purification of a Truncated Macrophage Colony Stimulating Factor in *Kluyveromyces lactis. Biochem. Mol. Biol. Int.* **1994,** *34*(2), 419–427.

76. Huang, B.; Guo, J.; Yi, B.; Yu, X.; Sun, L.; Chen, W. Heterologous Production of Secondary Metabolites as Pharmaceuticals in *Saccharomyces cerevisiae. Biotechnol. Lett.* **2008,** *30*(7), 1121–1137.

77. Hussein, L.; Elsayed, S.; Foda, S. Reduction of Lactose in Milk by Purified Lactase Produced by *Kluyveromyces lactis*. *J. Food Protect.* **1989**, *52*(1), 30–34.

78. Ijah, U. Studies on Relative Capabilities of Bacterial and Yeast Isolates from Tropical Soil in Degrading Crude Oil. *Waste Manage.* (*Oxford*) **1998**, *18*(5), 293–299.

79. Izgu, F.; Altinbay, D. Isolation and Characterization of the K5-type Yeast Killer Protein and its Homology with an Exo-Beta-1,3-Glucanase. *Biosci., Biotechnol., Biochem.* **2004**, *68*(3), 685–693.

80. Jacobs, J. Z.; Ciccaglione, K. M.; Tournier, V.; Zaratiegui, M. Implementation of the CRISPR-Cas9 system in fission yeast. *Nat. Commun.* **2014**, *5*, 5344.

81. Jenson, I. Bread and Baker's Yeast. *Microbiology of Fermented Foods*; Springer, US, 1998; pp 172–198.

82. Jiang, P.; Mizushima, N. Autophagy and Human Diseases. *Cell Res.* **2014**, *24*(1), 69–79.

83. Jullesson, D.; David, F.; Pfleger, B.; Nielsen, J. Impact of Synthetic Biology and Metabolic Engineering on Industrial Production of Fine Chemicals. *Biotechnol. Adv.* **2015**, *33*(7), 1395–1402.

84. Jyoti, S.; Singh, D. P. Fungi as Biocontrol Agents in Sustainable Agriculture, In *Microbes and Environmental Management*; Studium Press (India) Pvt. Ltd., New Delhi, India, 2016.

85. Kabak, B.; Dobson, A. D.; An Introduction to the Traditional Fermented Foods and Beverages of Turkey. *Crit. Rev. Food Sci. Nutr.* **2011**, *51*(3), 248–260.

86. Kaczorek, E.; Chrzanowski, L.; Pijanowska, A.; Olszanowski, A. Yeast and Bacteria Cell Hydrophobicity and Hydrocarbon Biodegradation in the Presence of Natural Surfactants: Rhamnolipides and Saponins. *Bioresour. Technol.* **2008**, *99*(10), 4285–4291.

87. Kashyap, P.; Sabu, A.; Pandey, A.; Szakacs, G.; Soccol, C. R. Extra-Cellular L-glutaminase Production by *Zygosaccharomyces rouxii* Under Solid-State Fermentation. *Process Biochem.* **2002**, *38*(3), 307–312.

88. Klein, S. M.; Elmer, G. W.; McFarland, L. V.; Surawicz, C. M.; Levy, R. H., Recovery and Elimination of the Biotherapeutic Agent, *Saccharomyces boulardii*, in Healthy Human Volunteers. *Pharm. Res.* **1993**, *10*(11), 1615–1619.

89. Kobayashi, K.; Kuwae, S.; Ohya, T.; Ohda, T.; Ohyama, M.; Ohi, H.; Tomomitsu, K.; Ohmura, T. High-Level Expression of Recombinant Human Serum Albumin from the Methylotrophic Yeast *Pichia pastoris* with Minimal Protease Production and Activation. *J. Biosci. Bioeng.* **2000**, *89*(1), 55–61.

90. Kondo, T.; Tezuka, H.; Ishii, J.; Matsuda, F.; Ogino, C.; Kondo, A. Genetic Engineering to Enhance the Ehrlich Pathway and Alter Carbon Flux for Increased Isobutanol Production from Glucose by *Saccharomyces cerevisiae*. *J. Biotechnol* **2012**, *159*(1), 32–37.

91. Kumar, A.; Chandel, D.; Bala, I.; Muwalia, A.; Mankotiya, L. Managing Water Pollution All the Way Through Well-Designed Environmental Biotechnology: A Review. *IUP J. Biotechnol.* **2010**, *4*(2).

92. Kumar, L.; Gangadharan, V.; Rao, D. R.; Saikia, T.; Shah, S.; Malhotra, H.; Bapsy, P.; Singh, K.; Rao, R. Safety and Efficacy of an Indigenous Recombinant Interferon-Alpha-2b in Patients with Chronic Myelogenous Leukaemia: Results of a Multicentre Trial from India. *Natl. Med. J. India* **2005**, *18*(2), 66.

93. Kunze, G.; Kang, H. A.; Gellissen, G. Hansenula polymorpha (Pichia angusta): *Biology and Applications*. *Yeast Biotechnology: Diversity and Applications*; Springer; Berlin, Germany, 2009; pp 47–64.

94. Kužela, L.; Mašek, J.; Zalabak, V.; Kejmar, J. *Some Aspects of the Biomass of Torula in Human Nutrition. A New Protein Source. Early Signs of Nutritional Deficiencies*; Karger Publishers; Basel, Switzerland, 1976; pp 169–173.

95. Lea, A. G.; Piggott, J. *Fermented Beverage Production*; Springer Science and Business Media, Springer US, 2012.

96. Leclercq-Perlat, M-N.; Latrille, E.; Corrieu, G.; Spinnler, H. E. Controlled Production of Camembert-Type Cheeses. Part II. Changes in the Concentration of the More Volatile Compounds. *J. Dairy Res.* **2004**, *71*(3), 355–366.

97. Lee, W. H.; Seo, S. O.; Bae, Y. H.; Nan, H.; Jin, Y. S.; Seo, J. H. Isobutanol Production in Engineered *Saccharomyces cerevisiae* by Overexpression of 2-Ketoisovalerate Decarboxylase and Valine Biosynthetic Enzymes. *Bioprocess Biosyst. Eng.* **2012**, *35*(9), 1467–1475.

98. Legras, J. L.; Merdinoglu, D.; Cornuet, J. M.; Karst, F. Bread, Beer and Wine: *Saccharomyces cerevisiae* Diversity Reflects Human History. *Mol. Ecol.* **2007**, *16*(10), 2091–2102.

99. Leiva-Candia, D. E.; Pinzi, S.; Redel-Macías, M. D.; Koutinas, A.; Webb, C.; Dorado, M. P. The Potential for Agro-Industrial Waste Utilization Using Oleaginous Yeast for the Production of Biodiesel. *Fuel* **2014**, *123*, 33–42.

100. Li, Y.; Zhao, Z.; Bai, F. High-Density Cultivation of Oleaginous Yeast *Rhodosporidium toruloides* Y4 in Fed-Batch Culture. *Enzyme Microb. Technol.* Elsevier; Amsterdam, Netherlands, **2007**, *41*(3), 312–317.

101. Logar, R. M.; Vodovnik, M. The Application of Microbes in Environmental Monitoring. *Communicating Current Research and Educational Topics and Trends in Applied microbiology*; 2007; 577–585.

102. Lourens-Hattingh, A.; Viljoen, B. C. Yogurt as Probiotic Carrier Food. *Int. Dairy J.* **2001**, *11*(1), 1–17.

103. Magalhães, K. T.; Pereira, G. V. d. M.; Campos, C. R.; Dragone, G.; Schwan, R. F., Brazilian Kefir: Structure, Microbial Communities and Chemical Composition. *Braz. J. Microbiol.* **2011**, *42*(2), 693–702.

104. Mannazzu, I.; Landolfo, S.; Da Silva, T. L.; Buzzini, P. Red Yeasts and Carotenoid Production: Outlining a Future for Non-conventional Yeasts of Biotechnological Interest. *World J. Microbiol. Biotechnol.* **2015**, *31*(11), 1665–1673.

105. Margesin, R.; Schinner, F. Bioremediation of Diesel-Oil-Contaminated Alpine Soils at Low Temperatures. *Appl. Microbiol. Biotechnol.* **1997**, *47*(4), 462–468.

106. Margesin, R.; Gander, S.; Zacke, G.; Gounot, A. M.; Schinner, F. Hydrocarbon Degradation and Enzyme Activities of Cold-Adapted Bacteria and Yeasts. *Extremophiles* **2003**, *7*(6), 451–458.

107. Martens, T.; Gram, L.; Grossart, H-P.; Kessler, D.; Müller, R.; Simon, M.; Wenzel, S. C.; Brinkhoff, T.Bacteria of the *Roseobacter* clade Show Potential for Secondary Metabolite Production. *Microb. Ecol.* **2007**, *54*(1), 31–42.

108. Mattanovich, D.; Branduardi, P.; Dato, L.; Gasser, B.; Sauer, M.; Porro, D. Recombinant Protein Production in Yeasts. *Methods Mol. Biol.* Humana Press, Totowa, NJ, **2012**, *824*, 329–358.

109. Mattanovich, D.; Sauer, M.; Gasser, B. Yeast Biotechnology: Teaching the Old Dog New Tricks. *Microb. Cell Fact.* **2014**, *13*(1), 34.
110. Mehta, G. D.; Agarwal, M. P.; Ghosh, S. K. Centromere Identity: A Challenge to be Faced. *Mol. Genet. Genomics* **2010**, *284*(2), 75–94.
111. Mellor, J.; Dobson, M.; Roberts, N.; Tuite, M.; Emtage, J.; White, S.; Lowe, P.; Patel, T.; Kingsman, A.; Kingsman, S. Efficient Synthesis of Enzymatically Active Calf Chymosin in *Saccharomyces cerevisiae. Gene* **1983**, *24*(1), 1–14.
112. Meng, D. M.; Zhao, J. F.; Ling, X.; Dai, H. X.; Guo, Y. J.; Gao, X. F.; Dong, B.; Zhang, Z. Q.; Meng, X.; Fan, Z. C. Recombinant Expression, Purification and Antimicrobial Activity of a Novel Antimicrobial Peptide PaDef in *Pichia pastoris. Protein Expression Purif.* **2017**, *130*, 90–99.
113. Menon, V.; Rao, M. Trends in Bioconversion of Lignocellulose: Biofuels, Platform Chemicals and Biorefinery Concept. *Prog. Energy Combust. Sci.* **2012**, *38*(4), 522–550.
114. Miller-Fleming, L.; Giorgini, F.; Outeiro, T. F. Yeast as a Model for Studying Human Neurodegenerative Disorders. *Biotechnol. J.* **2008**, *3*(3), 325–338.
115. Mishra, S.; Baranwal, R. Yeast Genetics and Biotechnological Applications. In *Yeast Biotechnology: Diversity and Applications*; Satyanarayana, T.; Kunze, G., Eds.;Springer Netherlands: Dordrecht, 2009; pp 323–355.
116. Mosier, N.; Wyman, C.; Dale, B.; Elander, R.; Lee, Y. Y.; Holtzapple, M.; Ladisch, M. Features of Promising Technologies for Pretreatment of Lignocellulosic Biomass. *Bioresour. Technol.* **2005**, *96*(6), 673–686.
117. Müller, S.; Sandal, T.; Kamp-Hansen, P.; Dalbøge, H. Comparison of Expression Systems in the Yeasts *Saccharomyces cerevisiae, Hansenula polymorpha, Klyveromyces lactis, Schizosaccharomyces pombe* and *Yarrowia lipolytica.* Cloning of Two Novel Promoters from *Yarrowia lipolytica. Yeast* **1998**, *14*(14), 1267–1283.
118. Mulligan, C. N.; Yong, R. N.; Gibbs, B. F. An Evaluation of Technologies for the Heavy Metal Remediation of Dredged Sediments. *J. Hazard Mat.* **2001**, *85*(1–2), 145–163.
119. Musch, E.; Malek, M.; von Eick, H.; Chrissafidou, A. Successful Application of Highly Purified Natural Interferon Alpha (multiferon) in a Chronic Hepatitis C Patient Resistant to Preceding Treatment Approaches. Hepato-gastroenterology **2003**, *51*(59), 1476–1479.
120. Nassar A. H.; El-Tarabily, K. A.; Sivasithamparam, K.Promotion of Plant Growth by an Auxin-Producing Isolate of the Yeast Williopsis Saturnus Endophytic in Maize (*Zea mays* L.) Roots. *Biol. Fertil. Soils* **2005**, *42*(2), 97–108.
121. Nel, S.; Labuschagne, M.; Albertyn J. Advances in Gene Expression in Non-Conventional Yeasts. In *Yeast Biotechnology: Diversity and Applications*; Satyanarayana, T.; Kunze, G., Eds.; Springer Netherlands: Dordrecht, 2009; pp 369–403.
122. Nielsen, J.; Larsson, C.; van Maris, A.; Pronk, J.Metabolic Engineering of Yeast for Production of Fuels and Chemicals. *Curr. Opin. Biotechnol.* **2013**, *24*(3), 398–404.
123. Ockerman, H. W.*Sausage and Processed Meat Formulations*; Van Nostrand Reinhold, New York, 1989.
124. Overton, T. W. Recombinant Protein Production in Bacterial Hosts. *Drug Discovery Today* **2014**, *19*(5), 590–601.

125. Ozmihci, S.; Kargi, F. Comparison of Yeast Strains for Batch Ethanol Fermentation of Cheese–whey Powder (CWP) Solution. *Lett. Appl. Microbiol.* **2007,** *44*(6), 602–606.

126. Paradise, E. M.; Kirby. J.; Chan, R.; Keasling, J. D.Redirection of Flux Through the FPP Branch-Point in *Saccharomyces cerevisiae* by Down-Regulating Squalene Synthase. *Biotechnol. Bioeng.* **2008,** *100*(2), 371–378.

127. Peralta-Yahya, P. P.; Ouellet, M.; Chan, R.; Mukhopadhyay, A.; Keasling, J. D.; Lee, T. S. Identification and Microbial Production of a Terpene-Based Advanced Biofuel. *Nat. Commun.* **2011,** *2*, 483.

128. Pereira, C.; Bessa, C.; Soares, J.; Leão, M.; Saraiva, L. Contribution of Yeast Models to Neurodegeneration Research. *J. Biomed. Biotechnol.* **2012,** *2012,* 941232.

129. Perez, J.; Munoz-Dorado, J.; de la Rubia, T.; Martinez, J. Biodegradation and Biological Treatments of Cellulose, Hemicellulose and Lignin: An Overview. *Int. Microbiol.* **2002,** *5*(2), 53–63.

130. Porro, D.; Gasser, B.; Fossati, T.; Maurer, M.; Branduardi, P.; Sauer, M.; Mattanovich, D.. Production of Recombinant Proteins and Metabolites in Yeasts. *Appl. Microbiol. Biotechnol.* **2011,** *89*(4), 939–948.

131. Poutanen, K.; Flander, L.; Katina, K. Sourdough and Cereal Fermentation in a Nutritional Perspective. *Food Microbiol.* **2009,** *26*(7), 693–699.

132. Pozo-Bayón, M.; Guichard, E.; Cayot, N. Flavor Control in Baked Cereal Products. *Food Rev. Int.* **2006**, *22*(4), 335–379.

133. Pray, L. L. H. Hartwell's Yeast: A Model Organism for Studying Somatic Mutations and Cancer. *Nat. Educ.* **2008,** *1*(1), 183.

134. Pretorius, I. S. Tailoring Wine Yeast for the New Millennium: Novel Approaches to the Ancient Art of Winemaking. *Yeast* **2000**, *16*(8), 675–729.

135. Priest, F. G.; Stewart, G. G. Eds. *Handbook of Brewing*; CRC Press, Boca Raton, Florida, United States, 2006.

136. Radecka, D.; Mukherjee, V.; Mateo, R. Q.; Stojiljkovic, M.; Foulquié-Moreno, M. R.; Thevelein, J. M. Looking Beyond *Saccharomyces*: The Potential of Non-conventional Yeast Species for Desirable Traits in Bioethanol Fermentation. *FEMS Yeast Res.* **2015,** *15*(6), fov053.

137. Randez-Gil, F.; Sanz, P.; Prieto, J. A. Engineering Baker's Yeast: Room for Improvement. *Trends Biotechnol.* **1999,** *17*(6), 237–244.

138. Rantsiou, K.; Urso, R.; Iacumin, L.; Cantoni, C.; Cattaneo, P.; Comi, G.; Cocolin, L. Culture-Dependent and Independent Methods to Investigate the Microbial Ecology of Italian Fermented Sausages. *Appl. Environ. Microbiol.* **2005,** *71*(4), 1977–1986.

139. Ratledge, C. Fatty Acid Biosynthesis in Microorganisms Being Used for Single Cell Oil Production. *Biochimie* **2004**, *86*(11), 807–815.

140. Ratledge, C.; Wynn, J. P. The Biochemistry and Molecular Biology of Lipid Accumulation in Oleaginous Microorganisms. *Adv. Appl. Microbiol.* **2002,** *51*, 1–51.

141. Ravindra, P. Value-Added Food: Single Cell Protein. *Biotechnol. Adv.* **2000,** *18*(6), 459–479.

142. Renninger, N. S.; Newman, J.; Reiling, K. K.; Regentin, R.; Paddon, C. J. Production of Isoprenoids. Google Patents.

143. Robin, S.; Petrov, K.; Dintinger, T.; Kujumdzieva, A. Tellier, C.; Dion, M. Comparison of Three Microbial Hosts for the Expression of an Active Catalytic scFv. *Mol. Immunol.* **2003,** *39*(12), 729–738.

144. Rodríguez, B.; Valdivié, M.; Lezcano, P.; Herrera, M. Evaluation of *Torula* Yeast (*Candida utilis*) Grown on Distillery Vinasse for Broilers. *Cuban J. Agric. Sci.* **2013**, *47*(2), 183–187.

145. Rosa-Magri, M. M.; Tauk-Tornisielo, S. M.; Ceccato-Antonini, S. R. Bioprospection of Yeasts as Biocontrol Agents Against Phytopathogenic Molds. *Braz. Arch. Biol. Technol.* **2011**, *54*(1), 1–5.

146. Rufino, R. D.; Rodrigues, G. I. B.; Campos-Takaki, G. M.; Sarubbo, L. A.; Ferreira, S. R. M. Application of a Yeast Biosurfactant in the Removal of Heavy Metals and Hydrophobic Contaminant in a Soil Used as Slurry Barrier. *Appl. Environ. Soil Sci.* **2011**, *2011,* 7.

147. Ryder. J. A. Jet Fuel Compositions and Methods of Making and Using Same. Google Patents

148. Rymowicz, W.; Rywińska, A.; Żarowska, B.; Juszczyk, P. Citric Acid Production from Raw Glycerol by Acetate Mutants of *Yarrowia lipolytica*. *Chem. Pap.* **2006**, *60*(5), 391–394.

149. Saha, B. C.; Sakakibara, Y.; Cotta, M. A.. Production of D-Arabitol by a Newly Isolated *Zygosaccharomyces rouxii*. *J. Ind. Microbiol. Biotechnol.* **2007**, *34*(7), 519–523.

150. Sakai, Y.; Tani, Y. Formaldehyde Production by Cells of a Mutant of *Candida boidinii* S2 Grown in Methanol-limited Chemostat Culture. *Agric. Boil. Chem.* **1986**, *50*(10), 2615–2620.

151. Saliola, M.; Mazzoni, C.; Solimando, N.; Crisà, A.; Falcone, C.; Jung, G.; Fleer, R. Use of the KlADH4 Promoter for Ethanol-dependent Production of Recombinant Human Serum Albumin in *Kluyveromyces lactis*. *Appl. Environ. Microbiol.* **1999**, *65*(1), 53–60.

152. Sanchez, S.; Bravo, V.; Castro, E.; Moya, A. J.; Camacho, F. The Fermentation of Mixtures of D-glucose and D-xylose by *Candida shehatae*, *Pichia stipitis* or *Pachysolen tannophilus* to Produce Ethanol. *J. Chem. Technol. Biotechnol.* **2002**, *77*(6), 641–648.

153. Sauer, M.; Porro, D.; Mattanovich, D.; Branduardi, P. Microbial Production of Organic Acids: Expanding the Markets. *Trends Biotechnol.* **2008**, *26*(2), 100–108.

154. Schein, C. H. Production of Soluble Recombinant Proteins in Bacteria. *Nat. Biotechnol.* **1989**, *7*(11), 1141–1149.

155. Schmitt, M. J.; Breinig, F. The Viral Killer System in Yeast: from Molecular Biology to Application. *FEMS Microbiol. Rev.* **2002**, *26*(3), 257–276.

156. Schneiter, R. *Genetics, Molecular and Cell Biology of Yeast*; Universität Freiburg Schweiz: Schweiz, 2004; pp 4–18.

157. Seiler, H.; Busse, M. The Yeasts of Cheese Brines. *Int. J. Food Microbiol.* **1990a,** *11*(3–4), 289–303.

158. Seiler, H.; Busse, M. The Yeasts of Cheese Brines. *Int. J. Food Microbiol.* **1990b,** *11*(3), 289–303.

159. Senthilraja, P.; Kathiresan, K.; Saravanakumar, K. Comparative Analysis of Bioethanol Production by Different Strains of Immobilized Marine Yeast. *J. Yeast Fungal Res.* **2011**, *2*(8), 113–116.

160. Shalaby, M.; El-Nady, M. F. Application of *Saccharomyces cerevisiae* as a Biocontrol Agent Against Fusarium Infection of Sugar Beet Plants. *Acta Biologica Szegediensis* **2008**, *52*(2), 271–275.

161. Sharma, A.; Diwevidi, V.; Singh, S.; Pawar, K. K.; Jerman, M.; Singh, L.; Singh, S.; Srivastawav, D. Biological Control and its Important in Agriculture. *Int. J. Biotechnol. Bioeng. Res.* **2013**, *4*(3), 175–180.

162. Shi, S.; Valle-Rodríguez, J. O.; Khoomrung, S.; Siewers, V.; Nielsen, J. Functional Expression and Characterization of Five Wax Ester Synthases in *Saccharomyces cerevisiae* and Their Utility for Biodiesel Production. *Biotechnol. Biofuels* **2012**, *5*(1), 7.

163. Shivaji, S.; Prasad, G. S. Antarctic Yeasts: Biodiversity and Potential Applications. In *Yeast Biotechnology: Diversity and Applications*; Satyanarayana, T., Kunze, G., Eds.; Springer Netherlands: Dordrecht, 2009; pp 3–18.

164. Sibirny, A. A.; Boretsky, Y. R. *Pichia guilliermondii Yeast Biotechnology: Diversity and Applications*; Eds.; Springer: 2009; pp 113–134.

165. Sibirny, A. A.; Nielsen, J.; Yeast Peroxisomes: Structure, Functions and Biotechnological Opportunities. *FEMS Yeast Res.* **2016**, *16*(4), fow038.

166. Siegel, R. S; Brierley, R. A. Methylotrophic Yeast *Pichia pastoris* Produced in High-cell-Density Fermentations with High Cell Yields as Vehicle for Recombinant Protein Production. *Biotechnol. Bioeng.* **1989**, *34*(3), 403–404.

167. Sivakumar, T.; Ravikumar, M.; Prakash, M.; Shanmugaraju, V. Production of Extracelluar Invertase from *Saccharomyces cerevisiae* Strain Isolated from Grapes. *Int. J. Curr. Res. Acad. Rev.* **2013**, *1*(2), 72–83.

168. Smith, E. A.; Myburgh, J.; Osthoff, G.; de Wit, M. Acceleration of Yoghurt Fermentation Time by Yeast Extract and Partial Characterisation of the Active Components. *J. Dairy Res.* **2014**, *81*(4), 417–423.

169. Soares, E. V.; Soares, H. M.. Bioremediation of Industrial Effluents Containing Heavy Metals Using Brewing Cells of *Saccharomyces cerevisiae* as a Green Technology: A Review. *Environ. Sci. Pollut. Res.* **2012**, *19*(4), 1066–1083.

170. Songpim, M.; Vaithanomsat, P.; Vanichsriratana, W.; Sirisansaneeyakul, S.. Enhancement of Inulinase and Invertase Production from a Newly Isolated *Candida guilliermondii* TISTR 5844. *Kasetsart J.* **2011**, *45*(4), 675–685.

171. Spohner, S. C.; Schaum, V.; Quitmann, H.; Czermak, P. *Kluyveromyces lactis*: An Emerging Tool in Biotechnology. *J. Biotechnol.* **2016**, *222*, 104–116.

172. Steen, E. J.; Chan, R.; Prasad, N.; Myers, S.; Petzold, C. J.; Redding, A.; Ouellet, M.; Keasling, J. D. Metabolic Engineering of *Saccharomyces cerevisiae* for the Production of n-butanol. *Microb. Cell Fact.* **2008**, *7*(1), 36.

173. Strong, P. J.; Burgess, J. E. Treatment Methods for Wine-related and Distillery Wastewaters: A Review. *Biorem. J.* **2008**, *12*(2), 70–87.

174. Suzuki, C.; Ando, Y.; Machida, S. Interaction of SMKT, a Killer Toxin Produced by *Pichia farinosa*, with the Yeast Cell Membranes. *Yeast* **2001**, *18*(16), 1471–1478.

175. Swennen, D.; Paul, M-F.; Vernis, L.; Beckerich, J-M.; Fournier, A.; Gaillardin, C. Secretion of Active Anti-Ras Single-chain Fv Antibody by the Yeasts *Yarrowia lipolytica* and *Kluyveromyces lactis*. *Microbiology* **2002**, *148*(1), 41–50.

176. Swiegers, J.; Pretorius, I.; Modulation of Volatile Sulfur Compounds by Wine Yeast. *Appl. Microbiol. Biotechnol.* **2007**, *74*(5), 954–960.

177. Tai, M.; Stephanopoulos, G. N. Metabolic Engineering: Enabling Technology for Biofuels Production. *Advances in Bioenergy: The Sustainability Challenge*; 2015; pp 1–10.

178. Teoh, A. L.; Heard, G.; Cox, J. Yeast Ecology of Kombucha Fermentation. *Int. J. Food Microbiol.* **2004,** *95*(2), 119–126.

179. Thammakiti, S.; Suphantharika, M.; Phaesuwan, T.; Verduyn, C. Preparation of Spent Brewer's Yeast β-glucans for Potential Applications in the Food Industry. *Int. J. Food Sci. Technol.* **2004,** *39*(1), 21–29.

180. Thevenieau, F.; Nicaud, J.-M.; Gaillardin, C. *Applications of the Non-conventional Yeast* Yarrowia lipolytica. *Yeast Biotechnology: Diversity and Applications*; Springer: 2009; pp 589–613.

181. Tschopp, J. F.; Brust, P. F.; Cregg, J. M.; Stillman, C. A.; Gingeras, T. R. Expression of the lacZ Gene from Two Methanol-regulated Promoters in *Pichia pastoris. Nucleic Acids Res.* **1987,** *15*(9), 3859–3876.

182. Türker, M. *Yeast Biotechnology: Diversity and Applications*; 2014.

183. Úbeda, J.; Gil, M. M.; Chiva, R.; Guillamón, J. M.; Briones, A. Biodiversity of Non-Saccharomyces yeasts in Distilleries of the La Mancha Region (Spain). *FEMS Yeast Res.* **2014,** *14*(4), 663–673.

184. Ubersax, J. A.; Platt, D. M. Genetically Modified Microbes Producing Isoprenoids, 2010, Google Patents.

185. van den Berg, J. A.; van der Laken, K. J.; van Ooyen, A. J. J.; Renniers, T. C. H. M.; Rietveld, K.; Schaap, A.; Brake, A. J.; Bishop, R. J.; Schultz, K.; Moyer, D.; Richman, M.; Shuster, J. R. *Kluyveromyces* as a Host for Heterologous Gene Expression: Expression and Secretion of Prochymosin. *Nat. Biotechnol.* **1990,** *8*(2), 135–139.

186. Van Dijk, R.; Faber, K. N.; Kiel, J. A.; Veenhuis, M.; van der Klei, I. The Methylotrophic Yeast *Hansenula polymorpha*: A Versatile Cell Factory. *Enzyme Microbial. Technol.* **2000,** *26*(9), 793–800.

187. van Ooyen, A. J. J.; Dekker, P.; Huang, M.; Olsthoorn, M. M. A.; Jacobs, D. I.; Colussi, P. A.; Taron, C. H. Heterologous Protein Production in the Yeast *Kluyveromyces lactis. FEMS Yeast Res.* **2006,** *6*(3), 381–392.

188. Verpoorte, R.; Van Der Heijden, R.; Ten Hoopen, H.; Memelink, J. Metabolic Engineering of Plant Secondary Metabolite Pathways for the Production of fine Chemicals. *Biotechnol. Lett.* **1999,** *21*(6), 467–479.

189. Vieira, E. F.; Ferreira, I. M. Antioxidant and Antihypertensive Hydrolysates Obtained from By-products of Cannery Sardine and Brewing Industries. *Int. J. Food Prop.* **2016,** *20,* 662–673.

190. Viljoen, B. C. Yeast Ecological Interactions: Yeast/Yeast, Yeast/Bacteria, Yeast/Fungi Interactions and Yeasts as Biocontrol Agents. In *Yeasts in Food and Beverages*; Querol, A., Fleet, G., Eds.; Springer Berlin Heidelberg: Berlin, Heidelberg, 2006; pp 83–110.

191. Walker, G. M. *Yeast Physiology and Biotechnology*; John Wiley & Sons, 1998.

192. Weatherholtz, W.; Holsing, G. Evaluation of Torula Yeast for Use as a Food Supplement. *Fed. Proc.* **1975,** *20814–3998,* 890–890.

193. Wegner, G. H. Emerging Applications of the Methylotrophic Yeasts. *FEMS Microbiol. Rev.* **1990,** *7*(3–4), 279–283.

194. Westfall, P. J.; Pitera, D. J.; Lenihan, J. R.; Eng, D.; Woolard, F. X.; Regentin, R.; Horning, T.; Tsuruta, H.; Melis, D. J.; Owens, A. Production of Amorphadiene in Yeast, and its Conversion to Dihydroartemisinic Acid, Precursor to the Antimalarial Agent Artemisinin. *Proc. Natl. Acad. Sci.* **2012,** *109*(3), E111–E118.

195. Wu, W.; Lu, M.; Yu, L. Expression of Carotenogenic Genes and Astaxanthin Production in *Xanthophyllomyces dendrorhous* as a Function of Oxygen Tension. *Z. Naturforschung C* **2011**, *66*(5–6), 283–286.

196. Xia, C.; Zhang, J.; Zhang, W.; Hu, B. A New Cultivation Method for Microbial Oil Production: Cell Pelletization and Lipid Accumulation by *Mucor circinelloides*. *Biotechnol. Biofuels* **2011**, *4*(1), 15.

197. Yuzbashev, T. V.; Yuzbasheva, E. Y.; Sobolevskaya, T. I.; Laptev, I. A.; Vybornaya, T. V.; Larina, A. S.; Matsui, K.; Fukui, K.; Sineoky, S. P. Production of Succinic Acid at Low pH by a Recombinant Strain of the Aerobic Yeast *Yarrowia lipolytica*. *Biotechnol. Bioeng.* **2010**, *107*(4), 673–682.

198. Zhang, A.; Lee, B.; Lee, S.; Lee, K.; An, G.; Song, K.; Lee, C. Effects of Yeast (*Saccharomyces cerevisiae*) Cell Components on Growth Performance, Meat Quality, and Ileal Mucosa Development of Broiler Chicks. *Poult. Sci.* **2005**, *84*(7), 1015–1021.

199. Zhao, J.; Fleet, G. Yeasts are Essential for Cocoa Bean Fermentation. *Int. J. Food Microbiol.* **2014**, *174*, 72–17487.

200. Zhou, J.; Liu, L.; Chen, J. Improved ATP Supply Enhances Acid Tolerance of *Candida glabrata* During Pyruvic Acid Production. *J. Appl. Microbiol.* **2011**, *110*(1), 44–53.

201. Zhou, Y. J.; Buijs, N. A.; Zhu, Z.; Qin, J.; Siewers, V.; Nielsen, J. Production of Fatty Acid-derived Oleochemicals and Biofuels by Synthetic Yeast Cell Factories. *Nat. Commun.* **2016**, *7*, 11709.

202. Zhuge, J.; Fang, H. Y.; Wang, Z. X.; Chen, D. Z.; Jin, H. R.; Gu, H. L. Glycerol Production by a Novel Osmotolerant yeast *Candida glycerinogenes*. *Appl. Microbiol. Biotechnol.* **2001**, *55*(6), 686–692.

203. Zinjarde, S.; Apte, M.; Mohite, P.; Kumar, A. R. *Yarrowia lipolytica* and Pollutants: Interactions and Applications. *Biotechnol. Adv.* **2014**, *32*(5), 920–933.

PART II

Microbes in the Environment

CHAPTER 8

ANTIMICROBIALS FOR TEXTILE FINISHES

SAGARIKA DEVI*

*DTU Bioengineering, Department of Biotechnology and Biomedicine, Technical University of Denmark, 2800 Kgs. Lyngby, Denmark, *E-mail: sadevi@dtu.dk, devi.sagarika@gmail.com*

CONTENTS

ABSTRACT

The use of elemental silver, ionizable silver, nanoparticles, chitosan and natural dyes as antimicrobial textile agents is well known for a variety of microbes including *Staphylococcus aureus*, methicillin-resistant

Staphylococcus aureus (MRSA), *Escherichia coli*, *Streptococcus pyogenes*, *Proteus vulgaris*, *Bacillus subtilis*, *Pseudomonas* and *Aspergillus niger*. Clothing which served as mere protection has now changed to health-based clothing. There is a good demand not only for fabrics having functional/special finishes in general but also with antimicrobial finishes in particular. Such fabrics not only serve the wellness sector but due to their antimicrobial finishes suffice prevention and healing also. The application of antimicrobial textile finishes includes a wide range of textile products for industrial, home furnishing, technical, medicinal and apparel sectors. Textiles with biocidal finishes, despite their commercial potential, until recently, were not widely accepted in the market. To avoid such menaces, cautious and steadfast in vitro as well as in vivo test systems need to be developed. The upsurge of recent technologies and products should earnestly study the subtle skin microflora balance for selection and consideration of compounds and techniques. Although textile fabrics ought to sustain physiological functions, they should not become a threat to the human health. Standard protocols should be developed for evaluating the antimicrobial potential of the textile fabrics as well as for the determination of any other associated undesirable side effects. The present chapter focuses on developments and utilization of various antimicrobial agents for healthy textile finishes with improved performance and durability.

8.1 TEXTILES: AN INTRODUCTION

Textiles have an unceasingly important involvement in the cultural evolution of human beings by being at the forefront of both creative and technological development.[1] Apart from providing the basic necessities of life, textile industry conjointly plays a dynamic role through its input to industrial output, employment generation and also the export retributions of the country. Textiles are an ideal medium for adherence, transmission and proliferation of infection in the human body by virtue of their make and vicinity, instigating microbial proliferation.[2,3] The response to microbial growth thus varies greatly in natural and synthetic fibres. Each could act as a willing substrate for the microbial attack with different mechanism of action. Natural fibres such as cotton, wool, flax and jute, due to their porosity and hydrophilic nature, are additionally vulnerable to the attack of microorganism. A greater concern, however, is that textiles can also act as active agents for the microbial propagation and do not act as substrates merely. Microorganisms metabolize sweat and soiling present in the cloth as nutrients to grow and produce foul-smelling molecules. Fungi generate multiple snags in the textiles including discolouration, stains and fibre destruction, disagreeable odour and a slimy feel (Table 8.1).

TABLE 8.1   Microorganisms and the Ailments Caused by them.

Microorganism	Causative disease or conditions
Gram-negative bacteria	
Escherichia coli	Urinary tract infections
Pseudomonas aeruginosa	Infection of wounds and burns
Proteus mirabilis	Urinary infections
Gram-positive bacteria	
Staphylococcus aureus	Pyrogenic infections
Staphylococcus epidermis	Body odour
Corynebacterium ditheroides	Body odour
Brevibacterium ammoniagenes	Diaper rash
Streptococcus pneumoniae	Bacterial pneumonia
Fungi	
Candida albicans	Diaper rash
Epidermophyton floccosum	Infections of skin and nails
Aspergillus niger	Damage cotton

Although growth and propagation of microbes are not supported by synthetic fibres, compounds with low molecular weight used as textile finishes might provide adequate nutrient for the microbial proliferation. This results in the development of unpleasant odour, stains and discolouration in the fabric thereby reducing the strength of the fabric. The spiteful odour grows when perspiration in human beings is converted into foul-smelling substances, such as carboxylic acid, aldehydes and amines, by microbes. The metabolism of Gram-positive bacteria *S. aureus* produces E-3-methyl-2-hexenoic acid (E-3M2H) which causes the distinctive body odour.[4] The Gram-negative bacteria *P. vulgaris* is competent to metabolize urea to form ammonia and causing formation of odour. The manifestation of various microflora on human skin and their ability to grow and persist until a year in an individual proposes that skin supports normal microflora. The predominance of different microorganisms on different body parts differs in individuals of unrelated age groups. Microflora found on clothes was akin to the normal skin microflora.[5] Body odour is produced in undershirts by *Staphylococcus epidermis* and *Coryneform* bacteria. *Bacilli* and lesser amounts of *S. epidermis* and *Micrococci* are present in the trouser legs and pockets. The skin of the groin, perineum and feet being the moist and dark areas harbours skin infections producing *S. aureus*, Gram-negative bacteria, yeast, and *Candida albicans*. More notably, textile materials utilized in hotels and hospitals are favourable for cross-infection or transmission of diseases caused by microbes. For these reasons, it is remarkably indispensable that the microbial proliferation and propagation on textiles be curtailed during usage and storing. A large number of natural as well synthetic antimicrobial finishes are being used in the global market recently to achieve this.

8.2   ANTIMICROBIAL AGENTS

The Greek words, anti (against), mikros (little) and bios (life) mean antimicrobial and refer to any or all compounds that inhibit microbial growth. In contrast, antibiotics by definition strictly refer to microbial metabolite substances that inhibit or kill another microorganism. The term 'antimicrobials' includes all agents that prevent against bacteria (antibacterial), viruses (antiviral), fungi (antifungal) and protozoa (antiprotozoal). Antimicrobials are categorized in several ways with respect to their spectrum of activity, effect on bacteria and mode of action. Albeit the practice of antimicrobials usage has been known for several spans, it is only recently that numerous efforts are made on textile finishes with antimicrobial efficacy. Antimicrobials augment the functionality and value of textile products by preventing microbial growth that causes odour and fibre degradation.[6] An antimicrobial finish averts the growth of microbes, protecting health thereby preventing diseases.[7]

8.3   ANTIMICROBIAL TEXTILES—HISTORY AND DEVELOPMENT

Application of naturally derived antimicrobial agents on textiles goes back to the distant past, when spices and herbs were used by ancient Egyptians for the preservation of mummy wraps. Herbs were used to protect textiles and inhibit growth of bacteria.[8] Moulds, soil, and plants have been used in many ancient cultures to cure bacterial infections. Old mouldy bread was pressed against wounds to prevent infection in Ancient Serbia, China and Greece. The crusts of moldy wheat bread were applied on pustular scalp infections and 'medicinal earth' was conferred for its recuperative properties in Egypt.[9] Around 1550 BC, use of a concoction of honey, lard and lint for dressing wounds by the Egyptians has also been recorded. It is now evident that honey contains substantial amounts of hydrogen peroxide which can eliminate bacteria. These cures were believed to influence the spirits or gods liable for illness and suffering. The occasional efficacy of these early treatments could be attributed to the active metabolites and chemicals present in the concoctions which are better understood today. The concern and requirement for protection against microorganisms during World War II instigated the research race to find or make suitable antimicrobial finishes. One of the first antimicrobial finishes used during World War II was made to prevent cotton textiles such as tents, tarpaulins and vehicle covers from rotting.[10] There has been an increased attentiveness in antimicrobial finishes over the last few years.[11]

The beginning and promotion for use of silver foil in wound dressings were initiated by one of the first American surgeons, Halsted, and silver sutures were often used in surgical incisions to prevent infections. The utilization of elemental silver as an antimicrobial agent is also nearly as old as the history of mankind. The third metal known to be used by the ancients after gold and copper was metallic silver which was known to the Chaldeans as early as 4000 B.C.E.[12] To attain enhanced wound healing, silver plates was utilized by Macedonians, possibly the first attempt to avert or treat surgical infections. Hippocrates used silver preparations for the treatment of ulcers and to promote wound healing. The probability that silver nitrate also was utilized medically is evident from citations in a pharmacopeia published in Rome in 69 B.C.E.[12] After observing Halsted applying silver foil to wounds to treat infections, B.C. Crede, a surgeon, is credited to be the first to use colloidal silver for wound antisepsis in 1891.[12,13]

Furthermore, silver nitrate was effectively used to treat compound fractures, skin ulcers and pus-forming wounds, well ahead the time of Lister. Evolving biotechnology has facilitated the integration of ionizable silver into textiles for medical use to shrink the menace of infections and for personal cleanliness.[3] Creation of metal oxides and nano-sized metals, mostly silver (Ag), zinc oxide (ZnO), copper II oxide (CuO) and titanium dioxide (TiO_2) has facilitated emergence of a new class of biocides. Silver, amongst all antimicrobial agents, shows strong biocidal effects against many pathogenic microbes and thus has been extensively utilized in various fields. The antimicrobial potential of nanosilver finish is demonstrated for a variety of bacteria such as *Escherichia coli, Pseudomonas, S. aureus* and methicillin-resistant *Staphylococcus aureus* (MRSA).[14]

Among natural antimicrobial agents, plant products encompass the foremost section. The healing capacity of some plants has been tapped ever since early times. Useful phytochemicals with pronounced antimicrobial efficacy can be divided into more than a few categories as summarized in Table 8.2.[15]

TABLE 8.2   Major Classes of Antimicrobial Compounds from Plants.

Class	Subclass	Example	Mechanism
Phenolics	Simple phenols	Catechol	Substrate deprivation
		Epicatechin	Membrane disruption
	Phenolic acids	Cinnamic acid	Bind to adhesins, complex
	Quinones	Hypericin	with cell wall, inactivate enzymes
	Flavones	Chrysin	Bind to adhesins
	Flavonoids		Complex with cell wall
		Abyssinone	Inactivates enzymes
			Inhibits HIV reverse transcriptase
		Totarol	Not known
	Flavonols	Ellagitannin	Binds to proteins
	Tannins		Binds to adhesins
			Enzyme inhibition
			Substrate deprivation
			Complex with cell wall
			Membrane disruption
			Metal ion complexation
	Coumarins	Warfarin	Interaction with eukaryotic DNA (antiviral activity)
Terpenoids Essential oils		Capsaicin	Membrane disruption

TABLE 8.2   *(Continued)*

Class	Subclass	Example	Mechanism
Alkaloids		Berberin Peperine	Intercalate into cell wall and/or DNA
Lectins and polypeptides		Mannose-specific agglutinin	Block viral fusion or adsorption
		Fabatin	Form disulphide bridges
Polyacetylenes		8S Heptadeca-2(Z),9(Z)-diene-4,6-diyne-1,8 diol	Not known

Source: Adapted From Joshi, M.; Wazed Ali, S.; Purwar, R.; Rajendran, S. Ecofriendly Antimicrobial Finishing Of Textiles Using Bioactive Agents Based On Natural Products. *Indian J. Fibre Text. Res.* **2009,** *34,* 295–304.

The finishing on cotton fabric using Aloe vera extract by utilization of glyoxal (100 g/l), an ecofriendly cross-linker by pad-dry-cure technique has been demonstrated at different concentrations (5, 10, 15, 20 and 25 g/l).[16] The antibacterial potential of Aloe vera gel applied on cotton fabric as a textile finish has additionally been evaluated by both qualitative and quantitative tests such as American Association of Textile Chemists and Colorists (AATCC)-147-1998 and AATCC-100-1998, respectively.[17]

Recent advances in various treatments have been made in resistance of shrinkage, felt proofing, ant pilling, antimicrobial, surface properties (hydrophilic, soil resistance, water and oil repellency), self-cleaning, anti-odour and flame retardation to confer functional characteristics on wool and synthetic fibres such as acrylic, polyamide and polyester. Enhancement of these properties can offer the fibres a vital position between the textile fibres making them more suitable for varied uses.[30]

The utilization of natural products such as natural dyes and chitosan for antimicrobial finishes in textiles has been extensively described.[18–22] Natural chitosan obtained from crabs, shrimps and other crustaceans is ecofriendly, safe and biocompatible polymer. It has been employed as a natural biopolymer material in drug, health, papermaking and food processing industries. Due to its polycationic nature, chitosan holds potent antibacterial efficacy against various microbes through ionic cell surface interactions eventually killing the cell. Chitosan and derivatives of chitosan have been attributed a lot of importance as antimicrobial agents in textile finishing.[17] Chitosan oligomers made by acid degradation for finishing nonwoven polypropylene fabrics have been stated to disseminate potential antimicrobial activity against *P. vulgaris, S. aureus* and *E. coli* at 0.01–0.05% level.[23] Chitosan has also been woven into wool after acylation using succinic anhydride and phthalic anhydride.[24] The antimicrobial potential was determined against *S. aureus* and *E. coli* by a qualitative method. The antimicrobial activity was more pronounced against *E. coli*.

Two different cross-linking agents, namely butane tetracarboxylic acid and Arkofix NEC (by Clariant), were utilized to bind chitosan to cotton fabrics chemically.[25] Antimicrobial potential of the treated textiles was assessed against *Bacillus subtilis, Bacillus cereus, E. coli, Pseudomonas aeruginosa, S. aureus* and *Candida albicans* by a disc diffusion method. The maximum antimicrobial activity was achieved on the treatment of cotton fabrics with 0.5–0.75% chitosan with 1.5–5 kDa molecular weight, and curing at 160°C for 2–3 min. Dimethylol dihydroxy ethylene urea was used to covalently cross-link chitosan to cotton.[26] Fabrics were evaluated against *E. coli* and *S. aureus* according to 'diffusion agar test' of the American Society for Microbiology, and 'shake flask test' ASTM E 2149-01. Chitosan-treated fabrics exhibited 100% inhibition of *E. coli* and *S. aureus*. In order to evaluate fastness to laundering, the fabrics were washed at 60°C in an automatic washing machine for 45 min using soap and sodium carbonate. The test results indicated a decline in antimicrobial activity after 5 and 10 washing cycles to 90% and 83%, respectively.

The upsurge in contemporary antimicrobial fibre know-hows in the last few years with developing awareness about cleaner environment and healthy lifestyle has led to the development of a range of textile products centred on synthetic antimicrobial agents such as triclosan, organometallics, phenols, metal and their salts and quaternary ammonium compounds of which relatively a few are available commercially.[2] Recognized to be ideal biocidal groups, halamine structures are chemically integrated to polymers and textiles for delivering powerful biocidal functions through chemical modification processes. Halamine chemistry is specified to have inhibited an array of microorganisms without hosting drug resistance.[27] Biguanides, amines, glucoprotamine and quaternary ammonium compounds demonstrate polycationic, porous and absorbent properties. Textiles finished with these substances bind to the cell membrane of the microorganisms, disrupt the lipopolysaccharide structure leading to breakdown of the cell. In this regard, abundant lines of evidence advocate that heterocyclic compounds which are extensively disseminated in nature and essential to life are also potential antimicrobial agents.[28]

Exploration of antimicrobial potential of a number of 1N-substituted-2-methyl benzimidazole has revealed notable antibacterial activity against *S. aureus* (38 mm) and *B. subtilis* (28 mm) as well as potent antifungal activity (28 mm) against *A. niger*.[29] Biological screening result of activities of 3-chloro-1-{4-[5- (substituted phenyl)-4, 5-dihydro-pyrazol-3-yl] phenyl}-4-(4-hydroxyphenyl) azetidin-2-one-based derivatives have shown better activity against *E. coli* and *S. pyogenes* at 100–225 µg/ml.[30]

Poly(p-vinylbenzyl tetramethylenesulfonium tetrafluoroborate) and poly(p-ethylbenzyl tetramethylenesulphonium tetrafluoroborate) were synthesized with different molecular weights. The antimicrobial efficacies of the polymers were pronounced against Gram-positive *S. aureus*, whereas against Gram-negative *E. coli* the activity was less. The activity of polymeric sulphonium salts was found to be much higher than that of corresponding monomers, and with the increase in molecular weight, the efficacy increased.[31] Polymerisation of (benzofuran-2-yl) (3-mesityl-3 methyl cyclobutyl)-*O*-methacryl ketoxime monomer resulted in effective inhibition of the growth of microorganisms, such as *P. aeruginosa, E. coli, C. albicans* and *S. aureus*. This could have been due to the inhibition of enzyme production by the oxime esters and carbonyl groups.[32]

A series of polymeric silver sulphadiazine with durable and rechargeable antibacterial and antifungal activities were developed.[33] Acryloyl sulphadiazine was copolymerized with methyl methacrylate to produce poly(methyl methacrylate)-based polymeric silver sulphadiazine. The antibacterial potential of the polymeric silver sulphadiazine was assessed according to AATCC Test Method 100–1999 against *S. aureus* (ATCC 6538) and *E. coli* (ATCC 15 597). The antifungal activity was tested against *Candida tropicalis* (ATCC 62690). The biocidal activity of the polymeric silver sulphadiazine was against bacteria and fungi. Further confirmation was done by Kirby–Bauer tests to indicate that there was no leaching of silver ions to the polymers.

Novel acrylamide-type monomer (*N*-(4-hydroxy-3-methoxy-benzyl)-acrylamide was effectively polymerized by conventional radical polymerization technique and the polymers showed promising antibacterial activity against *B. subtilis* (ATCC 6633).[34]

Ever since the origin of the concept of 'antibacterial finishing for textiles' in 1941, the textile industry has had ample progress in developing fibres and agents with antibacterial properties. Man has embraced substances with antimicrobial properties since olden times and fabric protection and preservation, also, have extended a prominent role. In several textile uses today, the need to shield and preserve is still fundamental. People had not comprehended until recently that it is equally more significant to protect wearers also against the spread of microbes and diseases. Consequently, great attention in the utilization of agents with antimicrobial properties for textiles or antibacterial finishing of fibres and fabrics for practical applications has been witnessed in the recent years.[35,36] Textile materials, today, are used extensively in countless milieus, and treatment with antimicrobials is promptly becoming a precondition for textiles utilized in clinics, hotels, sports and personal care industries.

Textile with antimicrobial properties can be utilized to produce many goods such as sportswear, outdoor apparels, undergarments, shoes, furnishings, upholstery, hospital linens, wound care wraps, towels and wipes. Self-sterilizing fabrics could possibly have potential benefits to lessen transmission of diseases in hospitals, bio-warfare prevention and other bids. The insight of market demand for finishes and additives with antimicrobial properties is evolving and an improved evolution projected in future.[37] The business development strategies in antimicrobial finished fibrous products have been ascertained by the commercial gurus. Some examples would include the following:[38]

1. Products such as masks, scrubs, towels and linens are anticipated to be preferred in substantial quantity in the United States of America where there are more than 6000 hospitals.
2. Development of socks and underwear for military personnel with antimicrobial and antifungal finishes. Unavailability of such product has also been indicated in health care sector to provide sustained regulation of bacteria and fungi causing foot ulcer.
3. Market comprising socks, gloves, bras, caps and so forth for sports.

The global consumer market for antimicrobial wipes was approximately 5.8 billion and 1.2 billion US dollars, respectively, in 2007. The market was anticipated to expand and estimated to reach 8.5 and 1.7 billion US dollars, respectively, by 2012.[39] A substantial increase in the diaper market for babies is projected around the world. The global diaper market for babies is estimated to reach 52.2 billion US dollars in 2017, according to the latest report by Transparency Market Research (Albany, NY, USA).[40]

The major products were anticipated to be iodophors, nitrogen compounds, phenolic compounds and organometallics. A progressive growth from 2009 to 2014 is forecasted in the USA after a recent study evaluated the worth of 2 billion dollars disinfectant and antimicrobial chemical industry.[41] The chemo diversity of an assortment of vital antimicrobials which are commercially available in the market is summarized in Table 8.3.[42]

TABLE 8.3 Some Commercially Available Antimicrobial Finishes and Their Chemical Nature.

Sl. no.	Finish name (Company)	Chemical nature
1	Ecosy (Unitika)	Based on chitosan
2	Irgaguard (B5000,B6000,B7000) (BASF (Ciba))	Silver based for synthetic fibres
3	Irgaguard F3000 (BASF (Ciba))	Benzimidazole
4	Irgaguard 1000 (BASF (Ciba))	Triclosan
5	Irgasam (Sigma Aldrich)	Triclosan
6	iSys (CHT)	Based on silver chloride
7	Microban (Microban International)	Triclosan
8	Reputex 20 (Arch Chemicals)	Based on PHMB
9	Ruco-Bac AGL; EPA (Rudolf Chemie)	Hygienic finish containing salts and surfactants
10	Ruco-Bac AGL; AGP (Rudolf Chemie)	Silver chloride
11	Ruco-Bac AGL; CID (Rudolf Chemie)	Antifungal based on triazole compound
12	Ruco-Bac AGL; EXE (Rudolf Chemie)	Antimycotic finish based on quaternary ammonium compounds
13	Ruco-Bac AGL; MED (Rudolf Chemie)	Diphenyl alkane
14	Sanigard (DC, 7500, 500) (LN Chemical Industries)	Based on organosilicon quaternary compounds
15	Sanigard (CHF) (LN Chemical Industries)	Based on Triclosan
16	Sanigard (KC) (LN Chemical Industries)	Cationic surface active agent based on quaternary ammonium group
17	Silpure (Thompson Research Associates)	Based on fine silver particles
18	Smart silver (Nanohorizon Inc.)	Based on silver nanoparticles
19	Tinosan HP 100 (BASF(Ciba))	Based on halogenated phenols
20	Utex (Nantec Textile Co Ltd.)	Natural chitin rayon fibre with antimicrobial performance
21	Vantocil IB (Zeneca Ltd.)	Based on PHMB

Source: Adapted from Simoncic, B.; Tomsic, B. Structures of Novel Antimicrobial Agents for Textiles—A Review. *Text. Res. J.* **2010,** *80*(16), 1721–1737.

8.4   CLASSIFICATION OF ANTIMICROBIAL FINISHES

Antimicrobial textiles inhibit the growth and proliferation of microorganism. Antimicrobials can be categorized into two types, namely leaching and non-leaching type. The antimicrobial compound disperses from the clothing and is in interaction with microbe in the leaching type. Antimicrobials drift off the clothing, establishing a province of activity and destroy any microbes coming into contact. Eventually, the product is utilized by the microbes and gradually loses effectiveness. In the non-leaching type, the antimicrobials do not drift off the clothing. The antimicrobials destroy microbes by interaction with the cell membrane. The finishing effect of these products is mostly permanent retaining functionality through the life of the textile. These types of textile finish can endure more than 40 laundry washes. Based on its activity, it is suitable to again regroup a general type of finish into three main groups, namely rot-proofing, hygienic and aesthetic finishes. [43] The rot-proofing antimicrobial finishes are applied to endure protection of the materials either short- or long-term against physical wear and tear. Hygiene finishes are concerned with control of infection and undesirable bacteria. And, antimicrobials with aesthetic finishes are employed to prevent the development of odour and staining.

8.5   FUNCTIONS OF ANTIMICROBIAL TEXTILES

Some of the utmost significant functions of any antimicrobial textile finishes are to control infestation and prevent from cross infections caused by pathogenic microorganisms. The antimicrobial finish must impart durability by arresting microbial metabolism and reduce odour formation. It must also safeguard the textile from getting stains, discoloured and quality deterioration. Additionally, the antimicrobials should limit the growth of microbes to their extinction and protect the textile user against pathogens causing diseases.

8.6   APPLICATION METHODOLOGIES

The antimicrobials can be augmented into the textiles by utilizing the methodologies described herein. The use of spun in additives is a common practice for imparting antibacterial properties to synthetic fibres. The bioactive agents are incorporated into melt and spinning dope solution. To provide fibres with enduring antimicrobial properties, numerous agents with antimicrobial properties can be unified in the polymer matrix during the fibre/yarn manufacturing process. This also helps to avoid further tough subsequent operations. In the padding process however the textiles can be padded with antimicrobials with nearly 70–80% of expression. Certain cross-linking and binding agents can be amended together with the antimicrobials. The padding process is generally followed by drying in the air or curing in stenter. Methodologies such as spraying solutions of active antimicrobials are customarily not recommended taking into account the creation and subsequent inhalation of drops of irrespirable size. Predominantly apt for nonwoven fabrics, spraying treatments can be applied if there are provisions of appropriate containment facilities. In polymer modification methods, copolymerization of monomers along with biologically active functional groups can be accomplished. By being an integral part of the fibre, the bioactive elements result in durable effects which are an added advantage. The requirement for special polymerization units makes the technology expensive, for example copolymerization of monomers with bioactive to modify acrylic polymer which gives cationic amines or quaternary ammonium salts or even carboxylic group in the polymer which are able to react with antibiotics.

For achieving good antimicrobial durable synthetic fibres, in microencapsulation, the controlled release of antimicrobials from the fibres appears to be an established and workable technology. This technique however is unsuitable for cotton textiles. In the fixation and controlled release process, the antimicrobial agents persist on cotton fabrics after extensive launderings. This is because the capsules with bioactive ingredients are covalently fixed on the fibres. Then also it is crucial to ensure, the regulated release of any antimicrobial in an individual capsule system. Without a controlled release, the treated cotton fabric would not demonstrate good antimicrobial effectiveness. And the wash fastness of the treated fabric will be poor if the release is too fast. Additionally, the capsules should be firm enough to endure the processes commonly involved in fabric treatment. Also, they should be small enough so that no external change in the other properties of the treated fabrics is caused. Cotton fabrics treated with this system indeed present good antimicrobial properties also after 100 launderings. More recently, innovative approaches on the utilization of fabrics with various coatings have been reported. To improve and influence the reflectivity, conductivity, insulation or for allocating decorative purposes coatings can be applied onto fabrics. Enhanced biological activity when in contact with soil or in a humid environment is a great asset for fabrics with antimicrobial properties. In recent years, know-hows such as physical vapour deposition (PVD) have been efficient to modify textiles because of its characteristic advantages of being ecofriendly and a solvent-free process. Amongst the most frequently used techniques in PVD, sputter coating produces very thin metallic or ceramic coatings on to a wide range of substrates, be it either metallic or nonmetallic. For technical applications, the technique is also utilized to coat textile materials.

8.7 CHARACTERISTICS OF ANTIMICROBIAL AGENTS FOR TEXTILES

Antimicrobial agents if bacteriostatic/fungistatic or bactericidal/fungicidal and are heat, light and ultraviolet rays stable, can be considered for textile finishes. As a compound, the antimicrobial agent must retain its stability for the estimated shelf life of the finished goods as well as throughout the storage. The microbes before they can grow to deteriorate the fabric get destroyed essentially by a large number of products. The antimicrobials moreover should be active at a relatively lesser percentage enabling cost effectiveness and allowing the user to retain low weight add-on. Apart from being nontoxic, the compounds must not cause discolouration or unpleasant odour formation in the end item, particularly if it is in wear-apparel class. While most antimicrobials are completely free from this distasteful feature, some agents possess unpleasant odour features. Also importantly, the antimicrobial compound must not considerably alter the hand/feel of the fabric, especially in fabrics used for the manufacture of wear-apparel. After treatment with the antimicrobial, the fabric must not get a harsh hand. No adverse chemical effects on the processed fabric would be an added advantage. The tensile strength maintained under all types of conditions for long usage periods will also be advantageous.[44]

The antimicrobials should also be tested for their durability and efficacy after incorporation into the textiles. Proper evaluation systems to assess the effectiveness as well as the safety of antimicrobial textiles are imperative. These tests could be performed by following the standard procedures of textile performance tests. A detailed knowledge of the technology involved and the bioactive substances used as a finish helps in deciding a proper test method to evaluate its efficacy. The in vitro tests such as agar diffusion and suspension tests can be employed to record qualitative and quantitative data on the potential and effectiveness of antimicrobial textiles and quantify microbial reduction by the antimicrobial textiles. It is past the scope of this article to compare all the most important accepted international test methods for antimicrobial textiles as summarized in Table 8.4.[3]

TABLE 8.4 Overview of the Standard Performance Tests of Textiles for Antimicrobial Efficacy.

Designation	Title	Principle
SN 195920-1992	Textile fabrics: determination of the antibacterial activity	Agar diffusion test
SN 195921-1992	Textile fabrics: determination of the antimycotic activity	Agar diffusion test
EN 14119:2003-12	Textiles evaluation of the action of microfungi	Agar diffusion test
ASTM E 2149-01	Standard test method for determining the antimicrobial activity of immobilized antimicrobial agents under dynamic contact conditions	Suspension test
JIS Z 2801	Antimicrobial products—test for antimicrobial activity and efficacy	Suspension test
JIS L 1902–2002	Testing for antibacterial activity and efficacy on textile products	Suspension test

Source: Adapted from Wollina, U.; Abdel Naser, M.; Verma, S.; Hipler, U.C.; Elsner, P. (eds.). Skins physiology and textiles- consideration of basic interactions, biofunctional textiles and the skin. Curr Probl Dermatol, 2006, 33, 1-16.

The Kirby–Bauer agar diffusion test has been a time-honoured traditional testing method in microbiology for antibiotic sensitivity studies. The test can be performed to screen against various microorganisms. The method can be commended as a rapid and preliminary qualitative method to discriminate between active (i.e. clear zone of growth inhibition) and passive antimicrobial principles (no zone of inhibition), in textile research. A measurement of the diameter of the clearance zone indicates the dimension of the antimicrobial activity. Nevertheless, the agar diffusion test is also recommended to test textiles with passive antifungal activity. However, the method is limited to textile fabrics and for materials other than fabrics the test conclusions can be imprecise.

With the assistance of suspension tests such as the JIS 1902–2002, it is probable to determine the maximum viable in vitro "degree of effectiveness" of the finished textiles against various test microbes. This standard method is an excellent method to determine the antimicrobial activity of textiles. Suspension tests could also be modified to record the activity of textiles with passive principles.

8.8 ANTIMICROBIAL TEXTILE FINISHES—CONCERN AND CHALLENGES

An extensive array of textile products is now available for the consumers benefit. The primary objective of such textile finishes initially was to safeguard textiles such as uniforms, tents and military garments from being deteriorated by microbes. Later, followed antimicrobial finishes for home textiles such as curtains, towels and mats. The application of different finishes is now explored in textiles for healthcare, sports, outdoor sector, as well as leisure. While a lot of research investigation has been done to explore antimicrobial finishes for textile substrates, several concerns still need to be investigated.[45-47] Current available antimicrobial products, such as triclosan and silver, suffer from precarious flaws of short active shelf life or high price. Such low-molecular weight antimicrobial agents commonly leach out from the fabrics to the milieu and sometimes to the wearer's skin. High molecular weight antimicrobial polymers could outwit these complications by reducing or averting the leaching of bioactive substances. These polymers are thus considered a striking way for the "non-leaching" approach in the production of fabrics with bactericidal finishes. This approach is interesting for many applications in textiles owing to improved ecofriendliness, stability, lack of diffusion to the wearers' skin, low toxicity, enhanced biocompatibility, low corrosion of metals and long residence time with potent biological activity.

Though several bacteriostatic textile finishes already exist for personal wear, outerwear fabrics, hosiery, inner-wears, carpeting, throw rugs, filter fabrics and so forth, their utilization has not gained ready acceptance. The pursuit to discover a natural antimicrobial agent still persists. Also with the increase in the microbial-resistance, the urge to find new groups of antimicrobial agents with enhanced performance is imperative. The mechanism of action of many natural agents with antimicrobial property whether bactericidal or static is unknown still. Most of the products are insoluble in water and dissolution of antimicrobial agents used for textile finishing is a major challenge. As a result of which the preliminary screening methods require proper optimization to establish the accurate inhibitory concentrations of these antimicrobial agents. The adherence of substances which are bioactive to several composite textile substrates for extended stability with antimicrobial performance is also an innovative avenue of research. Poor activity against mold and mildew, lack of wash durability, inadequate safety data to encounter current requirements, or a blend of these factors has hindered their utilization. Consequently, a safe, wash resistant textile finish capable of inhibiting the microbial growth is required. Trendsetting development in textile finishes with antimicrobials underlines the requirement for evaluating ecofriendliness of antimicrobial finishes. Hence, there is persistent interest in the expansion of new antimicrobial agents ascribed to the increasing emergence of bacterial resistance as well as to meet the global challenges of providing improved value-added products.

8.9 CONCLUSION

In the pyramid of human requirements, it can be believed that clothing stands second top most in preference next to food. The demand for ecofriendly antimicrobial textiles to effectively lessen the unpleasant effects of growth of microbes on textile materials is expanding. But an antimicrobial textile finish that involves contact with skin will require added data concerning the safety aspect. The prospective is substantial if the economics and detailed performance can be accomplished. Benefits could embrace better durability and enhanced safety. The ill effects of textile finish contents released in water bath, occupational hazards concerned with the skin of user upon contact/inhalation/intake by other living species entail exhaustive investigation. The quest for safer and cost-effective testing methods thus will endure. Altogether refined enactment and cost-effectiveness, while fulfilling the environmental and toxicity requirements, will linger to test all those operational in the field.

KEYWORDS

- **antimicrobial**
- **textiles**
- **performance**
- **durability**

REFERENCES

1. Zacharia, J. Developments of Kerosene Vapour Recovery System to Control Air Pollution in India. *Text. Print. Ind.* **1999,** 16–17.
2. Abdel Naser, M. B.; Verma, S.; et al. *Bifunctional Text. Skin* **2006,** *33,* 1.
3. Wollina, U.; Abdel Naser, M.; Verma, S.; Hipler, U. C.; Elsner, P. Eds. Skins Physiology and Textiles—Consideration of Basic Interactions, Biofunctional Textiles and the Skin. *Curr. Probl. Dermatol.* **2006,** *33,* 1–16.
4. Wasif, A. I.; Laga, S. K. Use of Nano Silver as an Antimicrobial Agent for Cotton. *AUTEX Res. J.* **2009,** *9*(1), 5–13.
5. Barnes, C.; Warden, J. Microbial Degradation: Fiber Damage from *Staphylococcus aureus. Text. Chem. Color.* **1971,** *3,* 52–56.
6. Thiry, M. C. Testing Antimicrobial Performance. *AATCC Review: Mag. Text. Dyeing, Printing, Finishing Industry,* **2010,** *10*(6), 26–37.
7. Krishnaveni, V.; Rajkumar, G.; Shanmugam, S. Antimicrobial Finish in Textiles. *Colourage* **2007,** *46.*
8. Ammayappan, L.; Jeyakodi Moses, J. Study of Antimicrobial Activity of Aloe Vera, Chitosan, and Curcumin on Cotton, Wool, and Rabbit Hair. *J. Fibers Polym.* **2009,** *10*(2), 161–166.
9. Keyes, K.; Lee, M. D.; Maurer, J. J. *Microbial Food Safetry in Animal Agriculture Current Topics*; Torrence, M. E., Isaacson, R. E., Eds., Iowa State Press: Ames, Iowa, USA, 2003.
10. Mudnoor, V.; Laga, S. K. Application of Antimicrobial Finish in Medical Textiles-An Overview. *Colourage,* **2012,** *59*(10), 40.
11. Pannu, S. Investigation of Natural Variants for Antimicrobial Finishes in Innerwear—A Review Paper for Promotion of Natural Hygiene in Innerwear. *Int. J. Eng. Trends Technol.* **2013,** *4*(5), 2168–2170.

12. Hill, W. R.; Pillsbury, D. M. *Argyria—The Pharmacology of Silver*. Williams & Wilkins: Baltimore, 1939.

13. Grier, N. In *Disinfection, Sterilization and Preservation;* Block, S. S., Ed.; Lea & Febiger: Philadelphia, 1968; pp 375–398.

14. Kanokwan, S.; Pranee, R.; Supin, S. A Study on Antimicrobial Efficacy of Nanosilver Containing Textile. *CMU J. Nat. Sci.* **2008**, *7*, 134–136.

15. Joshi, M.; Wazed Ali, S.; Purwar, R.; Rajendran, S. Ecofriendly Antimicrobial Finishing of Textiles Using Bioactive Agents Based on Natural Products. *Indian J. Fibre Text. Res.* **2009**, *34*, 295–304.

16. Wasif, S. K.; Rubal, S. *Proceedings of 6th International Conference TEXSCI*. Czec Republic, Liberec, 2007.

17. http://www.fibre2fashion.com/industry-article/textile-industry-articles/an-eco-friendly-herbal-herbal-antimicrobial-finish/an-eco-friendly-herbal-antimicrobial-finish1.asp (accessed Sept 6, 2008).

18. Anderson, K. F.; Sheppard, R. W. Dissemination of Staphylococcus aureus from woollen blankets. *Lancet* **1959**, *1*, 514–515.

19. Lim, S. H.; Hudson, S. M. Review of Chitosan and its Derivatives as Antimicrobial Agents and Their Uses as Textile Chemicals. *J. Macromol. Sci.: Polym. Rev.* **2003**, *43*, 223–269.

20. Gupta, D.; Khare, S. K.; Laha, A. Antimicrobial Properties of Natural Dyes Against Gram-Negative Bacteria. *Color. Technol.* **2004**, *120*, 167–171.

21. Singh, R.; Jain, A.; Panwar, S.; Gupta, D.; Khare, S. K. Antimicrobial Activity of Some Natural Dyes. *Dyes Pigm.* **2005**, *66*(2), 99–102.

22. Gupta, D.; Jain, A.; Panwar, S. Anti-UV and Anti-Microbial Properties of Some Natural Dyes on Cotton. *Ind. J. Fibre Text. Res.* **2005**, *30*(2), 190–195.

23. Shin, Y.; Yoo, D. I.; Min, K. Antimicrobial Finishing of Polypropylene Nonwoven Fabric by Treatment with Chitosan Oligomer. *J. Appl. Polym. Sci.* **1999**, *74*(12), 2911–2916.

24. Ranjbar-Mohammadi, M.; Arami, M.; Bahrami, H.; Mazaheri, F.; Mahmoodi, N. M. Grafting of Chitosan as a Biopolymer onto Wool Fabric Using Anhydride Bridge and its Antibacterial Property. *Colloids Surf., B: Biointerfaces* **2010**, *76*, 397–403.

25. El-tahlawy, K. F.; El-bendary, M. A.; Elhendawy, A. G.; Hudson, S. M. The Antimicrobial Activity of Cotton Fabrics Treated with Different Crosslinking Agents and Chitosan. *Carbohydr. Polym.* **2005**, *60*, 421–430.

26. Öktem T. Surface Treatment of Cotton Fabrics with Chitosan. *Color. Technol.* **2003**, *119*, 241–246.

27. Gang, Sun.; Worley, S. D. Environmental Concern in Antimicrobial Textiles Chemistry of Durable and Generable Biocidal Textiles. *J. Chem. Educ.* **2005**, *82*(1), 60.

28. Danie`le, M.; Garnaud, C.; Thierry, C.; Cornet, M. Resistance of *Candida* Spp. to Antifungal Drugs in the ICU: Where are we Now? *Intensive Care Med.* **2014**, *40*(9), 1–15.

29. Ansari, K. F.; Lal, C. Synthesis and Biological Activity of Some Heterocyclic Compounds Containing Benzimidazole and Beta-Lactam Moiety. *J. Chem. Sci.* **2009**, *121*(6), 1017–1025.

30. Shah, S. H.; Patel, P. S. Synthesis and Antimicrobial Activity of Azetidin-2-one Containing Pyrazoline Derivatives. *Res. J. Chem. Sci.* **2012,** *2*(7), 62–68.
31. Kanazawa, A.; Ikeda, T.; Endo, T. Antibacterial Activity of Polymeric Sulfonium Salts. *J. Polym. Sci., Part A: Polym. Chem.* **1993,** *31*, 2873–2876.
32. Erol, I. Synthesis and Characterization of a New Methacrylate Polymer with Side Chain Benzofurane and Cyclobutane Ring: Thermal Properties and Antimicrobial Activity. *High Perform. Polym.* **2009,** *21*, 411–423.
33. Cao, Z.; Sun, X.; Sun, Y.; Fong, H. Rechargeable Antibacterial and Antifungal Polymeric Silver Sulfadiazines. *J. Bioact. Compat. Polym.* **2009,** *24*, 350–367.
34. Liu, H.; Lepoittevin, B.; Roddier, C.; Guerineau, V.; Bech, L.; Herry, J. M.; Bellon-Fontaine, M. N.; Roger, P. Facile Synthesis and Promising Antibacterial Properties of a New Guaiacol-Based Polymer. *Polymer* **2011,** *52*, 1908–1916.
35. Worley, S. D.; Sun, G. Biocidal Polymers. *Trends Polym. Sci.* **1996,** *4*, 364–370.
36. Sun, G. *Bioactive Fibres and Polymers, Chapter 14, ACS Symposium Series 792*; Edwards, J. V., Vigo, T. L. American Chemical Society: Washington, D.C, 2001.
37. http://www.hkc22.com/antimicrobials.html.The Market for Antimicrobial Additives/Biocides in Plastics and Textiles Worldwide 2011–2025: Development, Strategies, Markets, Companies, Trends, Nanotechnology, Home page (accessed Oct 31, 2012).
38. http://www.infectioncontroltoday.com/news. Infection Control Today: Hospital, Hospitality, Consumer Markets-Driving Market for Antimicrobial Textiles (September 16, 2010), Home page (accessed Nov 1, 2012).
39. http://www.simithersapex.com.The Future of Antimicrobial and Antimicrobial Wipes—Strategic Five-Year Forecasts, Market Reports (September 24, 2007), Home page (accessed Oct 31, 2012).
40. Baby Diaper Market Projected to Reach US$52.2 Billion Globally by 2017. *Medical Textiles*, 2012, 10.
41. http://www.freedoniagroup.com. Disinfectant and Antimicrobial Chemicals to 2009—Demand and Sales Forecasts, Market Share, Market Size, Market Leaders, Study—1975, (September 2005), Home page (accessed Oct 31, 2012).
42. Simoncic, B.; Tomsic, B. Structures of Novel Antimicrobial Agents for Textiles—A Review. *Text. Res. J.* **2010,** *80*(16), 1721–1737.
43. Dorugade, V. A.; Bhsilveryashri, K. *Man Made Textiles in India*; 2010, p 89.
44. Naresh, M. S.; Ashok, G. S.; Vaishali, Rane. Antimicrobial Finish—Strengthening the Next Generation. *Int. Dyer* **2011,** 34–36.
45. McNeil, E.; Greenstein, M. Control of Transmission of Bacteria by Textiles and Clothing. In *Proceedings of Chemical Specialties Manufacturers Association, Chemical Specialties Manufacturers Association Proceedings*, Reilly, F. J. Ed.; New York, 1961; pp 134–141.
46. McNeil, E. Dissemination of Microorganisms by Fabrics and Leather. *Dev. Ind. Microbiol.* **1964,** *5*, 30–35.
47. Wiksell, J. C.; Pickett, M. S.; Hartman, P. A. Survival of Microorganisms in Laundered Polyester-Cotton Sheeting. *Appl. Microbiol.* **1973,** *25*, 431–435.

CHAPTER 9

INFLUENCE OF WASTEWATER USE IN AGRICULTURE: ADVANCES IN HUMAN AND PLANT HEALTH

KAVINDRA KUMAR KESARI[1], POOJA TRIPATHI[2], and DR. VIJAY TRIPATHI[3],*

[1]*Department of Applied Physics & Department of Bioproduct and Biosystem, Aalto University, Espoo, Finland*

[2]*Center of Bioinformatics, IIDS, University of Allahabad, Allahabad, UP, India*

[3]*Department of Molecular and Cellular Engineering, Jacob Institute of Biotechnology and Bioengineering, Sam Higginbottom University of Agriculture, Technology and Sciences, Naini, Allahabad, India, *E-mail: vijay.tripathi@shiats.edu.in*

CONTENTS

ABSTRACT

The use of wastewater in agriculture is increasingly important as the world's population increases. The influence of treated or untreated wastewater disposal for irrigation is one of the main concerns of this chapter. From several studies, it has been confirmed that direct use of wastewater for crop irrigation is very injurious to health. Wastewater contains many kinds of contaminants sch as, metal, microbial pathogens, chemical, antibiotics or medicinal and many more. These contaminants come in contact with plants, crops or farmers and cause severe health issues by consuming wastewater-irrigated crops or vegetables. This chapter explores the several types of contaminantes and their role in causing diseases. Another important issue is of open burning of agricultural waste on filed after harvesting crop. This is even more serious concern that wastewater-irrigated crops may burnet after harvesting on the field. Due to open burning, it may release many toxic compounds in the air and soil, which could also participate in ground water contamination through leaching process. So, instead of burning on the field, it might be useful to utilize them as energy produces as biofuel production. In this chapter, an occurrence and detection of antibiotic resistance in wastewater treatment plants (WWTPs) have been considered in association with the use of wastewater in agriculture and possible effects. These are the issues, which are very useful to discuss and spread the awareness in farmers and consumers especially in developing countries because they generally use untreated wastewater for crop irrigation.

9.1   INTRODUCTION

Seventy per cent of freshwater including all the water diverted from rivers and pumped from underground is used for agricultural irrigation[1] and 93% of water consumption worldwide[2,3] which has serious concern for the countries facing drinking water crisis. Wastewater use is increasingly seen as an option to meet these growing needs for water. The fact that an estimated 20 million hectares worldwide are irrigated with wastewater suggests that the wastewater is potentially a major source of irrigation. However, much of the wastewater that is used for irrigation is untreated.[4,5] This is especially so in developing countries, where the partially treated or untreated wastewater is used for the purpose.[6] Therefore, treated wastewater reuse in agriculture is a common practice in many countries and an ideal resource to replace freshwater use in agriculture. The use of treated wastewater may be useful in land reclamation, improving agricultural growth and reducing fertilization costs. The use of treated wastewater for irrigation is now widely used since decades.[7]

Untreated wastewater often contains a large range of chemical contaminants from waste sites, chemical wastes from industrial discharges, heavy metals, fertilizers, textile, leather and paper industry, sewage waste, food processing waste and pesticides. There are significant health implications associated with the use of wastewater for agriculture where consumption of wastewater-irrigated crops and direct exposure to wastewater are linked to many viral, bacterial, protozoan, diarrhoeal diseases.[8] Moreover, the application of wastewater to soil, particularly untreated wastewater, followed by its infiltration, poses a significant risk of pollution, not only to soil and crops but also to the surface and subterranean water sources surrounding the irrigated area.[9,10] Pathogenic bacteria or helminths agents present in wastewater are the main factors for the cause of soil contamination. Not only soil but the air also gets polluted due to untreated wastewater used for irrigation. This chapter aims to describe what it is known and what it is unknown regarding the positive and negative impacts of the reuse of treated/untreated wastewater in agricultural irrigation. It will be shown in detail how this practice can benefit soil and farmers, while at the same time posing a risk of contamination to the ecosystem as also represented in Figure 9.1.

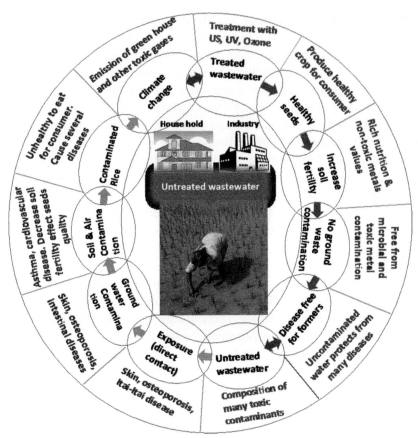

FIGURE 9.1 A schematic showing the overall conceptual framework based on the impact of treated or untreated wastewater use in irrigation.

9.2 INFLUENCE OF UNTREATED WASTEWATER

Wastewater use in agriculture has substantial benefits for agriculture and water resources management, but can also pose substantial risks to public health, especially when used untreated for crop irrigation. Factors, which affects the direct use of wastewater for agriculture is the pollution of soil and air, contamination of crops and groundwater, where these are directly connected with human health risks. However, the positive aspects of untreated or partially treated water benefit agriculture with reliable and possibly less costly irrigation water supply; increased crop yields, often with larger increases than with freshwater due to the wastewater's nutrient content. The characteristics of the pollutants received by soil via wastewater may be different in developing and developed countries. There are certain microorganisms, which may differ in different types of wastewater. In developed countries, it might be treated before use, but opposite to this in developing countries it may be used directly for irrigation. In consideration to above, following factors will be considered in this chapter for the evaluation of wastewater use in agriculture as also shown in Figure 9.1.

9.3   UNHEALTHY CROP IRRIGATION AND HEALTH

Crops are polluted by direct contact with wastewater during irrigation. The microbial contamination of crops in wastewater irrigation systems is closely related to the survival of microorganisms. Untreated wastewater often contains a large range of chemical contaminants from waste sites, chemical wastes from industrial discharges, heavy metals, fertilizers, textile, leather and paper industry, sewage waste, food processing waste and pesticides. Serious health concerns arise from the presence of large amounts of pathogens, skin irritants and toxic chemicals in wastewater. The farmer and their family members who work in the field poured with untreated wastewater may face several kinds of diseases mainly skin disease and infection. Wastewater-irrigated crops majorly consume heavy metals, which originate from mining, foundries and smelters metal-based industries.[11] Exposure to heavy metals including arsenic, cadmium, lead, and mercury in wastewater-irrigated crops is a cause for many health problems. For example, the consumption of high amounts of cadmium causes osteoporosis in humans.[12] The uptake of heavy metals by the rice crop irrigated with untreated effluent from a paper mill has been shown to cause serious health concerns.[13] Pandey et al. (2012)[14] found that while root vegetables showed greater uptake of heavy metals over leaves and fruits irrigated with wastewater, when the inputs of atmospheric deposition were considered together with wastewater irrigation, leafy vegetables showed greater contamination levels. Therefore, owing to these widespread health risks, the World Health Organization (WHO) published the third edition of its guidelines for the safe use of wastewater in irrigating agricultural crops.[8]

9.3.1   CONTAMINATIONS AND HEALTH

The use of waste vwater in agriculture may produce several types of contaminations such as soil, ground water contamination, air contamination, heavy metal contamination, microbial contamination and many more. This is an interesting aspect to know that how wastewater-irrigated crop may affect human health by different sources, which has been discussed here.

9.3.2   GROUND WATER CONTAMINATION

The use of untreated wastewater contains several types of contaminants, which may get activated during croup irrigation. Direct release of untreated effluents to land and water bodies can potentially contaminate surface and groundwater as well as soils and eventually the crops grown on these soils, which affect the quality of the food produced. Contamination of groundwater resources with faecal bacteria or viruses from wastewater poses a threat to the portability and the use of water resources.[15,16] Soil and groundwater pollution is the main risk of using wastewater in agriculture; the microbiological pollution of groundwater is a lesser risk, as most soils will retain pathogens in the top few meters of soil except in certain hydrogeological situations like limestone formations. Chemical risks include, among others, nitrates in groundwater from sewage irrigation, salination of soils and aquifers, and changes in soil structure from, for example, boron compounds common in industrial and domestic detergents. The most important anthropogenic factor responsible for groundwater pollution is urban and industrial wastewater.

9.3.3   SOIL CONTAMINATION

Direct use of wastewater in agriculture leads to contamination of soil and crops by pathogenic agents. This effect of wastewater reuse in agriculture has received the most attention from environmentalists and scientists. Wastewater irrigation may lead to transport of heavy metals to fertile soils, affecting soil flora and fauna and may result in crop contamination. Some of these heavy metals may bio-accumulate in the soil while other such as

Cd and Cu may be redistributed by soil fauna such as earthworms.[17,18] Soils irrigated by wastewater accumulate heavy metals such as Cd, Zn, Cr, Ni, Pb and Mn in surface soil. When the capacity of the soil to retain heavy metals is reduced due to repeated use of wastewater, soil can release heavy metals into ground water or soil solution available for plant uptake. Heavy metals can also contaminate the food chain and reduce crop yields.[19,20] Soil has traditionally been an important medium for waste disposal.[21,22] Municipal waste dump soils are always higher in organic matter. In the long term, the use of municipal solid waste compost may also cause a significant accumulation of Zn, Cu, Pb, Ni and Cd in the soil and plants.

9.4   CLIMATE CHANGE AND ENVIRONMENT

The use of wastewater is also emerging as a form of climate change adaptation because it provides a consistent source of water in variable or dry conditions.[23] Reuse increases the total available water supply and reduces the need to develop new water resources and therefore provides an adaptation solution to climate change. Wastewater reuse could provide a mitigation solution to climate change through the reduction in greenhouse gases by using less energy for wastewater management compared to that for importing water, pumping deep groundwater, seawater desalination, or exporting wastewater.[24] The use of wastewater and its application to soil is known to increases greenhouse gas (GHG) emission.[25] It also contributes to GHG emission directly through the release of CO_2, CH_4 and N_2O from C and N compounds present in the wastewater. Several researchers reported that use of wastewater for irrigation increases GHG emissions like N_2O (due to microbial transformation of the N contained in wastewater);[26] CO_2 (influenced by treated wastewater application on a silt loam soil).[27] Zou et al. (2009)[28] investigated that sewage-irrigated paddy cultivated land significantly increased CH_4 emission.

9.5   INFLUENCE OF TREATED WASTEWATER IN AGRICULTURE

Treated wastewater not only offers an alternative water irrigation source but also the opportunity to recycle plant nutrients.[29] Treated wastewater, which in many ways is ideal for agricultural water,[30–32] is already being widely used throughout the world,[33] and previous studies suggest that more than 10% of the world's population is consuming agricultural products cultivated by wastewater irrigation.[8] When using treated wastewater for irrigation, water quality must be strictly controlled considering factors such as the possible accumulation of substances, which are harmful to crop growth, the potential damage to soil by the transformation of its physical and chemical characteristics[34,35] and microbe infection.[36,37] Angelakis et al. (1999), Brissaud (2008), Kalavrouziotis et al. (2015), and Paranychianakis et al. (2015)[38–41] researched the current status of wastewater reuse for agriculture, assessed its standards and guidelines for water quality, and asserted the necessity for a common criterion based on scientific evidence. Treated wastewater may increase several properties of crops, soil fertility rate, healthy seeds and vegetable plants. Wastewater is a valuable source of plant nutrients and organic matter needed for maintaining fertility and productivity of arid soils.[42]

Treated wastewater, which has been used for crop irrigation, has a high nutritive value that may improve plant growth, reduce fertilizer application rates, and increase the productivity of poor fertility soils.[43] Several other studies have reported that treated wastewater irrigation increases soil organic matter[43–45] as well as the concentrations of different nutrients involved in plant growth such as nitrogen (N), phosphorus, iron, manganese, potassium, calcium, magnesium, and others[46–51] Wastewater, which comes from medicinal or pharmaceutical industries, has typical characteristic due to the occurrence of antibiotic resistance. The detection of antibiotic is important before disposal, though several methods used to detect these.

9.6 OCCURRENCE AND DETECTION OF ANTIBIOTIC RESISTANCE IN WASTEWATER TREATMENT PLANTS (WWTPs)

Antibiotics are the most effective group of antimicrobials drugs used for the improvement of human, animal and plant health and the treatment of microbial infections caused by pathogenic bacteria. Most of the antibiotics are also used as a growth promoter to increase the feed efficiency in animals. Therefore uncontrolled production and unmonitored consumptions of antibiotics have induced the proliferation and development of antibiotic resistance bacteria (ARB) and antibiotic resistance genes (ARG) and introduced these bacteria into the different compartments of the environments.[52–55] Horizontal gene transfer (HGT) is believed to be a major route for the widespread dissemination and exchange of antibiotic resistance genes between similar and different species and also responsible for the transfer of plasmid carrying antibiotic resistance genes in bacteria. Mobile genetic elements such as integrons, plasmid and transposons are the most helpful in the transferring antibiotic resistance genes amongst bacterial populations. Most of the studies prove that the wastewater is the main breeding ground, reservoir and suppliers of both ARBs and ARGs,[56–58] because since the late 1990s, several classes of antibiotics have been reported in sewage and wastewater treatment plants including β-lactams, sulphonamides, trimethoprim, macrolides, fluoroquinolones and tetracyclines. Wastewater treatment plants (WWTPs) receive high daily loads of sewage from diverse anthropogenic environments, including residential areas, hospitals, farms and industry.

During the past decade, various methods have been applied for the detection of antibiotic resistance genes in WWTPs. These methods include culture-based, molecular techniques and culture-independent methods, which provide both independently or combined useful insight into the dynamics of antibiotic resistance determinants.

Culture-based methods are mainly isolating target bacteria on general or selective media and evaluating growth in response to specific concentrations of antibiotics. Culture-based assessments of bacteria in WWTPs are generally limited to selected commensal and pathogenic genera include *E. coli, Enterococcus, Salmonella* and *Staphylococcus*, while less commonly targeted microbial groups *Aeromonas, Klebsiella pneumoniae* and *Pseudomonas aeruginosa*. Because they can only be applied to a small fraction of cultivable bacteria they do not provide a comprehensive overview of environmental microbiomes.[59,60] However, these methods can be combined with molecular-based techniques and high throughput sequencing to identify and observe pathogens carrying ARG within specific populations.

Quantitative polymerase chain reaction (qPCR) array is a research tool used for the detection and relative profiling of antibiotic resistance genes. This array provides a convenient way to easily and rapidly detect the presence of antibiotic resistance genes that may be present in isolated bacterial colonies, metagenomic samples. The array contains assays for 87 antibiotic resistance genes belonging to macrolide-lincosamide-streptogramin B, β-lactam, vancomycin, erythromycin, aminoglycoside, tetracycline and fluoroquinolone. The qPCR assays use PCR amplification primers and hydrolysis-probe detection, increasing the specificity of each assay. This pivotal technique can be applied also to track Mobile genetic element (MGEs) in complex environments in order to estimate the potential of HGT.

Metagenomics, the sequencing of whole genomes from complex microbial communities, overcome the limitations of culture-based methods[61,62] and this approach has the potential to identify all the known resistance genes in WWTPs on a more comprehensive way.[63] Analysis of metagenomic data needs extensive bioinformatic algorithms, necessitating expertise and strong computer infrastructure. However, metagenomics has revolutionized understanding of environmental microbiomes and has been effectively applied to WWTPs and allows the isolation of completely novel antibiotic resistance genes. Metagenomic analysis performed on different platforms and various technologies are available, including 454 pyrosequencing, Illumina MiSeq or HiSeq and the Applied Biosystems SOLiD system, Ion Torrent and Pacific Biosciences.[64] These techniques for sequencing DNA generate shorter fragments than Sanger sequencing.

9.7 BURNING AFTER HARVESTING OF WASTEWATER-IRRIGATED CROP

Due to lack of knowledge regarding the significance of crop residues, they are often burned in the field.[65] After burning, this may convert into air and soil pollution by induced contamination. However, discharged to the environment, agricultural wastes can be both beneficial and detrimental to living matter. Therefore, the better options for the utilization of biomass are to reproduce bioenergy products instead of burning or leaving them on agricultural land after harvesting. Open burning increases soil temperature decreases soil water content and bulk density and, consequently, leads to soil compaction, higher surface water runoff, and soil erosion.[66-68] Moreover, the quantity of the crop residues burned and the fire intensity strongly influence the amount of carbon and nutrients released during the fire.[69] This burning process may decrease the soil fertility rate and due to leaching, wastewater comes in contact with several other types of contaminations like ground water contamination. More dangerously, burning of biomass either of sugar or rice may cause serious health issues. It pollutes the air and emission of particulate residues represents an environmental problem, which can trigger respiratory diseases such as asthma.[70,71] During open burning of rice straw, the emissions of fine particles are a major cause of concern due to their harmful effects on human health[72] as well as the earth's climate change.[73] Such climatic changes due to biomass burning are also a critical source of greenhouse gases such as methane (CH_4) and carbon dioxide (CO_2),[74] which contribute to global warming.[75]

KEYWORDS

- **wastewater**
- **groundwater**
- **soil contamination**
- **antibiotics**
- **antibiotic resistance genes**

REFERENCES

1. Pedrero, F.; Kalavrouziotis, I.; Alarcón, J. J.; Koukoulakis, P.; Asano, T. Use of Treated Municipal Wastewater in Irrigated Agriculture—Review of Some Practices in Spain and Greece. *Agric. Water Manage.* **2010,** *97*(9), 1233–1241

2. Turner, K.; Georgiou, S. R.; Clark, R.; Brouwer, R.; Burke, J. *Economic Valuation of Water Resources in Agriculture: From the Sectoral to a Functional Perspective of Natural Resource Management*, FAO Water Reports 27, Food Agriculture Organization of the United Nations, Rome, 2004.

3. Tanji, K. K.; Kielen, N. C. Agricultural Drainage Water Management in Arid and Semiarid Areas. FAO Irrigation and Drainage Paper 61. Food and Agriculture Organization of the United Nations, Rome, 2002.

4. Jiménez, B.; Asano, T. Water Reclamation and Reuse Around the World. In: *Water Reuse: An International Survey of Current Practice, Issues and Needs*; Jiménez, B, Asano T, Eds.; IWA Publishing: London, UK, 2003; Vol. 3, p 26.

5. Scott, C. A.; Faruqui, N. I.; Raschid-Sally, L. Wastewater use in Irrigated Agriculture: Management Challenges in Developing Countries. In *Wastewater Use in Irrigated Agriculture: Confronting the Livelihood and Environmental Realities*; Scott, C. A., Faruqui, N. I., Raschid-Sally, L. Eds.; CABI Publishing: Wallingford, UK, 2004; pp 1–10.

6. Scott C.A.; Drechsel, P.; Raschid–Sally, L.; Bahri A.; Mara, D.; Redwood, M. et al. Wastewater Irrigation and Health: Challenges and Outlook for Mitigating Risks in Low-Income Countries. In: *Wastewater irrigation and health: Assessing and*

mitigating risk in low-income countries; Drechsel, P.; Scott, C. A.; Raschid-Sally, L.; Redwood, M.; Bahri, A. Eds.; Earthscan: London, 2010; pp 381–394.

7. DHWA. Department of Health Western Australia. *Draft Guidelines for the Reuse of Greywater in Western Australia*. Department of Health: Perth, Australia, 2002, p 37.

8. WHO. *Guidelines for the Safe Use of Wastewater, Excreta and Greywater*; World Health Organization: Geneva, Switzerland, 2006.

9. Toze, S. Reuse of Effluent Water – Benefits and Risks. *Agric. Water Manage.* **2006,** *80*, 147–159.

10. Keraita, B.; Jiménez, B.; Drechsel, P. Extent and Implications of Agricultural Reuse of Untreated, Partially Treated and Diluted Wastewater in Developing Countries. *CAB Rev. Perspectives in Agriculture, Veterinary Science, Nutrition and Natural Resources.* **2008,** *3*(58), 1–15.

11. Fergusson, J. E. Ed. *The Heavy Elements: Chemistry, Environmental Impact and Health Effects*. Pergamon Press: Oxford, 1990.

12. Dickin, S. K.; Schuster-Wallace, C. J.; Qadir, M.; Pizzacalla, K. A Review of Health Risks and Pathways for Exposure to Wastewater Use in Agriculture. *Environ. Health Perspect.* **2016,** *124*, 900–909.

13. Fazeli, M. S.; Khosravan, F.; Hossini, M.; Sathyanarayan, S.; Satish, P. N. Enrichment of Heavy Metals in Paddy Crops Irrigated by Paper Mill Effluents Near Nanjangud, Mysore District, Karnataka, India. *Environ. Geol.* **1998,** *34*, 297–302.

14. Pandey, R.; Shubhashish, K.; Pandey, J. Dietary Intake of Pollutant Aerosols Via Vegetables Influenced by Atmospheric Deposition and Wastewater Irrigation. *Ecotoxicol. Environ. Saf.* **2012,** *76*, 200–208.

15. Crane, S. R.; Moore, J. A. Bacterial Pollution of Groundwater: A Review. *Water Air Soil Pollut.* **1984,** *22*, 67–83.

16. Kadam, A. M.; Oza, G. H.; Nemade, P. D.; Shankar, H. S. Pathogen Removal from Municipal Wastewater in Constructed Soil Filter. *Ecol. Eng.* **2008,** *33*(1), 37–44.

17. Dikinya, O.; Areola, O. Comparative Analysis of Heavy Metal Concentration in Secondary Treated Wastewater Irrigated Soils Cultivated by Different Crops. *Int. J. Environ. Sci. Tech.* **2010,** *7*(2), 337–346.

18. Kruse, E. A.; Barrett, G. W. Effects of Municipal Sludge and Fertilizer on Heavy Metal Accumulation in Earthworms. *Environ. Pollut., Ser. A* **1985,** *38*(3), 235–244.

19. Obrador, A.; Rico, M. I.; Mingot, J. I.; Alvarej, J. M. Metal Mobility and Potential Bioavailability in Organic Matter-Rich Soil-Sludge Mixtures: Effect of Soil Type and Contact Time. *Sci. Total Environ.* **1997,** *206*, 117–126.

20. Wang, Q. R.; Cui, Y. S.; Liu, X. M.; Dong, Y. T.; Christie, P. Soil Contamination and Plant Uptake of Heavy Metals at Polluted Sites in China. *J. Environ. Sci. Health.* **2003,** *38*, 823–838.

21. Nyles, C. B.; Ray, R. N. *The Nature and Properties of Soils*, 12th ed.; United States of America.1999, pp 743–785.

22. Anikwe, M. A. N.; Nwobodo, K. C. A. Long Term Effect of Municipal Waste Disposal on Soil Science Society of Nigeria held Nov. 21–25, 1999. Bein city, Nigeria, 2002.

23. Trinh, L. T.; Duong, C. C.; Van Der Steen, P.; Lens, P. N. Exploring the Potential for Wastewater Reuse in Agriculture as a Climate Change Adaptation Measure for Can Tho City, Vietnam. *Agr. Water Manage.* **2013,** *128*, 43–54.

24. De Wrachien, D.; Ragab, R.; Hamdy, A.; Trisorio Liuzzi, G. Global Warming Water Scarcitiy and Food Security in the Medeterranean Environment. In: *Food security under water scarcity in the Middle East: problems and solutions*; Hamdy, A., Monti, R. Eds.; CIHEAM: Bari, 2005, Vol. 65, pp 29–48. (Options Me'diterrane'ennes, Se'rie A n.).

25. Thangarajan, R.; Kunhikrishnan, A.; Seshadri, B.; Bolan, N. S.; Naidu, R. Greenhouse Gas Emission from Wastewater Irrigated Soils. Chapter 4; WIT Transactions on Ecology and The Environment, 2012, Vol. 168; pp 225–236.

26. Mackenzie, F. T.; Mackenzie, J. A. Our Changing Planet: An Introduction to Earth System Science and Global Environmental Change, Prentice Hall: NJ, 1998, p 486.

27. Xue, Y.; Yang, P.; Luo, Y.; Li, Y.; Ren, S.; Su, Y.; Niu, Y. Characteristics and Driven Factors of Nitrous Oxide and Carbon Dioxide Emissions in Soil Irrigated with Treated Wastewater. *J. Integr. Agric.* **2012**, *11*(8), 1354–1364.

28. Zou, J.; Liu, S.; Qin, Y.; Pan, G.; Zhu, D. Sewage Irrigation Increased CH_4 and N_2O Emissions from Rice Paddies in Southeast China. *Agric. Ecosyst. Environ.* **2009**, *129*(4), 516–522.

29. Chen, W.; Wu, L.; Frankenberger, W. T. Jr., Chang, A. C. Soil Enzyme Activities of Long-Term Reclaimed Wastewater-Irrigated Soils. *J. Environ. Qual.* **2008**, *37*, 36–42.

30. Chavez, A.; Rodas, K.; Prado, B.; Thompson, R.; Jimenez, B. An Evaluation of the Effects of Changing Wastewater Irrigation Regime for the Production of Alfalfa (Medicago sativa). *Agric. Water Manag.* **2012**, *113*, 76–84.

31. Jang, T. I.; Kim, H. K.; Seong, C. H.; Lee, E. J.; Park, S. W. Assessing Nutrient Losses of Reclaimed Wastewater Irrigation in Paddy Fields for Sustainable Agriculture. *Agric. Water Manag.* **2012**, *104*, 235–243.

32. Jeong, H. S.; Jang, T. I.; Seong, C. H.; Park, S. W. Assessing Nitrogen Fertilizer Rates and Split Applications Using the DSSAT Model for Rice Irrigated with Urban Wastewater. *Agric. Water Manag.* **2014**, *141*, 1–9.

33. Hamilton, A. J.; Stagnitti, F.; Xiong, X.; Kreidil, S. L.; Benke, K. K.; Maher, P. Wastewater Irrigation: The State of Play. *Vadose Zone J.* **2007**, *6*, 823–840.

34. Mohammad, M. J.; Mazahreh, N. Changes in Soil Fertility Parameters in Response to Irrigation of Forage Crops with Secondary Treated Wastewater. *Commun. Soil Sci. Plant Anal.* **2011**, *34*, 128–11294.

35. Xu, J.; Wu, L.; Chang, A. C.; Zhang, Y. Impact of Long Term Reclaimed Wastewater Irrigation on Agricultural Soils: A Preliminary Assessment. *J. Hazard. Mater.* **2010**, *183*, 780–786.

36. Rhee, H. P.; Yoon, C. G.; Jung, K. W.; Son, J. W. Microbial Risk Assessment Using E. Coli in UV Disinfected Wastewater Irrigation on Paddy. *Environ. Eng. Res.* **2009**, *14*, 120–125.

37. Forslund, A.; Ensink, J. H. J.; Battilani, A.; Kljujev, I.; Gola, S.; Raicevic, V.; Jovanovic, Z.; Stikic, R.; Sandei, L.; Fletcher, T.; et al. Faecal Contamination and Hygiene Aspect Associated with the Use of Treated Wastewater and Canal Water for Irrigation of Potatoes. *Agric. Water Manag.* **2010**, *98*, 440–450.

38. Angelakis, A. N.; Marecos do Monte, M. H. F.; Bontoux, L.; Asano, T. The Status of Wastewater Reuse Practice in the Mediterranean Basin: Need for Guidelines. *Water Res.* **1999**, *33*, 2201–2217.

39. Brissaud, F. Criteria for Water Recycling and Reuse in the Mediterranean Countries. *Desalination* **2008**, *218*, 24–33.
40. Kalavrouziotis, I. K.; Kokkinos, P.; Oron, G.; Fatone, F.; Bolzonella, D.; Vatyliotou, M.; Fatta-Kassinos, D.; Koukoulakis, P. H.; Varnavas, S. P. Current Status in Wastewater Treatment, Reuse and Research in Some Mediterranean Countries. *Desalin. Water Treat.* **2015**, *53*, 2015–2030.
41. Paranychianakis, N. V.; Salgot, M.; Snyder, S. A.; Angelakis, A. N. Water Reuse in EU States: Necessity for Uniform Criteria to Mitigate Human and Environmental Risks. *Crit. Rev. Environ. Sci. Technol.* **2015**, *45*, 1409–1468.
42. Weber, B.; Avnimelech, Y.; Juanico, M. Salt Enrichment of Municipal Sewage: New Prevention Approaches in Israel. *Environ. NIanag.* **1996**, *20*(4), 487–495.
43. Kiziloglu, F.; Tuean, M.; Sahin, U.; Angin, I.; Anapali, O.; Okuroglu, M. Effects of Wastewater Irrigation on Soil and Cabbage-Plant (*Brassica olereacea* var. capitate cv. Yavola-1) Chemical Properties. *J. Plant. Nutr. Soil. Sc.* **2007**, *170*, 166–172.
44. Mañas, P.; Castro, E.; De las Heras, J. Irrigation with Treated Wastewater: Effects on Soil, Lettuce (Lactuca sativa) Crop and Dynamics of Microorganisms. *J. Environ. Sci. Heal. A.* **2009**, *44*, 1261–1273.
45. Jueschke, E.; Marschner, B.; Tarchitzky, J.; Chen, Y. Effects of Treated Wastewater Irrigation on the Dissolved and Soil Organic Carbon in Israeli Soils. *Water Sci. Technol.* **2008**, *57*, 727–733.
46. Akponikpe, P.; Wima, K.; Yakouba, H.; Mermoud, A. Reuse of Domestic Wastewater Treated in Macrophyte Ponds to Irrigate Tomato and Eggplants in Semi-Arid West-Africa: Benefits and Risks. *Agr. Water Manag.* **2011**, *98*, 834–840.
47. Rezapour, S.; Samadi, A. Soil Quality Response to Long-Term Wastewater Irrigation in Inceptisols from a Semi-Arid Environment. *Nutr. Cycling Agroecosyst.* **2011**, *91*, 269–280.
48. Sacks, M.; Bernstein, N. Utilization of Reclaimed Wastewater for Irrigation of Field-Grown Melons by Surface and Subsurface Drip Irrigation. *Isr. J. Plant Sci.* **2011**, *59*, 159–169.
49. Gwenzi, W.; Munondo, R. Long-Term Impacts of Pasture Irrigation with Treated Sewage Effluent on Nutrient Status of a Sandy Soil in Zimbabwe. *Nutr. Cycling Agroecosyst.* **2008**, *82*, 197–207.
50. Kim, S.; Park, S.; Lee, J.; Benham, B.; Kim, H. Modeling and Assessing the Impact of Reclaimed Wastewater Irrigation on the Nutrition Loads from an Agricultural Watershed Containing Rice Paddy Fields. *J. Environ. Sci. Heal. A.* **2007**, *42*, 305–315.
51. Angin, I.; Yaganoglu, P.; Turan, M. Effects of Long-Term Wastewater Irrigation on Soil Properties. *J. Sustain. Agric.* **2005**, *26*, 31–42.
52. Ferreira da Silva, M.; Tiago, I.; Verissimo, A.; Boaventura, R. A. R.; Nunes, O. C.; Manaia, C. M. Antibiotic Resistance of Enterococci and Related Bacteria in an Urban Wastewater Treatment Plant. *FEMS Microbiol. Ecol.* **2006**, *55*, 322–329.
53. Figueira, V.; Serra, E.; Manaia, C. M. Differential Patterns of Antimicrobial Resistance in Population Subsets of Escherichia Coli Isolated from Waste- and Surface Waters. *Sci. Total Environ.* **2011**, *409*, 1017–1023.
54. Kümmerer, K. Antibiotics in the Aquatic Environment—A Review-Part II. *Chemosphere* **2009**, *75*, 435–441.

55. Lupo, A.; Coyne, S.; Berendonk T. Origin and Evolution of Antibiotic Resistance: The Common Mechanism of Emergence and Spread in Water Bodies. *Front. Microbiol.* **2012**, *3*, 18.
56. Auerbach, E. A.; Seyfried, E. E.; McMahon, K. D. Tetracycline Resistance Genes in Activated Sludge Wastewater Treatment Plants. *Water Res.* **2007**, *41*, 1143–1151.
57. Kim, S.; Park, H.; Chandran, K. Propensity of Activated Sludge to Amplify or Attenuate Tetracycline Resistance Genes and Tetracycline Resistant Bacteria: A Mathematical Modeling Approach. *Chemosphere* **2010**, *78*, 1071–1077.
58. Czekalski, N.; Berthold, T.; Caucci, S.; Egli, A.; Buergmann, H. Increased Levels of Multiresistant Bacteria and Resistance Genes After Wastewater Treatment and Their Dissemination into Lake Geneva, Switzerland. *Front. Microbiol.* **2012**, *3*, 106.
59. Daniel, R. The Metagenomics of Soil. *Nat. Rev. Microbiol.* **2005**, *3*, 470–478.
60. McNamara, C. J.; Lemke, M. J.; Leff, L. G. Culturable and Non-Culturable Fractions of Bacterial Populations in Sediments of a South Carolina Stream. *Hydrobiologia* **2002**, *482*, 151–159.
61. Hugenholtz, P.; Tyson, G. W. Microbiology: Metagenomics. *Nature.* **2008**, *455*, 481–483.
62. Sharpton, T. J. An Introduction to the Analysis of Shotgun Metagenomic Data. *Front. Plant Sci.* **2014**, *16*(5), 209.
63. Schmieder, R.; Edwards, R. Insights into Antibiotic Resistance Through Metagenomic Approaches. *Future Microbiol.* **2011**, *7*, 73–89.
64. Hodkinson, B. P.; Grice, E. A. Next-Generation Sequencing: A Review of Technologies and Tools for Wound Microbiome Research. *Adv. Wound Care* **2015**, *4*, 50–58.
65. Samra, J. S.; Singh, B.; Kumar, K. Managing Crop Residues in the Rice-Wheat System of the indo Gangetic Plain. In *Improving the productivity and sustainability of rice-wheat system: issues and impact;* Ladha, J. K., Hill, J. E., Duxbury, J. M., Gupta, R. K., Buresh, R. J. Eds.; American Society of Agronomy: Madison, WI, USA, 2003; pp 173–195.
66. Dourado-Neto, D.; Timm, C.; Oliveira, J. C. M. et al. State-Space Approach for the Analysis of Soil Water Content and Temperature in a Sugarcane Crop. *Sci. Agric.* **1999**, *56*, 1215–1221.
67. Oliveira, J. C.; Reichardt, K.; Bacchi, O. O. S. et al. Nitrogen Dynamics in a Soil-Sugar Cane System. *Sci. Agric.* **2000**, *57*, 467–472.
68. Tominaga, T. T.; Cassaro, F. A. M.; Bacchi, O. O. S.; Reichardt, K.; Oliveira, J. C. M.; Timm, L. C. Variability of Soil Water Content and Bulk Density in a Sugarcane Field. *Aust. J. Soil Res.* **2002**, *40*, 604–614.
69. Sharma, P. K.; Mishra, B. Effect of Burning Rice and Wheat Crop Residues: Loss of N, P, K and S from Soil and Changes in Nutrient Availability. *J. Indian Soc. Soil Sci.* **2001**, *49*, 425–429.
70. Cancado, J. E. D.; Saldiva, P. H. N.; Pereira, L. A. A. et al. The Impact of Sugar Cane-Burning Emissions on the Respiratory System of Children and the Elderly. *Environ. Health Perspect.* **2006**, *114*, 725–729.
71. Dawson, L.; Boopathy, R. Use of Post-Harvest Sugarcane Residue for Ethanol Production. *Bioresour. Technol.* **2007**, *98*, 1695–1699.

72. Pope, C. A.; Ezzati, M.; Dockery, D. W. Fine-Particulate Air Pollution and Life Expectancy in the United States. *N. Engl. J. Med.* **2009,** *360*, 376–386.

73. Bond, T. C.; Streets, D. G.; Yarber, K. F.; Nelson, S. M.; Woo, J. H.; Klimont, Z. Technology-Based Global Inventory of Black and Organic Carbon Emissions from Combustion. *J. Geophys. Res.* **2004,** *109*, (D14203).

74. Andreae, M. O.; Merlet, P. Emission of Trace Gases and Aerosols from Biomass Burning. *Global Biogeochem. Cycles* **2001,** *15*, 955–966.

75. Sun, J. F.; Peng, H. Y.; Chen, J. M. et al. An Estimation of CO2 Emission Via Agricultural Crop Residue Open Field Burning in China from 1996 to 2013. *J. Cleaner Prod.* **2016,** *112*, 2625–2631.

CHAPTER 10

MICROBIAL FUEL CELLS FOR WASTEWATER TREATMENT, BIOREMEDIATION, AND BIOENERGY PRODUCTION

RAVINDER KUMAR[1] and PRADEEP KUMAR[2,*]

[1]*Faculty of Engineering Technology, Universiti Malaysia Pahang, Kuantan 26300, Malaysia*

[2]*Department of Forestry, North Eastern Regional Institute of Science and Technology, Nirjuli 791109, Arunachal Pradesh, India, *E-mail: pkbiotech@gmail.com*

CONTENTS

ABSTRACT

Microbial fuel cells (MFCs) are emerging as a versatile renewable energy technology. This is particularly because of the multidisciplinary applications of this eco-friendly technology. The technology depends on the electroactive bacteria, popularly known as exoelectrogens to simultaneously produce electric power and treat wastewater. The electrode modification with nanoparticles or some other pretreatment methods and use of genetically modified microorganisms have shown to enhance the MFC performance for electricity generation and wastewater treatment. The MFC technology has been also investigated for the removal of various heavy metals and toxic elements, and to detect the presence of toxic elements present in the wastewater. This chapter provides a comprehensive and the state-of-the-art information on prime applications (electricity generation, wastewater treatment and bioremediation) of the MFC technology and also some methods that can be employed to improve the MFC performance for these applications.

10.1 INTRODUCTION

Microbial fuel cell (MFC) technology has become an attractive technology today because of its capability to convert the chemical energy present in organic/inorganic wastes into electrical energy. MFC technology efficiently links microbial metabolism with electrochemical reactions.[1,2] Consequently, the technology can be used for electricity generation, wastewater treatment, and bioremediation of heavy metals/toxic compounds. The general principle of an MFC is given in Figure 10.1. MFCs are bioelectrochemical devices, typically consisting of two chambers that is the anode chamber (anaerobic; contains an electrode, microorganisms and anolyte) and the cathode chamber (aerobic/anaerobic; an electrode, electron acceptor and a catalyst), separated by a proton exchange membrane (PEM), for example, Nafion.[3-5] Microorganisms are used as catalysts to oxidize the substrate in the anode chamber and thus they have been denoted as the power house of MFCs. The electrons are transferred to the anodic (an electrode) surface, which are then directed to the cathode through an electrical connection.[6,7] In the cathode, the electrons combine with protons and oxygen to form water. A catalyst, for example platinum, is generally used to catalyse the reduction in the cathode; alternatively, a microorganism can also be used to replace such costly catalyst.[8,9] The exquisiteness of MFCs mainly lies in the use of microorganisms as the catalyst at the anode of MFCs. The unique characteristic of the microorganisms used in MFCs is their ability to transfer the electrons (resulted from their metabolism) from their outer cell surface to the electrode surface (in anode) and to accept the electrons from the electrode surface (in cathode) to catalyse the reduction of electron acceptors and such microorganisms are usually called as exoelectrogens. Due to this unique characteristic of exoelectrogens, the MFC technology has been experimented for various applications, mainly for electricity generation, wastewater treatment and bioremediation, which are discussed in next sections of this chapter.

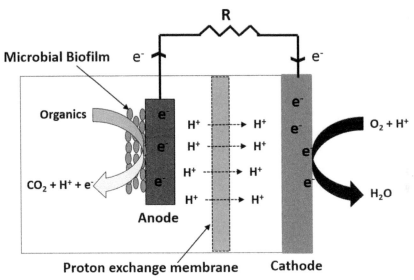

FIGURE 10.1   General principle of a microbial fuel cell.

10.2   MFCs FOR WASTEWATER TREATMENT

MFC technology has the potential to provide an effective platform for the treatment of highly polluted industrial wastewater or urban wastewater. Evidently, different types of MFCs have shown the ability to treat various industry and urban wastewaters.[10–13] On one side of the picture, MFC technology can be used to treat the wastewater while on the other side, the wastewater can be used to provide: (1) substrate as the carbon source for the bacterial growth and hence for the end products of the oxidation

process that is electrons and protons for sustainable bioelectricity generation, (2) primary wastewater from an industry such as chocolate industry
wastewater, palm oil mill effluent (POME) can be used to provide the
inoculum or the biocatalysts for the substrate oxidation. Moreover, defined
bacterial culture (pure or mixed) can be isolated from wastewater that can
be further used as inoculum for the MFCs, (3) the wastewater can be used
as catholyte as well though it may contain some minerals that can act as
electron acceptors.

10.2.1   MFC PERFORMANCE FOR WASTEWATER TREATMENT

Different parameters can be evaluated to measure the treatment capacity
of MFCs. These primarily include chemical oxygen demand (COD),
biochemical oxygen demand (BOD), total solids, total dissolved solids,
acidity and so forth Typically, COD test is performed (or is sufficient)
to examine the performance of MFC toward wastewater treatment. Some
examples of MFC studies demonstrated for wastewater treatment are
given in Table 10.1. Wastewater containing high BOD and COD can be
treated in MFCs to meet discharge regulations before it is released into
the environment; simultaneously, it also provides the organic substrate
for microbial metabolism and growth to extract the energy for electricity
generation. The MFCs have achieved up to 98% COD removal from
wastewater, which is a quite appreciable achievement for COD removal.
Almost all the studies regarding wastewater treatment are coupled with the
foremost application of MFCs that is electricity production. For example,
animal wastewaters contain high organic content and high concentrations
of phosphate and nitrate in wastewater that also cause the water pollution through eutrophication of surface water. A few studies have demonstrated the use of animal wastewater in different MFCs for its treatment
and bioenergy production. A study using swine wastewater in different
MFCs (two chambered MFC and single-chambered MFC [sMFC])
achieved maximum 92% COD removal and approximately 83% ammonia
reduction after 150 h of operation.[14] Animal carcass wastewater (ACW)
containing high organic content has been also treated by MFC (upflow
tubular MFC).[15] The disposed animal carcasses can be further hydrolysed
with alkaline treatment (sodium hydroxide or potassium hydroxide) into
smaller constituents like amino acids, sugars and minerals forming a sterile

solution referred as ACW (of BOD: 70, COD: 105 and ammonia: 1 g/l). The maximum COD reduction obtained in the demonstration was more than 50% and the nitrate removal efficiency of MFC was nearly 80%.[15]

TABLE 10.1 Performance of Microbial Fuel Cells for Wastewater Treatment.

Wastewater	Type of MFC	Electrode material	% COD reduction	Reference
Swine wastewater	Single-chamber MFC	Toray carbon paper as anode carbon cloth as cathode	92	[14]
Starch-processing wastewater	Single-chamber MFC	Carbon paper	98	[16]
Paper recycling wastewater	Single-chamber MFC	Graphite fibres-brush	76	[8]
Cassava mill wastewater	Double-chambered MFC	Graphite plates electrode	86	[13]
Food processing wastewater	Double-chambered MFC	Carbon paper electrodes	95	[11]
Domestic wastewater	Double-chambered MFC	Plain graphite electrodes	88	[12]
Chocolate industry wastewater	Double-chambered MFC	Graphite rods as electrodes	75	[18]
Biodiesel wastes	Single-chamber MFC	Carbon brush electrodes	90	[19]
Beer brewery wastewater	Single-chamber MFC	Carbon fibres	43	[20]
Potato Processing wastewater	Tubular MFC	Graphite particles as anode Graphite felt as cathode	91	[17]
Palm oil mill effluent	[a]UML-MFCs	Graphite granules, Carbon fibre felt	90	[21]
Animal carcass wastewater	Upflow tubular MFC	Graphite felt as anode Carbon cloth as cathode	52	[10]

[a]UML-MFCs = Up-flow membrane-less microbial fuel cell.

Food wastewater or food industry wastewater is rich in sugars and starch and thus contains higher BOD as compared to other industrial wastewaters. Food wastewaters can be treated by MFCs, simultaneously providing the electron donors for bioenergy production. A study using cereal wastewater in a double-chambered MFC achieved more than 95% COD removal. The COD values measured before and after treatment in the MFC were 595 and 30 mg/l, respectively.[11] The production of starch foodstuffs (e.g. potato chips) in food industries requires great usage of water, and consequently releases large quantities of wastewater in the environment. Such starch-processing wastewater (SPW) is highly energy rich resource that comprises high contents of proteins, carbohydrates, cellulose, vitamins and other nutrients. An MFC demonstration used SPW to evaluate the treatment efficiency of a double-chambered MFC. The results were significant, achieving a 98% COD reduction after 140 days of operation of the MFC. The maximum removal efficiency of ammonia-nitrogen in the study was nearly 91%.[16] In another study using potato-processing wastewater (PPW), 91% of COD reduction was achieved with a rate of 26.23 ± 1.71 mg l^{-1} h^{-1}.[17] Similarly, chocolate wastewater was used in a double-chambered MFC by.[18] The results showed that maximum 75% COD was removed after the MFC operation in batch mode. The BOD removal and total solid removal was ca. 65 and 68%, respectively.[18]

Wastewaters containing lignocellulosic biomass (e.g., cellulose, hemicellulose and lignin) are not completely treated by traditional wastewater treatment techniques. The bacteria in the MFC can not only degrade the lignocellulose biomass or convert it into simpler sugars but can also extract energy from such wastewaters to generate electricity. MFC technology has proved to treat lignocellulose-containing wastewater. Huang et al. used paper recycling wastewater in a sMFC for its treatment and electricity generation. The results suggested that the MFC after nearly 3 weeks of operation achieved more than 76% COD removal while ca. 96 % of cellulose was removed by the bacteria.[8]

Brewery wastewater has been widely used in different MFCs for its treatment and bioenergy production. The brewery wastewater exhibits high COD, up to 5000 mg/l; moreover, it contains high levels of carbohydrates or sugars that can be used as electron donors in the MFCs. For example, air cathode sMFC was used with different concentrations of the wastewater and was operated in fed-batch mode.[19] When the wastewater with less COD value was used in the MFC, low COD removal was obtained: for COD concentrations of 84 and 1600 mg/l, the COD removal was ~58 and 98%, respectively.[19] In another study, sMFC was operated in continuous mode with a hydraulic resistance time of 2.13 h. The wastewater COD ranged between 600 and 660 mg/l, and a 43%–46% COD removal was achieved.[20]

Wastewaters from different other industries (agro-industries and oil industries) have also been treated in MFCs. Such wastewaters are very toxic, having very high COD, organic content and carbohydrates. For example, cassava mill effluent has very high COD that is >16,000 mg/l and a cyanide concentration of ca. 86 mg/l.[13] A double-chambered MFC of a capacity of 30 l was used to treat the wastewater, and nearly 90% COD removal was achieved after 120 h of operation.[13] Palm oil industries release the large amount of highly toxic wastewater, referred as POME. POME exhibits high COD and BOD that is >50,000 and >25,000 mg/l, respectively.[21] Cheng et al. treated POME in upflow membrane-less MFC (UML-MFC) coupled with MFC and upflow anaerobic sludge blanket (UASB) reactors. This integrated system produced significant results for COD and nitrogen removal that is ca. 96 and 94%, respectively.[21]

The MFC technology, no doubt, has the potential to treat the various wastewaters but there are some critical issues as well that hinder the application of this technology at the industrial level. The first limiting issue is the low COD removal rates achieved in the studies. The literature suggests a wide range of organic removal rates ranging from 0.0053 to 5.57 g COD/l/d using different wastewaters of varying COD strengths,[22] which are quite low than the required COD removal rates for real-world applications. The second drawback in the MFC technology is the cathode biofouling. It has been identified that in the long-term MFC operation, the anolyte passes through the PEM to the cathode, which could be a possible reason for low performance of the MFCs. Alternatively, in the air-cathode MFCs, the activity of the anaerobic microbial community at the anode can be adversely affected due to the passage of oxygen from the cathode, consequently, which could have a negative influence on the MFC performance. Moreover, the MFC technology has been investigated for wastewater treatment under fed-batch conditions with very small working volumes. Therefore, the optimization of various vital parameters would be a challenging task for real-world applications. In addition, the high cost of materials cannot be ignored in scaling up the technology, which otherwise would make the technology uneconomical at a large scale.

10.2.2 APPROACHES TO IMPROVE TREATMENT EFFICIENCY OF MFCs

Usually, the MFCs produce more power density with wastewater of high COD values. However, the highly concentrated substrate can cause fouling of the PEM, resulting in the restriction of protons, which consequently lead to the accumulation of protons in the anode chamber (low pH) and less availability of protons in the cathode (high pH). Therefore, concentrated wastewaters with small dilutions are suitable to maintain the proper functioning of the MFC. Furthermore, some pretreatment methods can be employed to change the physiochemical or biological properties of the wastewater for enhanced performance of the MFCs. For example, wastewater can be autoclaved to kill the methanogens (the anaerobic bacteria that yield methane as a metabolic by-product) that otherwise use the organic matter to produce methane instead of protons and electrons. A study showed that MFC with the autoclaved wastewater produced ca. 5% more power density than with raw wastewater.[23] Another pretreatment method that is sonication is useful to increase the performance of the MFCs considerably. This approach was employed in a similar former study (with diluted wastewater) that produced ca. 16% more power density than with raw wastewater and increased the COD removal efficiency by nearly 5%. The sonication process may improve the performance of the MFC by altering the biodegradability of the organic matter present in the wastewater or changing the molecular weight or particle size spectra of the organic matter. Moreover, stirring the wastewater has also shown marginal improvement in the COD removal in the MFC.[23] In MFCs, the wastewater treatment efficiency can be further improved if the fuel cells are operated for longer periods. It has been found that the MFCs operated for shorter period resulted in less COD removal than the MFCs ran for the longer time. For example, an MFC (air-cathode) was operated for four cycles, each cycle lasting for approximately 35 days. The results suggested that the COD removal after the first cycle was ca. 95%, which increased to more than 98% after the end of four cycles (after 140 days of MFC operation).[16] This could be due to the microorganisms getting enough time to degrade the complex substrates into simpler substituents. However, the coulombic efficiency achieved in the demonstration was ca. 7%, indicating that most of the substrates did not convert to electricity, which could be due the following reasons: oxygen diffusion, production of fermented products, oxidization of other electron acceptors and biomass production.[16] The another approach that can be employed to increase the wastewater efficiency is the operation of MFCs at the mesophilic temperature. The effect of temperature on treatment efficiency of MFCs was investigated by[12] using air-cathode sMFC. They operated the fuel cell (batch mode and continuous mode) at two different temperatures that is ambient temperature ($23 \pm 30°C$) and mesophilic temperature ($30 \pm 10°C$). The results showed that the % COD removal, as well as the COD removal rate was higher in the MFCs operated at mesophilic temperature than the ambient temperature. Moreover, ca. 10% more nitrogen removal was achieved from the MFCs operated at higher temperature. Overall, the MFCs in the fed-batch mode removed more than 2.5 times COD as compared to MFCs operated in continuous mode.[12]

The integration of MFCs with other wastewater treatment technologies could be a better idea to extract more energy and to further improve the pollutant removal efficiency. Generally, the bacteria in MFCs degrade the simpler or low-strength wastewaters more effectively. While some other bioreactors such as anaerobic digester or UASB have the ability to treat high-strength wastewaters.[24] Therefore, the wastewaters with complex composition (e.g. POME) can be subjected to fermentation in UASB that can provide more suitable or simpler substrates for electricity generation in MFCs. Moreover, the residual organics present in the effluent of UASB can be further removed in the MFCs.

10.3 MFCs FOR BIOREMEDIATION

The exoelectrogens produce electrons from their metabolism in the anode chamber of an MFC, which need to be reduced at the cathode chamber. Therefore, an electron acceptor is provided at the cathode to overcome the potential losses. In addition, a catalyst can also be used to increase the reduction reaction rate. Usually, the electron acceptors that exhibit a high redox potential and faster kinetics, and are economical and easily available are of great interest in MFC applications. For example, oxygen is one of the promising and widely used electron acceptors in the MFCs. In MFC system, various organic and inorganic toxic elements or compounds can be utilized as the electron acceptor in the cathode chamber for its removal or reduction to less toxic form and simultaneously for the electric current generation. For examples, metal ions, perchlorate, nitrobenzene, azo dyes, nitrate (NO^{3-}) and so forth have been already used as electron acceptors in different MFCs to explore the bioremediation potential of this technology. Some examples of MFC performance for bioremediation application are given in Table 10.2.

TABLE 10.2   Performance of Microbial Fuel Cells for Bioremediation.

Heavy metals/ Wastewater	Type of MFC	Electrode material	% Removal	Reference
Chromium (VI)	Double-chambered MFC	Graphite granules-cathode	94	[32]
		Graphite brush-anode		
Chromium (VI)	Double-chambered MFC	Carbon fibre felt	76	[33]
Sulphide	Double-chambered MFC	Carbon fibre felt	85	[34]
Cadmium	Single-chambered MFC	Carbon cloth	90	[35]
Zinc	Single-chambered MFC	Carbon cloth	97	[35]
Vanadium	Double-chambered MFC	Carbon fibre felt		
Ammonia– copper (II)	Double-chambered MFC	Graphite felt-anode	96	[26]
		Graphite plate-cathode		
Mercury (Hg^{2+})	Double-chambered MFC	Graphite felt-anode	99.5	[28]
		Carbon felt-cathode		
Azo dye Congo red	Single-chambered MFC	Carbon brush	98.3	[28]
Cyanide	Double-chambered MFC	Carbon cloth	88.3	[27]
Copper (Cu^{2+})	Double-chambered MFC	Graphite felt electrodes	99.5	[15]
Chromium (VI)	Single-chambered MFC	Carbon brush-anode	99	[30]
		Carbon cloth-cathode		
Nitrate	Single-chambered MFC	Graphite rods	30	[31]
Nitrite	Single-chambered MFC	Graphite rods	37	[37]

10.3.1   GENERAL MECHANISM FOR BIOREMEDIATION OF HEAVY METALS/POLLUTANTS

Various heavy metals have been experimented in the anode chamber as well as in the cathode chamber of MFCs for their efficient removal. For anodic removal, generally, a specific concentration of a heavy metal or toxic element is added in the anolyte (supplemented with carbon source and bacterial inoculum). A heavy metal with a high redox potential can be used as the electron acceptor in the cathode chamber. Few mechanisms have been demonstrated that are responsible for the removal of heavy metals or other toxic elements during the MFC operation. The first mechanism that is biosorption has been widely recognized for the removal of toxic elements in the MFCs.[25] Biosorption is a combined term for the processes such as microprecipitation, complexation, chelation, coordination and ion exchange. The biomolecules like polysaccharides, proteins and lipids contain the functional groups such as amine, sulphate, carboxylate, hydroxyl and phosphate that help in the biosorption process to remove the heavy metals or toxic pollutants. These biomolecules containing the functional group may present in the anolyte or on the bacterial cell walls, which play a major role in the removal of toxic pollutants. Moreover, some processes like biological oxidation, chemical oxidation, volatilization and anode electrode adsorption have been found responsible for the sulphide removal during the MFC operation.

10.3.2   MFC PERFORMANCE FOR BIOREMEDIATION OF HEAVY METALS AND TOXIC POLLUTANTS

High concentration of toxic heavy metals (e.g. cadmium, mercury, lead, arsenic, chromium etc.) in the industrial effluents is harmful to the cellular metabolism of the aquatic flora and the fauna. Therefore, the wastewaters that contain the high concentration of toxic heavy metals are essential to be reduced to nontoxic form before they are discharged into the environment. MFCs have shown a great potential for the reduction of heavy metals and have achieved productive results when used in the anode as well as the electron acceptor in the cathode chamber.[25-27] Generally, the heavy metals with a higher redox potential are of great interest to act as the electron acceptor, to achieve higher power output from the cell.

A single-chambered air-cathode MFC showed high removal efficiencies of cadmium (90%) and zinc (97%).[28] Moreover, in a dual chamber MFC, vanadium containing wastewater was employed as the cathodic electron acceptor for its simultaneous removal. The fuel cell after 10 days' operation achieved ca. 70% removal of vanadium (V) with a maximum power density of ca. 970 mW/m^2.[25] A two-chamber MFC obtained more than 99.5% removal efficiency for Hg^{2+} (used as an electron acceptor in the fuel cell) and a power generation of ca. 431 mW/m^2.[25] Ammonia–copper (II) complexes from wastewater have been substantially recovered using MFC technology. Cu $(NH_3)^{2+}$ complexes can be reduced to Cu or Cu_2O. In a study by Wang et al., 96% copper was successfully removed after 12 h' operation of the MFC at pH-9.0.[27]

Different types of dyes are used in the textile industry, which result in the generation of millions of tons of dye wastewater per year around the world.[29] This type of wastewater contains many toxic and recalcitrant organic molecules and carcinogenic chemicals. The discharge of such wastewater is a threat to the environment, animals as well as to the plants. Therefore, treatment of such wastewater is essential before its discharge into the environment. MFC technology provides an eco-friendly alternative for the treatment of wastewater and simultaneous bioelectricity generation. MFCs use microorganisms, therefore, the dyes can be reduced by different decolourization mechanisms involving enzymes, low molecular weight redox mediators, and chemical reduction by biogenic reductants. In the MFCs, the dye decolourization occurs in the anode chamber biologically under anaerobic conditions, where the azo bond of the dye is broken into the intermediates such as aromatic amines that can be completely degraded abiotically in cathode chamber.[30]

An sMFC with bioanode and biocathode was demonstrated to achieve 98% decolourization of an azo dye effluent within 1 day.[30] Transfer of electrons from anode microorganisms and protons through PEM leads to the degradation of azo bond (–N=N–) in the cathode. Reduction of azo bond results in the formation of colourless and biodegradable aromatic amines.[30] Dechlorinating microorganisms can be used in MFCs for the bioremediation of pentachloroethane (PCE) and trichloroethene to reduce them into non-toxic end product ethene. A successful effort was accomplished by Strycharz et al.[31] using *Geobacter lovleyi* and graphite electrodes serving as the electron donor for reductive dechlorination of PCE.

A consortium of anaerobic and aerobic bacteria in the cathodic chamber of dual chamber MFC demonstrated efficient degradation of pentachlorophenol (PCP). In that study, degradation rate for PCP was investigated at different pH and temperatures. The most effective degradation rates achieved at a constant temperature of 500°C and a pH of 6.[31] In addition, *Geobacter* species have shown the tendency to reduce aqueous, soluble U (VI) into an insoluble form as U (IV). Multiple lines of evidence suggest that *G. sulfurreducens* entails the outer-surface c-type cytochromes for U (VI) reduction but do not require pili for the same purpose.[32] Further investigation revealed that *G. sulfurreducens* strain lacking the pilA gene reduced U (VI) to the parallel extent to wild type strain. Similarly, c-type cytochromes are also indispensable for S. oneidensis to reduce U (VI). Gene deletion studies demonstrated the importance of outer membrane, decaheme cytochrome MtrC in the electron transport to U (VI), similar to Mn (IV) and Fe (III) reduction, as the strains deficient in mtrC and/or omcA were incompetent for U (VI) reduction as compared to wild-type MR-1.[33] Moreover, MFCs utilizing anaerobic biocathodes have shown the ability to reduce highly toxic Cr (VI) to much less toxic Cr (III) and subsequent precipitation to Cr (OH)$_3$ with simultaneous electricity generation.[34] The MFC with set biocathode potentials reduced Cr (VI) with increased reduction rate of 19.7 mg/L^{-d}. Further, use of Shewanella oneidensis MR-1 (produced riboflavin, an electron shuttle mediator to transfer electrons) as a biocatalyst in the cathode under aerated conditions in the presence of lactate showed increased reduction rate for Cr (VI).[35] An MFC fed with sulphide and glucose, predominated by Firmicutes obtained sulphide removal efficiencies up to 84.7% and 572.4 mW/m^2 power output at the current density of 1094.0 mA/m^2.[36] Recently, analysis of 16S rRNA revealed that a strain showing similarity to *Klebsiella sp.* is capable of bioremediation of cyanide-containing wastewater in MFC (more than 99.5% of cyanide biodegradation rate and ca. 88% COD removal rate).[37] The investigations described in this section reflect that the MFC technology is a promising alternative for the bioremediation and wastewater treatment of hazardous contaminants, however, there are some more aspects that can be studied regarding this application. For example, most of studies utilized only electron acceptor in the MFCs, but the cathodic reaction with different electron acceptors should also be investigated since the wastewater contains a complex composition of different minerals or ions. Moreover, the investigation of the electron transfer mechanisms from an electrode to a microorganism would be highly interested when a biocathode is used in the MFC study.

10.4   MFCs FOR BIOELECTRICITY GENERATION

The MFCs are chiefly used for the application of electric current generation and many efforts have been made to ameliorate the current density such as electrode modifications, MFCs designs, use of metal catalysts at the anode as well as at the cathode and so forth.[5,9,38] The first step in MFCs towards current generation is the acclimatization of the exoelectrogens in the anode chamber and subsequent biofilm formation on the electrode surface (anode). Consequently, the exoelectrogens form a biofilm (30–50 µm) on the electrode surface.[39,40] Biofilm formation by exoelectrogens is a unique characteristic and differs from other bacteria or microorganisms. The development of biofilm on the electrode surface from the single bacterial cell is stimulated by the assembly of adhesins and extracellular matrix components.[41,42] Subsequently some pivotal proteins, specifically pili and outer membrane c-type cytochromes (OMC c-Cyts), for example, OmcZ, OmcS and so forth, also promote the biofilm formation.[43] Evidently, available studies showed that *Geobacter sulfurreducens* was unable to form biofilm in the absence of pili and OMC c-Cyts.[43] The formation of thick biofilm is taken as an important parameter in MFCs for efficient performance. Usually, optimal biofilm thickness is preferred in MFCs for higher current densities, as highly thick biofilms also confine the electron passage. In addition, the selection of suitable bacterial inoculum (pure culture of mixed culture) with preferred substrate can be highly beneficial to extract more energy for the current generation. For example, *Geobacter sulfurreducens* can reduce acetate with ~100% electron recovery to generate electricity.[44]

After the establishment of a suitable biofilm, the exoelectrogens transfer the metabolically generated electrons from their outer cell membrane to the anode surface. There are two known electron transfer mechanisms that is direct electron transfer (DET) and mediated electron transfer (MET), which are deeply and extensively studied in *Geobacter* species and *Shewanella* species.[45–47] In *Geobacter sulfurreducens*, DET involves OMC c-Cyts (e.g. OmcZ, OmcB) for the short-range electron transfer during the initial development of biofilms and pili (type IV) for long-range electron transfer in multilayer biofilms,[45] while MET process uses flavin molecules, for example, riboflavin (RF).[48] In *Shewanella oneidensis*, the complex of cytochromes-flavins mediates the exocellular electron transfer (EET) mechanism. For example, flavin mononucleotide acts as a cofactor for the cytochrome MtrC and RF for the cytochrome OmcA.[49]

The transferred electrons on the anode surface are transported to the cathode surface via an electrical connection. The electrons at the cathode surface react with protons and an electron acceptor. If the electron acceptor is oxygen, the end product will be water, resulting maximum open circuit voltage at the cathode of ca. 0.805 V. Generally, the cathode surface is bound with a catalyst to increase the oxygen reduction rate.[50] The most commonly and efficient catalyst used is platinum. The carbon/platinum electrodes are commercially available with different concentrations of platinum, for example, carbon cloth with $0.20.5$ mg/cm^2. Alternatively, a microorganism can also be used for the oxygen reduction to make the fuel cell more cost-effective. This also provides a tool to understand the electron transfer mechanisms from the cathode to microorganism. Electron acceptors other than oxygen, such as ferricyanide, potassium permanganate are useful alternatives.[6]

10.4.1 APPROACHES TO ENHANCE THE ELECTRICITY GENERATION

The researchers engaged in the MFC studies around the globe have endeavoured many innovative efforts to increase the power output of the fuel cells. Many of them are developing new MFC designs using different effective materials for the electrodes and membrane, operating MFCs at specific conditions (e.g. setting electrode potentials, maintaining pH of the electrolytes, pretreatment of membranes and electrodes), and nanomodification of the electrodes.

10.4.1.1 ELECTRODE MODIFICATION

Electrode modification with metal catalyst or nanoparticles or chemical treatment has become a new trend to improve the performance of MFCs. The main purpose to modify the electrodes in MFCs is to increase the power output, in the anode by providing high surface area for the biofilm formation and to increase the EET mechanisms while the cathode modifications are the centre of attraction to replace the highly costly platinum catalyst by cheaper catalysts of nearly or same catalytic properties. Most of the studies regarding electrode modifications also claimed to decrease the internal resistance of the system as well as start-up time of the reactor. In the anode, different approaches have been employed to modify the electrodes to increase the power outputs either by simple modifications methods such as heat-treated electrodes; nitrogen-doped electrodes or by some sophisticated tools such as by coating some highly effective catalysts (e.g. gold nanoparticles, graphene, carbon nanotubes (CNT) etc.) on the electrodes.[51–53] Interestingly, almost every kind of metal nanoparticles or other carbon nanoparticles has been used in the MFCs; therefore, the researchers are now using different composite materials (e.g. CNT–gold–titania nanocomposites) to modify the electrode.[52] A recent study shows that the heat treatment of the electrodes can effectively increase the power output. In this regard, the stainless steel felt electrodes (as anode) were heated at 600°C for 5 min in the furnace. The results found that the heat treatment generated a thick layer of iron oxide on the surface of the electrodes. Moreover, MFC with such heat-treated electrode produced approximately seven times more current density than with untreated electrodes, 1.5 mA/cm^2 and 0.22 mA/cm^2, respectively.[54] Another effective method includes the use of nitrogen-doped carbon nanoparticles to modify the electrode that increased the EET mechanism. Evidently, nitrogen-doped carbon nanoparticles were coated on carbon cloth electrodes in a *Shewanella oneidensis* MR-1 inoculated two chamber MFC. The findings of the study revealed that the treated electrodes absorbed more electron mediators (flavins) secreted by the organism that subsequently increased the electron transfer rate. Consequently, the power density also increased more than three times as compared to untreated electrodes.[53] The anode can also be modified with metal or non-metal nanoparticles (with different morphologies as well) to influence the EET and thus the performance of the MFCs. For example, CNT powder was directly added to the anode chamber to increase the biofilm growth of *G. sulfurreducens* in a two chamber MFC using plain carbon paper as the electrode material in both the chambers.[55] The addition of CNT powder in the anode chamber reduced the internal resistance of the system as well as the start-up time of the MFC. The decreased start-up time can be attributed to the promotion of the bacterial adhesion to the electrode material with the addition of CNT powder in the anode chamber.[55] The performance of the anode can be further improved by using different morphologies of the material that can provide more active sites and enhance biocompatibility with the electrode material. In a double-chambered MFC, the anode (carbon cloth) was modified with bamboo-like CNT that produced ca. four times higher power density than the MFC using plain carbon cloth as the anode.[51]

10.4.1.2 GENETIC ENGINEERING OF EXOELECTROGENS

The exoelectrogens exhibit specific proteins for electroactive biofilm formation and redox proteins/molecules that help the electrons to transfer from the microbial outer membrane to the electrode surface. The production of such proteins or the secretion of redox molecules can be enhanced in the exoelectrogens, consequently, the more electrons could be transferred at a faster rate and higher would be the current density. It can be achieved by inserting the genes encoding these redox proteins or molecules into the exoelectrogen's genetic material and it is feasible by using the genetic engineering technique. Therefore, such genetically engineered exoelectrogens, which has the ability to produce abundant redox proteins can be employed in the MFC anode to produce high power density. This approach has been already demonstrated in the MFCs. For example, Yang et al. introduced the genes encoding the flavin biosynthesis pathway in *S. oneidensis* MR-1. This genetically modified *S. oneidensis* MR-1 secreted approximately 25 times more flavins and subsequently ~13 times higher power density than the wild type exoelectrogen.[56] Similarly, the genes of biosynthetic pathway for RF synthesis were expressed in *E. coli* BL-21, which generated ca. 9.5 times more power density as compared to wild strain.[57] The proteins responsible for biofilm formation can also be altered to promote the biofilm formation for early start-up of the MFC and higher power output. Recently, a second messenger, bis-(3'-5')-cyclic dimeric guanosine monophosphate (c-di-GMP) has been identified to play a vital role in electroactive biofilm formation.[58] c-di-GMP promotes the expression of adhesive matrix components, and thus facilitates the bacterial biofilm formation. A protein named as YdeH catalyses the biosynthesis of c-di-GMP.[58] Liu et al. constructed two mutant strains of *S. oneidensis* MR-1, in one strain the expression of *ydeH* was under the control of isopropyl-β-D-1-thiogalactopyranoside (IPTG)-inducible promoter, and in the other strain *ydeH* was under the control of a constitutive promoter. Both the strains were inoculated in a two chamber MFC that produced ~2.8 times more power density as compared to wild type *S. oneidensis* MR-1.[59] Therefore, it is quite evident that genetic engineering is a powerful technique to significantly increase the power output of an MFC and hence it could be an advantageous approach to use genetically engineered exoelectrogens in scaling-up the technology.

10.5 CONCLUSION

The MFCs provide a suitable, eco-friendly alternative to produce energy and to treat the wastewater simultaneously since bacteria can degrade the organic matter present in the wastewater. Several wastewaters ranging from low-strength to high strength have been utilized in MFCs for their treatment and electricity generation simultaneously. However, the power outputs achieved in the MFCs are low and can be enhanced by the following approaches: (1) a suitable design that results in low internal resistance; (2) using metal nanoparticles that increase the electron transfer mechanisms; (3) use of genetically engineered microorganisms; (4) addition of pretreated inoculum or control inoculum; (5) decreasing the start-up time of the MFC. On the other hand, the wastewater treatment of MFCs can be further improved by operating the fuel cells at mesophilic temperatures and for longer periods. Moreover, the MFCs integrated with other anaerobic fermentation technologies such as with UASB, have shown to enhance the COD removal efficiency by manifolds. MFCs have shown a great potential for the reduction of heavy metals or toxic pollutants and have achieved encouraging results when used in the anode as well as in the cathode chamber as the electron acceptor.

10.6 ACKNOWLEDGEMENTS

The authors are grateful to the Universiti Malaysia Pahang (UMP) for Internal Research Scheme Grant (RDU 140379) for the financial support.

KEYWORDS

- **microbial fuel cell**
- **electricity generation**
- **wastewater treatment**
- **bioremediation**

REFERENCES

1. Van Eerten-Jansen, M. C. A. A.; Heijne, A. T.; Buisman, C. J. N.; Hamelers, H. V. M. Microbial Electrolysis Cells for Production of Methane from CO_2: Long-Term Performance and Perspectives. *Int. J. Energy Res.* **2012,** *36,* 809–819.

2. Xu, B.; Ge, Z.; He, Z. Sediment Microbial Fuel Cells for Wastewater Treatment: Challenges and Opportunities. *Environ. Sci.: Water Res. Technol.* **2015,** *1,* 279–284.

3. Li, W. W.; Yu, H. Q.; He, Z. Towards Sustainable Wastewater Treatment by Using Microbial Fuel Cells-Centered Technologies. *Energy Environ. Sci.* **2014,** *7,* 911–924.

4. Fornero, J. J.; Rosenbaum, M.; Angenent, L. T. Electric Power Generation from Municipal, Food, and Animal Wastewaters Using Microbial Fuel Cells. *Electroanal.* **2010,** *22,* 832–843.

5. Qian, Y.; Xu, Y.; Yang, Q.; Chen, Y.; Zhu, S.; Shen, S. Power Generation Using Polyaniline/Multi-Walled Carbon Nanotubes as an Alternative Cathode Catalyst in Microbial Fuel Cells. *Int. J. Energy Res.* **2014,** *38,* 1416–1423.

6. Yong, X. Y.; Feng, J.; Chen, Y. L.; Shi, D. Y.; Xu, Y. S.; Zhou, J.; Wang, S. Y.; Xu, L.; Yong, Y. C.; Sun, Y. M.; Shi, C. L.; OuYang, P. K.; Zheng, T. Enhancement of Bioelectricity Generation by Cofactor Manipulation in microbial Fuel Cell. *Biosens. Bioelectron.* **2014,** *56,* 19–25.

7. Yoshizawa, T.; Miyahara, M.; Kouzuma, A.; Watanabe, K. Conversion of Activated-sludge Reactors to Microbial Fuel Cells for Wastewater Treatment Coupled to Electricity Generation. *J. Biosci. Bioeng.* **2014,** *118,* 533–539.

8. Huang, L.; Logan, B. E. Electricity Generation and Treatment of Paper Recycling Wastewater Using a Microbial Fuel Cell. *Appl. Microbiol. Biotechnol.* **2008,** *80,* 349–355.

9. Ghoreishi, K. B.; Ghasemi, M.; Rahimnejad, M.; Yarmo, M. A.; Daud, W. R. W.; Asim, N.; Ismail, M. Development and Application of Vanadium Oxide/Polyaniline Composite as a Novel Cathode Catalyst in Microbial Fuel Cell. *Int. J. Energy Res.* **2014,** *38,* 70–77.

10. Li, X.; Zhu, N.; Wang, Y.; Li, P.; Wu, P.; Wu, J. Animal Carcass Wastewater Treatment and Bioelectricity Generation in up-Fow Tubular Microbial Fuel Cells: Effects of HRT and Non-Precious Metallic Catalyst. *Bioresour. Technol.* **2013,** *128,* 454–460.

11. Oh, S. E.; Logan, B. E. Hydrogen and Electricity Production from a Food Processing Wastewater Using Fermentation and Microbial Fuel Cell Technologies. *Water Res.* **2005,** *39*, 4673–4682.

12. Ahn, Y.; Logan, B. E. Effectiveness of Domestic Wastewater Treatment Using Microbial Fuel Cells at Ambient and mesophilic Temperatures. *Bioresour. Technol.* **2010,** *101*, 469–475.

13. Kaewkannetra, P.; Chiwes, W.; Chiu, T. Y. Treatment of Cassava Mill Wastewater and Production of Electricity Through Microbial Fuel Cell Technology. *Fuel* **2011,** *90*, 2746–2750.

14. Min, B.; Kim, J.; Oh, S.; Regan, J. M; Logan, B. E. Electricity Generation from Swine Wastewater Using Microbial Fuel Cells. *Water Res.* **2005,** *39*, 4961–4968.

15. Li, X. M.; Cheng, K. Y.; Wong, J. W. Bioelectricity Production from Food Waste Leachate Using Microbial Fuel Cells: Effect of NaCl and pH. *Bioresour. Technol.* **2013,** *149*, 452–458.

16. Lu, N.; Zhou, S.; Zhuang, L.; Zhang, J.; Ni, J. Electricity Generation from Starch Processing Wastewater Using Microbial Fuel Cell Technology. *Biochem. Eng. J.* **2009,** *43*, 246–251.

17. Durruty, I.; Bonanni, P. S.; Gonzalez, J. F.; Busalmen, J. P. Evaluation of Potato-Processing Wastewater Treatment in a Microbial Fuel Cell. *Bioresour. Technol.* **2012,** *105*, 81–87.

18. Patil, S. A.; Surakasi, A. P.; Koul, S.; Ijmulwar, S.; Vivek, A.; Shouche, Y. S.; Kapadnis, B. P. Electricity Generation using Chocolate Industry Wastewater and its Treatment in Activated Sludge Based Microbial Fuel Cell and Analysis of Developed Microbial Community in the Anode Chamber. *Bioresour. Technol.* **2009,** *100*, 5132–5139.

19. Feng, Y.; Wang, X.; Logan, B. E.; Lee, H. Brewery Wastewater Treatment Using Air-Cathode Microbial Fuel Cells. *Appl. Microbiol. Biotechnol.* **2008,** *78*, 873–880.

20. Wen, Q.; Wu, Y.; Cao, D.; Zhao, L.; Sun, Q. Electricity Generation and Modeling of Microbial Fuel Cell from Continuous Beer Brewery Wastewater. *Bioresour. Technol.* **2009,** *100*, 4171–4175.

21. Cheng, J.; Zhu, X.; Ni, J.; Borthwick, A. Palm oil Mill Effluent Treatment Using a Two-Stage Microbial Fuel Cells System Integrated with Immobilized Biological Aerated Filters. *Bioresour. Technol.* **2010,** *101*, 2729–2734.

22. Gude, V. G. Wastewater Treatment in Microbial Fuel Cells–an Overview. *J. Clean. Prod.* **2016,** *122*, 287–307.

23. Min, B.; Cheng, S.; Logan, B. E. Electricity Generation Using Membrane and Salt Bridge Microbial Fuel Cells. *Water Res.* **2005,** *39*, 1675–1686.

24. Papaharalabos, G.; Greenman, J.; Melhuish, C.; Ieropoulos, I. A. Novel Small Scale Microbial Fuel Cell Design for increased Electricity Generation and Wastewater Treatment. *Int. J. Hydrogen Energy* **2015,** *40*, 4263–4268.

25. Zhang, B.; Feng, C.; Ni, J.; Zhang, J.; Huang, W. Simultaneous Reduction of Vanadium (V) and Chromium (VI) with Enhanced Energy Recovery Based on Microbial Fuel Cell Technology. *J. Power Sources* **2012,** *204*, 34–39.

26. Abourached, C.; Catal, T.; Liu, H. Efficacy of Single-Chamber Microbial Fuel Cells for Removal of Cadmium and Zinc with Simultaneous Electricity Production. *Water Res.* **2014,** *51*, 228–233.

27. Wang, Z.; Lim, B.; Choi, C. Removal of Hg^{2+} as an Electron Acceptor Coupled with Power Generation Using a Microbial Fuel Cell. *Bioresour. Technol.* **2011**, *102*, 6304–6307.

28. Zhang, L. J.; Tao, H. C.; Wei, X. Y.; Lei, T.; Li, J. B.; Wang, A. J.; Wu, W. M. Bioelectrochemical Recovery of Ammoniaecopper (II) Complexes from Wastewater Using a Dual Chamber Microbial Fuel Cell. *Chemosphere* **2012**, *89*, 1177–1182.

29. Hai, F. I.; Yamamoto, K.; Fukushi, K. Hybrid Treatment Systems for Dye Wastewater. *Crit. Rev. Env. Sci. Tec.* **2007**, *37*, 315–377.

30. Kong, F.; Wang, A.; Cheng, H.; Liang, B. Accelerated Decolorization of Azo Dye Congo Red in a Combined Bioanode-Biocathode Bioelectrochemical System with Modified Electrodes Deployment. *Bioresour. Technol.* **2014**, *151*, 332–339.

31. Strycharz, S. M.; Woodard, T. L.; Johnson, J. P.; Nevin, K. P.; Sanford. R.; Löffler, F. E.; Lovley, D. R. Graphite Electrode as a Sole Electron Donor for Reductive Dechlorination of Tetrachlorethene by Geobacter Lovleyi. *Appl. Environ. Microbiol.* **2008**, *74*, 5943–5947.

32. Orellana, R.; Leavitt, J. J.; Comolli, L. R.; Csencsits, R.; Janot, N.; Flanagan, K. A.; Gray, A. S.; Leang, C.; Izallalen, M.; Mester, T. U(VI) Reduction by a Diversity of Outer Surface c-Type Cytochromes of *Geobacter sulfurreducens. Appl. Environ. Microbiol.* **2013**, *79*, 6369–6374.

33. Marshall, M. J.; Beliaev, A. S.; Dohnalkova, A. C.; Kennedy, D. W.; Shi, L.; Wang, Z. M.; Boyanov, M. I.; Lai, B.; Kemner, K. M.; McLean, J. S.; Reed, S. B.; Culley, D. E.; Bailey, V. L.; Simonson, C. J.; Saffarini, D. A.; Romine, M. F.; Zachara, J. M.; Fredrickson, J. K. c-Type Cytochrome-Dependent Formation of U(IV) Nanoparticles by *Shewanella oneidensis. PLos Biol.* **2006**, *4*, e268.

34. Huang, L.; Chai, X.; Chen, G.; Logan, B. E. Effect of Set Potential on Hexavalent Chromium Reduction and Electricity Generation from Biocathode Microbial Fuel Cells. *Environ. Sci. Technol.* **2011**, *45*, 5025–5031.

35. Xafenias, N.; Zhang, Y.; Banks, C. J. Enhanced Performance of Hexavalent Chromium Reducing Cathodes in the Presence of *Shewanella oneidensis* MR-1 and Lactate. *Environ. Sci. Technol.* **2013**, *47*, 4512–4520.

36. Zhang, B.; Zhang, J.; Liu, Y.; Hao, C.; Tian, C.; Feng, C.; Zhang, Z. Identification of Removal Principles and Involved Bacteria in Microbial Fuel Cells for Sulfide Removal and Electricity Generation. *Int. J. Hydrogen Energy* **2013**, *38*, 14348–14355.

37. Wang, W.; Feng, Y.; Tang, X.; Li, H.; Du, Z.; Yang, Z.; Du, Y. Isolation and Characterization of an Electrochemically Active and Cyanide-Degrading Bacterium Isolated from a Microbial Fuel Cell. *RSC Adv.* **2014**, *4*, 36458–36463.

38. Miran, W.; Nawaz, M.; Jang, J.; Lee, D. S. Sustainable Electricity Generation by Biodegradation of Low-Cost Lemon Peel Biomass in a Dual Chamber Microbial Fuel Cell. *Int. Biodeter. Biodegr.* **2016**, *106*, 75–79.

39. Prathap, P.; Cesar, I. T.; Tyson, B.; Sudeep, C. P.; Bradley, G. L.; Bruce, E. R. Kinetic, Electrochemical, and Microscopic Characterization of the Thermophilic, Anode-Respiring Bacterium *Therminocola ferriacetica. Environ. Sci. Technol.* **2013**, *47*, 4934–4940.

40. Kracke, F.; Vassilev, I.; Krömer, J. O. Microbial Electron Transport and Energy Conservation–the Foundation for Optimizing Bioelectrochemical Systems. *Front. Microbiol.* **2015**, *6*, 575.

41. Fazli, M.; Rybtke, M. L.; Almblad, H.; Tolker-Nielsen, T.; Givskov. M.; Eberl, L. Regulation of Biofilm Formation in *Pseudomonas* and *Burkholderia* Species. *Environ. Microbiol.* **2014**, *16*, 1961–1981.

42. Pamp, S. J.; Tolker-Nielsen, T. Multiple Roles of Biosurfactants in Structural Biofilm Development by *Pseudomonas aeruginosa*. *J. Bacteriol.* **2007**, *189*, 2531–2539.

43. Nevin, K. P.; Richter, H.; Covalla, S. F.; Johnson, J. P.; Woodard, T. L.; Orloff, A. L.; Jia, H.; Zhang, M.; Lovley, D. R. Power Output and Columbic Efficiencies from Biofilms of *Geobacter sulfurreducens* Comparable to Mixed Community Microbial Fuel Cells. *Environ. Microbiol.* **2008**, *10*, 2505–2514.

44. Richter, H.; McCarthy, K.; Nevin, K. P.; Johnson, J. P.; Rotello, V. M.; Lovley, D. R. Electricity Generation by *Geobacter sulfurreducens* Attached to Gold Electrodes. *Langmuir* **2008**, *24*, 4376–4379.

45. Clarke, T. A.; Edwards, M. J.; Gates, A. J.; Hall, A.; White, G. F.; Bradleya, J.; Reardonb, C. L.; Shib, L.; Beliaevb, A. S.; Marshallb, M. J.; Wangb, Z.; Watmougha, N. J.; Fredricksonb, J. K.; Zacharab, J. M.; Butta, J. N.; Richardson, D. J. Structure of a Cell Surface Decaheme Electron Conduit. *Proc. Natl. Acad. Sci. U. S. A.* **2011**, *108*, 9384–9389.

46. Coursolle, D.; Baron, D. B.; Bond, D. R.; Gralnick, J. A. The Mtr Respiratory Pathway is Essential for Reducing Flavins and Electrodes in *Shewanella oneidensis*. *J. Bacteriol.* **2010**, *192*, 467–474.

47. Inoue, K.; Leang, C.; Franks, A. E.; Woodard, T. L.; Nevin, K. P.; Lovley DR. Specific Localization of the c-Type Cytochrome OmcZ at the Anode Surface in Current-Producing Biofilms of *Geobacter sulfurreducens*. *Environ. Microbiol. Rep.* **2010**, *3*, 211–217.

48. Malvankar, N. S.; Lovley, D. R. Microbial Nanowires: A New Paradigm for Biological Electron Transfer and Bioelectronics. *Chem. Sus. Chem.* **2012**, *5*, 1039–1046.

49. Okamoto, A.; Kalathil, S.; Deng, X.; Hashimoto, K.; Nakamura, R.; Nealson, K. H. Cell-Secreted Flavins Bound to Membrane Cytochromes Dictate Electron Transfer Reactions to Surfaces with Diverse Charge and pH. *Sci. Rep.* **2014**, *4*, 5628.

50. Logan, B. E.; Wallack, M. J.; Kim, K. Y.; He, W.; Feng, Y.; Saikaly, P. E. Assessment of Microbial Fuel Cell Configurations and Power Densities. *Environ. Sci. Technol. Lett.* **2015**, *2*, 206–214.

51. Ci, S.; Wen, Z.; Chen, J.; He, Z. Decorating Anode with Bamboo-Like Nitrogen-Doped Carbon Nanotubes for Microbial Fuel Cells. *Electrochem. Commun.* **2012**, *14*, 71–74.

52. Wu, Y.; Zhang, X.; Li, S.; Lv, X.; Cheng, Y.; Wang, X. Microbial Biofuel Cell Operating Effectively Through Carbon Nanotube Blended with Gold–Titania Nanocomposites Modified Electrode. *Electrochim. Acta.* **2013**, *109*, 328–332.

53. Yu, Y. Y.; Guo, C. X.; Yong, Y. C.; Li, C. M.; Song, H. Nitrogen Doped Carbon Nanoparticles Enhanced Extracellular Electron Transfer for High-Performance Microbial Fuel Cells Anode. *Chemosphere* **2015**, *140*, 26–33.

54. Guo, K.; Soeriyadi, A. H.; Feng, H.; Prevoteau, A.; Patil, S. A.; Gooding, J. J.; Rabaey, K. Heat-Treated Stainless steel Felt as Scalable Anode Material for Bioelectrochemical Systems. *Bioresour. Technol.* **2015**, *195*, 46–50.

55. Liang, P.; Wang, H.; Xia, X.; Huang, X.; Mo, Y.; Cao, X.; Fan, M. Carbon Nanotube Powders as Electrode Modifier to Enhance the Activity of Anodic Biofilm in Microbial Fuel Cells. *Biosens. Bioelectron.* **2011**, *26*, 3000–3004.

56. Yang, Y.; Ding, Y.; Hu, Y.; Cao, B.; Rice, S. A.; Kjelleberg, S.; Song, H. Enhancing Bidirectional Electron Transfer of *Shewanella oneidensis* by a Synthetic Flavin Pathway. *ACS Synth. Biol.* **2015,** *7*, 815–823.

57. Tao, L.; Wang, H.; Xie, M.; Thia, L.; Chen, W. N.; Wang, X. Improving Mediated Electron Transport in Anodic Bioelectrocatalysis. *Chem. Commun.* **2015,** *51*, 12170–12173.

58. Chao, L.; Rakshe, S.; Leff, M.; Spormann, A. M. PdeB, a Cyclic Di-GMP Specific Phosphodiesterase that Regulates Shewanella Oneidensis MR-1 Motility and Biofilm Formation. *J. Bacteriol.* **2013,** *195*, 3827–3833.

59. Liu, T.; Yu, Y. Y.; Deng, X. P.; Ng, C. K.; Cao, B.; Wang, J. Y.; Rice, S. A.; Kjelleberg, S.; Song, H. Enhanced Shewanella Biofilm Promotes Bioelectricity Generation. *Biotechnol. Bioeng.* **2015,** *112*, 2051–2059.

CHAPTER 11

ROLE OF MICROORGANISMS IN THE REMOVAL OF FLUORIDE

SHUBHA DWIVEDI[1,2,*], PRASENJIT MONDAL[3], and
CHANDRAJIT BALOMAJUMDER[3]

[1]*Department of Biotechnology, Uttarakhand Technical University, Dehradun, India*

[2]*Department of Biotechnology, Meerut Institute of Engineering and Technology, Meerut, Uttar Pradesh-250005, India,*
*E-mail: shubha.dwivedi@rediffmail.com

[3]*Department of Chemical Engineering, Indian Institute of Technology Roorkee, Roorkee, Uttarakhand 247667, India*

CONTENTS

ABSTRACT

Fluorine exists as the p block element of group seventeen in periodic table and component of halogen family. Ionic form of the fluorine is fluoride. Various natural and industrial sources are responsible for the increased level of fluoride in groundwater. Fluoride serves as double, for example weapon for the human being. Below the prescribed level, that is 1.5 mg/l ((World Health Organization (WHO)), it is beneficiary for the tooth and bone health. But when the concentration increases, it possesses threatening effects on mankind and many kinds of fluorosis occur. There is the prime need to minimize the level of fluoride in drinking water as well as in wastewater. Various techniques are available to remove fluoride. Chemical and physical techniques have their own qualities and disadvantages. In the present scenario, bioremoval/bioadsorption is getting more attention to eliminate fluoride in eco-friendly manner. Live and dead cells of the microorganisms like *Shewanella putrefaciens*, *Spirogyra* species, *Aspergillus* sp., *Pleurotus* sp. etc. have potential to remove fluoride. Microbial biomass consists the high efficiency for the removal of contaminants from wastewater. The surface of the microorganism provides the enriched sites for bioadsorption process due to the existence of a variety of functional groups on it.

11.1 INTRODUCTION

11.1.1 GENERAL

Water is honeydew for sustaining life and precious resource provided by the nature for the existence of biotic organisms. Various socio-environmental factors either natural or manmade are responsible for the degradation of water quality.[17,2,34,3] Many hazardous contaminants like fluoride, arsenic, nitrate, sulphate and other heavy metals etc. are found in the underground water and has been reported from diverse parts of India.[20,7,31,32]

The contamination of surface and inland waters caused by the release of toxic chemicals from different industries can be dangerous to all classes of living organisms if discharged without proper treatment. Chemicals from industrial wastewater which produce toxicity include inorganic chemicals and organic chemicals. Both inorganic and organic chemicals, even in extremely low concentration, may be toxic to aquatic life as well as terrestrial life. Therefore, the treatment of these toxic materials is required before their discharge. Due to the detrimental and injurious effects of high fluoride concentration, there is an urgent prerequisite of methods for its removal from drinking as well as from industrial wastewater. Distillation, membrane separation processes, precipitation, ion exchange and adsorption are the methods used for fluoride removal. The adsorption method complies with the employ of physical or the biological adsorbents/microorganism. The present study highlights the contribution of microorganism which includes fungi, algae and bacteria for the removal of fluoride.

11.2 FLUORIDE, ITS PROPERTIES, OCCURRENCE, AND SOURCES

Fluoride belongs to halogen family, the main electronegative element which is the ionic form of fluorine and does not exist in the free state in nature because of its high reactivity. In the elemental form, fluorine is a combustible, infuriating, and is the most influential oxidizing agent. There are a number of sources from where fluoride is generally generated and added to the environment. The sources of fluoride are characterized into two categories: natural sources which include the fluorine already present in nature and distributed in Earth's crust, mainly as the minerals fluorspar, fluorapatite and cryolite;[22] and anthropogenic sources which include industrial activities. Further, many industrial processes such as coal combustion, steel production and other manufacturing processes (aluminium, copper and nickel production, phosphate ore processing, phosphate fertilizer production, glass, brick and ceramic manufacturing) etc. also contribute to increase fluoride levels in water. Many vegetables, fruits, and nuts consist organic fluorine. The concentration of fluoride found in some food products and agriculture crops are given in Figure 11.1 and Table 11.1.[40]

Properties of fluoride are given in Tab. 11.2. Fluoride compounds make up approximately 0.08% of the earth's outer layer.[14] Concentrations of fluoride are found nearly in all types of natural waters. Sea water contains about 1 mg/l whereas lakes possess a concentration of less than 0.5 mg/l. In groundwater, an elevated level of fluoride depends on the nature of rocks and the occurrence of fluoride-bearing minerals. In India, around 19 states are affected due to the occurrence of fluoride in water bodies. The Rajasthan and Andhra Pradesh are the states having highest fluoride concentration.[24] The Tab. 11.3 showing the record of all the states with the different concentrations of fluoride present. Figure 11.2 spotted all the places on the map of India where the fluoride has been found.

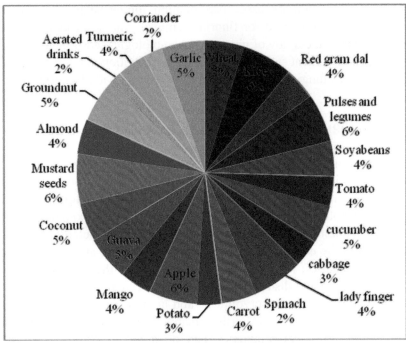

FIGURE 11.1   Fluoride concentration in mg/kg in different foodstuffs.

TABLE 11.1   Fluoride Concentrations in Other Food Products.

Food item	Fluoride concentration (mg/kg)
Beef	4.0–5.0
Pork	3.0–4.5
Mutton	3.0–3.5
Fishes	1.0–6.5
Tobacco	3.2–38
Beetle leaf (pan)	7.8–12.0
Rock salts	200.0–250.0
Areca but (supari)	3.8–12.0
Tea	60–112

TABLE 11.2 Properties of Fluoride.

Property	Sodium fluoride (NaF)	Hydrogen fluoride (HF)
Physical state	White crystalline powder	Colourless liquid or gas with biting smell
Melting point	993	−83
Boiling point	1695 at 100 kPa	19.5
Density	2.56	
Water solubility	42 g/l at 10°C	Readily soluble below 20°C
Acidity		Strong acid in liquid form; weak acid dissolved in water

TABLE 11.3 States Endemic to Fluoride in India.[24]

S. no.	States	Range of fluoride concentration (mg/l)
1	Andhra Pradesh	0.11–20.0
2	Assam	0.2–18.1
3	Bihar	0.6–8.0
4	Delhi	0.4–10.0
5	Gujarat	1.58–31.0
6	Haryana	0.17–24.7
7	Jammu and Kashmir	0.05–4.21
8	Karnataka	0.2–18.0
9	Kerala	0.2–2.5
10	Maharashtra	0.11–10.2
11	Madhya Pradesh	0.08–4.2
12	Odisha	0.6–5.7
13	Punjab	0.44–6.0
14	Rajasthan	0.2–37.0
15	Tamil Nadu	1.5–5.0
16	Uttar Pradesh	0.12–8.9
17	West Bengal	1.5–13.0

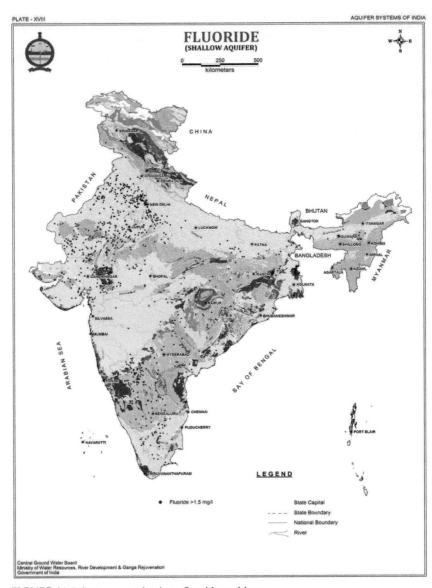

FIGURE 11.2 States are endemic to fluoride problem.

11.3   LIMITS OF FLUORIDE

Permissible limit of fluoride is set by various environmental organization and agencies and given in following sections.

11.3.1   LIMITS SET BY WORLD HEALTH ORGANIZATION (WHO)

While setting national standards or local guidelines for fluoride or to assess the probable health hazard due to the high concentration of fluoride, it is necessary to have a look on the consumption of water by the population residing in the fluoride prone areas where the concentration of fluoride is 6 mg/day. So, it is the requirement to set the standard or local guideline at a concentration lower than 1.5 mg/l. The WHO guideline set the limit for fluoride in drinking water is 1.5 mg/l. In developing countries, high fluoride concentrations are one of the major problems because of deficient infrastructure for treatment. Lacking in the technology advancement, WHO (1993)[48] also found difficulty in achieving the prescribed guideline for fluoride in fluoride prone areas (2004).[19]

11.3.2   LIMITS SET BY CPCB (CENTRAL POLLUTION CONTROL BOARD)

Central Pollution Control Board sets the limits for different forms of water presented in Tab. 11.4.[39]

TABLE. 11.4   Fluoride Concentration in Different Forms of Water.

Standards	Max. Fluoride concentration (as F) mg/l
Inland surface water	2.0 mg/l
Public Sewers	15 mg/l
Land for irrigation	–
Marine coastal areas	15 mg/l

11.3.3   LIMITS SET BY ENVIRONMENT PROCUREMENT AND CONTROL (EPC)

In 1974, Congress passed the Safe Drinking Water Act. This act directed Environmental Protection Agency (EPA) to examine the level of contaminants in drinking water at which no fluorosis is likely to occur. The maximum contaminant level goals (MCLGs) for fluoride are4.0 mg/l. EPA has also set a secondary standard (SMCL) for fluoride at 2.0 mg/l.

11.4   ADVANTAGEOUS PROPERTY OF FLUORIDE FOR HUMAN BEING

Following are the beneficiary effects of fluoride.

11.4.1   SHIELDING FROM DEMINERALIZATION

Bacteria present in the oral cavity combine with sugars and produce acid. This acid is responsible for the tooth enamel eroding. Fluoride can protect teeth from eroding due to the acid. Fluoride intake during the pre-eruptive development of the teeth has a cariostatic effect. It minimizes the risk of dental problems due to the uptake of fluoride by enamel crystallites and the formation of fluorhydroxyapatite.[5] Consumption of water that contains fluoride in a concentration of approximately 1 mg/l has been found to be efficient in reducing tooth decay. For this reason, fluoride compounds are mixed to water supplies as supplementary which contain less than the desired concentration.[19]

11.4.2   REMINERALIZATION

Remineralisation is the process which uses the fluoride as a supplementary at the affected areas which are damaged by the demineralization. Fluoride present in oral fluids, including saliva and dental plaque, also contributes to the cariostat fluorideic effect. This post-eruptive effect occurs to reduce acid production by plaque-forming bacteria and increases the rate of enamel remineralization during an acidogenic challenge. It also has the unique ability to stimulate new bone formation, and is also used in the treatment of osteoporosis.

11.5 DETRIMENTAL EFFECTS OF HIGH CONCENTRATION OF FLUORIDE ON HUMAN BEINGS, MICROORGANISMS, PLANTS AND ANIMALS

When the level of fluoride exceed the permissible limits in groundwater as well as in marine water, a number of harmful effects were seen in flora and fauna world.

11.5.1 EFFECT OF FLUORIDE ON HUMAN BEINGS

Fluoride poisoning and the biological response leading to ill effects depend on the following factors:

- Excess concentration of fluoride in drinking water
- Low calcium and high alkalinity in drinking water
- Total daily consumption of fluoride
- Period of exposure to fluoride
- Age of the individual

Expectant mothers and lactating mothers are the most vulnerable groups as fluoride crosses the placenta because there is no barrier and it also enters maternal milk. The hormonal disorder is also the result of fluoride poisoning.

11.5.1.1 MECHANISM OF FLUORIDE TOXICITY

There are a number of mechanisms of fluoride toxicity. Ingested fluoride initially acts locally on the intestinal mucosa. It can form hydrofluoric acid in the stomach which results in GI irritation. Fluoride also exerts effects on glucose metabolism. Results have revealed that fluoride exposure may contribute to impaired glucose intolerance. Fluoride inhibits Na^+/K^+-ATPase, which may result in hyperkalaemia by extracellular release of potassium. Fluorides influence the secretary pathway which leads to the dental fluorosis.

Several other cellular processes are also affected by the molecular mechanism of inorganic fluoride. Fluoride can induce oxidative stress, homeostasis and lipid peroxidation as well as also alters gene expression and cause cell death.

Redox status: Fluoride alters mitochondria-rich cells such as those of the human kidney.

Cellular respiration: Fluoride inhibits cellular respiration. Fluoride ions bind to the amino acid groups present on the active centre of an enzyme as in case of an enzyme of glycolytic pathways and Krebs cycle which is sensitive to fluoride.

Generation of ROS: Fluoride exposure increases the generation of anion superoxide (O_2^-), increased O_2^- concentration and downstream consequences such as H_2O_2. Fluoride also increases NO generation. Fluoride inhibits the antioxidant activity.

Apoptosis: Fluoride induces the apoptosis by uplifting the oxidative stress which further induced lipid peroxidation, thus causing mitochondrial dysfunction and activation of downward pathways observed as the primary mechanism of cell death in presence of relatively high fluoride concentrations. Fluoride acts as enzyme inhibitor but fluoride ions stimulate enzyme activity rarely.[4]

11.5.1.2 DEFICIENCY BY EXCESS OF FLUORIDE IN HUMAN BEING

As the fluoride is highly electronegative in nature, it possesses a strong affinity for the positively charged ions. The fluoride present in the drinking water firstly brings alternations in the tooth as the fluoride replaces tooth enamel hydroxyl group with fluoride resulting in the formation of more stable compound said to be fluorapatite and get deposited as calcium-fluorapatite crystals. There are many kinds of fluorosis occurs due to excess of fluoride in the body as summarized in Tab. 11.5. [33]

TABLE 11.5   Stages of Fluorosis in Human Beings.

Types of fluorosis	Responsible range of fluoride in mg/l [24]	Age	Symptoms
Dental fluorosis	1.0–3.0	Children (5–7years)	White opaque area on the tooth surface
Skeletal fluorosis	3.0–4.0	Children (5–7years) Adults (<18years)	Accumulation of fluoride in the joint of neck, knee, pelvic and shoulder bones
Crippling fluorosis	4.0–6.0	Children (5–7years) Adults (<18years)	Arthritis, sclerosis, low level of haemoglobin

11.5.1.2.1   Dental fluorosis

The disease generally is seen in the children in the age group of 5–8years due to excessive fluoride intake, enamel losses its lustre and fluorosis characterized by white, opaque areas on the surface of the tooth and in several forms as shown in Fig. 11.3. This sometimes results in the development of yellow browns to black stains and severe pitting of the teeth. This dental fluorosis develops at the age of 8–10. Thus, dental fluorosis was not observed in adults.

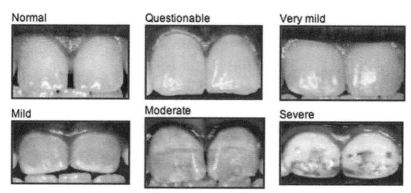

FIGURE 11.3   Stages of dental fluorosis.

11.5.1.2.2   Skeletal fluorosis

When an individual comes in contact with the high level of fluoride, chances of the bone disease known as skeletal fluorosis rises as shown in Fig. 11.4. The skeletal fluorosis occurs in all age group people i.e. in children also in adults. Fluoride assembles in the joints of neck, knee, pelvic and shoulder bones resulting in difficulty in movement or walking. The symptoms observed during this disease are very much alike to the symptoms seen in arthritis or spondylitis which are calcium deposits in bones, muscle weakness, etc. In the initial stages, known as 'pre-skeletal' fluorosis, a patient may suffer from a variety of symptoms with the lack of any detectable bone charges, including joint pain, joint stiffness and gastric distress.

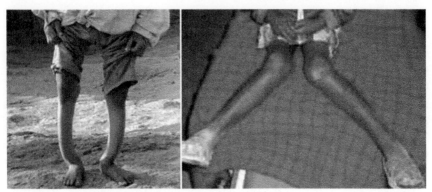

FIGURE 11.4   Stages of skeletal fluorosis.

11.5.1.2.3   Crippling fluorosis

Crippling fluorosis is the higher stage of fluorosis results in osteoporosis and bony outgrowths can also occur resulting rare bone cancer; osteosarcoma and finally spine, major joints, muscles and nervous system get damaged as shown in Fig. 11.5.

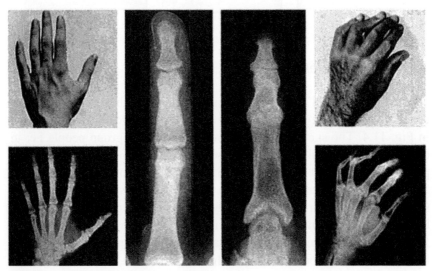

FIGURE 11.5 Stages of Crippling Fluorosis.

11.5.1.2.4   *Other problems*

Besides dental, skeletal and crippling fluorosis, excessive fluoride can result in muscle fibre degeneration, abnormalities in RBCs, low haemoglobin levels, excessive thirst, headache, skin rashes, nervousness, neurological man etc.

11.5.3   EFFECT OF FLUORIDE ON MICROORGANISMS

The microorganisms show different characteristic activity in the presence of fluoride. The effect of fluoride is different on different microbial fauna. Aerobic microorganisms are affected due to the presence of fluoride in anaerobic digestion. Review literatures point out that fluoride has a low inhibitory effect on the nitrifying bacteria in activated sludge process. Sodium fluoride did not showed any inhibition towards glucose utilizing bacteria in aerobic sewage sludge of fluoride concentrations of up to 540 mg/l. Fluoride presents inhibitory effects towards glucose fermenting microorganisms. Anaerobic microorganisms that utilize propionate and butyrate were found to be very sensitive to fluoride. Fluoride is an inhibitor of pyrophosphatase in various microorganisms. The toxicity of fluoride towards acetate, propionate and butyrate utilizing bacteria is not yet known.[14]

11.5.4   EFFECT OF FLUORIDE ON PLANTS

The effect of fluoride on plants depends upon many factors such as the concentration, time of exposure, type and age of plant, temperature, type of light and intensity, composition of the air, and its rate of circulation, concentrations of the gas (for an appropriate time interval) under controlled environmental conditions may produce one or more of the following manifestations: slight paling of normal green pigment at the tips or margins of the leaves, a pale green area at the margins which may gradually turn into a light buff colour and finally a reddish brown. Varying degrees of injury to leaf tissues may occur until the entire leaf is affected and it falls from the plant. Vital plant processes such as respiration and photosynthesis may be influenced (Fornasiero et al., 2001).

11.5.5   EFFECT OF FLUORIDE ON ANIMALS

The effects of fluorides on animals depend upon a number of factors such as the physical and chemical properties of the compounds, dosage or amount given, method, rate and length of administration and the presence of interfering substances. Among the factors which modify actions of potentially harmful substances in the body are age, state of health, and sensitivity. Young animals are generally more susceptible to harmful or toxic substances than older ones. Sick or inadequately nourished animals are sometimes more susceptible than healthy ones. Either natural or acquired sensitivity may render certain individual animals more sensitive than others. When large doses (several grams) of soluble inorganic fluorine compounds are given orally to animals in a short time interval (few minutes or hours) acute intoxication develops local, affecting the gastrointestinal tract (thirst, vomiting, abdominal pains, diarrhoea) and partly a result of absorption which causes alternate painful spasms and pareses, weakness, excessive salivation, perspiration, dyspnoea and weakened pulse. Death is attributed primarily to respiratory paralysis. One or more of the symptoms may be absent. Post-mortem examinations reveal a marked haemorrhagic gastro-enteritis with a tendency to necrosis.[45] The bones and joints of animals fed high amount of fluoride for extended periods gradually become enlarged and extra bone growths may appear in different parts of the skeleton. Calcification of ligaments and related structures attached to bones may occur in extreme cases.

11.5.6   EFFECT OF FLUORIDE ON AQUATIC ORGANISM

The concentration of fluoride in fresh water ranges from 0.01–0.3 mg/l and in uncontaminated marine water, it ranges from 1.2–1.5 mg/l. Due to many natural and manmade activity, like manufacturing brick, glass, ceramic etc., generation of fluoride through the wet process of phosphoric acid manufacture from phosphatic fertilizer industry, aluminium smelters etc., level of fluoride is elevated in water which causes serious health hazard to aquatic life. Population growth of algae may slow down or increase by the increase or decrease the level of fluoride. Some algae are able to bear high concentration of fluoride, that is up to 200 mg/l, for example *Cyclotella meneghiniana, Ankistrodesmus braunii* (freshwater algae), *Dunaliella tertiolecta, Spirogyra* sp. etc. Many of the aquatic algae seem to be effective in bioremediation process. Some of the algae have high negative effect at the high level of fluoride. Reduction by 50% in the rate of respiration of *Chlorella pyrenoidosa* with 570 mg/l concentration of fluoride. 25–27% growth inhibition was seen in *Amphidinium carteri* and in *Dunaliella tertiolecta*. Many of the aquatic animals like fishes and invertebrates directly get fluoride from water. Fluoride tends to accumulate in the exoskeleton and bone tissues of fishes. Toxicity of fluoride to aquatic invertebrates increases with the increasing level of fluoride. It exerts a poisonous effect on the health of aquatic animals by inhibiting the enzyme activity and finally interrupting metabolic process.[6]

11.6   TREATMENT METHODOLOGY

There are many processes available for the removal of fluoride from contaminated water. A brief description is given in the Tab. 11.6.

TABLE 11.6   Brief Description of Different Removal Methods.

S. no.	Removal methods	Basic principle	Merits	Demerits	References
1	Electrochemical	Fluoride can be removed from analyte by electrode potential. The oxidation reduction reaction takes place.	The process was operated under a wide pH range (pH 5–9). The process was highly promising and reliable. It was a little expensive.	It was a quite expensive.	[10]

TABLE 11.6 *(Continued)*

S. no.	Removal methods	Basic principle	Merits	Demerits	References
2	Ion-exchange	Fluoride can be removed from water supplies with the use of a strongly basic anion-exchange resin containing quaternary ammonium functional groups. The removal takes place according to the following reaction: Matrix-$NR3^+Cl^- +$ $F^- \rightarrow$ Matrix-$NR3^+F^- + Cl^-$	Removes fluoride up to 90–95%. Retains the taste and colour of water intact	The technique is expensive because of the cost of resin, pretreatment required to maintain the pH, regeneration and waste disposal.	[44]
3	Adsorption	Adsorption occurs quickly and may be monomolecular (unimolecular) layer or monolayer, or two, three or more layers thick (multi-molecular). Commonly used adsorbents are activated carbon fibres, alum impregnated activated alumina, activated carbon etc.	Treatment is cost-effective. This method can remove fluoride up to 90%.	Process is highly pH dependent. The process has low adsorption capacity, poor integrity. Pretreatment is required.	[47,29]

TABLE 11.6   *(Continued)*

S. no.	Removal methods	Basic principle	Merits	Demerits	References
4	Coagulation–precipitation	Coagulation is the destabilization of colloids by neutralizing the forces that keep them apart. Addition of lime leads to precipitation of fluoride as insoluble calcium fluoride and raises the pH value of water up to 11–12. $Ca\,(OH)_2 + 2F^- \rightarrow CaF_2 + 2OH^-$ Lime and alum are the most commonly used coagulants.	The bases of Nalgonda technique for defluoridation are the combined use of alum and lime in a two-step process.	It is not automatic process and Maintenance cost is high. Only (18–33%) of fluoride removes by this process.	Technical Digest, 1978
5	Reverse osmosis	*Molecules* and *ions* from solutions by applying pressure to the solution when it is on one side of a selective membrane.	The process is very much effective for fluoride removal. No chemicals are required and very little maintenance is needed. It works under wide pH range.	Remineralization is required after treatment. The method is expensive. in comparison to other options. The water becomes acidic and needs pH correction.	[43, 42]

TABLE 11.6 *(Continued)*

S. no.	Removal methods	Basic principle	Merits	Demerits	References
6	Membrane separation	Membrane separation involves the use of membrane for the separation of different chemicals. This method also uses some amount force for removal as in reverse osmosis and electrodialysis.	Life of membrane is long and process operates with minimal manpower. The process is highly reliable. The disinfection and treatment of water in a single step.	Other common ions are also removed. Remineralization of water was required after treatment so as to add the other essential components.	[24]
7	Electrodialysis	Electrodialysis is used to move saltions from one solution through *ion-exchange-membranes* to another solution under the influence of an applied *electric potential* difference.	It is used to purify Small and medium scale drinking water	Higher molecular weight, uncharged and less mobile ionic species will not typically be significantly removed.	[16]

11.6.1 BIOLOGICAL TREATMENT

Biological removal and biosorption is a promising technique for water treatment utilizing biomaterials abundantly available. Various bioadsorbents have been developed for fluoride removal. The effectiveness of bioadsorbents due to their low cost and presence of high contents of amino and hydroxyl functional groups which show significant adsorption potential.

Treatment using microorganisms: Conventional and less effective physicochemical methods are being replaced by the more effective biological methods, for example bioreduction/removal for the fluoride contamination in soil and groundwater by bacteria *Shewanella putrefaciens*etc., supports the use of eco-friendly ways for the bioremediation.[9] Simultaneous adsorption and bioaccumulation process (SABA) is gaining high momentum and widely used in the removal of contaminants from wastewater.[11] Some microorganisms were found very effective in bioremediation of heavy metals and iron. In present scenario, bioremediation/biotreatment/bioremoval of contaminants from wastewater has also proved to be a fresh technology for wastewater treatment.[23]

11.6.1.1   WORK DONE ON THE REMOVAL OF FLUORIDE BY MICROORGANISMS

Very few literatures are available on the removal of fluoride using microorganism. Comparative tabulation of removal efficiency for fluoride by various microorganisms is given in Tab. 11.7. Graphical representation of % removal is shown in Fig. 11.6.

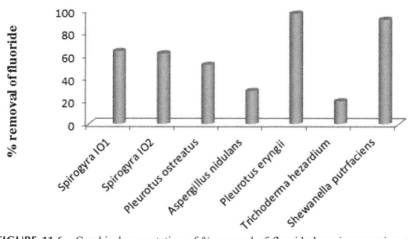

FIGURE 11.6   Graphical presentation of % removal of fluoride by microorganism on varying process parameters in batch study.[27, 27, 38, 25, 1, 18, 11]

TABLE 11.7   Comparative Tabulation of Removal Efficiency of Microorganisms on Different Process Parameters By Various Researchers.

Bioadsorbent	Initial Fluoride Concentration (IFC) (mg/l)	Temp.	pH	Adsorbent dose (g/l)	RPM	Time (hours)	% Removal	References
Spirogyra IO1	5	Room temp	2.0	10	100	2	64	[27]
Spirogyra IO2	5	30	7.0	0.1	100	2	62	27]
Pleurotus ostreatus 1804)	5	30	7	0.1	100	2	52	[38]
Aspergillus nidulans	–	40	4	1.5	–	20	29	[25]
Aspergillus,	35		2–10	5	150	2	65 91.	[30]
Ca-pretreated Aspergillus	5				250		96	
Pleurotus eryngii ATCC 90888	5	30	2	0.2	100	4	97.03	[1]
Trichoderma hezardium	4	30	7	0.4	100	1.0	20	[18]
Shewanella putrefaciens MTCC 8104	20	30	7	2.5 (NB media)	125	24	91.7	[11]

Dwivedi et al., (2016) investigated the potential of *Shewanella putre-faciens* in the bulk phase as well as immobilized phase in a batch reactor. The growth of the bacterial cells and their acclimatization in a fluoride media as well as under substrate stress has been investigated. Optimization of process parameters for the removal of fluoride in the bulk phase is investigated and the performance under optimum conditions is compared with the SABA process using *Citrus limetta* (mosambi peels) and *Ficus religiosa* (peepal) leaves as an adsorbent along with the bacteria. Optimum pH, nutrient broth (N. B.) media concentration, time and initial concentration were found 7, 2.5 g = L, 24 h and 20 mg/L, respectively. The SABA process shows maximum fluoride removal capacity followed by bioaccumulation in bulk phase and adsorption. 91.7% removal of fluoride occurred by bioaccumulation process. Both bioaccumulation and SABA are able to reduce the initial concentration of fluoride from 20 to below 1.5 mg/L.

Koshle et al., (2016)[18] investigated the possibility of *Trichoderma hezardium* for the removal of fluoride. Study was carried out at different pH ranges viz. 6.5, 7.5,8.5, 9.5 and initial fluoride concentration varied as 2,4,6 and 8 mg/l. Biosorption studies were carried out at different temperatures viz. 30, 40 and 50°C. Bioadsorbent dose varied as 0.4, 0.6, 0.8 and 1.6 g. Results showed the maximum removal of 38% at adsorbent dose 1.6 g. In this study, bioadsorption of fluoride on to fungus bioadsorbent followed the Freundlich isotherm which revealed the heterogeneous nature of surface binding sites. The value of Freundlich constant K_f was found to be 1.14 which shows the greater affinity for the fluoride-fungal system. Desorption studies indicate 82% desorption achieved.

Amin et al., (2015)[1] investigated the potential of white rot fungus *Pleurotus eryngii* ATCC 90888 for the fluoride removal in aqueous solution as a function of pH, initial fluoride concentration, biosorbent dose, temperature and contact time. Langmuir model showed better data interpretation than the Freundlich model. The monolayer biosorption capacity of *P. eryngii* biomass for fluoride ions was found to be 66.6 mg·g⁻¹. Biosorption study was carried out at different varying parameters viz. pH (2–7), initial fluoride concentration (5–25 mg/l) and contact time (60–300 min). At pH 2.0, initial fluoride concentration 5 mg/l and bioadsorbent dose 0.2 g, maximum fluoride removal, i.e. 97.03 % were achieved.

Mondal et al., (2013)[30] worked on bioadsorption of fluoride (F) by *Aspergillus* and Ca-pretreated *Aspergillus* biomass was conducted in an aqueous batch system and the maximum adsorption capacities of fluoride were found to be 8.09 mg/g (at pH 10) for *Aspergillus* and 4.80 mg/g (at pH 8) for Ca-pretreated *Aspergillus* biomass.The equilibrium data are best fitted to Freundlich isotherm for both adsorbents and the adsorption kinetics follow the pseudo-second order equation. Thermodynamic analysis observed that the adsorption process was exothermic and spontaneous for both adsorbents. Fourier transform infrared (FTIR) analysis of *Aspergillus* indicated the existence of potential fluoride-capturing functional groups.

Merugu et al., (2013) investigated that bioadsorbent prepared from *Aspergillus nidulans* was used for removal of fluoride from aqueous solution. Surface modified biomass (Ca^{2+} and alkali treated) was efficient in fluoride removal. The initial pH of fluoride-containing water plays an important role in defluoridation and it is decreased with increasing pH. The fluoride removal capacity was found to be 29% and 14% at pH 4.0 and pH 8 respectively. Fluoride removal was not affected by chloride and sulphate ions, however, the removal of fluoride decreased with increased bicarbonate concentration. The fluoride removal kinetics has shown a rapid phase of binding for a period of 1.5 h and a slower phase of binding during the subsequent time.

Merugu et al., (2013) investigated the fluoride removal by fungal bioadsorbent prepared from *Fusarium oxysporum*. This strain (*Fusarium oxysporum*) was isolated from Mahatma Gandhi University Campus, Nalgonda, A. P, India. In this study, the surface modified biomass (Ca^{2+} and alkali treated) was effective in the removal of fluoride. The fluoride removal capacity was found to be 19% and 9% at pH 5.0 and pH 9 respectively. The capacity of fluoride binding with bioadsorbent depends on the concentration of bicarbonate, but it is independent of chloride and sulphate. The effect of time on fluoride removal was also studied. It is observed that an equilibrium concentration of fluoride is achieved within 20 h, it is suggested that the 20 h of incubation could bind about 20% of the total fluoride present in water. However there was a little improvement in the sorption, it was not considerable. Hence, the time 20 h was used in further studies. The total amount of fluoride removal was found to be same still at the end of 20 h incubation period.

Chubar et al., (2008)[9] investigated the bioadsorption of fluoride and phosphate to viable cells of *Shewanella putrefaciens*. Here, the uptake is measured over a different concentration range. Comparative analysis has been done for the adsorption process for both anions (fluoride and phosphate), with viable cells and autoclaved cells. Batch potentiometric titration and Boehm method were applied to investigate the bacterial surfaces. FTIR absorption spectra and pH-dependent zeta potentials were similar for the viable and bacterial cells. Its physiological tolerance makes it a promising microorganism for bioremediation applications in a wide variety of environments.

Ramanaiah et al., (2007)[38] worked on the removal of fluoride from the aqueous phase using the waste fungus (*Pleurotus ostreatus*1804) and bioadsorbent obtained from laccase fermentation process. Batch bioadsorption studies were performed and the process followed pseudo-first-order rate equation. The adsorptive data fitted well with the Langmuir isotherm adsorption model. The fluoride sorption phenomena on fungal biosorbent might be attributed to the chemical sorption.

Mohan et al., (2007)[27] worked on nonviable algal *Spirogyra* IO1 for fluoride biosorption potential in batch studies. The results established the capacity of the bioadsorbent for fluoride removal. The bioadsorption process followed the pseudo-first-order rate equation. The intraparticle diffusion of fluoride molecules within the *Spirogyra* was the rate-limiting step. The adsorption isotherm followed the rearranged Langmuir isotherm adsorption model. Bioadsorption was highly dependent on the aqueous phase pH.

Mohan et al., (2007)[27] introduced the bioadsorption of fluoride onto algal *Spirogyra* sp.-*IO2*. The consequence of various physiochemical parameters on the removal of fluoride such as the fluoride concentration, the pH, the dosage of adsorbent, the temperature on the removal of fluoride, and the adsorption–desorption studies were investigated. The adsorption follows pseudo-first order rate equation. The adsorption data showed good fit Langmuir's adsorption isotherm model. The fluoride removal was varied due to the pH.

11.6.2 BIOLOGICAL PATHWAYS FOR FLUORIDE REMOVAL

Mainly two pathways such as bioadsorption on the bacterial cell surface and bioaccumulation in bacterial cells have been proposed for the removal of fluoride from water using bacterial whole cells.[9] The microorganism's cell wall primarily consists of polysaccharides, lipids and proteins, which have many binding sites for halides. This process is independent of the metabolism and metal binding is fast.

Bioaccumulation, in contrast, is an intracellular fluoride accumulation process which involves dehalogenation and it is mediated only by periplasmic and membrane fractions, not by cytoplasmic fractions of the cells.[37] Since it depends on the cell metabolism, it can be inhibited by metabolic inhibitors such as low temperature and lack of energy sources. Various processes of bioadsorption and bioaccumulation are shown in Fig. 11.7.

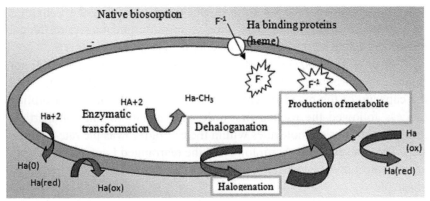

FIGURE 11.7   Various processes of bioadsorption and bioaccumulation in bacteria.

11.6.3   REVIEW ON SHEWANELLA PUTREFACIENS AND THE MECHANISM SELECTED BY THIS BACTERIA TO REMOVE FLUORIDE

This species has the tremendous potential in the field of bioremediation.[9] *Shewanella* species are Gram-negative, motile bacilli, phenotypically produce hydrogen sulphide. Previously, in 1941, this species should be kept to the genus *Pseudomonas* under the name *Pseudomonas putrefaciens* as proposed by Long and Hammer. But after, these organisms were placed in *Pseudomonas* group IV in the next three decades.[41] In the 1974, in the edition of Bergey's Manual of Systematic Bacteriology, it was observed that the G + C content of 43–55% was below the range for *Pseudomonas* spp. (58–70%), so, it was removed from the *Pseudomonas* group IV. In 1985, on the basis of phylogenetic analysis, these organisms were reclassified and were placed in the family *Vibrionaceae* and the new genus, *Shewanella*,[21] named after James Shewan in honour of his work in marine microbiology. Included in this new genus was *Shewanella putrefaciens*.[15]

Findings suggest that *Shewanella putrefaciens* is an obligate respiratory bacterium that can exploit a variety of terminal electron acceptor and it is capable of reductive dehalogenating activity by biotransformation the organic or inorganic halides. *Shewanella putrefaciens* consist of bacterial dehalogenase enzyme and halidohydrolases enzyme. *Shewanella putrefaciens* show the co-metabolic activity which is mediated by electron carriers of the respiratory electron-transport chains c-type cytochromes.[37,36] In case of halidohydrolase enzyme-mediated dehalogenation, the halogen substituent is replaced in a nucleophilic substitution reaction by a hydroxy group which is derived from water and reductive dehalogenation is associated with halide substituent transfer by hydrogen.[12] Bacteria possess the two processes simultaneously, that is dehalogenation and halogenations. In the procedure of dehalogenation, organic or inorganic halides are biotransformed and the liberated halide group itself is taken in the biosynthetic pathway of bacteria to produce some beneficial metabolite. However, very little information is available about those enzymes which catalyse the integration of halogen atoms during the biosynthesis of metabolites. Since the mechanism is not much known about the fluoride, that how it is incorporated into organic compounds during halogenations.[35]

11.7 CONCLUSION

The present investigation is conducted to appraise the efficiency of microorganisms for the removal of fluoride in a batch reactor. The comparative study was conducted by various researchers for removal of fluoride by using different algae, fungi and bacteria in batch reactors.

For the treatment of fluoride-containing wastewater, the bioadsorption and bioaccumulation are two significant methods. The bioadsorption process is better to other conventional treatment techniques in terms of efficiency and low cost. Bioadsorption systems are not affected by the toxicity of target compound(s). It is simple to operate and no need to use hazardous chemicals in this process. Biological processes are gradually getting high momentum due to the cheaper than other (physical, chemical) processes. It is eco-friendly, requires less energy input than the chemical and physical options. It can deal with widely dispersed contaminants and minimizes disturbance to the environment and no danger to workers. Further, the more investigation required to deal with real wastewater instead synthetic wastewater. Amongst various microorganisms, bacteria *Shewanella putrefaciens* has the applicability in bioremediation process.

KEYWORDS

- fluoride
- wastewater
- microorganism
- bioadsorption
- bioaccumulation
- simultaneous adsorption and bioaccumulation (SABA)

REFERENCES

1. Amin, F.; Talpur, F. N.; Balouch, A.; Surhio, M. A.; Bhutto, M. A. Biosorption of Fluoride from Aqueous Solution by White-Rot Fungus *Pleurotus eryngii* ATCC 90888. *Environ. Nanotech., Mon. & Mgt.* **2015,** *3*, 30–37.
2. Amina, C.; Lhadi, L. K.; Younsi, A.; Murdy, J. Environmental Impact of an Urban Landfill on a Coastal Aquifer, *J. Afr. Earth Sci.* **2004,** *39*(3–5), 509–516.
3. Anwar, F. Assessment and Analysis of Industrial Liquid Waste and Sludge Disposal at Unlined Landfill Sites in Arid Climate. *Waste Manage.* **2003,** *23*(9), 817–824.
4. Barbier, O.; Arreola-Mendoza, L.; Razo, L. M. D. Molecular Mechanism of Fluoride Toxicity. *Chem. -Biol Interact, 188*, 319–333.
5. Bennajah, M.; Maalmi, M.; Darmane, Y.; Touhami, E. M. Defluoridation of Drinking Water by Electrocoagulation/Electroflotation: Kinetic Study. *J. Urban. Environ. Eng.* **2010,** *4*(1), 37–45.
6. Camargo, A. J. 0 Fluoride Toxicity to Aquatic Organisms: A Review. *Chemosphere* **2003,** *50*, 251–264.
7. Charles, F. H.; Swartz, C. H.; Badruzzaman, A. B. M.; Nicole, K. B.; Yu, W.; Ali, A.; Jay, J.; Beckie, R.; Niedan, V.; Brabander, D.0 Groundwater Arsenic Contamination on the Ganges Delta: Biogeochemistry, Hydrology, Human Perturbations and Human Suffering on a Large Scale. *C. R. Geosci.* **2005,** *337*(1-2), 285–296.

8. Chaturvedi, A. K.; Yadava, K. P.; Pathak, K. C.; Singh, V. N. De Fluorination of Water. Water, Air, Soil Pollut. **1990**, *49*, 41–69.

9. Chubar, N.; Behrends, T.; Behrends, P. V. Biosorption of Metals (Cu_2^+, Zn_2^+) and Anions (F⁻, H_2PO_4) by Viable and Autoclaved Cells of the Gram-Negative Bacterium *Shewanella putrefaciens*. *J. Colloid Interface Sci*. **2008**, *65*, 126–133.

10. Cui, H.; Qian, Y.; An, H.; Sun, C.; Zhai, J.; Li, Q. Electrochemical Removal of Fluoride from Water by PAO Amodified Carbon Felt Electrodes in a Continuous Flow Reactor. *Water Res*. **2012**, *46*, 3943–3950.

11. Dwivedi, S.; Mondal, P.; Majumder, C. B. Bioremoval of Fluoride from Synthetic Water using the Gram-Negative Bacteria *Shewanella putrefaciens*. *J. Hazard. Toxic., Radioact.Waste*. **2016**, *10*.1061/(ASCE)HZ.2153-5515.0000341, pp04016023-1 to 04016023–04016026.

12. Fetzner, S.; Lingens, F. Bacterial Dehalogenases: Biochemistry, Genetics, and Biotechnological Application. *Microbiol. Rev*. **1994**, *58*, 641–685.

13. Fornasiero, R. B. Phytotoxic Effects of Fluorides. *Plant Sci*. **2001**, *161*(5), 979–985.

14. Herrera-Ochoa, V.; Banihani, Q.; Leon, G.; Khatri, C.; Field, J.; Sierra-alvarez, R. Toxicity of Fluoride to Microorganisms in Biological Wastewater Treatment Systems. *Water Res*. **2009**, *43*, 3177–3186.

15. Holt, H. M.; Gahrn-Hansenand, B.; Bruun, B. Shewanella Algae and *Shewanella putrefaciens*: Clinical and Microbiological Characteristics. *Clin. Microbiol. Infect*. **2005**, *11*, 347–352.

16. Kabay, N.; Arar, O.; Samatya, S.; Yuksel, U.; Yuksel, M. Separation of Fluoride from Aqueous Solution by Electrodialysis: Effect of Process Parameters and Other Ionic Species. *J. Hazard. Mater*. **2008**, *153*, 107–113.

17. Kass, A.; Yechieli, G. Y.; Vengosh, A.; Starinsky, A. The Impact of Freshwater and Wastewater Irrigation on the Chemistry of Shallow Groundwater: A Case Study from the Israeli Coastal Aquifer. *J. Hydrol*. **2005**, *300*(1–4), 314–331.

18. Koshle, S.; Mahesh, S.; Swami, N. S. Isolation and Identification of *Trichoderma harzianum* from Ground Water: An Effective Biosorbent for Defluoridation of Ground Water. *J. Environ. Biol*. **2016**, *37*, 135–140.

19. Lennon, M. A.; Whelton, H.; Mullane, D. O'.; Ekstrand, J. *Rolling Revision of the WHO Guidelines for Drinking-Water Quality Fluoride*, World Health Organization.

20. Liu, A.; Ming, J.; Ankumah, R. O. Nitrate Contamination in Private Wells in Rural Alabama, United States. *Sci. Total Environ*. **2005**, *346*(1–3), 112–120.

21. MacDonell, M. T.;, Colwell, R. R. Phylogeny of the Vibrionaceae, and Recommendation for Two New Genera, Listonella and Shewanella. *Syst. Appl. Microbiol*. **1985**, *6*, 171–182.

22. Mamilwar, B. M.; Bhole, A. G.; Sudame, A. M. Removal of Fluoride from Ground Water by Using Adsorbent. *Int. J. Eng. Res. Appl*. **2012**, *2*, 334–338.

23. Mandal,T.; Maity, S.; Dasgupta, D.; Datta, S. Advanced Oxidation Process and Biotreatment: Their roles in Combined Industrial Wastewater Treatment. *Desalination* **2010**, *250*, 87–94.

24. Meenakshi, M. R. C. Fluoride in Drinking Water and its Removal. *J. Hazard. Mater*. **2006**, *137*, 456–463.

25. Merugu, R.; Kudle, K. R.; Garimella, S.; Medi, N. K.; Prashanthi, Y. Factors Affecting the Defluoridation of Water using *Fusarium oxysporum* Bioadsorbent. *Int. J. Environ. Biol.* **2013,** 12–13.

26. Merugu, R.; Kudle, K. R.; Garimella, S.; Medi, N. K. Biodefluoridation of Water using *Aspergillus nidulans* Fungal Bioadsorbent. *Int. J. Environ.* Biol. **2013,** *3*(1), 9–11.

27. Mohan, S. V.; Ramanaiah, S. V.; Rajkumar, B.; Sarma, P. N. Removal of Fluoride from Aqueous Phase by Biosorption onto Algal Biosorbent Spirogyra sp.-IO2: Sorption Mechanism Elucidation, *J. Hazard. Mater.* **2007,** 465–474.

28. Mohan, S. V.; Ramanaiah, S. V.; Rajkumar, B.; Sarma, P. N. Biosorption of Fluoride from Aqueous hase onto Algal Spirogyra IO1 and Evaluation of Adsorption Kinetics. *Bio. Res. Technol.* **2007,** *985*, 1006–1011.

29. Mohapatra, M.; Anand, S.; Mishra, B. K.; Giles, D. E.; Singh, P. Review of Fluoride Removal from Drinking Water. *J. Environ. Management.* **2006,** *91*, 67–77.

30. Mondal, N. K.; Kundu, M.; Das, K.; Bhaumik, R.; Dattaa, J. K. Biosorption of Fluoride from Aqueous Phase onto *Aspergillus* and Its Calcium-Impregnated Biomass and Evaluation of Adsorption Kinetics. *Res. Rep.* **2013,** *46*(4), 239–245.

31. Muhammad, I. T.; Afzal, S.; Hussain, I. Pesticides in Shallow Groundwater of Bahawalnagar, Muzafargarh, D.G. Khan and RajanPur Districts of Punjab, Pakistan. *Environ Int.* **2004,** *30*(4), 471–479.

32. Mulligan, C. N.; Yong, R. N.; Gibbs, B. F. Remediation Technologies for Metal Contaminated Soils and Groundwater: An Evaluation. *Eng. Geol.* **2001,** *60*(1–4), 193–200.

33. Murray, J. J. A History of Water Fluoridation, *Br. Dent. J.* **1973,** *134*, 250–254, 299–302, 347–350.

34. Oren, O.; Yechieli, Y.; Bohlke, J. K.; Dody, A. Contamination of Groundwater under Cultivated Fields in an Arid Environment, Central Arava Valley, Israel. *J Hydrol.* **2004,** *290*(3/4), 312–328.

35. Van, P.; Heinz, K.; Unversucht, S. Biological Dehalogenation and Halogenation Reactions. *Chemosphere* **2003,** *52*, 299–312.

36. Picardal, F.; Arnold, R.G.; Huey, B. B. Effects of Electron Donor and Acceptor Conditions on Reductive Dehalogenation of Tetrachloromethane by *Shewanella putrefaciens* 200. *Appl. Environ. Microbiol.* **1995,** *61*, 8–12.

37. Picardal, F. W.; Arnold, R. G.; Couch, H.; Little, A. M.; Smith, M. E. Involvement of Cytochromes in the Anaerobic Biotransformation of Tetrachloromethane by *Shewanella putrefaciens* 200. *Appl. Environ. Microbiol.* **1993,** *59*, 3763–3770.

38. Ramanaiah, S. V.; Venkata, M. S.; Sarma, P. N. Adorptive Removal of Fluoride from Aqueous Phase using Waste Fungus (*Pleurotus ostreatus* 1804) Biosorbent: Kinetics Evaluation. *Ecol. Eng.* **2007,** *31*, 47–56.

39. Sanwal, M.; Schedule, V. I. *The Environment (Protection) Rules*, 1986.

40. Sengupta, S. R.; Pal, B. Iodine and Fluoride Content of Foodstuffs. *Indian J. Nutr. Diet.* **1971,** *8*, 66–71.

41. Shewan, J. M.; Hobbs, G.; Hodgkiss, W. A Determinative Scheme for the Identification of Certain Genera of Gram Negative Bacteria, with Special Reference to the *Pseudomonadaceae*. *J. Appl. Bacteriol.* **1960,** *23*, 379–390.

42. Simons, R. Trace Element Removal from Ash Dam Waters by Nanofiltration and Diffusion Dialysis. *Desalination* **1993,** *89*, 325–341.

43. Sourirajan, S.; Matsurra, T. Studies on Reverse Osmosis for Water Pollution Control. *Water. Res.* **1972,** *6*, 1073–1086.

44. Sundaram, C. S.; Viswanathan, N.; Meenakshi, S. Defluoridation Chemistry of Synthetic Hydroxyapatite at Nano Scale: Equilibrium and Kinetic Studies. *J. Hazard. Mater.* **2008,** *155*, 206–215.

45. Suttie, J. W. Effects of Inorganic Fluorides on Animals. *J. Air Pollut. Control Assoc.* **1964,** *14*, 461–480.

46. Digest, T. National Environmental Engineering Research Institute, Nagpur, NEERI Manual, 1978.

47. Tor, A.; Danaoglu, N.; Arslan, G.; Cengeloglu, Y. Removal of Fluoride from Water by Using Granular Red Mud: Batch and Column Studies. *J. Hazard. Mater.* **2009,** *164*, 271–278.

48. WHO. Guidelines for Drinking Water Quality. Geneva:World Health Organization. 1993, 1–3, 45–46.

CHAPTER 12

APPLICATIONS OF MICROORGANISMS IN BIODEGRADATION OF CYANIDE FROM WASTEWATER

NAVEEN DWIVEDI[1,2,*], CHANDRAJIT BALOMAJUMDER[3], and PRASENJIT MONDAL[4]

[1]*Uttarakhand Technical University, Dehradun 248007, India, Tel.: +91 9456456209, *E-mail: naveen.dwivedi@gmail.com*

[2]*Department of Biotechnology, S. D. College of Engineering & Technology, Muzaffarnagar, Uttar Pradesh 251001, India*

[3]*Professor and Head, Department of Chemical Engineering, Indian Institute of Technology Roorkee, Roorkee, Uttarakhand 247667, India*

[4]*Associate Professor, Department of Chemical Engineering, Indian Institute of Technology Roorkee, Roorkee, Uttarakhand 247667, India*

CONTENTS

ABSTRACT

Various procedures exist for treating cyanides from different industrial wastewaters, and comprise several physical, chemical and biological methods. It is still widely discussed and examined due to its potential toxicity and environmental impact. Physical and chemical methods are highly expensive and also cause secondary pollution. The treatment of cyanide by said method is rarely used, owing to cost-effectiveness. Thus, there is a pressing need for the development of an alternative treatment process capable of achieving high removal efficiency without troubling the environment. Several microbial species can degrade cyanide well into less toxic products. Biodegradation of cyanide compounds may take place through various enzymatic pathways, the enzymes of which are produced by microorganisms that utilize cyanides as substrate. The biological methods for the treatment of cyanide are not only cost-effective but also do not produce any secondary pollution. The present chapter describes the mechanism and ardent approaches of proficient biological methods for the removal of cyanide compounds and their advantages over other treatment processes.

12.1 INTRODUCTION

12.1.1 GENERAL

Cyanide is an extremely toxic nitrogenous compound for almost all living creatures found in wastewater. There are several bacteria, fungi, algae and plants that are able to produce cyanide as a defence mechanism against predation.[53] Cyanide is present in various foodstuffs consumed by humans such as cassava, apricots, bean sprouts, soya beans, cashews, almonds, cherries, potatoes, lentils, olives, sorghum and bamboo shoots. However, it is a common compound crucial to nature; it is extensively considered as a highly dangerous substance. In this drive it has been formed by its use as a genocidal agent during World War II, in mass suicides, and its continued use in some parts of the world for judicial executions.[17] Cyanide is included in the Comprehensive Environmental Response, Compensation and Liability Act (CERCLA) priority list of hazardous substances and it occupies 35th position in the priority list (revised in 2015) of most hazardous substances.[5] Cyanide has also been used worldwide in the extraction of gold and silver. The demand for sodium cyanide worldwide is about 360,000 tonnes per annum of which about one-third is used in the recovery of gold and silver.[64] The process was developed to extract gold from mines in Scotland in the year 1987, and commercially it was used on a large scale by the New Zealand Crown Mines Company at Karangahake in the year 1989.[42,28] Annually production of HCN has been reported to be around 1.4 million tonnes[63] whereby 13% is converted in sodium cyanide for the mining use. Different forms of cyanide have dreadful health effects on people as well as other living organisms.[23]

12.2 CYANIDE CONTAMINATION IN WATER, ITS SOURCES AND TOXIC EFFECT ON ENVIRONMENT

Cyanide compounds are present in environmental matrices and waste streams as free, simple and complex cyanides, cyanate and nitriles.[27] It enters into the water from natural sources and anthropogenic/man-made sources (industrial activities).

12.2.1 NATURAL SOURCES

Cyanide compounds are synthesized by many taxa including higher plants, arthropods, fungi and bacteria. Plants are an important source of cyanide compounds; they exploit the bitter tasting cyanogenic glucosides as defence against herbivores and pathogens.[88] Cyanides are produced by certain fungi, bacteria, algae and plants as well as particular foods such as lima beans and almonds. In certain plant foods, including almonds, millet sprouts, lima beans, soy, spinach, bamboo shoots and cassava roots (major source of food in tropical countries), cyanides occur naturally as part of sugars or other naturally occurring compounds. However, the edible parts of plants that are being eaten in the United States, including tapioca which is made from cassava roots, contain relatively low amounts of cyanide. Various food products may either contain cyanide naturally or it is used for their production.[59] In certain food products the cyanide concentration per kg observed is; cassava (104 mg/100 g plant tissue), lima beans (100–300 mg/100 g plant tissue), packed fruit juice (0.03–15.84 mg/l in sour cherry and orange), wild cherries (140–370 mg/100 g plant tissue), almond (297 mg/kg), almond products (50 mg/kg) and sorghum (250 mg/100 g plant tissue). Varying amounts of cyanide may be found in other vegetables and fruits such as apple, apricot, peach, cauliflower and strawberry.[51] Cyanide is also produced from forest fires and released into the atmosphere.[26] Much smaller amounts of cyanide may enter into water through storm water runoff where road salts are used that contain cyanide.

12.2.2 ANTHROPOGENIC/MAN-MADE SOURCES

Major sources of cyanide are discharged in large quantities by various industries such as electroplating and metal finishing, steel tempering, mining (extraction of metals such as gold and silver), automobile parts manufacturing, photography, pharmaceuticals and coal processing units.[3,60,72] Cyanide in landfills can contaminate underground water. However, cyanide has been detected in underground waters of a few landfills and industrial waste disposal sites. Cyanide becomes toxic to soil microorganisms at high concentrations found in some landfill leachates (water that seeps through landfill soil) and in the wastes stored in some disposal sites. Because these microorganisms can no longer change cyanide to other chemical forms, cyanide is able to pass through soil into the underground water. Generally, the concentrations of total cyanide in the effluent of such industries vary within the range of 0.01–10 mg/l.[89,33] The effluent of silver plating industry and coke plant industry containing total cyanide is around $5 \cdot 1 \pm 1 \cdot 5$ mg/l and 4–100 mg/l, respectively.[72] However, some industrial effluents from electroplating plants have been found to contain even higher cyanide levels of 100,000 mg/l.[89,47] One of the most significant cyanide complex species is thiocyanate (SCN), found in some coal conversion and coal coking effluents in the concentration range of 17–1500 mg/l SCN.[54]

Cyanide concentrations reported in the literature for several industrial wastewater effluents are presented in Table 12.1.

TABLE 12.1   Concentration of Cyanide Released from Various Industries/Industrial Effluent (mg/l).

Industrial wastewater source	Total cyanide	Reference
Oil refinery	2.25	[39]
Petroleum refining	0–1.5	[47]
Steel mill Coke plant liquor	7.5–396	[47]
Paint and ink formulation	0–2	[13]
Gold ore extraction	18.2–22.3	[47]
Coke plant waste	91–110	[47]
Coke oven plant	10–150	[54]
Coke plant	0.1–0.7	[54]
Coke plant	100–1000	[54]
Coke plant ammonia liquor	20–60	[47]
Explosives manufacture	0–2.6	[47]
Plating rinse	32.5	[10]
Plating industries (rising waste)	1.4–256	[47]
Electroplating plants	0.03–0.27	[89]
Electroplating plants	0.01–14.24	[89]
Chemical industry	10.4–50.9	[87]

Cyanide is a prominent metabolic inhibitor and inhibition is done by the inactivation of respiration due to its tight binding to cytochrome C oxidase. In this process, electron transport chain is blocked by the cyanide by binding with the iron ion in the terminal electron acceptor cytochrome C oxidase, consequently rapidly falling respiration rates, and ATP synthesis in mitochondria is inhibited.[36] Assimilation of cyanide can also result in either acute poisoning (including death) or chronic poisoning to human beings and animals.[24] The lowest reported oral lethal dose in humans is 0.54 mg/kg body weight; the average absorbed dose at the time of death has been estimated at 1.4 mg/kg body weight (calculated as hydrogen cyanide).[91] Aquatic organisms are very sensitive to cyanide. Fish have been shown to be the most sensitive aquatic organisms followed by invertebrates. Concentrations of free cyanide in the aquatic environment ranging from 5.0–7.2 µg/l, it reduces the swimming performance and inhibit reproduction in many species of fish. Other adverse effects include delayed mortality, pathology, susceptibility to predation, disrupted respiration, osmoregulatory disturbances and altered growth patterns. Concentrations of 20–76 µg/l free cyanide can cause the death of many species, and concentrations in the excess of 200 µg/l are rapidly toxic to the most species of fish. Algae and macrophytes can tolerate much higher environmental concentrations of free cyanide than fish and invertebrates, and do not exhibit adverse effects at 160 µg/l or more of free cyanide (Environmental and Health Effects of Cyanide, USA). Birds seem to have varying sensitivity to cyanide. It has been reported that oral lethal dose (LD50) for birds ranges from 0.8 mg^{-1}kg of body weight (American racing pigeon) to 11.1 mg^{-1}kg of body weight (domestic chickens). Workers in various professions may be exposed to cyanides primarily through inhalation and less frequently by skin absorption. Exposure to cyanide can cause breathing disorders, headaches, coma, heart pains, thyroid gland enlargement and even death. All forms of the cyanide can be toxic at high levels, but the most toxic form of cyanide is hydrogen cyanide. Various microorganisms and plants have shown the resistivity against cyanide poisoning since they have developed an alternate pathway for ATP production. Some of them have a diverse oxidase instead of cytochrome C oxidase.[80]

Cyanide containing wastewaters must be treated before being discharged into the environment to protect the water resources and environment. Hence, environmental regulations require reducing the cyanide concentration in wastewater and setting the permissible limit of cyanide in the surface water. Indian standard has set a minimal national standard (MINAS) limit for cyanide in effluent as 0.2 mg/l.[41] US Environmental Protection Agency (USEPA) standard for drinking and aquatic biota waters regarding total cyanide is 200 and 50 ppb, respectively.[90,85]

12.3 AVAILABLE TECHNIQUES OF CYANIDE REMOVAL FROM WASTEWATER

Several treatment processes such as physical, chemical and biological oxidation have been exploited for the reduction of cyanide levels in waste solutions/slurries in compliance with environmental regulations. Physicochemical methods for cyanide treatment can be accomplished using dilution, membranes, electrowinning and hydrolysis/distillation, chemical oxidation, alkaline chlorination,[56,4] electrochemical method,[68] capillary electrophoresis, adsorption and so forth.[1] Activated carbon has been reported to have cyanide adsorption capacity. However, in several cases the process is hampered with elevated funds, regent costs and royalty payments. In chemical process, various chemicals and reagents are used and secondary pollutants are created which need some additional treatment prior to their disposal. Adsorption is a simple and attractive method for the removal of toxic compounds from the effluents due to its high efficiency, easy handling, and economic feasibility. Adsorption systems are not affected by the toxicity of the target compound(s) and do not require hazardous chemicals. Moreover, adsorption facilitates concentrating and then recovering the adsorbed compounds if desired.[62] The biosorption of cyanide from aqueous solutions is quite a new process that has proven to be promising in the removal of contaminants from aqueous effluents.[25] Various agro-based adsorbents have been reported for cyanide removal from water and wastewater due to their abundant availability and low cost.[62] The adsorption of cyanide is, however, highly dependent on solution pH and initial cyanide concentration. Adsorbent concentration, agitation period and temperature also influence the cyanide adsorption process.

Biological treatments are feasible alternatives to chemical methods without creating or adding new toxic and biological persistent chemicals.[22,72] Cyanide is converted to carbon and nitrogen source by various enzymes present in the microorganisms. The metabolic pathway for the conversion of cyanide is influenced by its initial concentration, pH, temperature, availability of other energy sources in the form of organic carbon required for cell maintenance and growth, presence of oxygen, ammonia and various metal ions.[84,73] Microorganisms have the capability to degrade and assimilate cyanide into the form of amino acids, thiocyanate, β-cyanoalanine and vitamins.[84] White et al., (1988) reported the conversion of cyanide to formate and ammonia by *Pseudomonas* which is isolated from industrial wastewater. The strain of isolated *Bacillus pumilus* could survive in the solutions of potassium cyanide of up to 2.5 M and grow in the media containing 0.1 M KCN.[81] Biodegradation was performed in the presence of microbes either in mobilized or immobilized phase.[46] The immobilization of living microbial cells on a suitable adsorbent improves the removal efficiency.[61] There were some reports on the biodegradation of cyanide compounds by immobilized cells of *Pseudomonas* species,

Fusarium solani.[19,22,12] Adsorption and biodegradation are two significant methods for the treatment of cyanide-containing wastewater.[24] The
biological process is generally the preferred technique for treating wastewater owing to its cost-effectiveness and environmental easiness.

12.3.1 MICROBIAL MECHANISM FOR CYANIDE DEGRADATION

The biodegradation of cyanide compounds may take place through various
pathways. Generally, degradation of cyanide is induced by the presence of
cyanide in the media which is followed by the conversion of cyanide into
carbon and nitrogen. Various researchers have described various organisms
that use different pathways for cyanide degradation.[21] Sometimes, more
than one pathway can be applied for cyanide biodegradation in some organisms.[65] Five general pathways as reported in the literature for the biodegradation of cyanide are: hydrolytic pathway, oxidative pathway, reductive
pathway, substitution/transfer pathway and syntheses pathway.[52,79] First
three are degradation pathways in which enzymes catalyze the conversion
of cyanides into simple organic or inorganic molecules and convert them
into ammonia, methane, CO_2, formic acid and carboxylic acid. Last two
pathways are for the assimilation of cyanide in the microbe as nitrogen and
carbon source.[9] All these pathways depend on the mechanism of tolerance of cyanide in microbes and on the process it uses to dissociate the
cyanide metal complexes or for chelating metals.[36]

12.3.1.1 HYDROLYTIC PATHWAY

The hydrolytic pathway of cyanide degradation is catalyzed by various
enzymes present in microbial system, like cyanide hydratase, nitrile
hydratase, cyanidase and nitrilase.[27,52,9] First two enzymes have a specific
substrate and they direct hydrolyse and cleave the carbon–nitrogen triple
bond to form formamide, while the last two convert it to ammonia and
carboxylic acid, which are utilized in their metabolism activity.[36] Cyanide
compounds are degraded by the following enzymes through different reactions as discussed below:[27]

12.3.1.1.1 Cyanide hydratase

Cyanide hydratase is primarily a fungal enzyme; the most frequently encountered cyanide conversion takes place through this inducible enzyme, resulting in the formation of formamide which subsequently decomposes to carbon dioxide and ammonia by another enzyme formamide hydratase (FHL).

$$HCN + H_2O \circledR HCONH_2 \qquad (12.1)$$

This enzyme belongs to the family of lyases, specifically the hydrolyases which cleave the carbon–nitrogen bonds. The systematic name of this enzyme class is formamide hydro-lyase (cyanide-forming). Other names in common use include formamide dehydratase, and formamide hydro-lyase. This enzyme participates in cyanoamino acid metabolism.[36] Cyanide hydratase was first partially purified by *Stemphylium loti*, and is highly conserved between species.[8]

12.3.1.1.2 Cyanidase

Cyanidase is also known as cyanide dihydratases. It comprises a group of bacterial enzymes that are available in *Pseudomonas stutzeri* AK61, *Alcaligenes xylosoxidans* subsp. *Denitrificans* DF3 and *Bacillus pumilus* C1.[43,58,44] Cyanide dihydratases readily convert cyanide to relatively nontoxic formate directly as shown below:

$$RCN + H_2O \ll RCOOH + NH_3 \qquad (12.2)$$

12.3.1.1.3 Nitrile hydratase

Primarily aliphatic nitriles can be effectively degraded by nitrile hydratases. Nitrile hydratases convert cyanides to their corresponding amides as shown in the following reaction:

$$R\text{–}CN + H_2O \circledR R\text{–}CONH_2 \qquad (12.3)$$

The nitrile hydratase, isolated from *Pseudonocardia thermophila* shows high activity compared to other microorganisms known for the production of nitrile hydratase, that is *Rhodococcus rhodoclous, Pseudomonas, Corynebacterium, Klebsiella* and *Rhizobium*. Some new bacterial strains, *Pseudomonas putida* MA113 and *Pseudomonas marginales* MA32, containing nitrile hydratases were isolated from soil samples by an enrichment procedure. This isolated microbe could tolerate up to 50 mM cyanide and also has broad substrate range small substrates like acrylonitrile, nitriles with longer side chains and even nitriles with quaternary alpha-carbon atoms. *P. putida* MA113 and *Pseudomonas marginales* MA32 were used as whole cell biocatalysts for the hydration of acetone cyanohydrin to a hydroxyisobutyramide, which is a precursor of methacrylamide.[67] Nitrile hydratase is composed of two types of subunits, α and β, which are not related in amino acid sequence. Nitrile hydratase exists as αβ dimers or $\alpha_2 \beta_2$ tetramers and bind one metal atom per αβ unit.[34]

12.3.1.1.4 Nitrilase

Nitrilase enzymes catalyze the hydrolysis of nitriles to carboxylic acids and ammonia, without the formation of 'free' amide intermediates.

$$R-CN + 2H_2O ® R-COOH + NH_3 \qquad (12.4)$$

Nitrilases are involved in the biosynthesis of proteins and their post-translational modifications in plants, animals, fungi and certain prokaryotes. The structure of nitrilases is usually inducible enzymes composed of one or two types of subunits of different size and number, and these subunits of nitrilase self-associate to convert the enzyme to the active form. *Nocardia* sp. nitrilase was reported to be induced by enzonitrile.[35] Acetonitrile has been used as an inducer for the formation of nitrilase in *F. oxysporum*.[15]

12.3.1.2 OXIDATIVE PATHWAY

In oxidative pathway, cyanide conversion involves oxygenolytic conversion to carbon dioxide and ammonia. This pathway requires NADPH to catalyze this degradation pathway. Microorganisms that opt for cyanide conversion by this pathway also require additional carbon source with cyanide. There are two types of oxidative mechanisms involving three different enzymes:

i)   Cyanide monooxygenase and cyanase: Cyanide monoxy-genase[27,75] converts cyanide to cyanate which is further cata-lyzed by cyanase, resulting in the overall conversion of cyanate to ammonia and carbon dioxide as given in the following reaction:

$$HCN^- + O_2 + 2H^+ + NADPH \circledR CNO^- + NADP + 2H_2O$$

$$CNO^- + 3H^+ + HCO_3^- \circledR NH_4^+ + 2CO_2 \quad (12.5)$$

ii) Cyanide dioxygenase: This oxidative pathway utilizes cyanide dioxygenase to form ammonia and carbon dioxide directly.[27]

$$HCN + O_2 + 2H^+ + NADPH \circledR NADH + CO_2 + NH_3 \quad (12.6)$$

Immobilized cells of *P. putida* can effectively use oxidative pathway to produce ammonia and carbon dioxide.[12] The cyanide degradation in three white rot fungi, *Trametes Versicolor* ATCC 200801, *Phanerochaete chrysosporium* ME 496 and *Pleurotus sajor-caju,* was achieved by an oxidative reaction whose end products were ammonia and CO_2. In *T. versicolor,* this reaction was the most effective with 0.35 g dry cell/100ml degrading 2 mM KCN (130 mg/l) over 42 h, at 300 °C, pH 10.5 with stirring at 150 rpm.[50]

12.3.1.3   REDUCTIVE PATHWAY

The reductive pathways of degradation of cyanide are generally considered to occur under anaerobic conditions. This pathway is mediated by an enzyme nitrogenase. The enzyme utilizes HCN and produces methane and ammonia as the end products.

$$HCN + 2H^+ + 2e^- \circledR CH_2 = NH + H_2O \circledR CH_2 = O \quad (12.7)$$

$$CH_2 = NH + 2H^+ + 2e^-$$

$$\rightarrow CH_3 - NH + 2H^+ + 2e^- \rightarrow CH_4 + NH_3 \quad (12.8)$$

Klebsiella oxytoca is able to degrade cyanide compounds to methane and ammonia by this path.[45]

12.3.1.4   SUBSTITUTION/TRANSFER PATHWAY

The activity of this pathway involves cyanide assimilation and useably this tends to increase the growth of the microorganism by providing extra nitrogen source and preventing it from cyanide toxicity. There are two types of enzymes that catalyze cyanide assimilation through this pathway such as rhodanese and mercaptopyruvate sulfurtransferase. Both the enzymes are widely distributed in living organisms and catalyze the formation of pyruvate and thiocyanate from mercaptopyruvate.[78] Rhodaneses are extremely conserved and the prevalent enzymes are presently regarded as one of the mechanisms evolved for cyanide detoxification. In vitro rhodaneses catalyze the irreversible transfer of a sulphur atom from a suitable donor (that is thiosulfate) to cyanide, leading to the formation of less toxic sulfite and thiocyanate. The enzyme activity is modulated by phosphate ions and divalent anions are found to interact with the active site.[11]

$$\text{Thiosulfate} + \text{cyanide} \rightarrow \text{sulphite} + \text{thiocyanate} \qquad (12.9)$$

Rhodaneses have also been recognized in a variety of bacterial species including *Escherichia coli*[14] *Azotobacter vinelandii*[30] and several species of *Thiobacillus*.

The second enzyme mercaptopyruvate sulphurtransferase belongs to the family of transferases, specifically the sulphurtransferases, which transfer sulphur-containing groups. The systematic name of this enzyme class is 3-mercaptopyruvate: cyanide sulphurtransferase. This enzyme participates in cysteine metabolism. It catalyzes the following chemical reactions:

$$\text{HSCH}_2\text{COCOO}^- + E \ll \text{ CH}_2\text{COCOO}^- + ES \text{ (step1)} \qquad (12.10)$$

$$ES + CN^- \ll \text{ } E + SCN^- \text{ (step2)} \qquad (12.11)$$

3-Mercaptopyruvate is converted to thiosulfate by the two-step reaction as mentioned above. In step 1, the enzyme-sulphur (ES) intermediate is formed, and in step 2, intermediate reacts with cyanide to E (mercapto-pyruvate sulfurtransferase) and thiocyanide.[36]

12.3.1.5   SYNTHESES PATHWAY

This pathway is also a cyanide assimilation pathway by using enzyme β-cyanoalanine synthase and γ-cyano-α-aminobutyric acid synthase. It involves the synthesis of amino acid, β-cyanoalanine and γ-cyano-α-aminobutyric acid by using amino acid residues as precursors that react with cyanide compounds.[66] β-Cyanoalanine synthase is believed to play an important role in the removal of endogenous cyanide and is produced during a highly active growth period in the microbe.[66] This enzyme belongs to the family of lyases, specifically the class of carbon–sulphur lyases. The systematic name of this enzyme class is L-cysteine hydrogen-sulphide-lyase. *Bacillus magaterium* grows by converting cyanide to β-cyanoalanine and then to asparagines. β-Cyanoalanine synthase is induced by various amino acids such as serine, cysteine, asparagine and so forth. Cyanide produced by *C. violaceum* can first be converted to β-cyanoalanine then into asparagines as shown in the reaction below.[32]

$$HCN + HS-CH_2CH(NH_2)COOH \circledR H_2S + NC-CH_2CH(NH_2)COOH$$

$$\text{(Cysteine)} \qquad \qquad ^- \text{(β − cyanoalanine)} \qquad (12.12)$$

$$H_2NCO-CH_2CH(NH_2)COOH \text{ (Asparagine)}$$

γ-Cyano-α-aminobutyric acid synthase is an alternative pathway for cyanide assimilation. This pathway requires pyrodoxal phosphate for function, and is induced by glutamate or glycine. Once γ-cyano-α-aminobutyric acid is synthesized, it is slowly converted to glutamate. A thermophilic and cyanide ion-tolerant bacterium, *Bacillus stearothermophilus* CN$_3$, isolated from a hot spring in Japan, was found to produce thermostable γ-cyano-α-aminobutyric acid synthase.[69]

12.4   FACTORS RESPONSIBLE FOR THE BIODEGRADATION OF CYANIDES

Physiological and metabolic capabilities of microbes play a very essential role for the successful degradation of cyanide. Various parameters that play a major role in the degradation of cyanide include temperature, oxygen, pH, nutrients and so forth. Temperature is a significant factor for the determination of biodegradation rate. Cyanide degrading enzymes are generally produced by mesophilic microorganisms, often isolated from soil, with temperature optima typically ranging between 20 and 40°C.[9,45,2] pH of the solution is also a predominantly important factor in the bioremediation of cyanide. The optimum pH for bacterial growth is 6–8, and for fungal growth it is normally 4–5. Cyanide degrading enzymes usually have pH optima between 6 and 9; therefore, extremes of pH may have a significant effect on biodegradation. However, *F. solani* and mixed cultures of fungi including *F. solani*, *F. oxysporum*, *Trichoderma polysporum*, *Scytalidium thermophilum* and *Penicillin miczynski* have been found to be capable of degrading iron cyanides at pH 4.[7] Biodegradation rate of cyanide compounds may also be affected by the availability of nutrients. Carbon has been identified as a limiting factor in the microbial degradation of metal-cyanides, which may prevent the bioremediation of industrially contaminated soils.[9] Availability of oxygen is an important factor in the microbial mineralization of cyanide, as oxygen is consumed during several of the cyanide degrading pathways.[9] Physical and chemical processes employed to degrade cyanide and its related compounds are often high-priced and difficult to operate. A verified substitute to these processes is biological treatment, which typically relies upon the acclimation and enhancement of indigenous microorganisms. Biological degradation of cyanide has often been offered as a potentially economical and environmental-friendly alternative over conventional processes.[19]

12.5 ADVANTAGES OF THE BIOREMOVAL OF CYANIDE

The constant generation of enormous amount of cyanide-bearing wastes from various industries has resulted in the contamination of soil and water which recommends that novel processes are required to alleviate the serious problem of cyanide pollution in environment. The utilization of various chemical and physical treatment processes to rectify the cyanide problem from environment may be disagreeable because of the economic view point, less efficiency and secondary pollution. The biological treatment method is an eco-sociable and substitutional approach for removal of cyanide from wastewater. It can be less expensive than chemical and physical methods, and much faster than natural oxidation with efficiency equal or exceeding that of chemical or physical methods. Lower plant construction and operating cost, that is no chemical handling equipment or expensive control instruments are needed. Costs are relatively uniform and greater volumes of waste do not necessarily increase the costs proportionately. Efficiency wise, biodegradation provides superior resistance to shock loading, as well as recovery from such upsets and results in lower production of total dissolved solids and sludge, making it environment-friendly.

12.6 REVIEW ON THE TOPIC

The biodegradation of cyanide compounds by various microorganisms have been reported by a number of researchers. Table 12.2 presents the summary of literature on the biodegradation of various cyanide compounds.

Potivichayanon and Kitleartpornpairoat (2010)[74] isolated and identified cyanide-degrading bacteria *Agrobacterium tumefaciens* SUTS 1 and studied cyanide removal efficiency. The maximum growth rate of SUTS 1 obtained was 4.7×108 CFU/ml within 4 days. The cyanide removal efficiency was studied at 25, 50, and 150 mg/l cyanide. The cyanide removal efficiency at 25 and 50 mg/l cyanide, was approximately 87.50%. At 150 mg/l cyanide, SUTS 1 enhanced the cyanide removal efficiency up to 97.90%. Cell count of SUTS 1 increased when the cyanide concentration was set lower. Ammonia increased when the removal efficiency increased.

Ozel et al. (2010)[70] investigated to determine cyanide degradation characteristics in some basidiomycetes strains including *Polyporus arcularius* (T438), *Schizophyllum commune* (T 701), *Clavariadelphus truncatus* (T 192), *Pleurotus eryngii* (M 102), *Ganoderma applanatum* (M 105), *T. versicolor* (D 22), *Cerrena unicolor* (D 30), *S. commune* (D 35), and *Ganoderma lucidum* (D 33). The cyanide degradation activities of *P. arcularius*, *S. commune* and *G. lucidum* were found to be more than that of the other fungi examined. Maximum cyanide degradation was obtained after 48 h of incubation at 30°C by *P. arcularius* (T 438). The optimum pH and agitation rate were measured as 10.5 and 150 rev/min, respectively. The amount of biomass was found as 3.0 g for the maximum cyanide biodegradation with an initial cyanide concentration of 100 mg/l.

TABLE 12.2 Details of Different Microbial Species Involved in Biological Processes for Cyanide Removal.

Microorganisms	CN⁻ compound used	pH	Temp. (°C)	Time (h)	Initial conc.	% Removal of CN⁻	Remarks	Reference
Polyporus arcularius	KCN	10.5	30	48	100 mg/l	72.08	–	[70]
Pseudomonas pseudoalcaligenes	NaCN	9.5–10	30	57.6	45 mg/l	~60	–	[40]
Scenedesmus obliquus	NaCN	10.3	–	77	77.9	91	–	[37]
Bacillus sp. (Consortium)	KCN	9.5	37	192	200	65	–	[57]
Pseudomonas fluorenscens	ferrous (II) cyanide (ferrocyanide)	5	25	–	100 mg/l	78.9	–	[29]
Escherichia coli	KCN	6–8	20–40	–	50 mg/l	90	–	[13]
Pseudomonas fluorescens NCIMB 11764	KCN	7	30	6.3	50 mM KCN	80	in 6 h	[49]
Klebsiella oxytoca	KCN	7	30	80	0.58 mM	91.4	–	[45]
Rhodococcus species	KCN	7	30	10	12 mM	50	–	[55]
Pseudomonas fluorescens	FeCN	6	26	60	50 mg/l	99.9	–	[19]
Pseudomonas sp. (CM5, CMN2)	–	9.2–11.4	30	70	–	100	Bacteria remove cyanide between 45–70 h without high concentration of acclimatization	[2]
Fusarium solani	KCN	7.5	–	–	–	50	–	[7]
Pseudomonas putida	FeCN	7	30	–	100 mg/l	78.2	–	[20]

TABLE 12.2 (Continued)

Microorganisms	CN⁻ compound used	pH	Temp. (°C)	Time (h)	Initial conc.	% Removal of CN⁻	Remarks	Reference
Pseudomonas sp. Citrobacter sp.	K, Zn, Cu-cyanide	7.5	35	15	0.5 mM	68–93% 88–93%	Biodegradation in a rotating biological contactor	[72]
Mixed culture of Fusarium solani and T. polysporum	K₂Ni (CN)₄	7.0	–	–	0.75 mM	90%	–	[7]
Rhizopus oryzae	CN⁻	5.6	25	120	150 mg/l	83%	–	[18]
Stemphylium loti	CN⁻	7.2			150 mg/l	90%	–	
R. oryzae	CN⁻	5.6	25	120	150 mg/l	95.3%	Removal efficiency increased by using SAB process at same parameters	[18]
S. loti		7.2				98.6%		
Mixed (Anaerobic)	CN⁻	7.5	–	48	>100 mg/L	>70%	Methanogenesis	[31]

'–' Not reported

Gupta et al. (2010)[36] reported that there are many enzymes, which are produced by microorganisms that utilize cyanides as substrate to make alanine, glutamic acid, alpha-amino-butyric acid, beta-cyanoalanine and so forth. Five types of enzymatic pathways in various microorganisms have been reported, which are involved in cyanide degradation. All of them have their advantages and disadvantages. Cyanide hydratase is very substrate specific. The microorganisms containing this enzyme can degrade cyanide up to 200 ppm.

Gurbuza et al. (2009)[39] reported that the biological degradation of cyanide by *Scenedesmus obliquus* has shown a viable and healthy process for degrading cyanide in the mining process wastewaters. Gold mill effluents containing weak acid dissociable (WAD) cyanide concentration of 77.9 mg/l was fed into a batch unit, to examine the ability of *S. obliquus* for degrading cyanide. Cyanide was reduced to 6 mg/l in 77 h. Microbial growth and metal uptake of Zn, Fe and Cu was examined during cyanide degradation. The bio-treatment process was considered to be successful in degrading cyanide in the mine process water.

Kao et al. (2003)[45] reported cyanide degradation by *K. oxytoca*. They observed that resting cells could degrade, but cell-free extract was not able to degrade which happened as a result of inactivation of nitrogenase (an oxygen-labile enzyme) caused by the oxygen exposure after cell disruption. *K. oxytoca* also possessed enzymatic mechanisms to degrade nitriles. Nitriles can also be degraded by *Nocardia rhodochrous, Arthrobacter, Brevibacterium, P. putida, Pseudomonas marginalis, Pseudomonas aeruginosa, Rhodococcus erythropolis, Rhodococcus rhodochrous*.

Luque V. M. (2005)[53] studied the alkaline cyanide biodegradation by *Pseudomonas pseudoalcaligenes* CECT5344 at the pH of lower than 9.5, and found that cyanide compounds were used as the sole source of nitrogen under alkaline conditions, which prevents the volatile HCN formation. The cyanide consumed by this strain is stoichiometrically converted to ammonium. In addition, this bacterium grows with the heavy metal cyanide-containing wastewater generated by the jewellery industry. The strain CECT5344 was able to grow with the residue generated in the jewellery industry. This residue contains free cyanide and heavy metals such as Fe, Cu and Au complexed with cyanide, thus making the residue highly toxic.

Timur et al. (2005)[83] investigated a process for recovering gold from aqueous cyanide solutions using naturally occurring soil microorganisms.

In this research, microbes successfully liberated gold cations from auro-cyanide complexes. The gold was subsequently captured on iron and zinc electron donors. After cyanide degrading microorganisms populated microcosm reactors, the biomass was fed with a gold cyanide solution. Microorganisms consumed the CN^- moiety of aurocyanide complexes, releasing gold cations. A portion of the gold was recovered on steel wool or zinc which served as cathodes. Although gold recovery rates were low, methods for improving efficiency were investigated.

Akcil et al. (2003)[2] used two strains of *Pseudomonas sp.* isolated from a copper mine for biodegradation of cyanides at a concentration range of 100–400 mg/l. They compared their studies with chemical treatment methods and concluded that biological treatment methods are less expensive and environment-friendly but are as effective as chemical method.

Kwon et al. (2002)[52] investigated yeast, *Cryptococcus humicolus* MCN2, which was isolated from coke plant wastewater and grown on KCN as a sole nitrogen source, degraded concentrations of potassium tetracyanonickelate up to 65 mM when supplied with sufficient carbon.

Patil and Paknikar (2000)[72] used a mixed consortium of three *Pseudomonas species* and a *Citrobacter species* and obtained very high cyanide degradation capability as compared to the microbes used alone. Silver, nickel and zinc cyanide complexes were degraded successfully by the consortium.

Dursun et al. (1999)[23] investigated degradation of ferrous (II) cyanide complex (ferrocyanide) ions by free cells of *P. fluorescens*, in the presence of glucose and dissolved oxygen, as a function of initial pH, initial ferrocyanide, and glucose concentrations, and aeration rate in a batch fermenter. The microorganism used the ferrocyanide ions as the sole source of nitrogen. They found that 79% cyanide removal efficiency was achieved with maximum biodegradation rate at pH 5, and glucose concentration at 0.465 g/l.

Cowan et al. (1998)[16] reported that the cyanide degrading enzymes are generally produced by mesophilic microorganisms, often isolated from soil, with temperature optima usually ranging between 20 and 40°C.

Suh et al. (1994)[82] also used cyanide degrading bacteria, *Pseudomonas fluorescens* NCIB 11764 for the treatment of cyanide containing wastewater. After successive subculture for two months in cyanide containing medium, culture was completely adapted to grow in a medium containing up to 260 mg/l of cyanide.

Apart from the degradation, biosorption process also occurs, in which microorganisms adsorb the toxic compounds instead of degrading it.[71] Various fungal species such as *Aspergillus niger, Aureobasidium pullulans, Aspergillus fumigatus, Fusarium moniliforme, Fusarium oxysporum, Cladosporiumsp* and so forth, can act as biosorbents for removal of cyanide compounds.

12.7   ADVANCES ON BIODEGRADATION OF CYANIDE

Immobilization of cells improves the degradation rate[48,22,6,82] by preventing washing of cells, and it also increases the cell density. The immobilization technique was used by Kowalska et al. (1998). They used ultrafiltration membranes made of polyacronitrile[48] and carried out simultaneous degradation of cyanides and phenol using *Agrobacterium radiobacter, Staphylococcus sciuri,* and *Pseudomonas diminuta.* The efficiency of phenol and cyanide biodegradation was dependent on transmembrane pressure. The immobilization technique was earlier studied by Babu et al., (1992) where immobilized cells of *P. putida* were able to degrade sodium cyanide as a sole source of carbon and nitrogen. Various immobilization matrixes are available like granular activated carbon (GAC), alginate beads, zeolite, which have shown very high efficiency, and improved degradation rate.[6,12,23]

Biodegradation and biosorption processes can be used in combination for removal of cyanides as they can be very proficient as reported by Patil and Paknikar (1999). Simultaneous adsorption and biodegradation (SAB) is the latest development in cyanide removal.[19] The major mechanisms that are concerned with simultaneous adsorption and biodegradation are:

- Less biodegradable organics can be adsorbed on carbon at first, and are then slowly degraded by microorganisms[86,76]
- Adsorbent can be partially regenerated by microorganisms while the carbon bed is in operation[77,76]
- Biological reaction rate becomes higher on adsorbent due to an enrichment of the organics by carbon adsorption[86]

Dash et al. (2008)[19] reported high removal efficiency using *P. fluorescens* immobilized on granular activated carbon in SAB process and Dwivedi et al. (2016) reported cyanide removal using *Bacillus cereus* immobilized on natural bioadsorbent almond shell in SAB process. The degradation of cyanides has also been observed in several fungal strains. *F. solani, Trichoderma polysporum, Fusarium oxysporum, S. thermophilum, Penicillium miczynski* are able to grow on metallo-cyanide complex as a source of nitrogen.[19] A strain, *P. pseudoalcaligenes* CECT5344,[82] isolated from the Guadalquivir river (Cordoba), was found to use several nitrogen sources including cyanide, cyanate, b-cyanoalanine, cyanacetamide, and nitroferricyanide under alkaline conditions, which prevents volatile HCN (pKa 9.2) formation. This bacterium was also grown on heavy metal from cyanide containing wastewater generated by the jewellery industry.[40]

Besides bacteria and fungi, algae such as *Arthrospira maxima, Chlorella* sp. and *S. obliquus* could be used for the degradation of cyanides.[37] The detoxification of cyanide by algae was examined by exposing cultured suspensions of *A. maxima, Chlorella* sp. and *S. obliquus* at the pH of 10.3. The removal efficiency was 99%. This study has explored the use of algae in this field under extreme conditions with high removal efficiency.

Biosorption process is also used in cyanide removal. In biosorption processes, the microorganisms adsorb the toxic compounds instead of degrading it.[72] Several live fungal species (*A. fumigatus, A. niger, A. pullulans, Cladosporium cladosporioides, F. moniliforme, F. oxysporum, Mucor hiemalis*), can act as biosorbents for cyanide compounds. Biodegradation and biosorption process can be used in combination for removal of cyanides as they can be very efficient as shown by Patil and Paknikar (2000). Iron (III) cyanide complex is known to get adsorbed on *Rhizopus arrhizus* which is a filamentous fungus.[3] This fungus can adsorb the complex at very high alkaline conditions of pH 13 with very high loading capacity of 612.2 mg/g of cyanide.

One novel study has proposed a transcription-based assay for monitoring the biodegradation of both simple and metal cyanides.[8] The gene encoding cyanide hydratase (*chy*) in *F. solani* has been sequenced and primers utilized in reverse transcription-polymerase chain reaction (RT–PCR) to demonstrate transcription of this gene. The *chy* gene from *F. solani* displays significant homology to the corresponding cyanide hydratase gene from *Gloeocercospora sorghi, F. lateritium,* and *Leptosphaeria maculans.* This observation implies that the assay could be utilized in different contexts, provided that expression of the *chy* gene can be conclusively linked with the activity of cyanide hydratase. This assay could be an important tool to site managers, regulators and industries that generate cyanide wastes.[27]

12.8   CONCLUSION

From the above discussions the following conclusions are made:

- The development of microbial processes to be competent in severe environmental conditions, such as low pH, and contaminant toxicity, is required to ensure a technology that is competitive with the current physiochemical remediation strategies now practiced to reduce cyanide pollution.
- New microorganisms must be identified with respect to their degrading capabilities.
- Cyanide hydrolysis has an optimal pH range of 7–9, whereas thiocyanate biodegradation can occur at pH 10.
- The drawback of hydrolytic pathway is that, the microorganisms using cyanide hydratase to degrade cyanide do not use the carbon atom of cyanide as nutrient source; they require an additional carbon source. Cyanide hydratase is a really substrate specific enzyme.
- Microorganisms containing nitrile hydratase use cyanide as both nitrogen and carbon source.
- Biological processes have significant advantages over the physicochemical ones, and are attractive alternative methods for the cyanide removal.
- Biological treatment can be less expensive than chemical and physical methods, and much faster than natural oxidation with efficiency equal or exceeding that of chemical or physical methods.

KEYWORDS

- **bacteria**
- **biodegradation**
- **cyanide**
- **fungus**
- **wastewater**
- **SAB**

REFERENCES

1. Aguilar, M.; Farran, A.; Marti, V. Capillary Electrophoretic Determination of Cyanide Leaching Solutions from Automobile Catalytic Converters. *J. Chromatogr.* **1997**, *778*, 397–402.

2. Akcil, A.; Karahan, A. G.; Ciftci, H.; Sagdic, O. Biological Treatment of Cyanide by Natural Isolated Bacteria (Pseudomonas Sp.). *Miner. Eng.* **2003**, *16*, 643–649

3. Aksu, Z.; Calik, A.; Dursun, A. Y.; Demircan, Z. Biosorption of Iron (III)–Cyanide Complex Ions on *Rhizopus arrhizus*: Application of Adsorption Isotherms. *Process Biochem.* **1999**, *34*, 483–491

4. Alicilar, A.; Komurcu, M.; Guru, M. The Removal of Cyanides from Water by Catalytic Air Oxidation in a Fixed Bed Reactor. *Chem. Eng. Process.* **2002**, *41* 525–529.

5. ATSDR and USEPA. Comprehensive Environmental Response, Compensation, and Liability Act (CERCLA), priority list of hazardous substances, 2015.

6. Babu, G. R. V.; Wolfram, J. H.; Chapatwala, K. D. Conversion of Sodium Cyanide to Carbon Dioxide and Ammonia by Immobilized Cells of *Pseudomonas putida*. *J. Ind. Microbiol.* **1992**, *9*, 235–238.

7. Barclay, M.; Tett, V. A.; Knowles, C. J. Metabolism and Enzymology of Cyanide/ Metallocyanide Biodegradation by *Fusarium solani* under Neutral and Acidic Condition. *Enzyme Microb. Technol.* **1998**, *23*, 321–330.

8. Barclay, M.; Day, J. C.; Thompson, I. P.; Knowles, C. J.; Bailey, M. J. Substrate-Regulated Cyanide Hydratase (chy) Gene Expression in *Fusarium solani*: The Potential of a Transcription-Based Assay for Monitoring the Biotransformation of Cyanide Complexes. *Environ. Microbiol.* **2002**, *4*, 183–189.

9. Baxter, J.; Cummings, S. P. The Current and Future Applications of Microorganism in the Bioremediation of Cyanide Contamination. *Antonie van Leeuwenhoek* **2006**, *90*, 1–17.

10. Bernardin, F. E. Cyanide Detoxification using Adsorption and Catalytic Oxidation on Granular Activated Carbon. *J. Water. Pollut. Control. Fed.* **1973**, *45*(2), 221–231.

11. Bordo, D.; Forlani, F.; Spallarossa, A.; Colnaghi, R.; Carpen, A.; Bolognesi, M.; Pagani, S. A Persulfurated Cysteine Promotes Active Site Reactivity in *Azotobacter vinelandii* Rhodanese. *Biol. Chem.* **2001**, *382*, 1245–1252

12. Chapatwala, K. D.; Babu, G. R. V.; Vijaya, O. K.; Kumar, K.P.; Wolfram, J. H. Biodegradation of Cyanides, Cyanates and Thiocyanates to Ammonia and Carbon Dioxide by Immobilized Cells of *Pseudomonas putida*. *J. Ind. Microbiol.* **1998**, *20*, 28–33.

13. Ciminellei, F. M. M.; DeAndrade, M. C.; Linardi, V. R. Cyanide Degradation by an *Escherichia coli* Strain. *Can. J. Microbiol.***1996**, *42*, 519–523.

14. Cipollone, R.; Bigotti, M. G.; Frangipani, E.; Ascenzi, P.; Visca, P. Characterization of a Rhodanese from the Cyanogenic Bacterium *Pseudomonas aeruginosa*. *Biochem. Biophys. Res. Commun.* **2004**, *325*, 85–90.

15. Collins, P.; Knowles, C. The Utilization of Nitriles and Amides by *Nocardia rhodochrous*. *J. Gen. Microbiol.* **1983**, *129*, 248–254.

16. Cowan, D.; Cramp, R. A.; Periera, R.; Graham, D.; Almatawah, Q. Biochemistry and Biotechnology of Mesophilic and Thermophilic Nitrile Metabolizing Enzymes. *Extremophiles* **1998**, *2*, 207–216.

17. Environment Australia, Department of Environment. Best Practice Environmental Management in Mining Cyanide Management, 1998.
18. Dash, R. R.; Balomajumder, C.; Kumar, A. Cyanide Removal by Combined Adsorption and Biodegradation Process. *Iran. J. Environ. Sci. Eng.* **2006**, *3*, 91–96.
19. Dash, R. R.; Majumder, C. B.; Kumar, A. Treatment of Metal Cyanide Bearing Wastewater by Simultaneous Adsorption and Biodegradation (SAB). *J. Hazard. Mater.* **2008**, *152*, 387–396.
20. Dash, R. R.; Gaur, A, Majumder, C. B. Removal of Cyanide from Water and Wastewater using Granular Activated Carbon. *Chem. Eng. J.* **2009a**, *146*, 408–413
21. Dubey, S. K.; Holmes, D. S. Biological Cyanide Destruction Mediated by Microorganisms. *World J. Microbiol. Biotechnol.* **1995**, *11*, 257–265
22. Dursun, A. Y.; Aksu, Z. Biodegradation Kinetics of Ferrous (II) Cyanide Complex Ions by Immobilized *Pseudomonas fluorescens* in a Packed Bed Column Reactor. *Process Biochem.* **2000**, *35*(6), 615–622.
23. Dursun, A. Y.; Alik, A. C.; Aksu, Z. Degradation of Ferrous (II) Cyanide Complex Ions by *Pseudomonas fluorescens*. *Process. Biochem.* **1999**, 34, 901–908.
24. Dwivedi, N.; Majumder, C. B.; Mondal, P. Comparative Evaluation of Cyanide Removal Efficiency by Adsorption, Biodegradation and Simultaneous Adsorption and Biodegradation (SAB) in a Batch Reactor by using *Bacillus cereus* and Almond Shell. *J. Environ. Biol.* **2016**, *37*(4), 551–556.
25. Dwivedi, N.; Majumder, C. B.; Mondal, P. Comparative Investigation on the Removal of Cyanide from Aqueous Solution using Two Different Bioadsorbents. *Water Resour. Ind.* **2016**, *15*, 28–40
26. Dzombak, D. A.; Ghosh, R. S.; Wong-Chong, G. M. Cyanide in Water and Soil Chemistry, Risk and Management. Taylor and Francis Group, CRC press: NW, 2006.
27. Ebbs, S. Biological Degradation of Cyanide Compounds. *Curr. Opin. Biotechnol.* **2004**, *15,* 1–6.
28. Eisler, R.; Wiemeyer, S. N. Cyanide Hazards to Plants and Animals from Gold Mining and Related Water Issues. *Rev. Environ. Contam. Toxicol.* **2004**, *183*, 21–54.
29. Environmental and Health Effects of Cyanide, International cyanide management code for the gold mining industry, 888 16th Street, NW, Suite 303, Washington, DC 20006, USA.
30. Ezzi, I. M.; Pascual, A. J.; Gould, J. B.; Lynch, M. J. Characterisation of the Rhodanese Enzyme in *Trichoderma* spp. *Enzyme Microb. Technol.* **2003**, *32,* 629–634.
31. Fallon, R. D.; Cooper, D. A.; Speece, R.; Henson, M. Anaerobic Biodegradation of Cyanide under Methogenic Conditions. *Appl. Environ. Microbiol.* **1991**, *57*, 1656–1662.
32. Fowden, L.; Bell, E. A. Cyanide Metabolism by Seedling. *Nature* **1965**, *206*, 110–112.
33. Ganczarczyk, J. J.; Takoaka, P. T.; Ohashi, D. A. Application of Polysulfide for Pretreatment of Spent Cyanide Liquors. *J. Water Pollut. Control Fed.* **1985**, *57*, 1089–1093.
34. Gerasimova, T.; Novikov, A.; Osswald, S.; Yanenko, A. Screening, Characterization and Application of Cyanide-Resistant Nitrile Hydratases. *Eng. Life Sci.* **2004**, *4*, 543–546.

35. Goldhust, A.; Bohak, Z. Induction, Purification and Characterization of the Nitrilase of *Fusarium oxysporum*. *Biotechnol. Appl. Biochem.* **1989**, *11*, 581–601.
36. Gupta, N.; Majumder, C. B.; Agarwal, V. K. Enzymatic Mechanism and Biochemistry for Cyanide Degradation: A Review. *J. Hazard. Mater.* **2010**, *1*, 176.
37. Gurbuza, F.; Ciftci, H.; Akcil, A.; Karahan, A. G. Microbial Detoxification of Cyanide Solutions: A New Biotechnological Approach using Algae. *Hydrometallurgy* **2004**, *72*, 167–176.
38. Gurbuza, F.; Ciftci, H.; Akcil, A. Biodegradation of Cyanide Containing Effluents by *Scenedesmus obliquus*. *J. Hazard. Mater.* **2009**, *162*, 74–79.
39. Hachiya, H.; Ito, S.; Fushinuki, Y.; Masadome, T.; Asano, Y.; Imato, T. Continuous Monitoring for Cyanide in Wastewater with a Galvanic Hydrogen Cyanide Sensor using a Purge System. *Talanta* **1999**, *48*, 997–1004.
40. Huertas, M. J.; Sáez, L. P.; Roldán, M. D.; Luque-Almagro, V. M.; Martínez-Luque, M.; Blascoc, R.; Castillo, F.; Moreno-Vivián, C.; Garcia-Garcia, I. Alkaline Cyanide Degradation by *Pseudomonas pseudoalcaligenes* CECT5344 in a Batch Reactor, Influence of pH. *J. Hazard. Mater.* **2010**, *179*, 72–78.
41. Indian standard Institution, Guide for treatment of effluents of electroplating industry, IS: 7453–1974.
42. InfoMine, Summary Fact Sheet on Cyanide, 2012. http://www.infomine.com/publications/docs/SummaryFactSheetCyanide.pdf.
43. Ingvorsen, K.; Hojerpedersen, B.; Godtfredsen, S. E. Novel Cyanide-Hydrolysing Enzyme from *Alcaligenes xylosoxidans* Subsp. *Denitrificans*. *Appl. Environ. Microbiol.* **1991**, *57*, 1783–1789.
44. Ingvorsen, K.; Godtfredsen, S. E.; Hojer-Pedersen, B.; Nordisk, N.1992 Microbial Cyanide Converting Enzymes, their Production and Use, US Patent 5,116,744.
45. Kao, C. M.; Liu, J. K.; Lou, H. R.; Lin, C. S.; Chen, S. C. Biotransformation of Cyanide to Methane and Ammonia by *Klebsiella oxytoca*. *Chemosphere* **2003**, *50*, 1055–1061.
46. Kapadan, K. I.; Kargi, F. Simultaneous Biodegradation and Adsorption of Textile Dye Stuff in an Activated Sludge Unit. *Process Biochem.* **2002**, *37*(9), 973–981.
47. Kenfield, J. W.; Qin, R.; Semmens, M. J.; Cussler, E. L. Cyanide Recovery Across Hollow Fibre Gas Membranes. *Environ. Sci. Technol* **1988**, *22*, 1151–1155.
48. Kowalska, M.; Bodzek, M.; Bohdziewicz, J. Biodegradation of Phenols and Cyanides using Membranes with Immobilized Organisms. *Process Biochem.* **1998**, *33*, 189–197.
49. Kunz, D. A.; Nagappan, O.; Silva-Avalos, J.; DeLong, G. T. Utilization of Cyanide as a Nitrogenous Substrate by *Pseudomonas fluorescens* NCIMB 11764: Evidence for Multiple Pathways of Metabolic Conversion. *Appl. Environ. Microbiol.* **1992**, *58*, 2022–2029
50. Kunz, D. A.; Wang, C. S.; Che, J. L. Alternative Routes of Enzymic Cyanide Metabolism in *Pseudomonas fluorescens* NCIMB 11764. *Microbiology* **1994**, *140*, 1705–1712.
51. Kuyucak, N.; Akcil, A. Cyanide and Removal Options from Effluents in Gold Mining and Metallurgical Processes. *Miner. Eng.* **2013**, *50*, 13–29.
52. Kwon, H. K.; Woo, S. H.; Park, J. M. Degradation of Tetracyanonickelate (II) by *Cryptococcus humicolus* MCN2. *FEMS Microbiol. Lett.* **2002**, *214*, 211–216.

53. Luque-Almagro, V. M.; Huertas, M. J.; Martinez, L. M.; Moreno, V. C.; Huertas, M. J. Bacterial Degradation of Cyanide and its Metal Complexes under Alkaline Conditions. *Appl. Environ. Micrbiol.* **2005,** *7*, 940.

54. Luthy, R. G.; Bruce, S. G. Kinetics of Reaction of Cyanide and Reduced Sulfur Species in Aqueous Solution. *Environ. Sci. Technol.* **1979,** *13*, 1481–1487.

55. Maniyam, M. N.; Sjahrir, F.; Latif, I. A. *Biodegradation of Cyanide by Rhodococcus Strains Isolated in Malaysia, International Conference on Food Engineering and Biotechnology IPCBEE* Vol. 9; IACSIT Press: Singapore, 2011.

56. Mapstone, G. E.; Thorn, B. R. The Hypochlorite Oxidation of Cyanide and Cyanate. *J. Appl. Chem. Biotechnol.* **1978,** *28*, 135–143.

57. Mekuto, L.; Jackson, V. A.; Ntwampe, S. K. O. Biodegradation of Free Cyanide Using Bacillus Sp. Consortium Dominated by *Bacillus Safensis*, Lichenformis and Tequilensis Strains: A Bioprocess Supported Solely with Whey. *J. Biorem. Biodegrad.* **2013.** DOI: 10.4172/2155-6199.S18-004.

58. Meyers, P. R.; Rawlings, D. E.; Woods, D. R.; Lindsey, G. G. Isolation and Characterization of a Cyanide Dihydratase for *Bacillus pumilus* C1. *J. Bacteriol.* **1993,** *175*, 6105–6112.

59. Mohsen, A.; Neda, M.; Masoud, A. Negative Effects of Cyanide on Health and its Removal Options from Industrial Wastewater. *Int. J. Epidemiol. Res.* **2015,** *2*(1), 44–49.

60. Monser, L.; Adhoum, N. Modified Activated Carbon for the Removal of Copper, Zinc, Chromium and Cyanide from Wastewater. *Sep. Purif. Technol.* **2002,** *26*, 137–146.

61. Mordocco, A.; Kuek, C.; Jenkins, R. Continuous Degradation of Phenol at Low Concentration using Immobilized *Psedomonas putida*. *Enzyme Microb. Technol.* **1999,** *25*, 530–536.

62. Moussavi. G.; Khosravi, R. Removal of Cyanide from Wastewater by Adsorption onto Pistachio Hull Wastes: Parametric Experiments, Kinetics and Equilibrium Analysis. *J. Hazard Mater.* **2010,** *183*, 724–730.

63. Mudder, T. I.; Botz, M. A Global Perspective of Cyanide. A Background Paper of the UNEP/ICME Industry Codes of Practice Workshop: Cyanide Management Paris, France, 2000. http://www.mineralresfurcesforum.org/cyanide.

64. Mudder, T. I.; Botz, M. M. Cyanide and Society: A Critical Review. *Eur. J. Miner. Process. Environ. Prot.* **2004,** *4*(1), 62–74.

65. Mufaddal, I. E.; James, M. L. Cyanide Catabolizing Enzymes in Trichoderma spp. *Enzyme Microb. Technol.* **2002,** *31*, 1042–1047.

66. Nagahara, N.; Ito, T.; Minami, M. Mercaptopyruvate Sulfurtransferase as a Defense against Cyanide Toxication: Molecular Properties and Mode of Detoxification. *Histol. Histopathol.* **1999,** *14*, 1277–1286.

67. Nawaz, S. M.; Chapatwala, D. K.; Wolfram, H. J. Degradation of Acetonitrile by *Pseudomonas putida*. *Appl. Environ. Microbiol.* **1989,** *55,* 2267–2274.

68. Ogutveren, U. B.; Toru, E.; Koparal, S. Removal of Cyanide by Anodic Oxidation for Wastewater Treatment. *Water Res.* **1999,** *33*(8), 1851–1856.

69. Omura, H.; Ikemot, M.; Kobayash, M.; Shimizq, S.; Yoshida, T.; Nagasawa, T. Purification, Characterization and Gene Cloning of Thermostable O-acetyl-homoserine sulfhydrylase forming γ-Cyano-α-aminobutyric Acid. *J. Biosci. Bioeng.* **2003,** *96*, 53–58.

70. Ozel, Y. K.; Gedikli, S.; Aytar, P.; Unal, A.; Yamaç, M.; Ahmet, Ç.; Kolankaya, N. New Fungal Biomasses for Cyanide Biodegradation. *J. Biosci. Bioeng.* **2010**, *110*(4), 431–435.

71. Patil, Y. B.; Paknikar, K. M. Removal and Recovery of Metal Cyanides Using a Combination of Biosorption and Biodegradation Processes. *Biotechnol. Lett.* **1999**, *21*, 913–919.

72. Patil, Y. B.; Paknikar, K. M. Development of a Process for Biodetoxification of Metal Cyanides from Wastewater. *Process Biochem.* **2000**, *35*, 1139–1151.

73. Pereira, P. T.; Arrabaca, J. D.; Maral-Collaco, A. M. T. Isolation, Selection, and Characterisation of a Cyanide-Degrading Fungus from an Industrial Effluent. *Int. Biodeterior Biodegrad.* **1996**, *37*, 45–52.

74. Potivichayanon, S.; Kitleartpornpairoat, R. Biodegradation of Cyanide by a Novel Cyanide Degrading Bacterium. World Academy of Science. *J. Eng. Technol.* **2010**, *42*, 1362–1365.

75. Raybuck, S. A. Microbes and Microbial Enzymes for Cyanide Degradation. *Biodegradation* **1992**, *3*, 3–18.

76. Rice, R. G.; Robson, C. M. Biological Activated Carbon Enhanced Aerobic Biological Activity in GAC Systems. Ann Arbor Science Publishers, England. 1992.

77. Rodman, C. A. Factors Involved with Biological Regeneration of Activated Carbon Adsorbers. J. Water Pollut. Control Fed. **1973**, *55*, 1168–1173.

78. Roy, A. B.; Trudinger, P. A. The Biochemistry of Inorganic Compounds of Sulphur. *Microb. Ecol.* **1977**, *4*, 9–25.

79. Sexton, A. C.; Howlett, B. J. Characterisation of a Cyanide Hydratasegene in the Phytopathenogenic Fungus *Leptosphaeria maculans*. *Mol. Genet. Genomics.* **2000**, *263*, 463–470.

80. Siedow, J. N.; Umbach, A. L. Plant Mitochondrial Electron Transfer and Molecular Biology. *Plant Cell* **1995**, *7*, 821.

81. Skowronski, B.; Strobel, G. A. Cyanide Resistance and Utilization by a Strain of *Bacillus pumilus. Can. J. Microbiol.* **1969**, *15*, 93–98.

82. Suh, Y.; Park, J. M.; Yang, J. Biodegradation of Cyanide Compounds by *Pseudomonas fluorescens* Immobilized on Zeolite. *Enzyme. Microb. Technol.* **1994**, *16*, 529–533.

83. Timur, A.; Daniel, M. W.; Indranil, S. Biological Gold Recovery from Gold–Cyanide Solutions. *Int. J. Miner. Process.* **2005**, *76*, 33–42.

84. Trapp, S.; Larsen, M.; Pirandello, A.; Danquah, B. J. Feasibility of Cyanide Elimination Using Plants. *Eur. J. Miner. Process. Environ Prot.* **2003**, *3*, 128–137.

85. USEPA, 1985. Drinking Water Criteria Document for Cyanide. 84 192.

86. Weber, W. J. Jr.; Ying, W. C. Integrated biological and physiochemical treatment for reclamation of wastewater. In: Proc. IAWPRC Int. Conf. on Advanced Treatment and Reclamation of Wastewater, 1977.

87. White, D. M.; Schnabel, W. Treatment of Cyanide Waste in a Sequencing Batch Biofilm Reactor. *Water Res.* **1998**, *32*, 254–257.

88. White, J. M.; Jones, D. D.; Huang, D.; Gauthier, J. J. Conversion of Cyanide to Formate and Ammonia by a Pseudomonad Obtained from Industrial Wastewater. J. Ind. Microbiol. **1988**, *3*, 263.

89. Wild, S. R.; Rudd, T.; Neller, A. Fate and Effects of Cyanide during Wastewater Treatment Processes. Sci. Total Environ. **1994**, *156*(2), 93–107.

90. Young, C. A.; Jordan, T.S. Cyanide Remediation: Current and Past Technologies Proceedings of the 10th Annual Conference on Hazardous Waste Research, 2004, 104–129.
91. Zagury, G. J.; Oudjehani, K.; Deschenes, L. Characterization and Availability of Cyanide in Solid Mine Tailings from gold extraction plants. *Sci. Total Environ.* **2004,** *320*, 211–224.

PART III

Microbes and Human Health

CHAPTER 13

MICROBIAL INFECTIONS AND HUMAN HEALTH: WHAT CAN AYURVEDA OFFER?

CHETHALA N. VISHNUPRASAD*

*School of Life Sciences, Institute of Transdisciplinary Health Science and Technology (TDU), #74/2, Jarakabande Kaval, Yelahanka, Bangalore 560064, Karnataka, India, *E-mail: drcnvp@gmail.com, Tel.: +919481384872*

CONTENTS

ABSTRACT

Ayurveda, the traditional Indian medical practice, primarily focuses on maintaining an optimal health and improving natural resistance (immunity) to protect the body from various diseases. Though the Ayurvedic concepts of health and disease are different from the modern biology, there is an increasing interest in integrating both the streams for the better understanding of human wellness and illness. The present article is an attempt to compare the concepts of microbiology in *Ayurveda* and modern medicine and to see how scientists study the anti-microbial concepts with the guidance of *Ayurveda*. The article also presents a comparative description of various Ayurvedic disease symptoms that are now identified as typical microbial infections. Though the article is not giving a comprehensive list of all microbial diseases, it introduces the potentials of *Ayurveda* and herbal formulations in the management of infections and highlights the significance and challenges of bridging two unique health management systems, which are epistemologically different. The chapter also emphasize on the necessity of a novel integrative and trans-disciplinary approach for bioprospecting Ayurvedic concepts of Microbiology.

13.1 INTRODUCTION

13.1.1 AYURVEDA—*A CODIFIED SYSTEM OF MEDICINE*

Ayurveda, the 'science of life' (*Ayu* = Life; *Veda* = Science), is considered as one of the oldest traditional medicines in the world.[22] It is a holistic and logical health management system originated in India, which had progressively systematized and established as an efficient medical science over a period of time. The fundamental principles of *Ayurveda* were recorded in the form of three classical texts viz. *Charaka Samhita, Sushrutha Samhita* and *Ashtanga Hrudaya,* collectively known as *Bruhatrayis* (the Great Trilogy) of *Ayurveda*.[30] Several commentaries and interpretations for *Bruhatrayis* along with many independent works, based on the principles of *Bruhatrayis,* are also available in various regional languages of India. The medical knowledge acquired by people (individually or in groups) were systematically recorded and transferred to successive generations and established as a unique system of medicine in India.[21] Being a science that originated and established much before the development of modern biology, *Ayurveda* follows unique concepts and resultant methodologies for defining and treating human diseases. However, the spectrum of *Ayurveda* is not limited to the human wellness. It also covers plants and agricultural sciences (*Vrukshaayurveda*) as well as veterinary sciences (*Hasthyayurveda, Ashwayurveda* etc.). *Ayurveda* considers the human being as an inseparable part of the surrounding nature and it is in equilibrium with the composition of the nature. Events that afflict the equilibrium will lead to various diseases. Though the concepts of wellness and illness in *Ayurveda* are epistemologically different from the contemporary biological sciences, it is widely accepted and practiced all over India (also in many other parts of the globe) as a successful complementary or alternative medicine for the management various diseases. All types of diseases described in modern medicine such as metabolic diseases, lifestyle diseases, microbial infections, psychological disorders etc. are addressed and managed by *Ayurveda*. Constructing a logical and epistemological bridge between the biological concepts in *Ayurveda* and modern medicine would certainly help in deepening the biological understanding of disease pathogenesis and therapy. A reasonable and logical synthesis of *Ayurveda* with modern science is the need of the hour, but it is challenging as well.

13.1.2 *WELLNESS AND ILLNESS IN* AYURVEDA

Ayurveda defines life as a combination of body, sense organs, mind and the soul. Being a holistic medical system, *Ayurveda*'s primary focus is to maintain an optimal health and improve body's natural resistance to protect it from various diseases. This is achieved through an integrated regulation of day-to-day activities of body, mind, spirit and the emotions. Hence the concept of healthy living or wellness in *Ayurveda* is nothing but establishing a harmony of physical, mental, social and spiritual aspects of life.[23] *Ayurveda* describes this holistic balance in terms of three basic bodily humours (*Tridosha*) viz. *Vata* (wind), *Pitta* (bile) and *Kapha* (phlegm) where the body, mind and consciousness synergistically regulate their balance and maintain the human healthy. (*This holistic approach of wellness in Ayurveda can be compared to the Hippocratic theory of 'humorism', wherein health is considered as a harmonious balance of four humours viz. blood, phlegm, yellow bile and black bile*). Celebrated invocation of *Ashtangahrudaya*, one of the great Triologies of *Ayurveda*, pays salutations to the "physician unprecedented (*Apoorva vaidya*)" who can destroy the diseases '*Raga etc*'.[20] The word '*Raga*' translates as vehement passion or desire of the human mind. The expression '*etc*' in the verse implies all other *Raga* related attributes of mind such as *Kama* (lust),

Krodha (anger), *Lobha* (greed), *Mada* (arrogance), *Matsarya* (jealousy), *Dvesha* (hatred), *Bhaya* (fear) and other emotions, where the physiological or external disease manifestations are ingrained. *Ayurveda* defines disease principally as a condition that causes an imbalance in *tridoshas*. Agents or attributes that vitiate the *dosha* balance are the etiological agents for the disease. In short, the disequilibrium of *doshas* is 'illness' and the equilibrium is 'wellness'. The role of an Ayurvedic practitioner is primarily to understand the *Dosha Prakriti* (Whole body constitution) of an individual and its vitiations, if any. This forms the basis for disease diagnosis as well as for prescribing appropriate lifestyle modification, diet and medications to restore the balance. The motto of *Ayurveda* is thus to preserve the health of a healthy person and treat the diseased to cure the disease as well as to restore the health.

Ayurveda mentions inappropriate food and actions are the primary causes (etiological agents) that vitiate *doshas*. Based on its holistic perspective, *Ayurveda* describes a person who, enjoys wholesome food and activities every day, introspects on his actions, is generous, truthful and forgiving etc. shall remain free from illness always and people who are not following the above mentioned actions (physical or psychological) are affected with illness. In other words, the healthy life is defined as a condition where an equilibrium of *Doshas* (body humours), *Agni* (metabolic fire), *Dhathu* (tissues in the body) and *Malakriya* (various excretory processes) are maintained. In modern medicine, but, the etiological agents are more tangible and have qualitative and quantitative properties. While modern medicine follow a more reductionist approach focusing on etiological agents, specific targets and symptoms, the therapeutic approaches of *Ayurveda* are patient centred and are directed to restore the balance of *Doshas* in the body. As the definitions of illness and wellness in *Ayurveda* are from the human body perspective, a direct comparison of modern biological concepts of disease to that of *Ayurveda* is difficult. The best possible way of comparison is based on the symptoms. In the case of microbial infections also, only a symptomatic correlation can be drawn between modern medicine and *Ayurveda*. Constructing a logical and epistemological bridge between the biological concepts in *Ayurveda* and modern medicine would certainly help in deepening the biological understanding of pathogenesis and therapy.

13.1.3 BRIDGING AYURVEDA AND MODERN BIOLOGY— APPROACHES AND METHODOLOGIES

Traditional Indian knowledge systems are highly entangled with stories of imaginations, myths and legends that are deep rooted in the cultural and religious aspects of the country. Perhaps the situation is same with any other traditional knowledge systems in the world. This, in fact, makes it very difficult to segregate and fish out the gems of logical science from the ocean of mythical and fictitious narrations. Reading *Ayurveda* from modern biology perspective always faces this problem. Further, the unique and epistemologically different concepts of pharmacology and medicines in *Ayurveda* also form another major hurdle in making a logical synthesis of Ayurvedic knowledge to the contemporary scientific language. The differences start even with fundamental approaches that define disease and therapy. *Ayurveda* (and other similar CAMs) follow a holistic approach whereas modern medicine (or modern biology) has a reductionist approach for defining diseases. In holistic approach, disease is a manifestation observed in the body as a result of vitiation of the physiological balance, which are semi-quantitative in nature. The physiology and biochemistry underlying the disease manifestation are more of conceptual and qualitative. Conversely in reductionist approach, both disease and its etiological agents can be defined in a more tangible way with qualitative and quantitative measurements of physiological and biochemical parameters associated with disease progression.

These are validated and are globally accepted. Microbial infections are classical examples for comparison. *Ayurveda* does not talk about 'microorganisms' as such. Microbial infections are also considered as diseases resulting from the vitiation of *Tridosha*, and the treatments are basically to restore the *dosha* balance. Whereas in modern medicine microbial infections are explained on the basis of germ theory and Koch's postulates wherein disease specific microbes are identified and isolated. Therefore the treatments are target oriented like specific antimicrobial agents like antibiotics etc. However, from the patient's perspective, cure is not just the recovery from diseases it is also the prevention of any recurring incidents of diseases. While a reductionist approach helps in understanding the causative agents and their precise molecular and biochemical mode of action (which will expand the scientific insights about the disease), a holistic multi-targeted approach would be better for the maintenance of good health and long-term prevention of any recurrent diseases. By establishing a logical and reasonable integration of two different medical knowledge systems, it is expected to have a better understanding of the disease from multiple angles and it can open up specific as well as holistic management strategies without any adverse effects. The chapter here is an attempt to understand how *Ayurveda* recognize microbial infections and what it offers for the management of microbial infections.

13.2 CONCEPTS OF MICROBES IN *AYURVEDA* AND MODERN BIOLOGY

The revolutionary developments in modern microbiology started in the 17th century with the discovery of microscopes by Antony Van Leeuwenhoek. The invention of microscope had shed light into the world of microbiology that facilitated the morphological identification, classification, isolation and propagation of microorganisms. It further deepened the scientific understanding of microbial infections, nature of causative agents, mode of infection and antimicrobial strategies. These developments, in fact, resulted in establishing specialized branches like medical microbiology, agricultural microbiology, food microbiology etc. Although *Ayurveda* hadn't have a separate field like 'microbiology', the age-old medical science had described various diseases similar to microbial infections (matching with symptoms of microbial infections described in modern medicine) and prescribed preventive and curative therapies for them. To cite some of the examples include common fever (*Jwara*), malaria (*Atisara*), leprosy (*Kustha*), smallpox (*Masoorika*), infectious wounds (*Vruna*) etc. However, in line with the Ayurvedic concepts of *Dosha Prakriti* and diseases aetiology, *Ayurveda* consider all these microbial diseases (as defined in modern medicine) are manifestations of *dosha* vitiation in the body (Fig. 13.1). Thus, unlike the targeted antimicrobial treatments, *Ayurveda* focus on restoring the body's physiological balance and curing the symptoms of microbial infections. This is a holistic approach that not only helps in curing the infection but also helps in restoring the health and preventing any recurring infections.

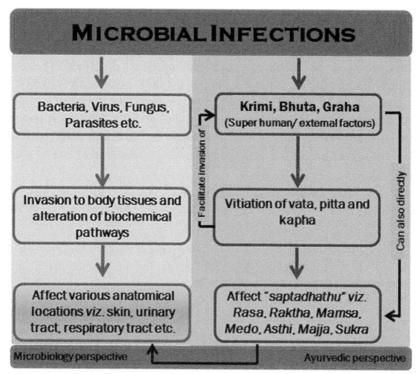

FIGURE 13.1   Pictorial representation of progression of microbial infection according to the concepts of modern biology and *Ayurveda*: Both the knowledge systems are converging at the level of anatomical locations where the manifestations are observed as diseases.

Modern microbiology classifies infections into different types based on the causative agents and their specific mode of action. Detailed understanding of the patho-physiology and biochemistry of infections helped us to identify the specific antigenic agents that elicit an adverse effect in the body. It is not based on the gross symptoms that one can observe. Diarrhoea is a classic example. Despite having symptomatic similarities, modern science clearly differentiates cholera from salmonellosis based on their patho-physiological and biochemical aspects. Whereas in *Ayurveda*, the disease classifications are mostly based on the symptoms seen in patient (e.g. dysentery, fever, cough etc.) and not based on the causative agents. Hence there is every chance that the symptoms resulting from infectious and non-infectious causes can overlap and can be considered as same or similar diseases. As in the case of cholera and salmonellosis,

Ayurveda classify both under dysentery (*atisara*) and the treatments are targeted to pacify the symptoms and restore the health. Table 13.1 lists out the major Ayurvedic diseases (or terminologies) that can be correlated to various types of specific microbial infections described in modern biology. Though the list is not exhaustive, the table helps readers to have a comparative understanding of 'infections' in *Ayurveda* and modern medicine. The table also lists out the major herbal and non-herbal substances used in infection management. The readers are also informed that the listed Ayurvedic symptoms can arise due to non-microbial agents as well.

TABLE 13.1 Comparison of Disease Terminologies in *Ayurveda* and Modern Medicine. Table compares the disease forms in *Ayurveda* with specific microbial infections in modern biology. The Ayurvedic information and herbal details in the table is prepared based on classical texts of *Ayurveda* as well as the health informatics software AyuSoft, Centre for Development of Advanced Computing (C-DAC), India. The table has not given the etiological agents from Ayurvedic perspective because *Ayurveda* basically see all these diseases are due to the vitiation of basic body humours (*Tridosha*). There is no individual etiological agent for the diseases in *Ayurveda*.

Sl. no.		Ayurveda perspective	Modern biology perspective
1	Disease	Abhishyanda	conjunctivitis (Ophthalmia neonatorum)
	Etiological agent		*Neisseria gonorrhoea,* herpes simplex virus, *Staphylococcus aureus, Streptococcus haemolyticus, Streptococcus pneumoniae, Chlamydia trachomatis*
	Treatment drug	Areca Catechu Linn. (Arecaceae). (Several other formulations are also used)	Antibiotics erythromycin, tetracycline, or silver nitrate
2	Disease	Amlapitta	*Gastritis,* Gastric ulcer
	Etiological agent		*Helicobacter pylori* and other bacterial and viral infections
	Treatment drug	Ricinus communis linn. (Euphorbiaceae), steel iron, calcium carbonate, conch shell, copper (cuprum), copper sulphate (blue vitriol)	'triple therapy' proton pump inhibitors (omeprazole) and the antibiotics clarithromycin and amoxicillin

TABLE 13.1   *(Continued)*

Sl. no.		Ayurveda perspective	Modern biology perspective
3	Disease	Athisara	Cholera, *colitis*, enterovirus infection, food poisoning, salmonellosis, enteritis
	Etiological agent		*Vibrio cholera, Salmonella typhi, Vibrio parahaemolyticus, Clostridium difficile* etc.
	Herbal/ non-herbal materials used in treatment	*Emblica officinalis* Gaertn. (Euphorbiaceae), *Mangifera indica* Linn. (Anacardiaceae), *Ficus religiosa* Linn. (Moraceae), *Aconitum heterophyllum* Wall. (Ranunculaceae), *Coleus vettiveroides* K.C. Jacob (Labiatae/ Lamiaceae), *Oxalis corniculata* Linn. (Oxalidaceae), *Pterocarpus santalinus* Linn. (Fabaceae), *Lepidium sativum* Linn. (Brassicaceae), *Plumbago zeylanica* Linn. (Plumbaginaceae), *Woodfordia fructicosa* Kurtz. (Lythraceae), *Anogeissus latifolia* Wall. Ex Bedd. (Combretaceae), *Scindapsus officinalis* Schott. (Araceae), *Callicarpa macrophylla* Vahl. (Verbenaceae), Cows urine, *Adiantum lunulatum* Brum. (Polypodiaceae), *Terminalia chebula* Retz. (Combretaceae), *Cuminum cyminum* Linn. (Umbelliferae/ Apiaceae), *Strychnos potatorum* Linn. (Loganiaceae), *Ichnocarpus fruitescens* R. Br. (Apocynaceae), *Holarrhena antidysenterica* Wall. (Apocynaceae), *Mimosa pudica* Linn. (Leguminosae/ Mimosaceae), *Symplocos racemosa* Roxb. (Symplocaceae), Honey, *Rubia cordifolia* Linn. (Rubiaceae), *Curculigo orchioides* Gaertn. (Amaryllidaceae), *Cyperus rotundus* Linn. (Cyperaceae), *Ficus bengalensis* Linn. (Moraceae), *Cissampelos pareira* Linn.(Menispermaceae), *Berberis aristata* D C. (Berberidaceae), *Asparagus racemosus* Willd. (Liliaceae), *Cuminum cyminum* Linn. (Umbelliferae/Apiaceae),	Antibiotics: doxycycline, cotrimoxazole, erythromycin, tetracycline, chloramphenicol, ciprofloxacin, etc.

TABLE 13.1   *(Continued)*

Sl. no.		Ayurveda perspective	Modern biology perspective
4	Disease	Kshaya	AIDS (acquired immunodefi- ciency syndrome), Tuberculosis (any chronic conditions with cachexia can be consid- ered under this)
	Etiological agent		AIDS virus, *Mycobacterium tubercu- losis,* any other infections
	Treatment drug	*Spondias mangifera* Willd. (Anacardiaceae), Mica, *Sesbania grandi- flora* Pers. (Fabaceae), *Alangium lamarckii* Thwaites (Alangiaceae), *Terminalia Arjuna* (Roxb.) Wt. & Arn. (Combretaceae), *Withania somnifera* (Linn.) Dunal (Solanaceae), *Dipterocarpus alatus* Roxb. (Dipterocarpaceae), *Coccinia indica* Wight. & Arn. (Cucurbitaceae), *Piper chaba* Hunter (Piperaceae), *Vitis vinifera* Linn. (Vitaceae), *Gmelina arborea* Roxb. (Verbenaceae), sulphur, gypsum, zircon (cinnamon stone), *Luffa echinata* Roxb. (Cucurbitaceae), *Bauhinia variegata* Linn. (Fabaceae), steel iron, *Saccharum spontanum* Linn. (Gramineae/Poaceae), calcium carbonate, *Pistacia integerrima* Stew. (Anacardiaceae), *Phoenix dacty- lifera* Linn. (Palmae), zinc carbonate, *Bauhinia purpurea* Linn. (Fabaceae), *Syzygium aromaticum* (Linn.) Merr. & Per. (Myrtaceae), ruby, honey, realgar, pearl (calcium carbonate), *Raphanus sativus* Linn. (Brassicaceae), *Piper longum* Linn. (Piperaceae), *Mimusops hexandra* Roxb. (Sapotaceae), silver (argentum), *Desmodium gangeticum* D C. (Fabaceae), *Asparagus racemosus* Willd. (Liliaceae), *Borassus flabellifer* Linn. (Arecaceae/ Palmae), *Abies webbiana* Lindl. (Pinaceae), *Adhatoda vasica* Nees. (Acanthaceae), *Glycerrhiza glabra* Linn. (Fabaceae)	

TABLE 13.1 *(Continued)*

Sl. no.		Ayurveda perspective	Modern biology perspective
5	Disease	Swasa roga	Anthrax, aspergillosis, bacterial pneumonia, viral pneumonia
			Group A & B streptococcal infection, pneumonia
	Etiological agent		*Bacillus anthracis, Aspergillus* sp., *Streptococcus pneumoniae, Klebsiella pneumoniae,* adenoviruses etc.
	Treatment drug	*Emblica officinalis* Gaertn. (Euphorbiaceae), *Zingiber officinale* Rosc. (Zingiberaceae), *Aquilaria agallocha* Roxb. (Thymelaeceae), goat's urine, *Garcinia pedunculata* Roxb. (Clusiaceae), *Calotropis gigantea* R. Br.Ex Ait. (Asclepiadaceae), *Holostemma rheedianum* Spreng. (Asclepiadaceae), *Dipterocarpus alatus* Roxb. (Dipterocarpaceae), *Psoralia corylifolia* Linn. (Leguminosae/Fabaceae), sheep urine, *Psoralia corylifolia* Linn. (Leguminosae/Fabaceae), *Citrus maxima* (Burm.) Merrill (Rutaceae), *Phyllanthus niruri* Linn. (Euphorbiaceae), *Eclipta alba* Hassk. (Compositae/Asteraceae), *Terminalia bellirica* Roxb. (Combretaceae), *Coccinia indica* Wight. & Arn. (Cucurbitaceae), *Solanum indicum* Linn. (Solanaceae), *Cassia tora* Linn. (Fabaceae), *Angelica archangelica* Linn. (Apiaceae), *Piper chaba* Hunter (Piperaceae), *Plumbago zeylanica* Linn. (Plumbaginaceae), *Angelica glauca* Edgw. (Umbelliferae/Apiaceae), *Coriandrum sativum* Linn. (Umbelliferae/Apiaceae), *Vitis vinifera* Linn. (Vitaceae), *Eletalaria cardamomum* Maton. (Zingiberaceae),	Antibiotics like amoxicillin, clarithromycin, azithromycin, fluoroquinolones etc.

TABLE 13.1 *(Continued)*

Sl. no.	Ayurveda perspective	Modern biology perspective
	Tribulus terrestris Linn. (Zygophyllaceae), cow's urine, *Terminalia chesbula* Retz. (Combretaceae), *Ferula foetida* Regel (Umbelliferae/Apiaceae), *Lagenaria vulgaris* Ser. (Cucurbitaceae), *Citrullus colocynthis* Schrad. (Cucurbitaceae), *Myristica fragrans* Houtt. (Myristicaceae), *Solanum surattense* Burm.F. (Solanaceae), *Curcuma zedoaria* Rosc., *Pistacia integerrima* Stew. (Anacardiaceae), *Costus speciosus* (Koeing) Sm. (Zingiberaceae), *Swertia chirata* (Roxb.Ex.Flem.) Karst. (Gentianaceae), *Ichnocarpus fruitescens* R. Br. (Apocynaceae), *Luffa acutangula* Roxb. (Cucurbitaceae), *Syzygium aromaticum* (Linn.) Merr. & Per. (Myrtaceae), *Madhuca longifolia* (Koen.), Honey, *Melia azedarach* Linn. (Meliaceae), *Piper nigrum* Linn. (Piperaceae), *Gymnema sylvestre* R. Br. (Asclepiadaceae), *Raphanus sativus* Linn. (Brassicaceae), Hart's horn, *Piper longum* Linn. (Piperaceae), Cassia *tora Linn.* (Fabaceae), *Inula racemosa* Hook.F. (Compositae/Asteraceae), *Allium sativum* Linn. (Liliaceae/Alliaceae), *Hedychium spicaticum* Ham.Ex Smith. (Zingiberaceae), *Zingiber officinale* Rosc. (Zingiberaceae), *Alstonia scholaris* R. Br. (Apocynaceae), *Marsilia minuta* Linn. (Marsileaceae), *Curcuma angustifolia* Roxb. (Zingiberaceae), *Ocimum sanctum* Linn. (Labiatae/Lamiaceae), *Adhatoda vasica* Nees. (Acanthaceae), *Bambusa arundinacea* (Retz.)Willd. (Poaceae), *Hordeum vulgare* Linn. (Gramineae/Poaceae)	

TABLE 13.1   *(Continued)*

Sl. no.		Ayurveda perspective	Modern biology perspective
6	Disease	Kushta	Leprosy and other skin disorders
	Etiological agent		Bacteria, fungus
	Treatment drug	*Ipomoea reniformis* Chois. (Convolvulaceae), *Merremia emarginata* (Burm. F.) Hallier F. (Convulvulaceae), *Emblica officinalis* Gaertn. (Euphorbiaceae), *Cassia fistula* Linn (Caesalpiniaceae), *Barleria strigosa* Willd. (Acanthaceae), *Cassia auriculata* Linn. (Leguminaceae/Fabaceae), Mica, *Aquilaria agallocha* Roxb. (Thymelaeceae), *Prunus cerasus* Linn. (Rosaceae), *Dipterocarpus turbinatus* Gaertn. (Dipterocarpaceae), *Laccifera lacca* (Kerr.) Lacciferidae, *Clitoria ternatea* Linn. (Fabaceae), *Acacia leucophloea* Willd. (Leguminosae/ Mimosaceae), *Calotropis procera* (Ait.) R.Br. (Asclepiadaceae), *Holostemma rheedianum* Spreng. (Asclepiadaceae), *Pterocarpus marsupium* Roxb. (Leguminosae/Fabaceae), *Linum usitatissimum* Linn. (Linaceae), *Psoralia corylifolia* Linn. (Leguminosae/ Fabaceae), *Semicarpus anacardium* Linn.F. (Anacardiaceae), *Betula utilis* D. Don (Betulaceae), *Eclipta alba* Hassk. (Compositae/Asteraceae), *Commiphora myrrha* (Nees) Engl. (Burseraceae), Bacopa monnieri (Linn.) Pennell (Scrophulariaceae), *Solanum indicum* Linn. (Solanaceae), *Oxalis corniculata* Linn. (Oxalidaceae), *Cassia tora* Linn. (Fabaceae), Angelica *archangelica* Linn. (Apiaceae), *Holoptelea integrifolia* Planch. (Ulmaceae), *Plumbago zeylanica* Linn. (Plumbaginaceae), *Angelica glauca* Edgw. (Umbelliferae/Apiaceae), *Berberis aristata* D C. (Berberidaceae),	Antibiotics

TABLE 13.1 *(Continued)*

Sl. no.		Ayurveda perspective	Modern biology perspective
		Baliospermum montanum Muell-Arg. (Euphorbiaceae), *Cedrus deodara* (Roxb.) Loud. (Pinaceae), *Datura metel* Linn. (Solanaceae), *Cynodon dactylon* (Linn.) Pers. (Gramineae/Poaceae), *Euphorbia hirta* Linn. (Euphorbiaceae), *Brassica campestris* Linn. Var Rapa (Linn.) Hartm. (Brassicaceae), *Tinospora cordofolia* (Willd.) Miers. (Meninspermaceae), *Commiphora mukul* Wight (Arn.) Bhandari (Burseraceae), *Abrus precatorius* Linn. (Fabaceae), *Terminalia chebula* Retz. (Combretaceae), *Curcuma longa* Linn. (Zingiberaceae), *Pongamia pinnata* Pierre (Leguminosae/Fabaceae), *Brassica campestris* Linn. Var Rapa (Linn.) Hartm. (Brassicaceae), chalcopyrite, iron pyrite, Euphorbia *Thomsoniana Boiss.* (Euphorbiaceae), copper (cuprum), *Lagenaria siceraria* (Mol.) Standl. (Cucurbitaceae), *Sesamum indicum* Linn. (Pedaliaceae), *Ougeinia dalbergioides* Benth. (Leguminaceae/Fabaceae), *Ocimum sanctum* Linn. (Labiatae/ Lamiaceae), *Hydnocarpus laurifolia* (Dennst.) Sleumer. (Flacourtiaceae), fluorspar, *Embelia ribes* Brum.F. (Myrsinaceae), *Hordeum vulgare* Linn. (Gramineae/Poaceae), *Alhagi pseudalhagi* (Bieb.) Desv. (Leguminosae/Fabaceae)	
7	Disease	Shotha	Cellulitis
	Etiological agent		*Staphylococcus* sp., *Streptococcus* sp.
	Treatment drug	*Emblica officinalis* Gaertn. (Euphorbiaceae), *Zingiber officinale* Rosc. (Zingiberaceae), *Premna mucronata* Roxb. (Verbenaceae), goat's urine, *Alangium lamarckii* Thwaites (Alangiaceae), *Clitoria ternatea* Linn. (Fabaceae), *Calotropis procera* (Ait.) R.Br. (Asclepiadaceae), *Withania somnifera* (Linn.) Dunal (Solanaceae), *Psoralia corylifolia* Linn. (Leguminosae/Fabaceae),	

TABLE 13.1 *(Continued)*

Sl. no.		Ayurveda perspective	Modern biology perspective
		Clerodendron serratum Spreng. (Verbenaceae), *Semicarpus anacardium* Linn.F. (Anacardiaceae), *Eclipta alba* Hassk. (Compositae/Asteraceae), *Aegle marmelos* Corr. (Rutaceae), *Bacopa monnieri* (Linn.) Pennell (Scrophulariaceae), *Solanum indicum* Linn. (Solanaceae), *Santalum album* Linn. (Santalaceae), *Plumbago zeylanica* Linn. (Plumbaginaceae), *Berberis aristata* D C. (Berberidaceae), *Baliospermum montanum* Muell-Arg. (Euphorbiaceae), *Cedrus deodara* (Roxb.) Loud. (Pinaceae), *Croton tiglium* Linn. (Euphorbiaceae), *Ricinus communis* Linn. (Euphorbiaceae), *Clitoria ternatea* Linn. (Fabaceae), cow's urine, *Commiphora mukul* Wight (Arn.) Bhandari (Burseraceae), *Terminalia chebula* Retz. (Combretaceae), *Curcuma longa* Linn. (Zingiberaceae), *Hygrophila spinosa* T. And. (Acanthaceae), *Solanum nigrum* Linn. (Solanaceae), *Corchorus capsularis* Linn. (Tiliaceae), steel iron, *Nelumbo nucifera* Gaertn. (Nymphaeaceae), *Solanum surattense* Burm.F. (Solanaceae), *Pongamia pinnata* Pierre (Leguminosae/Fabaceae), *Acacia catechu* Willd. (Leguminosae/Mimosaceae), *Swertia chirata* (Roxb. Ex.Flem.) Karst. (Gentianaceae)	
8	Disease	Masoorika	Smallpox (Variola), measles, chickenpox
	Etiological agent		Varicella-zoster, variola major and variola minor, measles virus
	Treatment drug	*Punica granatum* Linn. (Punicaceae), *Vitis vinifera* Linn. (Vitaceae), *Citrus limon* Linn. (Rutaceae), *Momordica charantia* Linn. (Cucurbitaceae), *Musa paradisiaca* Linn. (Musaceae), *Momordica dioica* Roxb. (Cucurbitaceae), *Trichosanthes dioica* Roxb. (Cucurbitaceae), *Moringa pterigosperma* Gaertn. (Moringaceae)	

TABLE 13.1   *(Continued)*

Sl. no.		Ayurveda perspective	Modern biology perspective
9	Disease	*Jwara, vata jwara, vishama jwara, vata raktha, ama vata*	Plague, meningitis, epidemic typhus, malaria, *influenzae*, chikungunya, rheumatoid fever etc.
	Etiological agent		Various microorganisms including bacteria, virus, fungi and protozoa
	Treatment drug	*Cassia fistula* Linn (Caesalpiniaceae), *Barleria strigosa* Willd. (Acanthaceae), *Michelia champaka* Linn. (Magnoliaceae), *Lepidium sativum* Linn. (Brassicaceae), *Ipomoea reniformis* Chois. (Convolvulaceae), *Merremia emarginata* (Burm. F.) Hallier F. (Convulvulaceae), *Cassia fistula* Linn (Caesalpiniaceae), *Gynandropsis gynandra* (Linn.) Briquet (Capparidaceae/Capparaceae), *Laccifera lacca* (Kerr.) Lacciferidae, *Sapindus trifoliatus* Linn. (Sapindaceae), *holostemma rheedianum* Spreng. (Asclepiadaceae), *Dipterocarpus alatus* Roxb. (Dipterocarpaceae), *Aconitum heterophyllum* Wall. (Ranunculaceae), *Psoralia corylifolia* Linn. (Leguminosae/Fabaceae), *Clerodendron serratum* Spreng. (Verbenaceae), *Semicarpus anacardium* Linn.F. (Anacardiaceae), *Terminalia bellirica* Roxb. (Combretaceae), *Coccinia indica* Wight. & Arn. (Cucurbitaceae), *Commiphora myrrha* (Nees) Engl. (Burseraceae), *Bacopa monnieri* (Linn.) Pennell (Scrophulariaceae), *Solanum indicum* Linn. (Solanaceae), *Oxalis corniculata* Linn. (Oxalidaceae), *Coscinium fenestratum* (Gaertn.) Colebr. (Menispermaceae), *Pterocarpus santalinus* Linn. (Fabaceae), *Angelica glauca* Edgw. (Umbelliferae/Apiaceae), *Punica granatum* Linn. (Punicaceae),	

TABLE 13.1   *(Continued)*

Sl. no.	Ayurveda perspective	Modern biology perspective
	Cedrus deodara (Roxb.) Loud. (Pinaceae), *Coriandrum sativum* Linn. (Umbelliferae/Apiaceae), *Datura metel* Linn. (Solanaceae), *Vitis vinifera* Linn. (Vitaceae), *Fagonia arabica* Linn. (Zygophyllaceae), *Ricinus communis* Linn. (Euphorbiaceae), *Cucumis utilissimus* Roxb. (Cucurbitaceae), *Tinospora cordofolia* (Willd.) Miers. (Meninspermaceae), *Abrus precatorius* Linn. (Fabaceae), *Hibiscus rosa-Sinensis* Linn. (Malvaceae), *Luffa echinata* Roxb. (Cucurbitaceae), *Cuminum cyminum* Linn. (Umbelliferae/Apiaceae), *Peristrophe bicalculata* Nees. (Acanthaceae), *Solanum nigrum* Linn. (Solanaceae), *Momordica charantia* Linn. (Cucurbitaceae), *Solanum surattense* Burm.F. (Solanaceae), *Momordica dioica* Roxb. (Cucurbitaceae), *Costus speciosus* (Koeing) Sm. (Zingiberaceae), *Acacia catechu* Willd. (Leguminosae/Mimosaceae), *Phoenix dactylifera* Linn. (Palmae), *Swertia chirata* (Roxb.Ex.Flem.) Karst. (Gentianaceae), *Ichnocarpus fruitescens* R. Br. (Apocynaceae), *Symplocos racemosa* Roxb. (Symplocaceae), *Centella asiatica* (Linn.) Urban (Umbelliferae/Apiaceae), *Rubia cordifolia* Linn. (Rubiaceae), *Polygonatum cirrifolium* Royle (Alliaceae), *Marsdenia tenacissima* W. & A. (Asclepiadaceae), *Selinum tenuifolium* Wall. Ex C. B. Clarke (Umbelliferae/Apiaceae), *Azadiracta indica* A. Juss. (Meliaceae), *Cissampelos pareira* Linn.(Menispermaceae), *Trichosanthes dioica* Roxb. (Cucurbitaceae), *Caesalpinia sappan* Linn. (Caesalpiniaceae), *Piper longum* Linn. (Piperaceae), *Inula racemosa* Hook.F. (Compositae/Asteraceae), *Allium sativum* Linn. (Liliaceae/Alliaceae),	

TABLE 13.1 *(Continued)*

Sl. no.	Ayurveda perspective	Modern biology perspective
	Cuminum cyminum Linn. (Umbelliferae/ Apiaceae), *Adhatoda vasica* Nees. (Acanthaceae), *Ricinus communis* Linn. (Euphorbiaceae), *Tinospora cordofolia* (Willd.) Miers. (Meninspermaceae), *Piper longum* Linn. (Piperaceae)	

Understanding the microbial diseases and their treatments from an Ayurvedic perspective can help in developing better holistic and integrated infection management strategies. However, the challenges remain with epistemological differences between *Ayurveda* and modern biology. *Ayurveda* describes the involvement of 'supernatural' and 'superhuman' entities (referred as *Graha, Bhootha, Rakshasa* etc.) as causative agents for various diseases particularly in diseases of children.

With careful analysis of the aetiology and symptoms of these diseases from a modern scientific viewpoint, the words '*Graha*', '*Bhuta*' etc. can be interpreted as invisible substances that may cause diseases in human beings. These include primarily microorganisms plus substances like enzymes, proteins, allergens and so on. The lack of advanced research and diagnostic techniques in the ancient Ayurvedic period might have resulted in giving a 'supernatural' hue to the various invisible disease-causing agents. Taking *Garha* as a special example, Table 13.2 summarises the classification of '*Garha*' according to *Ayurveda* with symptoms that they are known to cause. All the symptoms caused by *Garha* can be correlated to various microbial infections. A recent review by Kumar et al. from India had compared the male and female *Grahas* as Gram-positive and Gram-negative microbes.[17] Though the comparison seems to be logical, it is too premature to accept the claim and needs detailed studies to establish it. Even though the role of 'supernatural' agents in diseases pathogenesis is considered in *Ayurveda*, the management strategies are again focusing on restoring the *dosha* balance through lifestyle modifications and medications. As disease is considered as the cause of impartation of vitiated qualities of *doshas* on *dhatus*, treating microbial diseases through restoring *dosha* homeostasis could be considered as a process of empowering the body's immune system to fight the infections. *Ayurveda* has special emphasis on healthy lifespan that prepares the body to fight against diseases rather than targeted killing of the pathogens.

TABLE 13.2 Different Types of *Graha* and Various Symptoms Caused by Them. Ayurvedic classification of *Graha* and various disease symptoms that are caused by each *Graha*. The word *Graha* and the diseases caused by them are symptomatically correlated to various microorganisms and microbial infections. However, it is important to note that the same symptoms can also result from noninfectious diseases.

Name	Indicated symptoms comparable to microbial infections
Male grahas	
Skanda	Numbness, stiffness, sweating, vomiting, clenched fist
Visakha	Uncontrolled passage of stools and urine, froth from mouth, fever
Mesakhya	Distension of abdomen, froth from mouth, clenching fist, diarrhoea, cough, vomiting, fever, swelling of eye
Svagraha	Tremor, sweating
Pitrugraha	Fever, cough, diarrhoea, vomiting, clenching fist, stiffness
Female grahas	
Sakuni	Diarrhoea, ulcers of throat and tongue, joint pains, mouth ulcers, fever
Putana	Vomiting, diarrhoea, distensions of abdomen
Sitaputana	Abdominal gurgling, diarrhoea
Andhaputana	Vomiting, fever, cough, diarrhoea, discolouration, swelling, itching
Mukhamandita	Fever, loss of appetite, fatigue
Revati	Dark blue discolouration, cough, fever, green and watery faeces
Suskarevati	Emaciation of whole body

13.3 ATTEMPTS TO BRIDGE THE MICROBIAL DISEASES IN MODERN MEDICINE AND AYURVEDA—SELECTED EXAMPLES

A successful trans-disciplinary approach for bridging the concepts of microbiology in modern medicines and *Ayurveda* need to focus on the physiological and immunological aspects of diseases. However, unfortunately most of the attempts are focusing on understanding and/or establishing the antimicrobial effects of the prescribed Ayurvedic formulations and their component plants. Screening medicinal plants for antimicrobial activity has enormous potential in drug discovery through the identification of novel antimicrobial molecules, but has limited scope in understanding how *Ayurveda* cures infections. Nevertheless, such studies are important in understanding the molecular mechanism of antimicrobial action of medicinal plants as well as to understand the nature and type of bioactive molecules present in the plant extract. Scientists have also attempted to understand the various microbiological and antimicrobial concepts related to *Ayurveda* with an interest to correlate them with contemporary microbiology knowledge. The article highlights some of the important studies carried out in this direction.

13.3.1 ANTIVIRAL EFFECTS OF AYURVEDIC FORMULATIONS AND MEDICINAL PLANTS

Like any other disease of global importance, viral infections also demand the search of new antiviral compounds. The available antiviral drugs are often unsatisfactory due to the problems of viral resistance problems as well as viral latency and conflicting efficacy in recurrent infection in immunocompromised patients. Medicinal plants provide alternative sources to identify novel antiviral agents. One of the recent studies reported the potential use of *Cissampelos pariera* Linn (Cipa extract) against dengue virus where the study showed the Cipa extract could decrease viral titters as well as downregulate the production of TNF-α, a cytokine implicated in severe dengue disease. The study w carried out in both in vitro and in vivo models.[28] Similarly, the methanol bark extract of *Ficus religiosa*, used in traditional Ayurvedic and Unani medicine for treating respiratory tract infections, was found to be active against human rhinovirus. While the water extract of the same plant was found to be active against human respiratory syncytial virus.[4] *Ashwagandha* (*Withania somnifera*), an important plant widely used in Ayurvedic medicine as a nerve tonic and memory enhancer, was found to reverse β-amyloid-42-induced toxicity in human neuronal cells indicating its implications in HIV-associated neurocognitive disorders.

The plant is also reported to have direct antiviral effects as well.[18] Neem (*Azadirachta indica*) is another wonder plant having diverse antimicrobial activity and studies have shown that the polysaccharides obtained from *A. indica* act against poliovirus (PV-1) by inhibiting the initial stage of viral replication.[5] Apart from herbal components of various formulations, Ayurvedic concepts such as *Rasayana* are also reported to have beneficial effect in treating syndromes like AIDS. *Rasayana* is one of the eight disciplines described in *Ayurveda* that deals with dietary recipes and regimen, herbal and mineral supplements and health-promoting lifestyle that are said to enhance the quality of life and delay ageing. A clinical study carried out using *Shilajitu Rasayana* showed that it decreases the recurrent resistance of HIV virus to antiretroviral therapy and improves the outcome of the therapy.[7]

13.3.2 ANTIBACTERIAL EFFECT OF AYURVEDIC FORMULATIONS AND MEDICINAL PLANTS

Though antibacterial therapy is relatively easier than antiviral, there is still a huge demand for identifying better bioactive molecules and formulations for controlling infections. Particularly, the demand is very high with respect to various drug-resistant species of bacteria. *Triphala* is a well-known preparation in *Ayurveda* consisting of *Emblica officinalis*, *Terminalia chebula* and *Terminalia belerica*. It is been shown to have a wide range of biological activities including antimicrobial activity. It is well known for its applications in dentistry.[29,24,19,16] Oral rinses made from these are used in periodontal therapy. As an antimicrobial plant, *E. officinalis* is known to be effective against *E. coli*, *Klebsiella ozaenae*, *Klebsiella pneumoniae*, *Proteus mirabilis*, *Pseudomonas aeruginosa*, *S. paratyphi A*, *S. paratyphi B* and *Serratia marcescens* etc.[27] *Triphala* is one of the wide spectrum antimicrobial agents wherein *Sushruta Samhita*, one of the foundation texts of *Ayurveda*, recommends its use as a gargling agent in dental diseases. *Triphala* mouthwash (0.6%) has been shown to have significant anti-caries activity, which is comparable to that of chlorhexidine without possessing any disadvantages as staining of teeth.[25]

Similarly, *Caesalpinia bonduc* (Lin.) Roxb. is a known drug in *Ayurveda* to treat various diseases specifically tumours, cysts and cystic fibrosis (CF). It is been studied for its antimicrobial activities using seed coat and seed kernel extracts. Results showed that *C. bonduc* can have the potential to be developed as a promising natural medicine, with other forms of treatments, for CF patients with chronic *P. aeruginosa* lung infections.[2] *Laghupanchamula* is another important formulation in *Ayurveda* prepared as a combination of the roots of five herbs. In Ayurvedic classics, besides four common herbs, that is *Kantakari, Brihati, Shalaparni* and *Prinshniparni,* the fifth one is either *Gokshura* (LPG) or *Eranda* (LPE), and both formulations have been documented to have wound healing (Vruna) activity as well as antimicrobial activity against skin pathogens.[6] Another study has shown a synergistic effect of (+)-pinitol from *Saraca asoca* with β-lactam antibiotics. Similarly, another study showed that aqueous extracts of raw seeds of Guñjā exert its antibacterial effect on both Gram-positive, as well as Gram-negative bacteria but none of the Śodhita Gunja (purified) seeds showed any bactericidal effect on any bacterial strains.[1]

13.3.3   ANTIFUNGAL AND ANTIPROTOZOAN EFFECTS OF AYURVEDIC FORMULATIONS AND MEDICINAL PLANTS

Similar to viral and bacterial infections, fungal and protozoal infections are also contributing to the major health concerns in the infectious diseases field. Although there are successful antifungal and antiprotozoal agents, such as fluconazole, itraconazole, eflornithine, furazolidone etc., several unpleasant side effects, slow therapeutic responses and poor efficacy necessitates the discovery of novel antifungal and antiprotozoal agents. Phytochemicals and medicinal plants have shown very successful results in discovering such agents and that these could be a better source of medicine as compared to synthetically produced drugs.

The famous discovery of Artemisinin from Chinese traditional medical plant *Artemisia annua* is a classic example in this context. Although studies highlighting the antifungal and antiprotozoal effects of Ayurvedic medicinal plants are less compared to antiviral and antibacterial studies, some of the reported studies have significant impact in the area of fungal and protozoal disease management. One of the recent review articles highlights the diverse potentials of Bergenin, one of the active ingredients in herbals and Ayurvedic formulations and has been analysed and estimated in different plant extracts, having very good antifungal, antiplasmodial and antiviral effects.[3] Similarly, *Ocimum sanctum,* the holy basil that is well known for medicinal properties, is shown to have anticandidial effects where the *O. sanctum* essential oil having a synergistic action with fluconazole and ketoconazole. Another recent study had reported four Ayurvedic formulations namely *Yastimadhu choorna, Amalaki choorna* and *Asanadi kwatha* for antifungal activity against *Aspergillus niger,*

Aspergillus oryzae and *Mucor sp.* wherein they showed differential, but significant activity against all the fungal species studied.[10] The aerial parts and roots of *Murraya koenigii* (a plant of homestead gardens known as curry leaf) are shown to have antiprotozoal effect. Likewise, the stem bark of *Ficus sp., Ricinus communis* (Castor), *Curcuma longa* (turmeric), *Coccinia grandis*, is shown to have antiprotozoan activity.[9,13]

13.3.4 CONCEPT OF PROBIOTICS IN AYURVEDA

Apart from studying merely the antimicrobial activity of medicinal plants, researchers had also attempted to understand various other concepts related to microbiology in *Ayurveda*. One among them is the concepts of probiotics in *Ayurveda*. Probiotics are live bacteria and yeasts that are good for your health, especially your digestive system. Probiotics are often called 'good' or 'helpful' bacteria because they help to keep our gut healthy and are naturally found in our body. Generally, the use of antibiotics or similar substances will affect the 'good' bacterial population also along with the bad ones. Probiotics are good choices in refurbishing the good microbes in the gut. It is very important also, as gut microflora had shown to influence various biological activities in the body. It is well established in modern medicine that the gut microbes play several important roles in body's homeostasis and the overuse and abuse of antibiotics and other drugs are reported to alter and disturb the normal gut microflora. This leads to an increase in harmful microorganisms in the intestine, particularly pathogenic bacteria and yeast, and can end up with serious diseases such as irritable bowel syndrome, colitis and so on.

World Health organization (WHO) define probiotics as live microorganisms which when administered in adequate amounts confer a health benefit on the host and are accepted as nutraceuticals or dietary supplements. According to *Ayurveda*, *Takra* (butter milk or yogurt) is a great probiotic and is an adjunct food. Consumption of *Takra* is a simple and effective remedy as per *Ayurveda*. *Takra* has less fat compared to milk and is rich in calcium, potassium and vitamin B12. Hence in *Ayurveda* prime importance is given to buttermilk. Here again, the word 'probiotic' or microorganisms are not described in the Ayurvedic texts, but had given immense importance to the use of milk and fermented milk products for several health benefits.[26,15] *Ayurveda* recommends curd as a good therapeutic for diarrhoea and other acute/chronic gastero-intestinal disorders from time immemorial.[14] Another study had also observed that nutritional interventions and ingestion of probiotics assist the parasite clearance in giardiasis and highlights Ayurvedic formulation *Pippali rasayana* currently have the most clinical evidence supporting their use.[8]

13.3.5   FUMIGATION—A NOVEL ANTIMICROBIAL STRATEGY FROM AYURVEDA

Ayurveda is also known for unique methodologies for disease management. Fumigation is a therapeutic procedure in *Ayurveda*, wherein the drug is administered in a gaseous state. This is described as *Dhoopana* (*Dhoopa* = fumes; *Dhoopana* = fumigation) or *Dhumapana* (*Dhuma* = fume; *Dhumapana* = inhalation of fume) in *Ayurveda*. Fumigation is used for treating various types of diseases including infections of ear, nose and throat, and diseases of the female reproductive system, paediatric diseases and other common symptoms such as fever, asthma and cough. The spectrum covers microbial infections as well.[31] Besides *Ayurveda*, some other international studies also support the potentials of herbal fumigation as a potential antimicrobial strategy. Rice hull smoke was found to inactivate *Salmonella typhimurium* in laboratory media and to protect infected mice against mortality.[11,12]

In *Ayurveda*, herbal and non-herbal materials such as *Acorus calamus* L., *Abrus precatorius* L., *Ferula assa-foetida* L., *Brassica rapa* L. (*Brassica campestris* L.), *Hordeum vulgare* L., clarified butter, cast off skin of snake, hairs of camel, goat, sheep or cow are used in the preparation of medicated fumes. Similar materials, with several additions, are used for diseases such as fever, asthma, cough etc. Similarly herbal and non-herbal materials such as *Linum usitatissimum* L., *Commiphora mukul* (Hook.ex Stocks) Engl., *Acorus calamus* L. *Angelica glauca* Edgew., *Bacopa monnieri* (L.) Wettst., *Saraca indica* L., *Hordeum vulgare* L., *Brassica rapa* L. slough of snake mixed with *Ghee* are used for sterilization of cloths and bed spreads used for infants and children, which also can be considered as a technique of removal of potential microbial (bacterial, viral, fungal, protozoan) as well as insect contaminations.[31]

13.4 CONCLUSION

Microbiology is one of the most important and fascinating fields of biology and there is no life without microorganisms. However, at the same time it also represents the deadliest agents of human diseases. The importance of microorganisms in human diseases justifies the increasing research interest in understanding the pathogenesis of infections as well as in developing better therapeutic/management strategies for microbial infections. The holistic and patient centred therapeutic approaches in complementary and alternative medicines like *Ayurveda* can offer a lot in the area of microbial infections. *Ayurveda*, being a science originated and established long before the development of modern microbiology, does not have specific references to microorganisms like bacteria or virus. Nevertheless, diseases manifestations that are now understood as microbial infections were described in *Ayurveda* along with their ethology, progression, prevention and cure. These descriptions in *Ayurveda* are purely patient centred and the management strategies are, thus, emphasizing more on restoring the physiological homeostasis and wellness of the patient, rather than a targeted destruction of microbes. A logical and constructive synergism of the concepts of infections in *Ayurveda* and modern medicine can uncover novel concepts and strategies in understanding and managing microbial infections.

13.5 ACKNOWLEDGEMENTS

The author acknowledges Dr. Subramanya Kumar, assistant professor, School of Life Sciences, Transdisciplinary University (TDU), Bangalore for providing valuable suggestions in the manuscript preparation.

KEYWORDS

- *Ayurveda*
- microbiology
- tridosha
- bruhatrayis
- dosha Prakriti
- graha and Bhuta

REFERENCES

1. Ahmad, F.; Misra, L.; Gupta, V. K.; Darokar, M. P.; Prakash, O.; Khan, F.; Shukla, R. Synergistic Effect of (+)-Pinitol from *Saraca asoca* with β-Lactam Antibiotics and Studies on the in silico Possible Mechanism. *J. Asian Nat. Prod. Res.* **2016**, *18*, 172–183.

2. Arif, T.; Mandal, T. K.; Kumar, N.; Bhosale, J. D.; Hole, A.; Sharma, G. L.; Padhi, M. M.; Lavekar, G. S.; Dabur, R. In vitro and in vivo Antimicrobial Activities of Seeds of *Caesalpinia bonduc* (Lin.) Roxb. *J. Ethnopharmacol.* **2009**, *123*, 177–180.

3. Bajracharya, G. B. Diversity, Pharmacology and Synthesis of Bergenin and its Derivatives: Potential Materials for Therapeutic Usages. *Fitoterapia* **2015**, *101*, 133–152.

4. Cagno, V.; Civra, A.; Kumar, R.; Pradhan, S.; Donalisio, M.; Sinha, B. N.; Ghosh, M.; Lembo, D. Ficus Religiosa L. Bark Extracts Inhibit Human Rhinovirus and Respiratory Syncytial Virus Infection in vitro. *J. Ethnopharmacol.* **2015**, *176*, 252–257.

5. Faccin-Galhardi, L. C.; Yamamoto, K. A.; Ray, S.; Ray, B.; Carvalho Linhares, R. E.; Nozawa, C. The in vitro Antiviral Property of *Azadirachta indica* Polysaccharides for Poliovirus. *J Ethnopharmacol.* **2012**, *142*, 86–90.

6. Ghildiyal, S.; Gautam, M. K.; Joshi, V. K.; Goel, R. K. Wound Healing and Antimicrobial Activity of Two Classical Formulations of Laghupanchamula in Rats. *J. Ayurveda Integr. Med.* **2015**, *6*, 241–247.

7. Gupta, G. D.; Sujatha, N.; Dhanik, A.; Rai, N. P. Clinical Evaluation of Shilajatu Rasayana in Patients with HIV Infection. *Āyurvedāloka* **2010**, *31*, 28–32.

8. Hawrelak, J. Giardiasis: Pathophysiology and Management. *Altern. Med. Rev.* 2003, 8:129–142.

9. Husain, A.; Virmani, O. P.; Popli, S. P.; Misra, L. N.; Gupta, M. M.; Srivastava, G. N.; Abraham, Z.; Singh, A. K. *Dictionary of Indian Medicinal Plants;* CIMAP: Lucknow, India; 1992, p 546.

10. Kekuda, T. R. P.; Kavya, R.; Shrungashree, R. M.; Suchitra, S. V. Screening of Selected Single and Polyherbal Ayurvedic Medicines for Antibacterial and Antifungal Activity. *Ancient Sci. Life* **2010**, *29*, 22–25.

11. Kim, S. P.; Yang, J. Y.; Kang, M. Y.; Park, J. C.; Nam, S. H.; Friedman, M. Composition of Liquid Rice Hull Smoke and Anti-Inflammatory Effects in Mice. *J. Agric. Food Chem.* **2011**, *59*, 4570–4581.

12. Kim, S. P.; Kang, M. Y.; Park, J. C.; Nam, S. H.; Friedman, M. Rice Hull Smoke Extract Inactivates *Salmonella typhimurium* in Laboratory Media and Protects Infected Mice Against Mortality. *J. Food Sci.* **2012**, *77*, 80–85.

13. Kiso, Y.; Suzuki, Y.; Watanabe, N.; Oshima, Y.; Hikino, H. Antihepatotoxic Principle of *Curcuma longa*. *Planta Med.* 1983, 48:45–49.

14. Kore, K. B.; Pattanaik, A. K.; Sharma, K.; Mirajkar, P. P. Effect of Feeding Traditionally Prepared Fermented Milk Dahi (Curd) as a Probiotic on Nutritional Status, Hindgut Health and Haematology in Dogs. *Indian J. Tradit. Knowl.* **2012**, 11, 35–39.

15. Kukkupuni, S. K.; Shashikumar, A.; Venkatasubramanian, P. Fermented Milk Products: Probiotics of Ayurveda. *J. Med. Nutr. Nutraceuticals* **2015**, 4:14–21.

16. Kumar, M. S.; Kirubanandan, S.; Sripriya, R.; Sehgal, P. K. Triphala Promotes Healing of Infected Full-Thickness Dermal Wound. *J. Surg. Res.* **2008**, *144*, 94–101.

17. Kumar, S. H.; Prasad, G. G.; Neetu, S.; Jitesh, V. A Review on Clinical Application of the Concept of Balagrha in Modern Times. *J. Biol. Sci. Opin.* **2014**, *2*, 117–120.

18. Kurapati, K. R.; Samikkannu, T.; Atluri, V. S.; Kaftanovskaya, E.; Yndart, A.; Nair, M. P. β-Amyloid1-42, HIV-1Ba-L (clade B) Infection and Drugs of Abuse Induced Degeneration in Human Neuronal Cells and Protective Effects of Ashwagandha (*Withania somnifera*) and its Constituent Withanolide A. *PLoS One* **2014**, *9*, e112818.

19. Madani, A.; Jain, S. K. Anti-Salmonella Activity of *Terminalia belerica*: In vitro and in vivo Studies. *Indian J. Exp. Biol.* **2008**, *46*, 817–821.

20. Murthy, K. R. S. *Astanga Hrudayam*: (Text, English Translation, Notes, Appendix and Indices); Chowkhambha Krishnadas Academy: Varanasi, 2014.

21. Narayanaswamy, V. Origin and Development of Ayurveda: (A Brief History). *Ancient Sci. Life* **1981**, *1*, 1–7.

22. Pandey, M. M.; Rastogi, S.; Rawat, A.K. Indian Traditional Ayurvedic System of Medicine and Nutritional Supplementation. *Evidence-Based Complementary Alternat. Med.* 2013, **2013**, 376327.

23. Patwardhan, B. Bridging Ayurveda with Evidence-Based Scientific Approaches in Medicine. *EPMA J.* **2014**, *5*, 19.

24. Prabhakar, J.; Balagopal, S.; Priya, M. S.; Selvi, S.; Senthilkumar, M. Evaluation of Antimicrobial Efficacy of Triphala (an Indian Ayurvedic Herbal Formulation) and 0.2% Chlorhexidine Against Streptococcus Mutans Biofilm Formed on Tooth Substrate: An in vitro Study. *Indian J. Dent. Res.* **2014**, *25*, 475–479.

25. Prakash, S.; Shelke, A. U. Role of Triphala in Dentistry. *J. Indian Soc. Periodontol.* **2014**, *18*, 132–135.

26. Rajeshwari, P. N.; Verma, A.; Verma, J. P.; Pandey, B. B. Probiotics, in Ayurveda. *Int. Ayurvedic Med. J.* **2013**, *1*, 1–4.

27. Saeed, S.; Tariq, P. Antibacterial Activities of *Emblica officinalis* and *Coriandrum sativum* Against Gram Negative Urinary Pathogens. *Pak. J. Pharm. Sci.* **2007**, *20*, 32–35.

28. Sood, R.; Raut, R.; Tyagi, P.; Pareek, P. K.; Barman, T. K.; Singhal, S.; Shirumalla, R. K.; Kanoje, V.; Subbarayan, R.; Rajerethinam, R.; Sharma, N.; Kanaujia, A.; Shukla, G.; Gupta, Y. K.; Katiyar, C. K.; Bhatnagar, P. K.; Upadhyay, D. J.; Swaminathan, S.; Khanna, N. *Cissampelos pareira* Linn: Natural Source of Potent Antiviral Activity Against All Four Dengue Virus Serotypes. *PLoS Neglected Trop. Dis.* **2015**, *9*, e0004255.

29. Srinagesh, J.; Krishnappa, P.; Somanna, S. N. Antibacterial Efficacy of Triphala Against Oral Streptococci: An in vivo Study. *Indian J. Dent. Res.* **2012**, *23*, 696–701.

30. Tawalare, K. A.; Nanote, K. D.; Gawai, V. U.; Gotmare, A. Y. Contribution of Ayurveda in Foundation of Basic Tenets of Bioethics. *Āyurvedāloka* **2014**, *35*, 366–370.

31. Vishnuprasad, C. N.; Pradeep, N. S.; Cho, Y. W.; Gangadharan, G. G.; Han, S. S. Fumigation in Ayurveda: Potential Strategy for Drug Discovery and Drug Delivery. *J. Ethnopharmacol.* **2013**, *149*, 409–415.

CHAPTER 14

BACTERIA BIOFILMS AND THEIR IMPACT ON HUMAN HEALTH

MUKESH KUMAR YADAV*

*Department of Otorhinolaryngology, Head and Neck Surgery and Institute for Medical Device Clinical Trials, Korea University College of Medicine, Seoul, South Korea, *E-mail: mukiyadavgmail.com*

CONTENTS

ABSTRACT

Biofilm, a counterpart of the planktonic (free floating) form is a unique phenotype of bacterial growth in which bacteria adhere to each other, and to the biotic or abiotic surface. Bacterial biofilms have important implication in human health because these biofilms cause various chronic and recurrent infections in humans. National Institute of Health (NIH) has reported that more than 75% of microbial infections in human body occur due to biofilms. The bacteria within biofilms encased in extracellular polymeric substance (EPS) matrix have greatly reduced metabolic and divisional rates that confer resistance against antibiotic killing and host defense. Bacteria within biofilms are up to 1000 times more resistant to common antibiotics compared to planktonic type, which makes biofilm eradication even more challenging. The biofilm formation results in persistent infection, poor recovery, and higher cost of disease management. Bacterial biofilms are frequently found on indwelling medical implants such as prostheses, endotracheal tubes, stents and catheters that result in the replacement of medical implant or a second surgery, thus increasing cost and recovery time. In this chapter, we present an overview of bacterial biofilm formation on human tissue and medical devices, and their impact on human health.

14.1   INTRODUCTION

Bacteria exist in one of the two forms, in free-floating state called planktonic growth and in attached or aggregated forms. Bacteria attached to biotic or abiotic surfaces or to each other and surrounded by the self-produced extracellular polymeric substance (EPS) are called bacterial biofilms. Bacterial biofilms are composed of 75–95% EPS and only 5–25% bacteria. Biofilm mode of growth is considered a bacterial response for adaptation to hostile environments.[25,31] The biofilm lifestyle helps bacteria to survive and persist within the hostile environment. Bacterial biofilms are found in various types of habitats and cause infectious diseases in humans. According to the National Institutes of Health (NIH), more than 75% of microbial infections in human body occur due to biofilms. Cystic fibrosis (CF), native valve endocarditis, otitis media (OM), periodontitis, wound infections, superficial skin infections and chronic prostatitis all appear to be caused by biofilm-associated microorganisms. Moreover, bacterial biofilms have been detected on various implanted medical devices such as heart implants, bloodline catheters, peritoneal dialysis catheters, contact lenses, dental area caries and so forth.[2,12,46,53,82,19] Many researchers reported that the bacterial biofilms possess significantly enhanced antibiotic resistantance in comparison to their counterpart planktonic cells, and also possess resistance to the host immune defense.[37,36,34,35] Similarly, a significant higher minimum inhibitory concentration (MIC) as well as the minimum bactericidal concentration (MBC) have been detected for bacteria within biofilm (approximately 10–1000 times) in comparison to the planktonic bacteria.[37,34,35] Furthermore, the biofilms have greatly reduced metabolic and divisional rates.[45] Due to reduced metabolic and divisional rates, bacteria can escape antibiotic killing and failure of diagnosis by a conventional culture technique.[22] The bacteria also form poly-microbial biofilms that give rise to persistent and more virulent strains. Treatment of biofilm-related infection with antibiotics is a very effective measure for the control of bacterial infection; however, it is difficult to eradicate pre-established biofilms infections. Moreover, overdose of antibiotics give rise to biofilm progeny, and the situation is complicated by the emergence of antibiotic-resistant strains. Consequently, search for a new strategy and novel antimicrobial compounds that are safe to use and limit the emergence of antibiotic-resistant strains are required to combat biofilm-related infections. Many studies showed that development of antimicrobial compounds that could disrupt the biofilm bacteria (dormant bacteria) membrane bilayer of proteins could be one of the strategies.[39] Newly developed antibiotics such as daptomycin and telavancin have demonstrated considerable activity against bacterial biofilms.[30] Furthermore, many natural or plant-derived compounds such as tea tree oil (*Melaleuca alternifolia*), 220D-F2 obtained from plant *Rubus ulmifolius*, magnolol and carvacrol (a constituent of oregano) showed a significant effect against biofilms. However, their clinical application is not known.[88,9,92,63] In this chapter, we present an overview of bacterial biofilm formation on human tissues and medical devices and their impact on human health.

14.2 BIOFILMS ON MEDICAL INDWELLING DEVICES

The growth of bacterial biofilms on indwelling medical devices is a major problem in medical settings. It increases the cost of treatment and recovery time. With the advancement in medical science, many types of medical devices or artificial organs have been developed and their implantation has been increased to treat human diseases. This increases the frequency of biofilm-related infections. Bacterial biofilm formation has been detected on a vast majority if not all of the indwelling medical devices (Table 14.1). Both Gram-negative and Gram-positive bacteria can grow on medical devices and form biofilms. *Staphylococcus aureus, Staphylococcus epidermidis, Streptococcus viridans, Enterococcus faecalis,* along with *Pseudomonas aeruginosa, Escherichia coli, Klebsiella pneumoniae* and *Proteus mirabilis* have been isolated from indwelling medical devices. The bacteria may be originated from other infected patients, from patients' own skin and other parts, maybe transmitted from health care workers or from other environmental sources. Bacterial biofilms on the indwelling medical devices may be poly-microbial or single species, depending on the duration and insertion site in the patient's body. Many if not all of the prostheses or medical devices may result in biofilm-related infections, which include vascular prosthesis,[83] intravenous catheters,[84] cerebrospinal fluid shunts,[26] prosthetic heart valves,[19] urinary catheters,[19] joint prostheses and orthopedic fixation devices,[74] cardiac pacemakers,[72] peritoneal dialysis catheters,[14] intrauterine devices,[4] biliary tract stents,[18] dentures,[49] breast implants[67] and contact lenses.[1] The initial bacterial attachment to medical devices is reversible, and later the bacteria establish mature and irreversible biofilms. The biofilm establishment by *S. aureus* on biotic or abiotic surface consists of at least four stages (Fig. 14.1): the first step is initial attachment to a surface, and then aggregation of many bacteria or species and maturation step consists of EPS formation by bacteria within biofilm, and final step is dispersal.

TABLE 14.1   List of Bacteria and Biofilm Growth on Human Tissue and Medical Devices.

Bacteria	Biofilms human tissue or medical devices
Streptococcus pneumoniae, Haemophilus influenza and *Moraxella catarrhalis*	Tympanostomy tubes, cochlear implants or artificial ossicles
Staphylococcus aureus	Artificial hip prosthesis, artificial voice prosthesis
Escherichia Coli (E. coli), Pseudomonas aeruginosa	Urinary catheter
Coagulase-negative staphylococci	Central venous catheter, Intrauterine device, Prosthetic heart valve

TABLE 14.1 *(Continued)*

Bacteria	Biofilms human tissue or medical devices
Enterococcus spp.	Artificial hip prosthesis, central venous catheter, intrauterine device, prosthetic heart valve, urinary catheter
Klebsiella pneumonia	Central venous catheter, urinary catheter
P. aeruginosa	Artificial hip prosthesis, central venous catheter, urinary catheter
Staphylococcus aureus	Artificial hip prosthesis, central venous catheter, intrauterine device, prosthetic heart valve
P. aeruginosa	Contact lens
E. coli	Acute and recurrent urinary tract infection, biliary tract infection
P. aeruginosa	Cystic fibrosis lung infection, chronic wound infection, chronic rhinosinusitis, chronic otitis media
S. aureus	Chronic osteomyelitis, chronic rhinosinusitis, endocarditis, chronic otitis media
Staphylococcus epidermidis	Chronic osteomyelitis
S. pneumoniae	Colonization of nasopharynx, chronic rhinosinusitis, chronic otitis media, chronic obstructive pulmonary disease
Streptococcus pyogenes	Colonization of oral cavity and nasopharynx, recurrent tonsilitis
Porphyromonas gingivalis, Prevotella intermedia, Streptococcus mutans	Dental biofilms

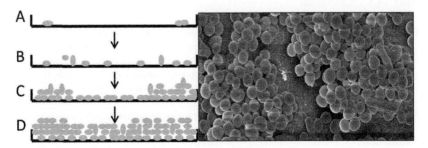

FIGURE 14.1 Systematic representation of Staphylococcus aureus biofilm formation and maturation. (A) Attachment of bacteria to substratum, (B) Micro-colony formation and firm attachment, (C) the bacteria spread out on the substratum, (D) the mature irreversible biofilm formation. The right panel shows scanning electron microscopic image of staphylococcus aureus biofilms.

14.3 BIOFILMS AND PERIODONTAL DISEASES

The periodontal disease caused by the tooth biofilm formed by pathogenic microorganisms is commonly known as dental plaque. Within one day of discontinuation of oral hygiene, dental biofilms begin to form and develop as gingivitis within 10–21 days.[57] Bacterial biofilm formation initiates with the attachment phase and within minutes host-derived molecules such as proline-rich proteins, phosphate-rich proteins (statherin), agglutinins, mucins, and enzymes such as alpha-amylase serve as the source of receptors for the primary colonizers of dental plaque.[43] Initial interactions among these bacteria are very crucial for biofilm development, and many have been elucidated from dental biofilms.[43] In dental biofilms, *Streptococcus spp.* represent 60–90% of early colonizers followed by secondary colonizing bacteria such as *Porphyromonas gingivalis*, *Treponema denticola*, *Eubacterium* spp., *Prevotella intermedia*, *Aggregatibacter actinomycetemcomitans*, *Tannerella forsythensis* and *Selenomonas flueggei*.[51,43] The sequential binding of bacteria results in the formation of nascent surfaces that act as a bridge for the next co-aggregating bacteria. In periodontal disease, an increased plague mass with higher microbial diversity has been detected, and a rise of obligate anaerobic and Gram-negative species has also been observed.[47] A significantly increased number of *Streptococcus mutans* and *P. gingivalis* have been detected in caries or periodontitis.[38]

To inhibit bacterial biofilms and to eradicate oral biofilms many strategies have been developed. One method is the dispersion of bacterial biofilms using extracellular enzymes such as deoxyribonucleases, proteases and glycosidases which could degrade the adhesive components matrix. The periodontal pathogen *A. actinomycetemcomitans* produces the glycoside hydrolase dispersin B, which degrades biofilm matrix components such as polysaccharide poly-N-acetyl glucosamine.[41] The advancement in biofilm detection techniques and knowledge of colonized bacteria in periodontal diseases paved the way for treatment and to develop new strategies such as vaccination for oral pathogens.[65] It has been reported that administration of vaccines targeting *Fusobacterium nucleatum* bacteria decreased gum inflammation and biofilm formation. *F. nucleatum* is an important bacterium that acts as a bridge between the primary and secondary colonizers.[44] Another approach to combat oral biofilm is the use of probiotics (containing beneficial bacteria) that eliminate pathogens with normal oral microflora in the mouth through an antagonistic mechanism.[93]

14.4 BIOFILM IN EAR INFECTIONS

Middle ear infections are the most common reason for children to visit doctors and are usually cleared up with antibiotics. But occasionally the infection persists, and becomes a chronic illness that only a surgical procedure can fix, and the bacterial biofilms are the culprit of this chronic form of ear infections. Chronic ear diseases like chronic OM, chronic suppurative otitis media (CSOM), otitis media with effusion (OME) and cholesteatoma are thought to be biofilm-related.[61,85] Although the acute infections are caused by the planktonic form of bacteria, the chronic ear infections or persistent effusions of the middle ear are biofilm-related infections. Otitis media with effusion is a chronic infection which is characterized by fluid discharge called middle ear fluid (MEF). Using rat OM experimental model, biofilms have been detected on middle ear mucosa suggesting that biofilm formation might be an important factor in the pathogenesis of chronic OME.[21] Yadav et al.[91] by using OM rat model demonstrated in vitro biofilm formation and in vivo colonization of *S. pneumoniae* in rat middle ear.[91] The *Streptococcus pneumoniae*, *Haemophilus influenzae* and *Moraxella catarrhalis* are major biofilm-forming bacteria in chronic OM along with *pseudomonas* and *staphylococci*. Furthermore, Hall-Stoodley et al.[32] detected bacterial biofilms on middle ear mucosa biopsy specimens from children with OME and recurrent OM, these results support the hypothesis that the chronic middle ear infections are related to bacterial biofilms.[32] Although, the traditional culturing methods have been proven inadequate to detect many viable bacteria present in OME,[59] molecular methods have detected quiescent dormant bacteria. In addition, bacterial DNA along with endotoxin in culture-negative middle ear effusion has been detected using PCR indicating the presence of bacteria in a dormant state.[16,60] In addition, biofilms in CSOM using non-human primate model infected by *P. aeruginosa* have been detected.[17] Exacerbations of COM in patients with cholesteatomas may also be associated with biofilms; the keratin matrix of a cholesteatoma appears an ideal environment for biofilm growth, and *P. aeruginosa* adheres to keratinocytes.[86] The bacterial biofilm has been detected on tympanostomy tubes, cochlear implants or artificial ossicles used to treat the abnormality of middle ear.[59,7] Biofilm formation on medical devices results in the extrusion of cochlear implants or the recalcitrant infection of implanted ears, forced to remove the implant device and loss of function.[54]

14.5 BIOFILMS IN NOSE AND THROAT INFECTIONS

The bacterial biofilms have been detected in chronic sinusitis,[13] and in chronic rhinosinusitis (CRS) patients during functional endoscopic sinus surgery (FESS).[64,71] A correlation between in vitro biofilm formation ability of *P. aeruginosa* and *S. aureus* and the unfavourable evolution after FESS indicates biofilm formation in chronic sinusitis.[6] Bacterial biofilms have been detected on frontal sinus stents in patients with chronic sinusitis who underwent FESS. These stents may actually serve as biofilm reservoirs. Biofilms have also been detected on adenoid tissue removed from children with CRS, and the nasopharynx may act as a chronic reservoir for bacterial pathogens.[95] Chronically diseased tonsils also evidence the presence of bacterial biofilms.

14.6 BIOFILMS IN URINARY TRACT INFECTIONS (UTI)

According to the National Institute of Health (NIH), among all the bacterial infections, 80% are due to biofilm-forming bacteria and has become a serious threat in the field of urology. Bacterial biofilm has been reported on the prostate stones, urothelium and medical implants.[80] Biofilms on uroepithelium act as a reservoir, and the bacteria in planktonic form could escape from those biofilms and invade the renal tissue causing pyelonephritis and chronic bacterial prostatitis.[50] The diagnosis of prostatitis is difficult because the colonized bacteria may not be present in the prosthetic secretion or the urine samples.[11] Bacteria biofilms can also form on catheters causing their blockage; therefore, catheter-associated urinary tracts infection (CAUTI) is one of the most common care-associated infections around the world.[78] It is estimated that more than 40% healthcare-associated infections in the United States are associated with CAUTI.[78] The most frequent biofilm forming microbes involved in CAUTI are *E. coli, Candida, Klebsiella, P. aeruginosa, Enterococci* or *Enterobacter* spp.[79] The catheters are an ideal surface for biofilm growth and the environmental conditions at that niche further enhance biofilm commodities.[52] Microorganisms produce hydrolyzing enzymes such as urease that hydrolyze urea to ammonium ions, that could cause encrustation, the formation of infected bladder calculi, and blockage of the urinary tract, on the indwelling medical devices such as urinary catheters. Due to ammonium ion formation, the pH of urine increases and finally causes precipitation of magnesium and calcium phosphate crystals.[76,77] Those crystals can form a protective layer for bacteria to reach antimicrobial agents and, thus increases bacterial resistance to antibiotics.[48] *Proteus vulgaris* and *Providencia rettgeri* also have the capacity to produce crystalline biofilms.[76] *P. mirabilis* produces several virulence factors such as mannose-resistant fimbriae, capsules, and urease that enhance biofilm formation.[73]

14.7 BIOFILMS IN CYSTIC FIBROSIS

Chronic bacterial infections are a key feature of a variety of lung conditions. *P. aeruginosa* is an opportunistic bacterium that possesses skills to both colonize and persist in the airways of patients with lung damage. The upper airways such as paranasal sinuses and nasopharynx play an important role as a silent reservoir of *P. aeruginosa*.[23] Cystic fibrosis (CF) is a life-threating lung infection leading to life-long airway infections and eventual respiratory failure.[27] The major pathogens of CF are *P. aeruginosa* and *S. aureus*.[29] The other most common multidrug-resistant pathogen infecting the airways of CF patients is *Stenotrophomonas maltophilia*.[27,5,75] *S. maltophilia* grows as biofilms on airway epithelial cells.[58,42] The mucoid forming strains (alginate-producing) of *P. aeruginosa* form biofilms in lungs of CF patients. The biofilm formation and maturation steps of *P. aeruginosa* are showed in Figure 14.2. The biofilm formation begins with the reversible attachment of planktonic bacteria with the help of flagella and pili to a substratum (step1). In the second step, the bacteria with the help of flagella and type IV pili form firm attachment (step 2). Thereafter, the attached bacterium loses flagella and pili and starts exopolysaccharides synthesis (step 3). The mature biofilm is formed with the aggregation of many bacteria embedded in EPS matrix with micro-colonies (step 4). As the biofilm matures a decreased concentration in the deep inside biofilm give rise to anaerobic phenotypes with the more persistent property. Once the biofilms established, the mature bacteria in form of planktonic cell detached and invade new site (step 5). Bacteria with-in biofilms are surrounded by self-produced EPS made up of DNA, protein and polymers. In CF patient's lungs, the biofilm forming *P. aeruginosa* is embedded in the polysaccharide alginate matrix. These biofilms prevent the bacteria phagocytosis and increase antibiotic tolerance, as well as resistance to both host innate and the adaptive immune response.[89] As a result, enhanced antibody response develops that leads to the immune complex-mediated chronic inflammation, increasing polymorphonuclear leukocytes. Biofilm formation in CF lungs is also associated with slow bacterial growth, increased frequency of mutations and adaptation to the conditions in the lungs. In addition, the bacterial metabolic activity and the increased doubling time in CF lungs are responsible for some antibiotic tolerance.[12] The *P. aeruginosa* survival in CF lungs is also attributed to chromosomal β-lactamase, upregulated efflux pumps and mutations of antibiotic target molecules in the bacteria.[33,90]

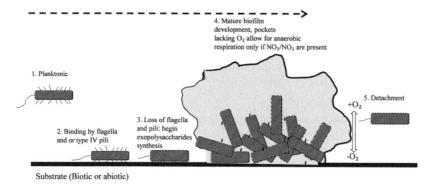

FIGURE 14.2 Systematic representation of biofilm formation, maturation and detachment of *Pseudomonas aeruginosa* on biotic or abiotic surface.

14.8 BIOFILMS IN WOUND INFECTION

The diabetic and cardiovascular patients are vulnerable to bacterial infection and develop chronic wound infections which could be due to the colonization of bacteria.[40,15,28,20] Chronic wound (CW) infections largely owe their chronicity to the inability of the host to clear a biofilm, with or without clinical intervention. Moreover, bacterial biofilms have been detected in non-healing CW infections in humans.[40,69,10] The poly-microbial biofilms of *P. aeruginosa* and *S. aureus* are highly resistant to antibiotics and to the host's defense. In poly-microbial biofilms, *P. aeruginosa* and *S. aureus* are competitive rather than antagonistic that give rise to more persistent and antibiotic resistant strains of *S. aureus* called small colony variants (SCVs). The poly-microbial colonization modifies the antibiotic selection and response of the host. In the CW infections, the biofilms act as a mechanical barrier against hostderived and exogenous antimicrobial agents as well as impede the re-epithelialization process. And this leads to a perpetual state of inflammation that delays wound healing.[87,94] More than 90% of all CW infections are due to bacterial biofilms,[3] and those biofilms make up to 51 to 1,000-fold more tolerant to antibiotics and the host immune response.[68] Thus, taking into account the increase in antibiotic-resistant bacteria and the biofilm formation ability of pathogen often makes effective treatment of these infections impossible. Many studies detected the presence of multiple bacterial species deep inside the dermal tissue in CW patients.[28,20,24,62,81] *S. aureus* and *P. aeruginosa* are the most commonly found organisms in wound infections along with other bacterial species.[28,81] In CW patients, the aggregation of *S. aureus* and *P. aeruginosa* was detected using microscopy. Later, the biofilm-like structures were confirmed by specific peptide nucleic acid (PNA) and fluorescence in situ hybridisation (FISH) probes in combination with confocal and scanning electron microscopy.[40] The *P. aeruginosa* wound infections are significantly larger in area than wound infections by other bacteria.

14.9   STRATEGIES TO COMBAT BIOFILM-RELATED INFECTIONS

Prevention of bacterial biofilm formation on human tissue or on medical devices is the most efficient measure to combat with biofilms. Human skin is the first barrier that prevents bacterial invasion of the body. However, during surgery and insertion of implants, this natural barrier is compromised that provides entry to the pathogen. To prevent the introduction of pathogens, surgeons should take extreme caution to maintain sterile condition while inserting implants or injecting dermal fillers. Ultraclean operation theatres should be used and the implants, surgical instruments and surgical garments should be prevented from coming into direct contact with the surgical site.[8] The use of implants made up of antibiofilm material should be encouraged along with the use of antibiotics or drug that inhibits bacteria quorum sensing. The following are strategies to combat bacterial biofilm.[66]

- Development of antimicrobial surfaces that inhibit biofilm growth
- Prevention of biofilm attachment to abiotic surfaces such as tissue or implant devices
- Physical treatments of biofilms
- Photodynamic treatment of biofilms
- Degradation of biofilm matrix to dismantle and eradicate biofilms
- Detachment induction
- Introduction of signal blockage
- Novel cell killing strategies
- Interference with biofilm regulation such as quorum sensing

14.10 CONCLUSION

Biofilm lifestyle helps bacteria to survive and persist within the hostile environment and increases bacterial resistance to antibiotics. Bacterial biofilms are key contributors to many chronic and recurrent infections in human tissues and medical indwelling devices. Biofilm formation is a major problem in a clinical setting that increases the cost of treatment and recovery time. The conventional antibiotic treatments are often inadequate to eradicate biofilm infections, and the sub-MIC antibiotics may promote biofilm formation and, thus compromise treatment effectiveness. To combat with biofilms, various strategies are being developed such as inhibition or disruption of biofilms using small molecules and enzymes targeting matrix, bactericidal and anti-adhesive coatings. The coating of modified indwelling medical devices' surface could prevent bacterial attachment and aggregation. Future approaches for controlling biofilms will also require a better understanding of bacterial interaction with each other and with host and their antibiotic activity.

KEYWORDS

- biofilm
- planktonic
- antibiotic resistance
- extracellular polymeric substance
- medical device

REFERENCES

1. Abidi, S. H.; Sherwani, S. K.; Siddiqui, T. R.; et al. Drug Resistance Profile and Biofilm Forming Potential of *Pseudomonas aeruginosa* Isolated from Contact Lenses in Karachi-Pakistan. *BMC Ophthalmol.* **2013**, *13*, 57.
2. Agarwal, A.; Singh, K. P.; Jain, A. Medical Significance and Management of Staphylococcal Biofilm. *FEMS Immunol. Med. Microbiol.* **2010**, *58*, 147–160.
3. Attinger, C.; Wolcott, R. Clinically Addressing Biofilm in Chronic Wounds. *Adv. Wound Care (New Rochelle).* **2012**, *1*, 127–132.

4. Auler, M. E.; Morreira, D.; Rodrigues, F. F.; et al. Biofilm Formation on Intrauterine Devices in Patients with Recurrent Vulvovaginal Candidiasis. *Med. Mycol.* **2010,** *48*(1), 211–216.

5. Ballestero, S.; Virseda, I.; Escobar, H.; Suarez, L.; Baquero, F. *Stenotrophomonas maltophilia* in Cystic Fibrosis Patients. *Eur. J. Clin. Microbiol. Infect. Dis.* **1995,** *14*, 728–729.

6. Bendouah, Z.; Barbeau, J.; Hamad, W. A.; Desrosiers, M. Biofilm Formation by *Staphylococcus aureus* and *Pseudomonas aeruginosa* is Associated with an Unfavorable Evolution After Surgery for Chronic Sinusitis and Nasal Polyposis. *Otolaryngol. Head Neck Surg.* **2006,** *134*, 991–996.

7. Berry, J. A.; Biedlingmaier, J. F.; Whelan, P. J. In vitro Resistance to Bacterial Biofilm Formation on Coated Fluoroplastictympanostomy Tubes. *Otolaryngol. Head Neck Surg.* **2000,** *123*, 246–251.

8. Bjarnsholt, T. The Role of Bacterial Biofilms in Chronic Infections. *APMIS* **2013,** *121*(Suppl. 136), 1–54.

9. Burt, S. A.; Ojo-Fakunle, V. T. A.; Woertman, J.; Veldhuizen, E. J. A. The Natural Antimicrobial Carvacrol Inhibits Quorum Sensing in *Chromobacterium violaceum* and Reduces Bacterial Biofilm Formation at Sub-Lethal Concentrations. *PLoS One* **2014,** *9*(4), e93414.

10. Chole, R.; Faddis, B. Evidence for Microbial Biofilms in Cholesteatomas. *Arch. Otolaryngol. Head Neck Surg.* **2002,** *28*, 129–1133.

11. Choong, S.; Whitfield, H. Biofilms and Their Role in Infections in Urology. *BJU Int.* **2000,** *86*(8), 935–941.

12. Costerton, J. W.; Stewart, P.S.; Greenberg, E. P. Bacterial Biofilms: A Common Cause of Persistent Infections. *Science* **1999,** *284*, 1318–1322.

13. Cryer, J.; Schipor, I.; Perloff, J. R.; Palmer, J. N. Evidence of Bacterial Biofilms in Human Chronic Sinusitis. *ORL J. Otorhinolaryngol. Relat. Spec.* **2004,** *66*, 155–158.

14. Dasgupta, M. K. Biofilms and Infection in Dialysis Patients. *Semin. Dial.* **2002,** *15*(5), 338–346.

15. Davies, C. E.; Hill, K. E.; Wilson, M. J.; Stephens, P.; Hill, C. M.; Harding, K. G.; et al. Use of 16S Ribosomal DNA PCR and Denaturing Gradient Gel Electrophoresis for Analysis of the Microfloras of Healing and Nonhealing Chronic Venous Leg Ulcers. *J. Clin. Microbiol.* **2004,** *42*, 3549–3557.

16. Dingman, J. R.; Rayner, M. G; Mishra, S.; et al. Correlation between Presence of Viable Bacteria and Presence of Endotoxin in Middle-Ear Effusions. *J. Clin. Microbiol.* **1998,** *36*, 3417–3419.

17. Dohar, J. E.; Hebda, P. A.; Veeh, R.; et al. Mucosal Biofilm Formation on Middle-Ear Mucosa in a Nonhuman Primate Model of Chronic Suppurative Otitis Media. *Laryngoscope* **2005,** *8*, 1469–1472.

18. Donelli, G.; Vuotto, C.; Cardines, R.; et al. Biofilm-Growing Intestinal Anaerobic Bacteria. *FEMS Immunol. Med. Microbiol.* **2012,** *65*(2), 318–325.

19. Donlan, R. M. Biofilms and Device-Associated Infections. *Emerging Infect. Dis.* **2001,** *7*(2), 277–281.

20. Dowd, S. E.; Sun, Y.; Secor, P. R.; Rhoads, D. D.; Wolcott, B. M.; James, G. A.; et al. Survey of Bacterial Diversity in Chronic Wounds Using Pyrosequencing, DGGE, and Full Ribosome Shotgun Sequencing. *BMC Microbiol.* **2008,** *8*, 43.

21. Ehrlich, G. D.; Veeh, R.; Wang, X.; et al. Mucosal Biofilm Formation on Middle-Ear Mucosa in the Chinchilla Model of Otitis Media. *JAMA* **2002,** *287,* 1710–1715.
22. Fitzgerald, G.; Williams, L. S. Modified Penicillin Enrichment Procedure for the Selection of Bacterial Mutants. *J. Bacteriol.* **1975,** *122*(1), 345–346.
23. Fothergill, J. L.; Neill, D. R.; Loman, N.; Winstanley, C.; Kadioglu, A. *Pseudomonas aeruginosa* Adaptation in the Nasopharyngeal Reservoir Leads to Migration and Persistence in the Lungs. *Nat. Commun.* **2014,** *5,* 4780.
24. Frankel, Y. M.; Melendez, J. H.; Wang, N. Y.; Price, L. B.; Zenilman, J. M.; Lazarus, G. S. Defining Wound Microbial Flora: Molecular Microbiology Opening New Horizons. Arch. Dermatol. **2009,** *145,* 1193–1195.
25. de Fuente-Núñez, C.; Reffuveille, F.; Fernandez, L.; et al. Bacterial Biofilm Development as a Multicellular Adaptation: Antibiotic Resistance and New Therapeutic Strategies. *Curr. Opin. Microbiol.* **2013,** *16*(5), 580–589.
26. Fux, C. A.; Quigley, M.; Worel, A. M.; et al. Biofilm-Related Infections of Cerebrospinal Fluid Shunts. *Clin. Microbiol. Infect.* **2006,** *12*(4), 331–337.
27. Gibson, R. L.; Burns, J. L.; Ramsey, B. W. Pathophysiology and Management of Pulmonary Infections in Cystic Fibrosis. *Am. J. Respir. Crit. Care Med.* **2003,** *168,* 918–951.
28. Gjodsbol, K.; Christensen, J. J.; Karlsmark, T.; Jorgensen, B.; Klein, B. M.; Krogfelt, K. A. Multiple Bacterial Species Reside in Chronic Wounds: A Longitudinal Study. *Int. Wound J.* **2006,** *3,* 225–231.
29. Govan, J. R.; Deretic, V. Microbial Pathogenesis in Cystic Fibrosis: Mucoid *Pseudomonas aeruginosa* and *Burkholderia cepacia. Microbiol. Rev.* **1996,** *60,* 539–574.
30. Guskey, M. T.; Tsuji, B. T. A Comparative Review of the Lipoglycopeptides: Oritavancin, Dalbavancin, and Telavancin. *Pharmacother. J. Human Pharmacol. Drug Therapy.* **2010,** *30,* 80–94.
31. Hall-Stoodley, L.; Costerton, J. W.; Stoodley P. Bacterial Biofilms: From the Natural Environment to Infectious Diseases. *Nat. Rev. Microbiol.* **2004,** *2*(2), 95–108.
32. Hall-Stoodley, L.; Hu, F. Z.; Gieseke, A.; et al. Direct Detection of Bacterial Biofilms on the Middle-Ear Mucosa of Children with Chronic Otitis Media. *JAMA* **2006,** *296,* 202–211.
33. Hassett, D. J.; Cuppoletti, J.; Trapnell, B.; Lymar, S. V.; Rowe, J. J.; Yoon, S. S.; Hilliard, G. M.; Parvatiyar, K.; Kamani, M. C.; Wozniak, D. J.; Hwang, S. H.; McDermott, T. R.; Ochsner, U. A. Anaerobic Metabolism and Quorum Sensing by *Pseudomonas aeruginosa* Biofilms in Chronically Infected Cystic Fibrosis Airways: Rethinking Antibiotic Treatment Strategies and Drug Targets. *Adv. Drug Delivery Rev.* **2002,** *54,* 1425–1443.
34. Hengzhuang, W.; Wu, H.; Ciofu, O.; et al. Pharmacokinetics/Pharmacodynamics of Colistin and Imipenem on Mucoid and Monmucoid *Pseudomonas aeruginosa* Biofilms. *Antimicrob. Agents Chemother.* **2011,** *55*(9), 4469–4474.
35. Hengzhuang, W.; Wu, H.; Ciofu, O.; et al. In vivo Pharmacokinetics/Pharmacodynamics of Colistin and Imipenem in *Pseudomonas aeruginosa* Biofilm Infection. *Antimicrob. Agents Chemother.* **2012,** *56*(5), 2683–2690.
36. Hoiby, N.; Bjarnsholt, T.; Givskov, M.; et al. Antibiotic Resistance of Bacterial Biofilms. *Int. J. Antimicrob. Agents* **2010,** *35*(4), 322–332.
37. Hoiby, N.; Ciofu, O.; Johansen, H. K.; et al. The Clinical Impact of Bacterial Biofilms. *Int. J. Oral Sci.* **2011,** *3*(2), 55–65.

38. Huang, R.; Li, M.; Gregory, R. L. Bacterial Interactions in Dental Biofilm. *Virulence* **2011,** *2,* 435–444.

39. Hurdle, J. G; O'Neill, A. J.; Chopra, I.; Lee, R. E. Targeting Bacterial Membrane Function: An Underexploited Mechanism for Treating Persistent Infections. *Nat. Rev. Micro.* **2011,** *9,* 62–75.

40. James, G. A.; Swogger, E.; Wolcott, R.; Pulcini, E. D.; Secor, P.; Sestrich, J.; et al. Biofilms in Chronic Wounds. *Wound Repair Regen.* **2008,** *16,* 37–44.

41. Kaplan, J. B. Biofilm Dispersal: Mechanisms, Clinical Implications, and Potential Therapeutic Uses. *J. Dent. Res.* **2010,** *89,* 205–218.

42. Keays, T.; Ferris, W.; Vandemheen, K. L.; Chan, F.; Yeung, S. W.; Mah, T. F.; Ramotar, K.; Saginur, R.; Aaron, S. D. A Retrospective Analysis of Biofilm Antibiotic Susceptibility Testing: A Better Predictor of Clinical Response in Cystic Fibrosis Exacerbations. *J. Cystic Fibrosis.* **2009,** *8,* 122–127.

43. Kolenbrander, P. E.; Andersen, R. N.; Blehert, D. S.; Egland, P. G.; Foster, J. S.; Palmer, R. J., Jr. Communication Among Oral Bacteria. *Microbiol. Mol. Biol. Rev.* **2002,** *66,* 486–505.

44. Liu, P. F.; Shi, W.; Zhu, W.; Smith, J. W.; Hsieh, S. L.; Gallo, R. L.; Huang, C. M. Vaccination Targeting Surface FomA of *Fusobacterium nucleatum* against Bacterial Co-Aggregation: Implication for Treatment of Periodontal Infection and Halitosis. *Vaccine* **2010,** *28,* 3496–3505.

45. Mah, T. F.; O'Toole, G. A. Mechanisms of Biofilm Resistance to Antimicrobial Agents. *Trends Microbiol.* **2001,** *9*(1), 34–39.

46. Maki, D. G.; Kluger, D. M.; Crnich, C. J. The Risk of Bloodstream Infection in Adults With Different Intravascular Devices: A Systematic Review of 200 Published Prospective Studies. *Mayo Clin. Proc.* **2006,** *81,* 1159–1171.

47. Marsh, P. D.; Devine, D. A. How is the Development of Dental Biofilms Influenced by the Host? *J. Clin. Periodontol.* **2011,** *38*(Suppl. 11), 28–35.

48. Morgan, S. D.; Rigby, D.; Stickler, D. J. A Study of the Structure of the Crystalline Bacterial Biofilms that can Encrust and Block Silver Foley Catheters. *Urol. Res.* **2009,** *37*(2), 89–93.

49. Murakami, M.; Nishi, Y.; Seto, K.; et al. Dry Mouth and Denture Plaque Microflora in Complete Denture and Palatal Obturator Prosthesis Wearers. *Gerodontology* **2013.** DOI: 10.1111/ger.12073.

50. Nickel, J. C.; Ruseska, I.; Wright, J. B.; Costerton, J. W. Tobramycin Resistance of *Pseudomonas aeruginosa* Cells Growing as a Biofilm on Urinary Catheter Material. *Antimicrob. Agents Chemother.* **1985,** *27*(4), 619–624.

51. Nyvad, B.; Kilian, M. Microbiology of the Early Colonization of Human Enamel and Root Surfaces In Vivo. *Scand. J. Dent. Res.* **1987,** *95,* 369–380.

52. Ong, C. Y.; Ulett, G. C.; Mabbett, A. N.; et al. Identification of Type 3 Fimbriae in Uropathogenic *Escherichia coli* Reveals a Role in Biofilm Formation. *J. Bacteriol.* **2008,** *190*(3), 1054–1063.

53. Otto, M. Looking Toward Basic Science for Potential Drug Discovery Targets Against Community-Associated MRSA. *Med. Res. Rev.* **2010,** *30,* 1–22.

54. Pawlowski, K. S.; Wawro, D.; Roland, P. S. Bacterial Biofilm Formation on a Human Cochlear Implant. *Otol. Neurotol.* **2005,** *26,* 972–975.

55. Perloff, J. R.; Palmer, J. N. Evidence of Bacterial Biofilms on Frontal Recess Stents in Patients with Chronic Rhinosinusitis. *Am. J. Rhinol.* **2004,** *18,* 377–380.
56. Perloff, J. R.; Palmer, J. N. Evidence of Bacterial Biofilms in a Rabbit Model of Sinusitis. *Am. J. Rhinol.* **2005,** *19,* 1–6.
57. Pihlstrom, B. L. Michalowicz, B. S.; Johnson, N. W. Periodontal Diseases. *Lancet* **2005,** *366,* 1809–1820.
58. Pompilio, A.; Crocetta, V.; Confalone, P.; Nicoletti, M.; Petrucca, A.; Guarnieri, S.; Fiscarelli, E.; Savini, V.; Piccolomini, R.; Di Bonaventura, G. Adhesion to and Biofilm Formation on IB3-1 Bronchial Cells by Stenotrophomonas Maltophilia Isolates from Cystic Fibrosis Patients. *BMC Microbiol.* **2010,** *10,* 102.
59. Post, J. C. Direct Evidence of Bacterial Biofilms in Otitis Media. *Laryngoscope* **2001,** *111,* 2083–2094.
60. Post, J. C.; Preston, R. A.; Aul, J. J.; et al. Molecular Analysis of Bacterial Pathogens in Otitis Media with Effusion. *JAMA* **1995,** *273,* 1598–1604.
61. Post, J. C.; Hiller, N. L.; Nistico, L.; Stoodley, P.; Ehrlich, G. D. The Role of Biofilms in Otolaryngologic Infections: Opdate 2007. *Curr. Opin. Otolaryngol. Head Neck Surg.* **2007,** *15,* 347–351.
62. Price, L. B.; Liu, C. M.; Melendez, J. H.; Frankel, Y. M.; Engelthaler, D.; Aziz, M.; et al. Community Analysis of Chronic Wound Bacteria Using 16S rRNA Gene-Based Pyrosequencing: Impact of Diabetes and Antibiotics on Chronic Wound Microbiota. *PLoS One* **2009,** *4,* e6462.
63. Quave, C. L.; Estévez-Carmona, M.; Compadre, C. M.; Hobby, G.; Hendrickson, H.; et al. Ellagic Acid Derivatives from Rubus Ulmifolius Inhibit *Staphylococcus aureus* Biofilm Formation and Improve Response to Antibiotics. *PLoS One* **2012,** *7,* e28737.
64. Ramadan, H. H.; Sanclement, J. A.; Thomas, J. G. Chronic Rhinosinusitis and Biofilms. *Otolaryngol. Head Neck Surg.* **2005,** *132,* 414–417.
65. Ramage, G.; Culshaw, S.; Jones, B.; Williams, C. Are We Any Closer to Beating the Biofilm: Novel Methods of Biofilm Control. *Curr. Opin. Infect. Dis.* **2010,** *23,* 560–566.
66. Ramasamy, M.; Lee, J. Recent Nanotechnology Approaches for Prevention and Treatment of Biofilm-Associated Infections on Medical Devices. *BioMed Res. Int.* **2016,** *17.* Article ID 1851242, DOI:10.1155/2016/1851242.
67. Rieger, U. M.; Mesina, J.; Kalbermatten, D. F.; et al. Bacterial Biofilms and Capsular Contracture in Patients with Breast Implants. *Br. J. Surg.* **2013,** *100*(6), 768–774.
68. Rogers, S. A.; Huigens, R. W., 3rd; Cavanagh, J.; Melander, C. Synergistic Effects Between Conventional Antibiotics and 2-aminoimidazole-derived Antibiofilm Agents. *Antimicrob. Agents Chemother.* **2010,** *54,* 2112–2118.
69. Roland, P. S. Chronic Suppurative Otitis Media: A Clinical Overview. *Ear Nose Throat J.* **2002,** *81,* /8–11.
70. Sanclement, J. A.; Webster, P.; Thomas, J.; Ramadan, H. H. Bacterial Biofilms in Surgical Specimens of Patients with Chronic Rhinosinusitis. *Laryngoscope* **2005,** *115,* 578–582.
71. Sanderson, A. R.; Leid, J. G.; Hunsaker, D. Bacterial Biofilms on the Sinus Mucosa of Human Subjects with Chronic Rhinosinusitis. *Laryngoscope* **2006,** *116,* 1121–116.
72. Santos, A. P.; Watanabe, E.; Andrade, D. Biofilm on Artificial Pacemaker: Fiction or Reality? *Arq. Bras. Cardiol.* **2011,** *97*(5), e113–120.

73. Siddiq, D. M.; Darouiche, R. O. New Strategies to Prevent Catheter-Associated Urinary Tract Infections. *Nat. Rev. Urol.* **2012,** *9*(6), 305–314.

74. Song, Z.; Borgwardt, L.; Høiby, N.; et al. Prosthesis Infections after Orthopedic Joint Replacement: The Possible Role of Bacterial Biofilms. *Orthop. Rev. (Pavia)* **2013,** *5*(2), 65–71.

75. Steinkamp, G.; Wiedemann, B.; Rietschel, E.; Krahl, A.; Gielen, J.; Barmeier, H.; Ratjen, F. Prospective Evaluation of Emerging Bacteria in Cystic Fibrosis. *J. Cystic Fibrosis* **2005,** *4*, 41–48.

76. Stickler, D.; Morris, N.; Moreno, M.; Sabbuba, N. Studies on the Formation of Crystalline Bacterial Biofilms on Urethral Catheters. *Eur. J. Clin. Microbiol. Infect. Dis.* **1998a,** *17*(9), 649–652.

77. Stickler, D. J.; Morris, N. S.; Mclean, R. J. C.; Fuqua, C. Biofilms on Indwelling Urethral Catheters Produce Quorum Sensing Signal Molecules In Situ and In Vitro. *Appl. Environ. Microbiol.* **1998b,** *64*(9), 3486–3490.

78. Tambyah, P. A. Catheter-Associated Urinary Tract Infections: Diagnosis and Prophylaxis. *Int. J. Antimicrob. Agents* **2004,** *24*(1), S44–48.

79. Tambyah, P. A.; Halvorson, K. T.; Maki, D. G. A Prospective Study of Pathogenesis of Catheter-Associated Urinary Tract Infections. *Mayo Clin. Proc.* **1999,** *74*(2), 131–136.

80. Tenke, P.; Kovacs, B.; Jäckel, M.; Nagy, E. The Role of Biofilm Infection in Urology. *World J. Urol.* **2006,** *24*(1), 13–20.

81. Thomsen, T. R.; Aasholm, M. S.; Rudkjobing, V. B.; Saunders, A. M.; Bjarnsholt, T.; Givskov, M.; et al. The Bacteriology of Chronic Venous Leg Ulcer Examined by Culture-Independent Molecular Methods. *Wound Repair Regen.* **2010,** *18*, 38–49.

82. Thornton, R. B.; Wiertsema, S. P.; Kirkham, L.-A.; Rigby, P. J.; Vijayasekaran, S.; et al. Neutrophil Extracellular Traps and Bacterial Biofilms in Middle Ear Effusion of Children with Recurrent Acute Otitis Media—A Potential Treatment Target. *PLoS One* **2013,** *8*, e53837.

83. Tollefson, D. F.; Bandyk, D. F.; Kaebnick, H. W.; et al. Surface Biofilm Disruption. Enhanced Recovery of Microorganisms from Vascular Prostheses. *Arch. Surg.* **1987,** *122*(1), 38–43.

84. Tran, P. L.; Lowry, N.; Campbell, T.; et al. An Organoselenium Compound Inhibits *Staphylococcus aureus* Biofilms on Hemodialysis Catheters in vivo. *Antimicrob. Agents. Chemother.* **2012,** *56*(2), 972–978.

85. Vlastarakos, P. V.; Nikolopoulos, T. P.; Maragoudakis, P.; Tzagaroulakis, A.; Ferekidis, E. Biofilms in Ear, Nose, and Throat Infections: How Important are they? *Laryngoscope* **2007,** *117*, 668–673.

86. Wang, E. W.; Jung, J. Y.; Pashia, M. E.; et al. Otopathogenic *Pseudomonas aeruginosa* Strains as Competent Biofilm Formers. *Arch. Otolaryngol. Head Neck Surg.* **2005,** *131*, 983–989.

87. Watters, C.; DeLeon, K.; Trivedi, U.; Griswold, J. A.; Lyte, M.; Hampel, K. J.; Wargo, M. J.; Rumbaugh, K. P. *Pseudomonas aeruginosa* Biofilms Perturb Wound Resolution and Antibiotic Tolerance in Diabetic Mice. *Med. Microbiol. Immunol.* **2013,** *202*, 131–141.

88. Weiss, E. C.; Zielinska, A.; Beenken, K. E.; Spencer, H. J.; Daily, S. J.; et al. Impact of SarA on Daptomycin Susceptibility of *Staphylococcus aureus* Biofilms In Vivo. *Antimicrob. Agents Chemother.* **2009,** *53,* 4096–4102.

89. Winstanley, C.; O'Brien, S.; Brockhurst, M. A. *Pseudomonas aeruginosa* Evolutionary Adaptation and Diversification in Cystic Fibrosis Chronic Lung Infections. *Trends Microbiol.* **2016,** *24*(5), 327–337.

90. Worlitzsch, D.; Tarran, R.; Ulrich, M.; Schwab, U.; Cekici, A.; Meyer, K. C.; Birrer, P.; Bellon, G.; Berger, J.; Weiss, T.; Botzenhart, K.; Yankaskas, J. R.; Randell, S.; Boucher, R. C.; Doring, G. Effects of Reduced Mucus Oxygen Concentration in Airway *Pseudomonas* Infections of Cystic Fibrosis Patients. *J. Clin. Invest.* **2002,** *109*, 317–325.

91. Yadav, M. K.; Chae, S. W.; Song, J. J. *In Vitro Streptococcus pneumoniae* Biofilm Formation and *In Vivo* Middle Ear Mucosal Biofilm in a Rat Model of Acute Otitis Induced by *S. pneumoniae. Clin. Exp. Otorhinolaryngol.* **2012,** *5*(3), 139–144.

92. Yadav, M. K.; Chae, S. W.; Im, G. J.; Chung, J. W.; Song, J. J. Eugenol: A Phyto-Compound Effective against Methicillin-Resistant and Methicillin-Sensitive *Staphylococcus aureus* Clinical Strain Biofilms. *PLoS One* **2015,** *10*(3), e0119564.

93. Zahradnik, R. T.; Magnusson, I.; Walker, C.; McDonell, E.; Hillman, C. H.; Hillman, J. D. Preliminary Assessment of Safety and Effectiveness in Humans of ProBiora3, a Probiotic Mouth Wash. *J. Appl. Microbiol.* **2009,** *107*, 682–690.

94. Zhao, G.; Usui, M. L.; Lippman, S. I.; James, G. A.; Stewart, P. S.; Fleckman, P.; Olerud, J. E. Biofilms and Inflammation in Chronic Wounds. *Adv. Wound Care (New Rochelle)* **2013,** *2*(7), 389–399.

95. Zuliani, G.; Carron, M.; Gurrola, J.; et al. Identification of Adenoid Biofilms in Chronic rhinosinusitis. *Int. J. Pediatr. Otorhinolaryngol.* **2006,** *70*(9), 1613–1617 (Epub ahead of print).

CHAPTER 15

PREVENTION OF COLORECTAL CANCER THROUGH PROBIOTICS, PREBIOTICS, AND SYNBIOTICS: A CRITICAL REVIEW

SUSHRIREKHA DAS[1], SMITA H. PANDA[1,*], NILADRI B. KAR[1], NAKULANANDA MOHANTY[1], and HRUDAYANATH THATOI[2]

[1]Department of Zoology, North Orissa University, Baripada, Odisha 757003, India, *E-mail: panda.smita@gmail.com

[2]Department of Biotechnology, Baripada, Odisha 757003, India

CONTENTS

ABSTRACT

Colorectal cancer (CRC) is the third most prevalent cancer and among the current trends of treatments which include chemotherapy, radiotherapy and surgery and so forth. Such treatments are associated with high risk of complication with no guaranty for complete recovery. Currently probiotics, prebiotics and synbiotics represent a new therapeutic option. Numerous studies such as in vitro and animal model studies have indicated the potential role of probiotics including *Lactobacillus* and bifidobacteria, and prebiotics such as oligo-fructose and inulin, to exert anti-neoplastic effects. Probiotic cultures can also serve as live vehicles to deliver biologic agents (vaccines, enzymes and proteins) to targeted locations within the body. Synbiotics, which is a combination of probiotics and prebiotics, can synergistically in improving CRC. This chapter highlights different preventive action of probiotics, prebiotics and synbiotics against CRC demonstrating various animals and human case studies.

15.1 INTRODUCTION

Colorectal cancer (CRC), malignant disease among cancers, affects both men and women population. In 2012, 140,000 new cases of CRC has been estimated in the United States having disease-specific mortality of up to 600,000 in 2011[1,2] epidemiological studies reported that diet rich in fruits and vegetables appears to be more protective against CRC rather than red meat and animal fat. Usually CRC occurs mutational activation of oncogenes and loss of heterozygosity of tumour suppressor genes caused by many carcinogenic mutagens, chemicals and so forth.[3] Current treatment options for CRC include surgery, chemotherapy and radiation therapy, all of which can significantly reduce quality of life. Also some emerging therapeutic strategies for CRC include the non-steroidal anti-inflammatory drug (Sulindac). Some new chemotherapeutics including Capecitabine,[4] and antibodies such as the monoclonal epidermal growth factor receptor antibody, Cetuximab are used for preventing the CRC.

In recent years, probiotics, prebiotics and synbiotics emerged as another therapeutic option for prevention of CRC. Probiotic bacteria may be defined as live microorganisms which when administered in adequate amounts confer a health benefit on the hosts.[5] A prebiotic is 'a non-digest-ible food ingredient that beneficially affects the host by selectively stimu-lating the growth and/or activity of one or a limited number of bacteria in the colon that have the potential to improve host health'. Prebiotics are non-digestible food ingredients that stimulate the growth of bifidogenic and lactic acid bacteria (LAB) in the gastrointestinal tract. A synbiotic has been defined as 'a mixture of probiotics and prebiotics that benefi-cially affects the host by improving the survival and implantation of live microbial dietary supplements in the gastrointestinal tract.[6] Probiotics and prebiotics function in gut by increasing the level of beneficial bacteria such as *Lactobacillus* and *bifidobacteria*, simultaneously reducing the levels of pathogenic microorganisms such as *Clostridium*, *Salmonella*, and so forth.[7] There are several experimental evidences found in animal and human studies to suggest that probiotics and prebiotics may be beneficial in the prevention and treatment of colon cancer, which has been discussed in briefly later. This chapter highlights the possible mechanisms of probi-otics, prebiotics and synbiotics which exert their beneficial effects for CRC prevention. Further, various animal and human case studies between microbiota and development of CRC are well represented.

15.2 COLORECTAL CANCER

Cancer is a class of diseases characterized by prolonged growth and one of the major health problems in the world. According to the Canadian Cancer Society and the National Cancer Institute, in 2012, an estimated 23,300 Canadians and 143 460 Americans were diagnosed with CRC and 9200 and 51,690, respectively, died of it during the period CRC develops by the accumulation of mutations, starting in stem cells at the base of the crypts and usually begins as a non-cancerous polyp. Mostly colon cancers origi-nates from a small, noncancerous (benign) tumours, (adenomatous polyps) that form on the inner walls of the large intestine (Fig. 15.1) Which grow into malignant colon cancers over time if they are not removed during colonoscopy. After the formation of malignant tumours, the cancerous cells propagate through the blood and lymph systems, spreading to other parts of the body.[8] The overall is called as metastatis.

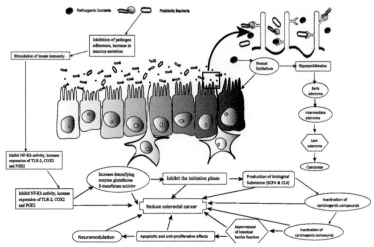

FIGURE 15.1   Mode of action of probiotics and pathogen towards CRC.

15.2.1 IMPACT OF DIFFERENT DIETS ON CRC PREVENTION

Mostly all probiotic bacteria in combination with dietary ingredients helps in detoxification and biotransformation of procarcinogens and carcinogens into less toxic metabolites, which helps in prevention of formation of tumours.[9] The consumption of prebiotics such as inulin, fructooligosaccharide (FOS) and galactooligosaccharide (GOS) causes a laxative effect on reaching the large intestine, which stimulated microbial growth resulting in increased bacterial biomass, faecal bulking and peristalsis. Rather, the faecal bulking helps in cancer prevention by reducing transit time of microbes and mutagens to intestinal lining of GI tract. The protective effect of dietary fibres against N-nitroso compounds (DMH), heterocyclic amines (HCA) and polycyclic aromatic hydrocarbons (PAH) were reported using animal models and in vitro assays.[10]

Mai et al.,[11] reported that diets which are rich in olive oil and extracts from freeze-dried fruits and vegetables considerably reduce the intestinal aenomas in mice indicating restriction in calorie (Table 15.1). Various studies in the colon of rats and mice have shown that inulin-type fructans can prevent chemically induced preneoplastic lesions, aberrant crypt foci (ACF) and tumours.[12–14] In a ACF mice model, Beneo Synergy1 (SYN1) is an oligofructose-enriched inulin has been shown to be particularly chemopreventive and also prevent the development of tumours in the Apc Min mice model.[15]

TABLE 15.1   Food Products and Their Potential Activity Against CRC.

Food products	Uses	CRC	Reference
Broccoli	Broccoli is a cruciferous vegetable and a good source of antioxidant lutein and sulforaphane, these are very potent antioxidant, and protective against stomach and intestinal cancers.	The *Journal of the American Dietetic Association* researchers, found that within the proximal and distal colon, brassica vegetables (brussels sprouts, cabbage, cauliflower and broccoli) were associated with decreased risk of these cancers	[41]
Spinach	The beta-carotene is found in spinach which may help to fight colon cancer.	The study from the *Journal of Nutrition and Cancer*, found that people who ate cooked green vegetables once a day had a 24% lower risk of colon cancer compared to people who did not.	
Ginger	Ginger *contains* very potent anti-inflammatory compound, called gingerols, which reduce colon inflammation.	In a study, 3 healthy participants who were at 'normal risk' for developing colorectal cancer, were not taking chronic medication, and had not taken aspirin or NSAIDs for 2 weeks prior were recruited; 16 were assigned to the ginger arm and 17 to a placebo arm. Data from all these were included in analysis, although 2 participants in the ginger and 1 in the placebo arm did not complete the full trial.	[47]

TABLE 15.1   *(Continued)*

Food products	Uses	CRC	Reference
White tea	White tea contains antioxidant, the antioxidant plays a good role to keep the cancer cells from growing and it is also specially powerful at controlling colon polyp growth.	Research from Oregon State University suggests that white tea is especially powerful at blocking colon polyp growth.	
Curry	Curry contains a yellow pigment called curcumin, can kill colon cancer cells.	In a controlled clinical trial, 41 human volunteers with a history of 8 or more aberrant crypt foci (ACF) were given doses of curcumin.	[48]
Mushroom	White button mushrooms are rich in antioxidant, called ergothioneine, it has cancer fighting properties.	Research is done on Proteoglycan from *P. linteus* to determine its possible anti-tumour effect on human cancer cells.	[49]
Black raspberries	Black raspberries contain higher levels of anthocyanins, having the capacity to fight against cancers.	Dr. Stoner's work examining preventive effects of whole raspberries against digestive tract cancers in humans.	[50]
Buttered corn	A compound is present in corn fibre called, inositol hexaphosphate, which prevents colon cancer growth.	Researchers wrote in the *American Journal of Epidemiology*, suggest that consumption of high amounts of trans-fatty acid may increase the risk of colorectal neoplasia and they provide additional support to recommendations to limit trans-fatty acid consumption	
Brown rice	Brown rice fibre contains a short-chain fatty acid that halts growth of cancer cells.	According to a study published in *Nutrition and Cancer*, people who ate brown rice at least once a week reduced their risk of developing colon polyps (growths that may lead to cancer) by 40%.	[51]

TABLE 15.1 *(Continued)*

Food products	Uses	CRC	Reference
Yogurt	Yogurt and other fermented milk products also contain vitamin K2 (menaquinone), which reduces colorectal cancer.	Epidemiological studies have shown that consumption of vitamin K2 through supplements or food can decrease cancer risk.	[52]
Barely	Resistant starch, similar to fibre, is a carbohydrate that is not broken down by fibre. Resistant starch acts as prebiotics, fuelling the growth of healthy bacteria (probiotics) in gut. The ferment fibre and resistant starch, forming short-chain fatty acids that have a number of anti-cancer effects.	In a research work of O'Keefe S.J., reported that when bacteria in the lower intestine break down fibre, a substance called butyrate is produced which may inhibit the growth of tumours of the colon and rectum.[3]	[53]
Peanut	Unsaturated fats, certain vitamins and minerals, and too many bioactives that have cancer-preventative effects.	Gonzalez CA et al. showed that, Unsaturated fats, certain vitamins and minerals, and the bioactive components have shown to have cancer-preventative effects, which are all packaged into a peanut kernel.	[54]
Oats	*Oats contain* a specific type of fibre known as beta-glucan which significantly *reduce* the risk of colorectal cancer		[55]

15.2.2 MICROORGANSIMS RESPONSIBLE FOR CRC

Human gut comprises 10^7 to 10^{12} bacterial cells per gram of the intestinal content (Fig. 15.2).[16] Microbiota in the gut performs vital functions in the host which includes, immune and nutritional status, assisting in health maintenance.[17] It has also been shown that microorganism alleviate lactose intolerance, lower serum cholesterol level, exert anticancer effect, improve constipation, enhance immunity, regulate obesity and relieve of vaginitis,

Imbalance of these gut microbiota leads to pathogenesis through chronic inflammation, immune evasion and suppression. In most of the diversity of bacterial community has been found by colonoscopic studies by significant elevation of bacteroides/Prevotella population, simultaneously increasing the level of IL-17 producing cells in the mucosa. *Clostridium* spp. has also been associated with metabolic and physiological changes in CRC patients. Hence the concept of probiotics, prebiotics and synbiotics, is becoming a revolution for health promotion in every individuals.[18]

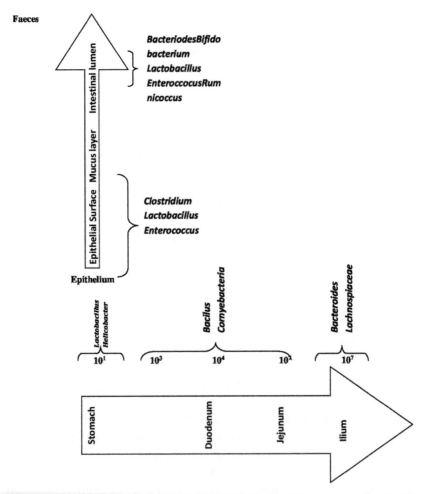

FIGURE 15.2   Variations in microbial numbers and composition across the length of the gastrointestinal tract.

15.3 PROBIOTICS

Probiotics are defined as 'live micro-organisms that confer a health benefit on the host when administered in adequate amounts'.[19,20] LAB are widely available, in yogurts and other functional foods such as cheese, fermented and unfermented milks, juices, smoothies, cereal, nutrition bars and infant formula. Probiotics prove their potential in preventing and treating disease, treatment of food allergy including atopic eczema, diarrhoea in children, delaying the effects of aging in humans and modulating blood lipid levels.[21] Probiotic bacteria should have potential features relevant to the development of CRC biotherapeutics. For example, LAB have shown protective effects against CRC by reinforcing and modulating the host's natural defence mechanisms.[20]

15.3.1 PROBIOTICS ADOPT DIFFERENT MECHANISMS TO PREVENT CRC

a) Adhesion of probiotics and competitive exclusion of pathogenic microflora

b) Alteration of enzyme activity of intestinal pathogenic microflora.

c) Reduction of carcinogenic bile acids and mutagens.

d) Upregulation of Short chain fatty acids SCFA by probiotics.

e) Reduction of DNA damage by probiotics.

f) Inhibition of Promotion phase of CRC.

g) Immune system improvement and intestinal barrier function.

h) Inhibition effect of probiotics on TLR4 and COX-2 expression.

i) Enhancement of innate immune function by probiotics.

15.3.2 PROBIOTICS IN COLORECTAL SURGERY

In an invited commentary, Martindale et al.[25] reported several studies on post-operative surgery using Probiotics. In a study four probiotics regimen reduces postoperative complication after colorectal surgery. Probiotics play a significant role in preserving intestinal microbiota in surgical and ICU patients. Further, supplements of probiotics have shown benefits in surgical, trauma and ICU patients suffering from antibiotic-associated diarrhoea, *Helicobacter pylori* infections, necrotizing enterocolitis, and so forth. Probiotics mainly responsible in perioperative setting include *Lactobacillus acidophilus*, *Lactobacillus plantarum*, *Bifidobacterium lactis* and one fungus *Saccharomyces boulardii*, which are responsible in decreasing the injection caused by *Clostridium difficile*. The concept of using probiotics in post-operative period is quite interesting but it is consistently confirmed, it may become the future standard for bioecological control.

15.3.3   PREBIOTICS

Prebiotics, defined as 'non-digestible substances that provide a beneficial physiological effect on the host by selectively stimulating the favourable growth or activity of a limited number of indigenous bacteria'. There are different types of prebiotics, such as trans-GOS, inulin, arabinogalactan, pectin, beta-glucans and Xylooligosaccharides (XOS). For a prebiotic to survive it must transit through the upper gastrointestinal tract undigested and then be metabolised primarily by probiotic bacteria in the lower intestine as an energy source where glucose is limited.[7] The most common prebiotics are FOS, inulin and GOS, which are synthetic and natural in origin and have gained interest in prevention of CRC. Inulin, is commonly used as a prebiotic for both *Lactobacilli* and *Bifidobacteria* because of the proven ability to increase microbial mass and lower the pH of the colonic content.[22] Most human prebiotic studies however have focussed on the use of inulin and FOS.

15.3.3.1   PREBIOTICS ADOPT DIFFERENT MECHANISMS TO PREVENT CRC

- a)   Stimulation of beneficial indigenous gut bacteria.
- b)   Production of SCFAs and lactic acid, as fermentation products.
- c)   Modification of gene expression in cecum, colon and faeces.
- d)   Enhanced micronutrient absorption in the colon.
- e)   Modulation of xenobiotic metabolizing enzymes.
- f)   Modulation of immune response.

15.4   SYNBIOTICS

Synbiotic is the co-administration of probiotic and prebiotics with the expectation that the prebiotics will enhance the survival and growth of the probiotics.[23] Presently numerous synbiotic combinations are available commercially, however there exists limited evidence in the literature as to their effectiveness in enhancing probiotic survival and growth in vitro or in vivo.[24] Recent times, several studies have demonstrated the capacity for synbiotics to reduce CRC.

15.4.1   SYNBIOTICS ADOPT DIFFERENT MECHANISMS TO PREVENT CRC

a)   Facilitating apoptotic response to carcinogen-induced DNA damage in the colon.

b)   Enhancing colonization, stimulate growth, survival and activity of probiotics in the presence of selective prebiotic substrate.

c)   Increase SCFA production, anti-proliferative activity of synbiotics and downregulation of inducible NO-synthaseandcycloxygenase-2enzymes, involved in colon carcinogenesis.

d)   Immunomodulation.

e)   Modification of colonic bacterial ecosystem, leading to an overall improvement in metabolic activity of the colon and cecum.

15.4.1.1   CLINICAL STUDIES INVOLVED IN PREVENTION OF CRC BY PROBIOTICS, PREBIOTICS AND SYNBIOTICS USING ANIMAL MODEL AND HUMAN MODEL

Food carcinogens such as PAH, HCA, NOC, mycotoxins (aflatoxins) and acrylamide are responsible for causing cancers such as CRC, breast and prostate cancer which are reported and briefly discussed in preclinical and clinical studies given below.

Campbell et al.[26] reported that, probiotics such as *Bifidobacterium-producing* metabolites, could suppress colon cancer, with eventually affecting the mixed-function of cytochrome P450s and conversion of azoxymethane (AOM) from proximate to ultimate carcinogen. Another human cell line study was done by Baricault et al.[27] *They used* cultured human colon cancer cell line (HT-29) and milk, fermented with individual strains of *Lactobacillus helveticus, Bifidobacterium, L. acidophilus* or a mix of *Streptococcus thermophilus* and *L. delbrueckii* subsp. *Bulgaricus and then* cancer cell line (HT-29) were added into fermented milk, and Baricault et al.[27] found that 10–50% of the HT-29 cells showed a decrease in growth with increased in dipeptidyl peptides (specific marker for HT-29 cell differentiation).

In a study, it is shown that the mice bearing colon cancer in the large intestine by the feeding of yoghurt can reduce the level of β-glucuronidase and nitroreductase. De Moreno et al.[28] took a fermented dairy product containing *L. acidophilus, Bifidobacterium bifidum* and mesophilic cultures *Streptococcus lactis* and *Streptococcus cremoris*, for their study in human to observe the increasing and decreasing activity of nitroreductase and β-glucuronidase, after 3 weeks period of ingesting this fermented dairy product on the nitroreductase activity.

In a study, *Lactobacillus casei* DN 114001 has been shown to grow and survive in the presence of metabolize toxic compounds such as, IQ (2-amino-3-methylimidazo[4,5-f]quinoline), MelQx (2-amino-3,8-dimethylimidazo[4,5-f] quinoxaline and PhIP (2-amino-1-methyl-6-phenylimidazo[4,5-b]pyridine), the probiotics have the ability to bind these toxic compounds, which are depends on pH and other physicochemical conditions, all this above result indicate that LAB may be antagonized the onset of CRC.[29] Another study carried out by O'Mahony et al.[30] *with* knockout mice, and reported that there are certain changes of gut micro flora in interleukin-10 [IL-10], the administration of probiotic *Lactobacillus salivarius* UCC118, resulted in reduced of colon cancer, from this above studies it is also clarified that the probiotics may act against CRC development. Further in an animal model, Lee et al.[31] reported that probiotics have remarkable immunoprotective activity and also have antitumour effects, on this basis, the authors administered *L. acidophilus* SNUL, *L. casei* YIT9029 and *B. longum* HY8001 for 4 weeks, after that they found survivability rate of mice injected with tumour cells was increased with an increase in cellular immunity by an augmentation in the number of total T cells, NK cells and MHC class II+ cells, and CD4−CD8+ T cells. These findings suggest that the treatment with probiotics have the potential role to prevent CRC.

In an animal model study, prebiotic feed supplements as long-chain inulin-type fructans, has found to have increased in bifidogenic effect, lowered pH and modulated immunity; and reduced the AOM-induced colonic pre-neoplastic ACF, and has shown a profound effect in prevention and treatment of CRC.[14]

Similarly in a study, potential effects of FOS and inulin were reported on ACF-induced rat models. XOS and FOS were observed to inhibit colonic ACF in dimethylhydrazine (DMH)-treated rats by lowering cecal pH and serum triglyceride concentration.

In an another study, soymilk fermented with four strains such as, *S. thermophilus, L. acidophilus, B. infantis, B. longum these strains* showed higher AMA against 3,2-dimethyl-4-amino-biphenyl due to the production of anti-mutagenic molecules during bacterial fermentation of milk.[32] In a study, lactobacilli and bifidobacteria were attributed to the production of butyrate, irreversible mutagen binding and AMA.[33] This emphasizes the importance of viable probiotic bacteria consumption. In a clinical study by[34] human strain *L. rhamnosus* 231, known as potential probiotic strain, was shown to possess antimicrobial activity against several human pathogens.

In a human study, it was observed that administration of 10 g trans-GOS, which have increased the bifidobacterial count and modifying the fermentative activity of colonic flora.[35]

In a rat model,[36] Combined probiotic strain *B. lactis* and resistant starch have shown to facilitate the acute apoptotic response towards genotoxic carcinogen and colonic fermentative events. In a clinical study, comparing the composition between bacteria in stool of CRC patients and healthy individuals, it was found that a significant dysbiosis. In case of CRC patients there was significant decrease in probiotic bacteria including *Bifidobacteria* and abundant firmicutes in intestinal lumen, when compared with healthy individuals.[8] In a clinical study, involving 398 CRC patients, showed a daily consumption of *L. casei*, 1 g after each meal were effective in preventing the development of colorectal tumours.[37] Thus intake of probiotics may represent a logical strategy to restore the normal microflora and reducing the risk of CRC.

Another study was done by double blind placebo controlled trail of 26 healthy adults. Considering them to intake probiotics strain (*L. plantarum*/100 ml) through fermented rose hip drink (0.7 g probiotics/100 ml), for three week. After 1 week there is a significant increase in the population of faecal *bifidobacteria* and *lactobacilli* with decrease in pathogenic sulphite-reducing bacteria clostridia.[38,39] In the SHIME model, *L. acidophilus* LA5 or *L. casei* 01 mixed with a pressurized longan juice product, shown significantly increase *Lactobacillus* and bifidobacteria, which led to enhanced the fermentative capacity of the ecosystem, as indicated by increase in production of propionate, acetate and butyrate.[40] Similarly in a human cancer cell line study reported by Oberreuther-Moschner et al.,[41] daily consumption of 300 g probiotic yogurt containing *L. acidophilus* 145 and *B. longum* 913 for 6 weeks. Resulted in reduction of genotoxic activity in colon cancer cells HT29clone 19 A.[39]

In a clinical study Shaughnessy et al.[42] has reported probiotics played a protective role against rectal biopsies subjects. Combination of some agents such as Crucifera, Chlorophyllin, and a yogurt drink contain *L. casei*, *L. bulgaricus* and *S. thermophilus*, in healthy subjects. These subjects were led to feed a diet which contains meat, the meat cooked at high temperature 200°C. After that, the combined treatment significantly reduced the HCA-induced DNA damage in CRC cells.[39] In a DMH-induced colon cancer model of mice, Chang et al.[43] has investigated *L. acidophilus* KFRI342, isolated from a traditional Koren food, called as Kimchi, which reduces the number of DMH-induced ACF.[39]

A recent study showed that the effect of two functional meat products that is, cooked ham (CH) and traditional chorizo sausage (CS) containing the prebiotic FOS insulin with respect to prevention of neoplasm. CRC was induced in 5 groups of 7 fisher, 344 rats by using AOM. During 5 months, rats were fed with regular feed means control group or with control meat products together with feed (CH or CS) or functional meat products together with feed (functional CS and functional CH). Arthur et al.[44] has demonstrated monocolonization with mouse commensal adherent invasive *E.coli* NC 101 and human commensal *Enterococcus faecalis* OGIRF causing severe colitis. When mouse treated with *E. coli*, NC101 developed invasive carcinoma, but interestingly when mice treated with *E. faecalis* rarely tumours develop. The authors found the presence of polyketide synthases in *E. coli* NC101, but not in *E. faecalis*. Vermaand Shukla[45] reported a treatment of *L.GG* or *L. acidophilus,* reduction in beta glucosidase activity, and also administration of *L. casei* or *L. plantarum* decreased in the nitroreductase activity in DMH (1,2-DMH dihydrochloride)-treated rats. Goldin et al.[46] has investigated a reduction in the DMH-induced tumours in the colon of F344 rats, with supplementation of *L.GG*. Another study was done by administration of probiotic bacteria, *Bacillus polyfermenticus*, significantly reduced the DMH-induced DNA damage and number of ACF in F344 rats, with less blood lipid peroxidation, and increase in total radical trapping antioxidant potential.

15.5 CONCLUSION AND FUTURE PROSPECTIVE

Recent research continues to support, that is,. that probiotic consumption may reduce tumour growth, modulate the host immune response and re-establish healthy gut conditions in CRC subjects. Probiotic formulations have shown the potential effect to protect the gut and colon epithelial cells against toxic substances digested or produced within the intestine. Numerous in vitro and animal model studies have indicated the potential for probiotics including lactobacillus and bifidobacteria, and prebiotics such as oligofructose and inulin, to exert anti-neoplastic effects. Still, further human studies are needed to guide the decision of their establishment as complementary treatment in CRC. Further, most studies were performed experimentally on animal models while large clinical studies are required to determine the exact effects and mechanism of probiotics in CRC. In the recent years, there is a growing interest in the genetic manipulation of probiotic which are designed to act as a delivery system for anti-proliferative or proapoptoic factors such as IL-10, TCF-*B,* superoxide dismutase, catalase (antioxidant) in the gastrointestinal tract. Therefore, well-designed randomized, double blind, placebo-controlled human studies using probiotics or prebiotics will adequate follow-up are necessary for prevention and therapy.

KEYWORDS

- **colorectal cancer**
- **probiotics**
- **prebiotics**
- **synbiotics**

REFERENCES

1. Antonic, V.; Stojadinovic, A.; Kester, K. E.; Weina, P. J.; Brucher, B. L.; Protic, M.; Avital, I.; Izadjoo, M. Significance of Infectious Agents in Colorectal Cancer Development. *J. Cancer* **2013,** *4,* 227–240.
2. Huxley, R.; Woodward, M.; Clifton, P. The Epidemiologic Evidence and Potential Biological Mechanisms for a Protective Effect of Dietary Fiber on the Risk of Colorectal Cancer. *Curr. Nutr. Rep.* **2013,** *2,* 63–70.

3. Sankpal, U. T.; Pius, H.; Khan, M.; Shukoor, M. I.; Maliakal, P.; Lee, C. M.; et al. Environmental Factors in Causing Human Cancers: Emphasis on Tumorigenesis. *Tumour Biol.* **2012,** *33,* 1265–1274.

4. Yerushalmi, R.; Idelevich, E.; Dror, Y.; Stemmer, S. M.; Figer, A.; Sulkes, A.; Brenner, B.; Loven, D.; Dreznik, Z.; Nudelman, I.; Shani, A.; Fenig, E. Preoperative Chemoradiation in Rectal Cancer: Retrospective Comparison Between Capecitabine and Continuous Infusion of 5-Fluorouracil. *J. Surg. Oncol.* **2006,** *93,* 529–533.

5. Bandyopadhyay, B.; Narayan, C. M.; Probiotics, Prebiotics and Synbiotics – In Health Improvement by Modulating Gut Microbiota. *Int. J. Curr. Microbiol. Appl. Sci.* **2014,** *3,* 410–420.

6. Pieścik-Lech, M.; Shamir, R.; Guarino, A.; Szajewska, H. D. The Management of Acute Gastroenteritis in Children. *Aliment. Pharmacol. Ther.* **2013,** *37*(3), 289–303.

7. Aachary, A. A.; Gobinath, D.; Srinivasan, K.; Prapulla, S. G. Protective Effect of Xylooligosaccharides from Corncob on 1,2-Dimethylhydrazine Induced Colon Cancer in Rats. *Bioact. Carbohydr. Diet. Fibers* **2015,** *5,* 146–152.

8. Zhu, Q.; Gao, R.; Wu, W.; Qin, H. The Role of Gut Microbiotata in the Pathogenesis of Colorectal Cancer. *Tumour Biol.* **2013,** *34,* 1285–1300.

9. Pool-Zobel, B.; Veeriah, S.; Böhmer, F. D. Modulation of Xenobiotic Metabolising Enzymes by Anticarcinogens – Focus on Glutathione S-Transferases and Their Role as Targets of Dietary Chemoprevention in Colorectal Carcinogenesis. *Mutat. Res.* **2005,** *591,* 74–92.

10. Smith-Barbaro, P.; Hanson, D.; Reddy, B. S. Carcinogen Binding to Various Types of Dietary Fiber. *J. Natl. Cancer Inst.* **1981,** *67,* 495–497.

11. Mai, V.; Colbert, L. H.; Perkins, S. N.; Schatzkin, A.; Hursting, S. D. Intestinal Microbiota: A Potential Diet-Responsive Prevention Target in ApcMin Mice. *Mol. Carcinog.* **2007,** *46,* 42–48.

12. Reddy, B. S.; Hamid, R.; Rao, C. V. Effect of Dietary Oligofructose and Inulin on Colonic Preneoplastic Aberrant Crypt Foci Inhibition. *Carcinogenesis* **1997,** *18,* 1371–1374.

13. Bolognani, F.; Rumney C. J.; Coutts, J. T.; Pool-Zobel, B. L.; Rowland, I. R. Effect of Lactobacilli, Bifidobacteria and Inulin on the Formation of Aberrant Crypt Foci in Rats. *Eur. J. Nutr.* **2001,** *40,* 293–300.

14. Verghese, M.; Rao, D. R.; Chawan, C. B.; Shackelford, L. Dietary Inulin Suppresses Azoxymethane-Induced Preneoplastic Aberrant Crypt Foci in Mature Fisher 344 Rats. *J. Nutr.* **2002,** *132,* 2804–2808.

15. Femia, A. P.; Luceri, C.; Dolara, P. Antitumorigenic Activity of the Prebiotic Inulin Enriched with Oligofructose in Combination with the Probiotics Lactobacillus Rhamnosus and Bifidobacteriumlactis on Azoxymethane-Induced Colon Carcinogenesis in Rats. *Carcinogen* **2002,** *23,* 1953–1960.

16. Eckburg, P. B.; Bik, E. M.; Bernstein, C. N.; Purdom, E.; Dethlefsen, L.; Sargent, M.; Gill, S. R.; Nelson, K. E.; Relman, D. A. Diversity of the Human Intestinal Microbial Flora. *Science* **2005,** *308,* 1635–1638.

17. Cerf-Bensussan, N.; Gaboriau-Routhiau, V. The Immune System and the Gut Microbiota: Friends or Foes? *Nat. Rev. Immunol.* **2010,** *10,* 735–744.

18. Delzenne, N. M.; Cani, P. D. Interaction Between Obesity and the Gut Microbiota: Relevance in Nutrition. *Annu. Rev. Nutr.* **2011,** *31,* 15–31.

19. Hill, C.; Guarner, F.; Gibson, G. R.; Merenstein, D. J.; Pot, B.; Sanders, M. E. Expert Consensus Document. The International Scientific Association for Pro Consensus Statement on the Scope and Appropriate Use of the Term Probiotic. *Gastroenterol. Herpetol.* **2014**, *11*(8), 506–514.

20. Faghfoori, Z.; Gargari, B. P.; Gharamaleki, A. S.; Bagherpour, H.; Khosroushahi, A. Y. Cellular and Molecular Mechanisms of Probiotics Effects on Colorectal Cancer. *J. Funct. Foods* **2015**, *18*, 463–472.

21. Omar, J. M.; Chan, Y.; Jones, M. L.; Prakash, S.; Jones, P. J. H. *Lactobacillus fermentum* and *Lactobacillus amylovorus* as Probiotics Alter Body Adiposity and Gut Microflora in Healthy Persons. *J. Funct. Foods* **2013**, *5*, 116–123.

22. Gobinath, D.; Madhu, A. N.; Prashant, G.; Srinivasan, K.; Prapulla, S. G. Benificial Effect of Xylo-Oligosaccharides and Fructo-Oligosaccharides in Streptozotocin-Induced Diabetic Rats. *Br. J. Nutr.* **2010**, *104*, 40–47.

23. Grimoud, J.; Duran, H.; de Souza, S.; Monsan, P.; Ouarne, F.; Theodorou, V.; Roques, C. In vitro Screening of Probiotics and Synbiotics According to Ant-Inflammatory and Anti-Proliferative Effects. *Int. J. Food Microbiol.* **2010**, *144*, 42–50.

24. Duncan, S. H.; Flint, H. J. Probiotics and Prebiotics and Health in Ageing Populations. *Maturitas* **2013**, *75*(1), 44–50.

25. Martindale, R.; Warren, M.; Tsikitis, V. L. Probiotics in Colorectal Surgery. *World J. Surg.* DOI: 10.2015 10007/s00268-015-3173-3177.

26. Campbell, T. C.; Hayes, J. R. The Effect of Quantity and Quality of Dietary Protein on Drug Metabolism. *Feder. Pro.* **1976**, *35*, 2470–2474.

27. Baricault, L.; Denariaz, G.; Houri, J. J.; Bouley, C.; Sapin, C.; Trugnan, G. Use of HT-29, a Cultured Human Colon Cancer Cell Line, to Study the Effect of Fermented Milks on Colon Cancer Cell Growth and Differentiation. *Carcinogenesis* **1995**, *16*, 245–252.

28. de Moreno de LeBlanc, A.; Perdigón, G. Reduction of Beta-Glucuronidase and Nitro-reductase Activity by Yoghurt in a Murine Colon Cancer Model. *Biochemistry* **2005**, *29*, 15–24.

29. Nowak, A.; Libudzisz, Z. Ability of Probiotic *Lactobacillus casei* DN 114001 to Bind or/and Metabolise Heterocyclic Aromatic Amines in vitro. *Eur. J. Nutr.* **2009**, *48*, 419–427.

30. O'Mahony, L.; Feeney, M.; O'Halloran, S. Probiotic Impact on Microbial Flora, Inflammation, and Tumour Development in IL-10 Knockout Mice. *Alimentary Pharmacol. Ther.* **2001**, *15*, 1219–1225.

31. Lee, J. W.; Shin, J. G.; Kim, E. H.; Kang, H. E.; Yim, I. B.; Kim, J. Y.; Joo, H. G.; Woo, H. J. Immunomodulatory and Antitumor Effects in vivo by the Cytoplasmic Fraction of *Lactobacillus casei* and *Bifidobacterium longum*. *J. Vet. Sci.* **2004**, *5*, 41–48.

32. Hsieh, M. L.; Chou, C. C. Mutagenicity and Antimutagenic Effect of Soymilk Fermented with Lactic Acid Bacteria and *Bifidobacteria*. *Int. J. Food Microbiol.* **2006**, *111*, 43–47.

33. Lankaputhra. W. E. V.; Shah, N. P. Antimutagenic Properties of Probiotic Bacteria and of Organic Acids. *Mutat. Res.* **1998**, *397*, 169–182.

34. Ambalam, P.; Dave, J. M.; Nair, B. M.; Vyas, B. R. In vitro Mutagen Binding and Antimutagenic Activity of Human *Lactobacillus rhamnosus* 231. *Anaerobe* **2011,** *17,* 217–222.

35. Bouhnik, Y.; Flourié, B.; D'Agay-Abensour, L.; Pochart, P.; Gramet, G.; Durand, M. Administration of Transgalacto- Oligosaccharides Increases Fecal Bifidobacteria and Modifies Colonic Fermentation Metabolism in Healthy Humans. *J. Nutr.* **1997,** *127,* 444–448.

36. Le Leu, R. K.; Brown, I. L.; Hu, Y.; Bird, A. R.; Jackson, M.; Esterman, A. A Synbiotic Combination of Resistant Starch and Bifidobacteriumlactis Facilitates Apoptotic Deletion of Carcinogen-Damaged Cells in Rat Colon. *J. Nutr.* **2005,** *135,* 996–1001.

37. Ishikawa, H.; Akedo, I.; Otani, T.; Suzuki, T.; Nakamura, T.; Takeyama, I.; Ishiguro, S.; Miyaoka, E.; Sobue, T.; Kakizoe, T. Randomized Trail of Dietary Fiber and *Lactobacillus casei* Administration for Prevention of Colorectal Cancer Tumors. *Int. J. Cancer* **2005,** *116,* 762–767.

38. Johansson, M.; Nobaek, S.; Berggren, A.; Nyman, M.; Bjorck, I.; Ahrne, S.; Jeppsson, M. G. Survival of *Lactobacillus plantarum* DSM 9843 (299V), and Effect on the Short Chain Fatty Acid Content of Faeces After Investigation of a Rose Hip Drink with Fermented Oats. *Int. J. Food Microbiol.* **1998,** *42,* 29–38.

39. Chong, E. S. L. A Potential Role of Probiotics in Colorectal Cancer Prevention: Review of Possible Mechanism Action. *World J. Microbial. Biotechnol.* **2014,** *30,* 351–374.

40. Chaikham, P.; Apichartsrangkoon, A.; Jirarattanarangsri, W.; Van de Wiele T. Influence of Encapsulated Probiotics Combined with Pressurized Longan Juice on Colon Microflora and Their Metabolic Activities on the Exposure to Simulated Dynamic Gastrointestinal Tract. *Food Res. Int.* **2012,** *49,* 133–142.

41. Oberreuther-Moschner, D.; Jahreis, G.; Rechkemmer, G.; Pool-zobel, B. Dietary Intervention with the Probiotics *Lactobacillus acidophilus* 145 and *Bifidobacterium longum* 913 Modulates the Potential of Human Faecal Water to Induce Damage in HT29clone 19A Cells. *Br. J. Nutr.* **2004,** *91,* 925–932.

42. Shaughnessy, D. T.; Gangarosa, L. M.; Schliebe, B.; Umbach, D. M.; Xu, Z.; Maclntosh, B.; Knize, M. G.; Matthews, P. P.; Swank, A. E.; Sandler, R. S.; DeMarini, D. M.; Taylor, J. A. Inhibition of Fried Meat Induced Colorectal DNA Damage and Altered Systemic Genotoxicity in Humans by Crucifera, Chlorophyllin and Yogurt. *PLosS One.* **2011,** *6,* 18707.

43. Chang, J. H.; Shim, Y. Y.; Cha, S. K.; Reaney, M. J. T.; Chee, K. M. Effect of *Lactobacillus acidophilus* KFRI342 on the Development of Chemically Induced Precancerous Growths in Rat Colon. *J. Med. Microbiol.* **2012,** *61,* 361–368.

44. Arthur, J. C.; Perez-Chanona, E.; Muhlbauer, M.; Tomkovich, S.; Uronis, J. M.; Fan, T. J.; Campbell, B. J.; Abujamel, T.; Dogan, B.; Rogers, A. B.; Rhodes, J. M.; Stintzi, A.; Simpson, K. W.; Hansen, J. J.; Keku, T. O.; Fodor, A. A.; Jobin, C. Intestinal Inflammation Targets Cancer Inducing Activity of the Microbiota. *Science* **2012,** *338,* 120–123.

45. Verma, A.; Shukla, G. Probiotics *Lactobacillus rhamonsus* GG, *Lactobacillus acidophilus* Suppresses DMH-Induced Procarcinogenic Fecal Enzymes and

Preneoplastic Aberrant Crypti Foci in Early Colon Carcinogenesis in Sprague Dawley Rats. *Nutr. Cancer* **2013b,** *65,* 84–91.

46. Goldin, B.; Gualtieri, L.; Moore, R. The Effect of Lactobacillus GG on the Initiation and Promotion of DMH-Induced Intestinal Tumours in the Rat. *Nutr. Cancer* **1996,** *25,* 197–204.

47. Annema, N.; Heyworthm, J. S.; McNaughton, S. A. Fruit and Vegetable Consumption and the Risk of Proximal Colon, Distal Colon, and Rectal Cancers in a Case-Control Study in Western Australia. *J. Am. Diet Assoc.* **2011,** *111,* 1479–1490.

48. Zick, S. M.; Ruffin, M. T.; Lee, J. Phase II Trial of Encapsulated Ginger As a Treatment for Chemotherapy-Induced Nausea and Vomiting. *Support Care Cancer* **2009,** *17*(5), 563–572.

49. Robert, E. C.; Richard, V. B.; Danielle, K. T.; Shaiju, V.; Malloree, N.; Luz, R.; Madhuri, K.; Philip, M. C.; Christine, M.; Frank, L. M.; Dean, E. B. Phase IIA Clinical Trial of Curcumin for the Prevention of Colorectal Neoplasia. *Cancer Prev. Res.* (Phila) **2011,** *4*(3), 354–364.

50. Li, Y. G.; Ji, D. F.; Zhong, S.; Zhu, J. X.; Chen, S.; Hu, G. Y. Anti-Tumor Effects of Proteoglycan from *Phellinus linteus* by Immunomodulating and Inhibiting Reg IV/EGFR/Akt Signalling Pathway in Colorectal Carcinoma. *Int. J. Biol. Macromol.* **2011,** *48,* 511–517.

51. Bi, X.; Fang, W.; Wang, L.; Stoner, G.; Yang, W. Black Raspberries Inhibit Intestinal Tumorigenesis in Apc1638 /– and Muc2–/– Mouse Models of Colorectal Cancer. *Cancer Prev. Res.* **2010,** *3*(11), 1443–1450.

52. Vargas, A. J.; Thompson, P. A. Diet and Nutrient Factors in Colorectal Cancer Risk. *Nutr. Clin. Pract.* **2012,** 27, 613–623.

53. Pala, V.; Sierim, S.; Berrino, F. Yogurt Consumption and the Risk of Colorectal Cancer in the Italian European Prospective Investigation into Cancer and Nutrition Cohort. *Int. J. Cancer* **2011,** *129,* 2712–2719.

54. O'Keefe, S. J.; Ou, J.; Aufreiter, S. Products of the Colonic Microbiota Mediate the Effects of Diet on Colon Cancer Risk. *J. Nutr.* **2009,** *139,* 2044–2048.

55. Gonzalez, C. A.; Jordi, S. S. The Potential of Nuts in the Prevention of Cancer. *Br. J. Nutr.* **2006,** *96,* 87–94.

CHAPTER 16

PURIFICATION AND CHARACTERIZATION OF MICROBIAL POLYGALACTURONASES: AN UPDATE

GAUTAM ANAND, SANGEETA YADAV, AIMAN TANVEER, and DINESH YADAV*

*Department of Biotechnology, Deen Dayal Upadhyay Gorakhpur University, Gorakhpur, Uttar Pradesh 273009, India, Tel.: 09411793038, *E-mail: dinesh_yad@rediffmail.com*

CONTENTS

ABSTRACT

Polygalacturonases (PGs) belong to a large group of pectinases, which synergistically mediate the complete decomposition of pectin substances that are abundantly present in plant tissues. Polygalacturonases represent an important member of pectin-degrading glycoside hydrolases of the family GH28 and is widely distributed among bacteria, fungi and yeasts.

Microbes secrete PGs to invade plant tissues during pathogenesis. Beside this, microbial PGs are widely used in food industry for clarification of fruit and vegetable juices. Alkaline PGs are used in textile industry for degumming and retting of plant fibres. This review focuses on comparative biochemical characterization of microbial PGs which will be useful for its application in different industries.

16.1 INTRODUCTION

The plant cell wall is a complex macromolecular structure that surrounds and protects the cell, and is a distinguishing characteristic of plants essential to their survival. As a consequence of limited mobility, plants show plasticity in their ability to withstand a variety of harsh environmental conditions and to survive attack by pathogens and herbivores. The structure formed by the polysaccharides, proteins, aromatic and aliphatic compounds of the cell wall enables plants to flourish in diverse environmental niches.

The plant cell lays down the middle lamella and the primary wall during initial growth and expansion of the cell. In many cells, the wall is thickened and further strengthened by the addition of a secondary wall. The primary wall is characterized by relatively less cellulose and greater pectin compared to secondary walls. The primary wall is thought to contribute to wall structural integrity, cell adhesion and signal transduction. The major fraction of primary wall non-cellulosic polysaccharides in the Type-I walls of dicot and non-graminaceous species are the pectic polysaccharides.

Pectin is a class of GalA-containing polysaccharides that are abundant in the plant cell wall; comprising as much as 30% of dicot, gymnosperm and non-Poales monocot walls.[79] It is mostly found in walls of cells surrounding soft parts of the plant, growing and dividing cells, middle lamella and corner of cells.[62] Pectin is also present in the junction zone between cells with secondary walls including xylem and fibre cells in woody tissue. Pectin polysaccharides are galacturonic acid polymers and are represented by three major types: homogalacturonan (HG), rhamnogalacturonan-I (RG-I) and RG-II.[6] Pectic polysaccharides are synthesized in the golgi and are delivered to the cell wall by secretory vesicles moving primarily along the actin cytoskeleton,[100,47] although there is recent evidence for kinesin-dependent pectin delivery via microtubules.[115] The pectin biosynthesis,[79,61,88] plant wall biosynthesis[112,87,53,24] and regulation of cell wall synthesis[37,113] has been extensively reviewed over the years.

16.1.1   PECTINASES

Pectinases or pectinolytic enzymes are a complex group of enzymes that degrade various pectic substances (pectin) present in the middle lamella of plant cell wall. Great complexity and diversity in the smooth and hairy regions of pectins require several kinds of degrading enzymes based on the specificity of substrate as well as type of reactions they catalyse.[30]

The groups of enzymes which are involved in the degradation of hairy region of pectins are RG hydrolase, RG lyase, RG rhamnohydrolase and RG galactohydrolase. There are only few reports about this group of enzymes.[89,94,50,64] There is a need for extensive studies on these groups of enzymes targeting their structural and functional aspects so as to explore its industrial applications. There are, however, other enzymes involved in degradation of side chains of pectins which include α-arabinofuranosidase [E. C 3.2.1.55], endoarabinase, [E. C 3.2.1.99] β-galactosidase, [E. C 3.2.1.23] endogalactanase, [E. C 3.2.1.89] and feruloyl and p-coumaroyl esterases.[102]

The group of enzymes which are associated with the degradation of 'smooth region' (HG) can be broadly categorized in two groups namely Esterases and Depolymerases.

Esterases are basically deesterifying enzymes which remove methoxyl and acetyl residues of pectin producing polygalacturonic acid (PGA). It includes pectin methyl esterases [PME, E. C 3.1.1.11] and pectin acetyl esterase [PAE, E. C 3.1.1.6]. The other group depolymerases act by breaking the α-1, 4 linkages either by hydrolysis that is Polygalactu-ronases [PG, E. C 3.2.1.15] or via transelimination mechanism namely pectate lyases [PL, E. C 4.2.2.2] and pectin lyases [PNL, E. C 4.2.2.10].

16.2   PURIFICATION OF MICROBIAL POLYGALACTURONASES (PGs)

PGs being of great industrial importance, knowledge of their biochemical properties are essential for their utilization in relevant industries. For carrying out the biochemical activity for its characterization and to study its industrial applications, the enzyme needs to be purified. Attempts have been made by several types of research to purify the PGs from several sources. Important purification methods used for purification of PG is briefly summarized in the Table 16.1.

TABLE 16.1 Purification Strategy and Biochemical Characterization of Microbial Polygalacturonases.

Source	Type	Purification strategy	pH optima	Temp. optima	Km	MW (kDa)	References
Zygoascus hellenicus	–	Ammonium sulphate precipitation, DEAE cellulose chromatography and Sephadex G-100 gel filtration	5.0	60	5.44	75.28	[55]
Penicillium occitanis	Endo-PG	MonoQ-FPLC purification	6.0	35		42	[86]
Aspergillus fumigatus	Endo-PG	Acetone precipitation and gel-filtration chromatography	10	30	2.4 mg/ml	43	[3]
Penicillium janthinellum	Exo-PG	Gel filtration chromatography	5	45	1.74 mg/ml	66.2	[107]
Penicillium oxalicum	Endo-PG	Anion exchange and gel filtration chromatography	5.0	60–70	1.27	38.0	[15]
Neosartorya fischeri	Endo-PG I	FPLC	5.0	–	0.73	37.9	[72]
	Endo-PGII		4.0	65	0.50	38.2	
	Exo-PGI		3.5	–	0.53	50.2	
Penicillium chrysogenum	–	–	5.0	43	–	–	[108]
Aspergillus niger	Endo-PG	Ammonium sulphate precipitation and gel filtration	5.0	50	88.54 μmol/ml	38.8	[114]
Caldicellulosiruptor bescii	Exo-PG	Ni²⁺-NTA agarose affinity column	5.2	72	0.31	50	[14]
Fusarium graminearum	Endo-PG	Gel filtration and ion exchange chromatography	5.0	50	–	40	[70]

TABLE 16.1 *(Continued)*

Source	Type	Purification strategy	pH optima	Temp. optima	Km	MW (kDa)	References
Ustilago maydis	Endo-PG	Ni²⁺-agarose affinity chromatography	4.5	34	57.84	37.9	[11]
Thermoascus aurantiacus	Endo-PG	Gel filtration and ion exchange chromatography	4.5–5.5	60–65	1.58	29.3	[59]
A. niger	–	Gel filtration chromatography	4.8	45	0.083	106	[39]
Rhizopus oryzae	Exo-PG	Ion exchange chromatography	5.0	50	2.7	75.5	[103]
Paecilomyces variotii	Exo-PG	Gel filtration and ion exchange chromatography	6.0	30	–	39.4	[74]
Klebsiella sp.	Exo-PG	Ni²⁺-NTA chromatography	6.0	50	–	72.2	[106]
Bacillus sp.	–	–	10.0	60–70	–	–	[4]
Aspergillus carbonarius	–	Integrated membrane process (IMP) and alginate affinity precipitation (AAP)	–	–	–	42	[66]
P. variotii	Exo-PG	Ion exchange and gel filtration chromatography	4.0	65	1.84	77.3	[55]
Mucor circinelloides	–	Gel filtration chromatography	5.5	42	2.2	66	[99]
Bacillus subtilis	Exo-PG	–	7.0	50	–	–	[95]
Fomes sclerodermeus	–	–	5.0	50	–	–	[85]
Aspergillus sojae	Exo-PG	Three-phase partitioning (TPP)	4.0	75	0.75	36 53	[21]

TABLE 16.1 (Continued)

Source	Type	Purification strategy	pH optima	Temp. optima	Km	MW (kDa)	References
Burkholderia glumae	—	Cation exchange and gel filtration chromatography	4.0	45	—	47	[18]
			3.5	45		50	
Streptomyces lydicus	Exo-PG	Ion exchange and gel filtration	6.0	50	1.63	43	[35]
T. aurantiacus	—	Gel filtration and ion exchange chromatography	5.5	60	1.46	30	[59]
Trichoderma harzianum	—	Anion exchange and gel filtration	4.4–7.0	55	1.42	29	[60]
Botrytis cinerea	Endo-PG	Ion exchange chromatography	4.2	—	0.60	34	[41]
			4.5		0.44	35	
			3.2–4.5		0.27	49	
			4.9		0.33	36	
			4.5		0.16	36	
Thermotoga maritime	Exo-PG	Anion exchange chromatography	6.4	80	—	50	[48]
R. oryzae	—	Gel filtration and ion exchange chromatography	—	—	—	37.4	[110]
Phytophthora parasitica	Endo-PG	—	—	—	—	39.7	[105]
Sporotrichum thermophile	—	—	7.0	55	0.416	—	[43]

TABLE 16.1 *(Continued)*

Source	Type	Purification strategy	pH optima	Temp. optima	Km	MW (kDa)	References
R. oryzae	Endo-PG	–	4.5	45	–	31	[82]
Aspergillus kawachii	Endo-PG I	Gel filtration	2.0–3.0	–	–	60	[17]
Aspergillus japonicas	Endo-PG I	Anion exchange chromatography and phenyl-Sepharose hydrophobic chromatography	–	–	–	38	[89]
	Endo-PG II					65	
B. cinerea	Exo-PGI	–	5.0	–	–	65	[10]
	Endo-PG II		5.2			52	
A. niger	Endo-PGIIs	–	3.8–4.3	43	0.12	61	[92]
	Endo-PG IV		3.0–4.6	46	0.72	38	
Kluyveromyces marxianus	Endo-PG	Ion exchange chromatography	4.0	55	–	41.7	[91]
Penicillium frequentans	–	–	3.9	50	0.68	74	[22]
A. niger	Exo-PG I	–	–	–	–	82	[84]
	Exo-PG II					56	
Fusarium moniliforme	Endo-PG	Ion exchange and gel filtration chromatography	4.8	–	0.12	38	[69]
Bacillus sp.	Exo-PG	Column chromatography	7.0	60	–	45	[49]
Fusarium oxysporum f. sp. Lycopersici	Exo-PG	–	–	–	–	47.9	[28]

TABLE 16.1 (Continued)

Source	Type	Purification strategy	pH optima	Temp. optima	Km	MW (kDa)	References
Aspergillus awamori	Endo-PG	Cation-exchange and size-exclusion chromatographies	–	–	–	41	[65]
Saccharomyces cerevisiae	Endo-PG	Anion exchange chromatography	3–5.5	25	–	42	[26]
Bacillus sp.	–	–	10.0	60	–	–	[40]
Lentinus edodes	–	–	5.0	50	–	–	[110]
Fusarium solani	Endo-PG	Gel filtration and cation exchange chromatography	6.0	–	1.34	38	[109]
Thermomyces lanuginosus	–	Ion exchange chromatography	5.5	60	0.67	59	[51]
Yersinia enterocolitica	Exo-PG	Cation exchange chromatography	–	–	–	66	[54]
A. niger	Endo-PG	–	3.8	–	2.40	35.5	[73]
Sclerotinia sclerotiorum	Endo-PG I	Ion exchange chromatography	3.5	50	0.5	41.5	[58]
	Endo-PG II		4.0	50	0.8	42.0	
Alternaria alternate	Endo-PG	Cation exchange HPLC and gel filtration chromatography HPLC	5.0	–	0.1	60	[34]
Sclerotinia borealis	Endo-PG	–	4.5	40–50	–	40	[97]
A. carbonarius	Endo-PGI	Molecular sieve chromatography, ion exchange chromatography followed by gel filtration	4.0	43	–	61	[19]
	Endo-PGII		4.1	46		42	
	Endo-PGIII		4.3	54		47	

TABLE 16.1 (Continued)

Source	Type	Purification strategy	pH optima	Temp. optima	Km	MW (kDa)	References
Aspergillus tubingensis	Exo-PG	—	4.2		3.2	78	[45]
Saccharomyces pastorianus	—	Ion exchange and gel filtration	4.2	50	0.625	43	[5]
F. oxysporum f. sp. Lycopersici	—	Preparative isoelectric focusing (IEF)	4.0	37	—	35 37.5	[20]
Aspergillus ustus	Endo-PG	—	5.0	—	0.82	36	[78]
Postia placenta	Endo-PG	—	3.2–3.9	25–37	—	34	[16]
P. frequentans	Endo-PG	Ion exchange chromatography	4.0–4.7	50	2.7	20	[25]
S. cerevisiae	Endo-PG	Gel filtration chromatography	5.5	45	4.7	39	[9]
B. cinerea	Endo-PG I	Isoelectric focusing, ion exchange chromatography and affinity chromatography	4–5	37	—	36	[36]
	Endo-PG II			34		36	
	Exo-PG I			48		65	
	Exo-PG II			50		70	
S. sclerotiorum	Exo-PG	Gel filtration and ion exchange	5.0	45	—	68	[80]
Neurospora crassa	Endo-PG	Ion exchange chromatography	6.0	45	5.0	37	[77]
Erwinia carotovora	Endo-PG	—	5.5	35–45	—	40	[81]
Rhizopus stolonifer	Endo-PG	Ion exchange and gel filtration chromatography	5.0	45	0.19	57.5	[57]

TABLE 16.1 *(Continued)*

Source	Type	Purification strategy	pH optima	Temp. optima	*Km*	MW (kDa)	References
Trichoderma koningii	Endo-PG I	Ion exchange and isoelectric focusing	–	–	–	32	[23]
	Endo-PG II					32	
Rhizoctonia fragariae	Endo-PG	Gel filtration	–	–	–	36	[12]
F. oxysporum f. sp. *Lycopersici*	Endo-PG I	Chromatography on DEAE-cellulose, CM-cellulose and hydroxyapatite	5.0	–	–	37	[93]
	Endo-PG II						
E. carotovora	–	Ion exchange chromatography	5.2–5.4	–	–	–	[67]

FPLC: Fast Protein Liquid Chromatography

Few examples of the strategy adopted by several groups for purification of PG are mentioned briefly. Gel filtration is the most commonly used purification method adopted by researchers for purification of PGs. For further purification of the enzyme ion exchange chromatography has been implemented by several research groups.

Before loading on to the gel filtration or ion exchange column subjecting the crude protein to ammonium sulphate precipitation is also a general practice.

Lu et al.[56] have purified the exo-PG from *Zygoascus hellenicus* V25 by initially concentrating on ammonium sulphate precipitation. The precipitate was dissolved in 50 mM sodium phosphate buffer (pH 7.0) and purified by using a Sephadex G-25 column. Fractions with high PG activity were further purified by DEAE cellulose column. The fractions were concentrated and again purified through Sephadex G-100 column. The final purified protein had 16.89-fold purification with a recovery of 18.46% and the specific activity of the final purified protein was 2469.77 U/ mg. An extracellular endo-PG has recently been purified by single step purification from a mutant strain of *Penicillium* occitanis. [86] This enzyme was unique in having a highest specific activity of 31397.26 U/mg. The method used by the group for purification was cation exchange chromatography on fast performance liquid chromatography (FPLC) column. Our research group has lately purified PG from *Aspergillus fumigatus* strain MTCC 2584 which is also probably the first report of alkaline PG from *Aspergillus fumigatus*.[3] The strategy used for purification was by first precipitation by chilled acetone and purification by Sephadex G-100 column. After these two steps of purification, 18.43 purified proteins were obtained which showed efficient retting of *Crotalaria juncea* fibres.

Apart from the purification of the native protein, the researchers have also attempted the purification of recombinant protein. Zhou et al. have cloned and expressed endo-PG gene of *Aspergillus niger*.[114] The over-expressed protein *Saccharomyces cerevisiae* EBY100 was purified by ammonium sulphate precipitation and gel filtration column chromatography. The enzyme was purified to 4.36 fold by the above approach and had specific activity of 1,448.48 U/mg. *Escherichia coli* origami (DE3) has been used for the expression of thermophilic PG from *Caldicellulosiruptor bescii* DSM 6725.[14] The gene was cloned in pET20b expression plasmid and the resultant recombinant protein (CbPelA) was affinity purified through Ni-NTA agarose column extraction. CbPelA exhibited very high activity for PGA and released monogalacturonic acid as its sole product.

PGases isolated from different microbial sources differ markedly from each other with respect to their physicochemical and biological properties and their mode of action. Among the PGases obtained from different microbial sources are acidic and most have the optimal pH range of 2–6. There are exceptional cases of PGs working in the alkaline range. An endo PG isolated from *A. fumigatus* had optimum pH 10.0.[3] Another PG from *Bacillus* spp. had pH optima of 10.[4] There is only one report of an exo-PG with optimum activity at pH 7.0.[49] Most the PGs work in the temperature range of 35–60. Enzymes working under extreme conditions have also been reported. Recombinant exo-PG from *Caldicellulosiruptor bescii* has been reported to have temperature optima of 72°C.[14] Another exo-PG from *A. sojae* has also been reported to have high-temperature optima of 75°C.[21] PG from thermophilic bacteria namely *Thermotoga maritima* and *Caldicellulosiruptor bescii* have temperature optima of 80 and 72°C respectively.[48,14]

The molecular mass of microbial PG ranges from 30–80 kDa[3,18,26,36,66,90,103,109] though molecular weight of exo-PG from *Bacillus* sp. was found to be 115 kDa.[49] Most of the PGs reported so far have their PI in the range between 3 and 9.[5,27,65,49]

The Km values reported for most of the microbial PGs is in the range of 0.1–5.0,[3,14,59, 68,77,92,99,103,109] with exception to exo-PG from the fungus *Penicillium frequentans* having very low Km value 0.059[7] and a high Km of 57.84 for PG from pathogenic fungus *Ustilago maydis*.[11] PG from *Aspergillus niger* also had a low Km of 0.083.[39]

16.3   APPLICATIONS OF POLYGALACTURONASES

Pectinase production shares about 10% of the overall manufacturing of enzyme preparations.[75] On the basis of chemical properties, pectinase has been utilized in a number of industries. Acid pectinases are used in clarification of fruit juices, maceration of vegetables in the production of pastes and purees, wine making and so forth.[29] Alkaline pectinase find sits application in the processing of natural textile fibres (such as jute, flax and hemp), treatment of pectic wastewater, coffee and tea fermentation, vegetable oil extraction, treatment of paper pulp and so forth.

Enzymes produced from the fungi *Aspergillus*, *Rhizopus* and *Penicillium* are generally regarded as safe (GRAS) and produces extracellular enzymes which can be easily recovered.[63] Commercial use of pectinases started in 1930 in various applications of food industries such as clarification of juice, mashing treatments and enhancing the yield and colour of the products.[44,13,38]

In vegetable and fruit juice industry, juices with high clarity and low viscosity have high commercial values. Microbial PGs have been shown to play role in viscosity reduction and clarification of juice.[39,8] Acidophilic PGs with optimum pH close to pH of fruit juices are preferably utilized for juice preparation and clarification. Pectinases, specifically PGs, find applications in the paper and textile industry, and in food processing especially in the production of fruit juice.[42] In the food industry, PGs are added to clarify fruit or vegetable juice. Besides commercial pectinases from *Aspergillus niger*[76] and *Bacillus subtilis*[96] predominantly used as enhancers of juice clarity, colour and yield,[71] several PGs from *Aspergillus carbonarius* and *Achaetomium* sp. Xz8 have been reported to have the capacities to improve the juice yield and clarity[66] and reduce the viscosity of papaya juice.[101] Polygalacturonase produced by *A. niger* using banana peel as a substrate has been used for clarification of banana juice.[8] Use of PGs from *Neosartorya fischeri* in clarification of apple and strawberry juice have also been reported.[72]

Bast fibres are the soft fibres formed in groups outside the xylem, phloem or pericycle, for example Ramie and sunn hemp. The fibres contain gum, which must be removed before its use for textile making.[33] Retting is a process employing the action of microorganisms and moisture on plants to dissolve or rot away much of the cellular tissues and pectins surrounding bast-fibre bundles, and so facilitating separation of the fibre from the stem. An alkaline endo-PG from *A. fumigatus* having a role in retting of *Crotalaria juncea* fibre has recently been reported.[3] Role of alkaline and thermostable PG from *Bacillus* sp. MG-cp-2 in degumming of ramie (*Boehmeria nivea*) and sunn hemp (*Crotalaria juncea*) bast fibre has been reported.[40] *Rhizopus oryzae* sb culture filtrate, which lacked xylanase and mannanase activities, has been successfully used for retting flax.[31] The *R. oryzae* filtrate was shown to have high pectinase activity and further work implicated a strong (endo) PGase as the primary retting agent.[1,2] It has been shown that PGase-induced degradation of low esterified pectin was the sole activity that correlated with the ability of the enzymatic mixture to perform retting. It has also been demonstrated that a purified (endo) PGase from *Aspergillus niger* was capable of retting flax.[111] Generally, alkaline PGs find application in retting and degumming of plant fibres but Yadav et al.[103] reported the role of acidic PG

from *R. oryzae* in retting of sunn hemp fibre. Polygalacturonase also find a
role in tea leaf fermentation. PG from *Mucor circinelloides* has been used
for the improvement of tea leaf fermentation by increasing the content of
phenolic compounds.[98] PGs have also been used in the development of
functional food. Oligogalacturonides, the digestion product of PGs were
shown to have antiulcer and anti-cancerous activity.[104,46] Non-digestible
oligosaccharides can also be used as health-promoting compounds analo-
gous to biologically active compound inulin. These oligosaccharides are
called prebiotics which promotes the growth of naturally residing bacteria
in the digestive tract.[52]

16.4   CONCLUSION

Microbial enzymes contribute immensely to the global market of industrial enzymes and several groups of enzymes have been reviewed over the
years. The application of microbial enzymes in different industries like
dairy, baking, beverages, animal feed, pulp and paper, polymer, detergent,
leather, cosmetics, organic synthesis and waste management is well documented. Polygalacturonases (PG) is an important member of pectinases
group of enzymes associated with degradation of pectin and its application
in fruit juice clarification and retting of fibres is of industrial importance.
The trend in enzyme research has undergone substantial changes over the
years marked by technological innovations and manifested in the form of
state-of-the-art technologies like directed evolution, metagenomics, bioinformatics and genomics. The technological advancement has influenced
the traditional enzyme research associated with isolation, purification and
characterization of microbial enzymes. The screening of microbial sources
from diverse habitats, developing appropriate purification protocols,
enzymatic characterization is still a prerequisite for elucidating diverse
industrial application. An attempt has been made in this review to update
the microbial sources, purification strategies, biochemical attributes and
diverse applications exclusively for the microbial PGs.

16.5 ACKNOWLEDGMENTS

S. Yadav sincerely acknowledges the financial grants received by DST, New Delhi, in the form of Women and Young Scientist Fellowships (SR/WOSA/LS-34/2004; SR/FT-LS-125/2008 and SB/FT/LS-430/2012). D. Yadav acknowledges the financial support of UGC, India, in the form of UGC-Major project (F. No. 37–133/2009-SR). UGC-Dr. D. S. Kothari Post-Doctoral fellowship to A Tanveer and JNM Scholarship for Doctoral Studies to G Anand is also sincerely acknowledged. The authors acknowledge the infrastructural support from the Head, Department of Biotechnology, D. D. U. Gorakhpur University, Gorakhpur.

KEYWORDS

- pectinases
- pectin
- polygalacturonases
- microbial
- purification
- application

REFERENCES

1. Akin, D. E.; Henriksson, G.; Slomczynski, D.; Eriksson, K. E. L. Retting of Flax with Endopolygalacturonase from *Rhizopus* Species. Two hundred-Eighteenth ACS National Meeting, New Orleans, LA, August 1999, 22–26.
2. Akin, D. E.; Foulk, J. A.; Dodd, R. B.; McAlister, D. D. Enzyme-Retting of Flax and Characterization of Processed Fibers. *J. Biotechnol.* **2001,** *89*(2–3), 193–203.
3. Anand, G.; Yadav, S.; Yadav, D. Purification and Characterization of Polygalacturonase from *Aspergillus fumigatus* MTCC 2584 and Elucidating its Application in Retting of *Crotalaria juncea* Fiber. *3 Biotech.* **2016,** *6*, 201.
4. de Andrade, M. V. V.; Delatorre, A. B.; Ladeira, S. A.; Martins, M. L. L. Production and Partial Characterization of Alkaline Polygalacturonase Secreted by Thermophilic *Bacillus* sp. SMIA-2 Under Submerged Culture Using Pectin and Corn Steep Liquor, Cienc Tecnol Aliment. *Campinas* **2011,** *31*(1), 204–208.
5. Astapovich, N. I.; Ryabaya, N. E. Specific Features of Growth and Biosynthesis of Polygalacturonase in *Saccharomyces pastorianus. Microbiology* **1996,** *65*(1), 32–35.

6. Atmodjo, M. A.; Hao, Z.; Mohnen, D. Evolving Views of Pectin Biosynthesis. *Annu. Rev. Plant Biol.* **2013**, *64*, 747–779.

7. Barense, R. I.; Chellegatti, M. A. S. C.; Fonseca, M. J. V.; Said, S. Partial Purification and Characterization of Exopolygalacturonase II and III of *Penicillium frequentans*. *Braz. J. Microbiol.* **2001**, *32*, 327–330.

8. Barman, S.; Sit, N.; Badwaik, L. S.; Deka, S. C. Pectinase Production by *Aspergillus niger* Using Banana (*Musa balbisiana*) Peel as Substrate and its Effect on Clarification of Banana Juice. *J. Food Sci. Technol.* **2015**, *52*(6), 3579–3589.

9. Blanco, P.; Sieiro, C.; Reboredo, N. M.; Villa, T. G. Cloning, Molecular Characterization, and Expression of an Endo Polygalacturonase-Encoding Gene from *Saccharomyces cerevisiae* IM1-8b. *FEMS Microbiol. Lett.* **1998**, *164*, 249–255.

10. Cabanne, C.; Doneche, B. Purification and Characterization of Two Isozymes of Polygalacturonase from *Botrytis cinerea*. Effect of Calcium Ions on Polygalacturonase Activity. *Microbiol. Res.* **2002**, *157*(3), 183–189.

11. Castruita-Domínguez, J. P.; González-Hernández, S. E.; Polaina, J.; Flores-Villavicencio, L. L.;Alvarez-Vargas, A.; Flores-Martínez, A.; Ponce-Noyola, P.; Leal-Morales, C. A. Analysis of a Polygalacturonase Gene of *Ustilago maydis* and Characterization of the Encoded Enzyme. *J. Basic Microbiol.* **2014**, *54*(5), 340–349.

12. Cervone, F.; Scala, A.; Foresti, M.; Cacace, M. G.; Noviello, C. Endopolygalacturonase from *Rhizoctonia fragariae*. Purification and Characterization of Two Isoenzymes. *Biochim. Biophys. Acta* **1977**, *482*(2), 379–385.

13. Chawanit, S.; Surang, S.; Lerluck, C.; Vittaya, P.; Pilanee, V.; Prisnar, S.; Screening of Pectinase Producing Bacteria and Their Efficiency in Biopulping of Paper Mulberry Bark. *Sci. Asia* **2007**, *33*, 131–135.

14. Chen, Y.; Sun, D.; Zhou, Y.; Liu, L.; Han, W.; Zeng, B.; Wang, Z.; Zang, Z. Cloning, Expression and Characterization of a Novel Thermophilic Polygalacturonase from *Caldicellulosiruptor bescii* DSM 6725. *Int. J. Mol. Sci.* **2014**, *15*(4), 5717–5729.

15. Cheng, Z.; Chen, D.; Lu, B.; Wei, Y.; Xian, L.; Li, Y.; Luo, Z.; Huang, R. A Novel Acid-Stable Endo-Polygalacturonase from *Penicillium oxalicum* CZ1028: Purification, Characterization and Application in the Beverage Industry. *J. Microbiol. Biotechnol.* **2016**, *26*(6), 989–998.

16. Clause, C. A.; Green, F. Characterization of Polygalacturonase from the Brown Rot Fungus *Postia placenta*. *Appl. Microbiol. Biotechnol.* **1996**, *45*, 750–754.

17. Contreras Esquivel, J. C.; Voget, C. E. Purification and Partial Characterization of an Acidic Polygalacturonase from *Aspergillus kawachii*. *J. Biotechnol.* **2004**, *110*(1), 21–28.

18. Degrassi, G.; Devescovi, G.; Kim, J.; Hwang, I.; Venturi, V. Identification, Characterization and Regulation of Two Secreted Polygalacturonases of the Emerging Rice Pathogen *Burkholderia glumae*. *FEMS Microbiol. Ecol.* **2008**, *65*, 251–262.

19. Devi, N. A.; Rao, A. G. A. Fractionation, Purification and Preliminary Characterization of Polygalacturonase Produced by *Aspergillus carbonarius*. *Enzyme Microb. Technol.* **1996**, *18*(1), 59–65.

20. Di Pietro, A.; Roncero, M. I. G. Cloning, Expression, and Role in Pathogenicity of pg1 Encoding the Major Extracellular Endopolygalacturonase of the Vascular Wilt Pathogen *Fusarium oxysporum*. *Mol. Plant Microbe. Interact.* J. **1998**, *11*(2), 91–98.

21. Dogan, N.; Tari, C. Characterization of Three Phase Partitioned Exo-Polygalacturo-nasefrom *Aspergillus sojae* with Unique Properties. *Biochem. Eng. J.* **2008,** *39*(1), 43–50.

22. dos Santos, C. C. M. A.; Fonseca, M. J.; Said, S. Purification and Partial Characterization of Exopolygalacturonase I from *Penicillium frequentans*. *Microbiol. Res.* **2002,** *157*(1),19–24.

23. Fanelli, C. Purification and Properties of Two Polygalacturonases from *Trichoderma koningii*. *J. Gen. Microbiol.* **1978,** *104*, 305–309.

24. Farrokhi, N.; Burton, R. A.; Brownfield, L.; Hrmova, M.; Wilson, S. M.; Bacic, A.; Fincher, G. B. Plant Cell Wall Biosynthesis: Genetic, Biochemical and Functional Genomics Approaches to the Identification of Key Genes. *Plant Biotechnol. J.* **2006,** *4*, 145–167.

25. De Fatima Borin, M.; Sais, S.; Fonseca, M. J. V. Purification and Biochemical Characterization of an Extracellular Endopolygalacturonase from *Penicillium frequentans*. *J. Agric. Food Chem.* **1996,** *44*, 1616–1620.

26. Gainvors, A.; Nedjaoum, A.; Gognies, S.; et al. Purification and Characterization of Acidic Endo-Polygalacturonase Encoded by PGL1-1 Gene from *Saccharomyces cerevisiae*, *FEMS Microbiol. Lett.* **2000,** *183*(1), 131–135.

27. Gao, S.; Choi, G. H.; Shain, L.; Nuss, D. L. Cloning and Targeted Disruption of *enpg-1*, Encoding the Major *in vitro* Extracellular Endopolygalacturonase from Chestnut Blight Fungus, *Cryphonectria parasitica*. *Appl. Environ. Microbiol.* **1996,** *62*(6), 1984-1990.

28. Garcia-Maceira, F. I.; Di Pietro, A.; Roncero, M. I. G. Cloning and Disruption of *pgx4* Encoding an *in planta* Expressed Exopolygalactuornases from *Fusarium oxysporum*. *Mol. Plant Microb. Interact. J.* **2000,** *13*(4), 359–365.

29. Garg, G.; Singh, A.; Kaur, A.; Singh, R.; Kaur, J.; Mahajan, R. Microbial Pectinases: An Ecofriendly Tool of Nature for Industries. *3Biotech.* **2016,** *6*, 47.

30. Gummadi, S. N.; Manoj, N.; Kumar, S. D. Structural and Biochemical Properties of Pectinases. In *Industrial enzymes: Structure, Function and Application*; Polaina, J., MacCabe, A. P., Eds.; Springer: Netherlands, 2007; pp 99–115.

31. Henriksson, G.; Akin, D. E.; Slomczynski, D.; Eriksson, K.-E. L. Production of Highly Efficient Enzymes for Flax Retting by *Rhizomucor pusillus*. *J. Biotechnol.* **1999,** *68*(2–3), 115–123.

32. Henrissat, B.; Coutinho, P. M. Carbohydrate-Active Enzymes: an Integrated Database Approach. In *Recent Advances in Carbohydrate Bioengineering;* Gilbert, H. J., Davies, G. J., Henrissat, B., Svensson, B., Eds.; Royal Society of Chemistry: Cambridge, 1999; pp 3–12.

33. Hoondal, G. S.; Tiwari, R. P.; Tewari, R.; Dahiya, N.; Beg, Q. K. Microbial Alkaline Pectinases and Their Industrial Applications: A Review. *Appl. Microbiol. Biotechnol.* **2002,** *59*, 409–418.

34. Isshiki, A.; Akimitsu, K.; Nishio, K.; Tsukamoto, M.; Yamamoto, H. Purification and Characterization of an Endopolygalacturonase from the Rough Lemon Pathotype of *Alternaria alternata*, the Cause of Citrus Brown Spot Disease. *Physiol. Mol. Plant Pathol.* **1997,** *51*, 155–167.

35. Jacob, N.; Prema, P. Influence of Mode of Fermentation on Production of Polygalacturonase by a Novel Strain of *Streptomyces lydicus*. *Food Technol. Biotechnol.* **2006**, *44*(2), 263–267.

36. Johnston, D. J.; Williamson, B. Purification and Characterization of Four Polygalacturonase from *Botrytis cinerea*. *Mycological Res*, **1992**, *96*, 343–349.

37. Johansen, J. N.; Vernhettes, S.; Hofte, H. The Ins and Outs of Plant Cell Walls. *Curr. Opin. Plant Biol.* **2006**, *9*, 616–620.

38. Jose, M.; Rodriguez, N.; Natividad, O.; Manuel, P.; Maria, D. B. Pectin Hydrolysis in a Free Enzyme Membrane Reactor: An Approach to the Wine and Juice Clarification. *Food Chem.* **2008**, *107*(112), 119.

39. Kant, S.; Vohra, A.; Gupta, R. Purification and Physicochemical Properties of Polygalacturonase from *Aspergillus niger* MTCC 3323. *Protein Expression Purif.* **2013**, *87*(1), 11–16.

40. Kapoor, M.; Beg, Q. K.; Bhushan, B.; Dadhich, K. S.; Hoondal, G. S. Production and Partial Purification and Characterization of a Thermo-Alkalistable Polygalacturonase from *Bacillus* sp. MG-cp-2. *Process Biochem.* **2000**, *36*, 467–473.

41. Kars, I.; Krooshof, G. H.; Wagemakers, L.; Joosten, R.; Benen, J. A. E.; van Kan, J. A. L. Necrotizing Activity of Five *Botrytis cinerea* Endopolygalacturonases Produced in *Pichia pastoris*. *Plant J.* **2005**, *43*, 213–225.

42. Kashyap, D. R.; Vohra, P. K.; Chopra, S.; Tewari, R. Applications of Pectinases in Commercial Sector: A Review. *Bioresour. Technol.* **2001**, *77*, 215–227.

43. Kaur, G.; Kumar, S.; Satyanarayana, T. Production, Characterization and Application of a Thermostable Polygalacturonase of a Thermophilic Mould *Sporotrichum thermophile* Apinis. *Bioresour. Technol.* **2004**, *94*(3), 239–243.

44. Kertesz, Z. A New Method of Enzymatic Clarification of Unfermented Apple Juice. New York State Agricultural Experimentation Station (Geneva) Bull. No. 689. U.S. Patent 1,932,833, 1930.

45. Kester, H. C. M.; Kusters-Van Someren, M. A.; Muller, Y.; Visser, J. Primary Structure and Characterization of an Exopolygalacturonase from *Aspergillus tubingensis*. *Eur. J. Biochem.* **1996**, *240*, 238–246.

46. Khan, M.; Nakkeeran, E.; Umesh-Kumar, S. Potential Application of Pectinase in Developing Functional Foods. *Annu. Rev. Food Sci. Technol.* **2013**, *4*, 21–34.

47. Kim, S.-J.; Brandizzi, F. The Plant Secretory Pathway: An Essential Factory for Building the Plant Cell Wall. *Plant Cell Physiol.* **2014**, *55*, 687–693.

48. Klusken, L. D.; van Alebeek, G. W. M.; Walther, J.; et al. Characterization and Mode of Action of an Exopolygalacturonase from the Hyperthermophilic Bacterium *Thermotoga maritima*. *FEBS J.* **2005**, *272*, 5464–5473.

49. Kobayashi, T.; Higaki, N.; Suzumatsu, A.; et al. Purification and Properties of a High-Molecular-Weight, Exo-Polygalacturonase from a Strain of *Bacillus*. *Enzyme Microb. Technol.* **2001**, *29*, 70–75.

50. Kofod, L. V.; Kauppinen, S.; Christgau, S.; Andersen, L. N.; Heldt-Hansen, H. P.; Dorreich, K.; Dalboge, H. Cloning and Characterization of Two Structurally and Functionally Divergent Rhamnogalacturonases from *Aspergillus aculeatus*. *J. Biol. Chem.* **1994**, *269*, 29182–29189.

51. Kumar, S. S.; Palanivelu, P. Purification and Characterization of an Extracellular Polygalacturonase from the Thermophilic Fungus, *Thermomyces lanuginosus*. *World J. Microbiol. Biotechnol.* **1999**, *15*, 643–646.

52. Lang, C.; Dörnenburg, H. Perspectives in the Biological Function and the Technological Application of Polygalacturonases. *Appl. Microbiol. Biotechnol.* **2000,** *53,* 366–375.

53. Lerouxel, O.; Cavalier, D. M.; Liepman, A. H.; Keegstra, K. Biosynthesis of Plant Cell Wall Polysaccharides—A Complex Process. *Curr. Opin. Plant Biol.* **2006,** *9,* 621–630.

54. Liao, C. H.; Revear, L.; Hotchkiss, A.; Savary, B. Genetic and Biochemical Characterization of an Exopolygalacturonase and a Pectate Lyase from *Yersinia enterocolitica. Can. J. Microbiol.* **1999,** *45*(5), 396–403.

55. de Lima Damasio, A. R.; da Silva, T. M.; Maller, A.; et al. Purification and Partial Characterization of an Exo-Polygalacturonase from *Paecilomyces variotii* Liquid Cultures. *Appl. Biochem. Biotechnol.* **2010,** *160*(5),1496–1507.

56. Lu, X.; Lin, J.; Wang, C.; Du. X.; Cai, J. Purification and Characterization of Exo-Polygalacturonase from *Zygoascus hellenicus* V25 and its Potential Application in Fruit Juice Clarification. *Food Sci. Biotechnol.* **2016,** *25,* 1379.

57. Manachini, P. L.; Fortina, M. G.; Parini, C. Purification and Properties of an Endopolygalacturonase Produced by *Rhizopus stolonifer. Biotechnol. Lett.* **1987,** *9*(3), 219–224.

58. Martel, M. B.; Letoublon, R.; Fevre, M. Purification and Characterization of Two Endopolygalacturonases Secreted During Early Stages of the Saprophytic Growth of *Sclerotinia sclerotiorum. FEMS Microbiol. Letts.* **1998,** *158*(1), 133–138.

59. Martins, E. S.; Leite, R. S. R.; Silva, R.; Gomes, E. Purification and Properties of Polygalacturonase Produced by Thermophilic fFngus *Thermoascus aurantiacus* CBMAI-756 on Solid-State Fermentation. *Enzyme Res.* **2013,** Doi:10.1155/2013/438645.

60. Mohamed, S. A.; Al-Malki, A. L.; Kumosani, T. A. Characterization of a Polygalacturonase from *Trichoderma harzianum* Grown on Citrus Peel with Application for Apple Juice. *Aust. J. Basic Appl. Sci.* **2009,** *3*(3), 2770–2777.

61. Mohnen, D. Biosynthesis of Pectins. In *Pectins and Their Manipulation;* Seymour, G. B., Knox, J. P., Eds.; Blackwell Publishing and CRC Press: Oxford, 2002; 52–98.

62. Mohnen, D. Pectin Structure and Biosynthesis. *Curr. Opin. Plant Biol.* **2008,** *11,* 266–277.

63. Mrudula, S.; Anitharaj, R. Pectinase Production in Solid-State Fermentation by *Aspergillus niger* Using Orange Peel as Substrate. *Global J. Biotechnol. Biochem.* **2011,** *6,* 64–71.

64. Mutter, M.; Colquhoun, I. J.; Schols, H. A.; Beldman, G.; Voragen, A. G. J. Rhamnogalacturonase B from *Aspergillus aculeatus* is a Rhamnogalacturonan α-L-rhamnopyranosyl-(1-4)-α-D-galactopyranosyluronide Lyase. *Plant Physiol.* **1996,** *110,* 73–77.

65. Nagai, M.; Katsuragi, T.; Terashita, T.; et al. Purification and Characterization of an Endo-Polygalacturonase from *Aspergillus awamori. Biosci. Biotechnol. Biochem.* **2000,** *64*(8), 1729–1732.

66. Nakkeeran, E.; Umesh-Kumar, S.; Subramanian, R. *Aspergillus carbonarius* Polygalacturonases Purified by Integrated Membrane Process and Affinity Precipitation for Apple Juice Production. *Bioresour. Technol.* **2011,** *102,* 3293–3297.

67. Nasuno, S.; Starr, M. P. Polygalacturonase of *Erwinia carotovora. J. Biol. Chem.* **1966,** *241*(22), 5298–5306.

68. Niture, S. K. Comparative Biochemical and Structural Characterizations of Fungal Polygalacturonases. *Biologia,* **2008,** *63*(1), 1–19.

69. Niture, S. K.; Pant, A.; Kumar, A. R. Active Site Characterization of Single Endo-polygalacturonase Produced by *Fusarium monoliforme NCIM 1276*. *Biologia* **2001**, *268*(3), 832–840.

70. Ortega, L. M.; Kikot, G. E.; Rojas, N. L.; Lopez, L. M.; Astoreca, A. L.; Alconada, T. M. Production, Characterization, and Identification Using Proteomic Tools of a Poly-galacturonase from *Fusarium graminearum*. *J. Basic Microbiol*. **2014**, *54*, S170–177.

71. Oszmianski, J.; Wojdylo, A.; Kolniak, J. Effect of Enzymatic Mash Treatment and Storage on Phenolic Composition, Antioxidant Activity, and Turbidity of Cloudy Apple Juice. *J. Agric. Food Chem*. **2009**, *57*, 7078–7085.

72. Pan, X.; Li, K.; Ma, R.; Shi, P.; Huang, H.; Yang, P.; Meng, K.; Yao, B. Biochemical Characterization of Three Distinct Polygalacturonases from *Neosartorya fischeri* P1. *Food Chem*. **2015**, *188*, 569–575.

73. Parenicova, L.; Benen, J. A. E.; Kester, H. C. M.; Visser, J. *pga*E Encodes a Fourth Member of the Endopolygalacturonase Gene Family from *Aspergillus niger*. *Eur. J. Biochem*. **1998**, *251*, 72–80.

74. Patil, N. P.; Patil, K. P.; Chaudhari, B. L.; Chincholkar, S. B. Production, Purification of Exo-Polygalacturonase from Soil Isolates *Paecilomyces variotii* NFCCI 1739 and its Application. *Indian J. Microbiol*. **2012**, *52*(2), 240–246.

75. Pedrolli, D. B.; Carmona, E. C. Pectin Lyase from *Aspergillus giganteus*: Compara-tive Study of Productivity of Submerged Fermentation on Citrus Pectin and Orange Waste. *Prikl Biokhim Mikrobiol*. **2009**, *45*(6), 677–683.

76. Perrone, G.; Mule, G.; Susca, A.; Battilani, P.; Pietri, A.; Logrieco, A. Ochratoxin A Production and Amplified Fragment Length Polymorphism Analysis of *Aspergillus carbonarius*, *Aspergillus tubingensis*, and *Aspergillus niger* Strains Isolated from Grapes in Italy. *Appl. Environ. Microbiol*. **2006**, *72*, 680–685.

77. Polizeli, M. L. T. M.; Jorge, J. A.; Terenzi, H. F. Pectinase Production by *Neurospora crassa*: Purification and Biochemical Characterization of Extracellular Polygalactu-ronase Activity. *J. Gen. Microbiol*. **1991**, *137*, 1815–1823.

78. Rao, M. N.; Kembhavi, A. A.; Pant, A. Implication of Tryptophan and Histidine in the Active Site of Endo-Polygalacturonase from *Aspergillus ustus*: Elucidation of the Reaction Mechanism. *Biochim. Biophys. Acta* **1996**, *1296*(2), 167–173.

79. Ridley, B. L.; O'Neill, M. A.; Mohnen, D. Pectins: Structure, Biosynthesis, and Oligogalacturonide-Related Signaling. *Phytochemistry* **2001**, *57*, 929–967.

80. Riou, C.; Freyssinet, G.; Fivre, M. Purification and Characterization of Extracellular Pectinolytic Enzymes Produced by *Sclerotinia sclerotiorum*. *Appl. Environ. Micro-biol*. **1992**, *58*(2), 578–583.

81. Saarilahti, H. T.; Heino, P.; Pakkanen, R.; et al. Structural Analysis of the *pehA* Gene and Characterization of its Protein Product, Endopolygalacturonase, of *Erwinia caro-tovora* Subspecies *carotovora*. *Mol. Microbiol*. **1990**, *4*(6), 1037–1044.

82. Saito, K.; Takakuwa, N.; Oda, Y. Purification of the Extracellular Pectinolytic Enzyme from the Fungus *Rhizopus oryzae* NBRC 4707. *Microbiol. Res*. **2004**, *159*(1), 83–86.

83. Sakamoto, T.; Sakai, T. Analysis of Structure of Sugar-Beet Pectin by Enzymatic Methods. *Phytochemistry* **1995**, *39*, 821–823.

84. Sakamoto, T.; Bonnin, E.; Quemener, B.; Thibault, J. F. Purification and Characterization of Two Exo-polygalacturonases from *Aspergillus niger* Able to Degrade Xylogalactu-ronan and Acetylated Homogalacturonan. *Biochim. Biophys. Acta*. **2002**, *1572*(1), 10–18.

85. Salariato, D.; Diorio, L. A.; Mouso, N.; Forchiassin, F. Extraction and Characterization of Polygalacturonase of *Fomes sclerodermeus* Produced by Solid-State Fermentation. *Rev. Argent. Microbiol.* **2010**, *42*, 57–62.

86. Sassi, A. H.; Tounsi, H.; Trigui-Lahiani, H.; Bouzouita, R.; Romdhane, Z. B.; Gargouri, A. A Low-Temperature Polygalacturonase from *P. occitanis*: Characterization and Application in Juice Clarification. *Int. J. Biol. Macromol.* **2016**, *91*, 158–164.

87. Scheible, W. R.; Pauly, M. Glycosyltransferases and Cell Wall Biosynthesis: Novel Players and Insights. *Curr. Opin. Plant Biol.* **2004**, *7*, 285–295.

88. Scheller, H. V.; Jensen, J. K.; Sørensen, S. O.; Harholt, J.; Geshi, N. Biosynthesis of Pectin. *Physiol. Plant* **2007**, *129*, 283–295.

89. Schols, H. A.; Geraeds, C. C. J. M.; Searle-van Leeuwen, M. F.; Kormelink, F. J. M.; Voragen, G. J. Rhamnogalacturonase: A Novel Enzyme that Degrades the Hairy Regions of Pectins. *Carbohydr. Res.* **1990**, *206*, 106–115.

90. Semenova, M. V.; Grishutin, S. G.; Gusakov, A. V.; Okunev, O. N.; Sinitsin, A. P. Isolation and Properties of Pectinases from the Fungus *Aspergillus japonicas*. *Biochemistry (Moscow)* **2003**, *68*(5), 559–569.

91. Serrat, M.; Bermudez, R. C.; Villa, T. G. Production, Purification and Characterization of a Polygalacturonase from a New Strain of *Kluyveromyces marxianus* Isolated from Coffee Wet-Processing Wastewater. *Appl. Biochem. Biotechnol.* **2002**, *97*(3), 193–208.

92. Singh, S. A.; Rao, A. G. A. A Simple Fractionation Protocol for, and a Comprehensive Study of the Molecular Properties of, Two Major Endopolygalacturonases from *Aspergillus niger*. *Biotechnol. Appl. Biochem.* **2002**, *35*(2), 115–123.

93. Strand, L. L.; Corden, M. E.; MacDonald, D. L. Characterization of Two Endopolygalacturonase *Isozymes Produced by Fusarium oxysporum f.* sp. *Lycopersici*. *BBA-Enzymol* **1976**, *429*(3), 870–883.

94. Suykerbuyk, M. E. G.; Schaap, P. J.; Musters, W.; Visser, J. Cloning, Sequence and Expression of the Gene for Rhamnogalacturonan Hydrolase of *Aspergillus aculeatus*. A Novel Pectinolytic Enzyme. *Appl. Microbiol. Biotechnol.* **1995**, *43*, 861–870.

95. Swain, M. R.; Ray, R. C. Production, Characterization and Application of a Thermostable Exo-Polygalacturonase from *Bacillus subtilis* CM5. *Food Biotechnol.* **2010**, *24*(1), 37–50.

96. Takao, M.; Nakaniwa, T.; Yoshikawa, K.; Terashita, T.; Sakai, T. Purification and Characterization of Thermostable Pectate Lyase with Protopectinase Activity from Thermophilic *Bacillus* sp. TS 47. *Biosci. Biotechnol. Biochem.* **2000**, *64*, 2360–2367.

97. Takasawa, T.; Sagisaka, K.; Yagi, K.; et al. Polygalacturonase Isolated from the Culture of the Psychrophilic Fungus *Sclerotinia borealis*. *Can. J. Microbiol.* **1997**, *43*(5), 417–424.

98. Thakur, J.; Gupta, R. Improvement of Tea Leaves Fermentation Through Fermentation. *Acta Microbiol. Immunol. Hung.* **2012**, *59*(3), 321–334.

99. Thakur, A.; Pahwa, R.; Singh, S.; Gupta, R. Production, Purification, and Characterization of Polygalacturonase from *Mucor circinelloides* ITCC 6025. *Enzyme Res.* **2010**, DOI:10.4061/2010/170549.

100. Toyooka, K.; Goto, Y.; Asatsuma, S.; Koizumi, M.; Mitsui, T.; Matsuoka, K. A Mobile Secretory Vesicle Cluster Involved in Mass Transport from the Golgi to the Plant Cell Exterior. *Plant Cell Online* **2009**, *21*, 1212–1229.

101. Tu, T.; Luo, H.; Meng, K.; Cheng, Y.; Ma, R.; Shi, P.; Huang, H.; Bai, Y.; Wang, Y.; Zhang, L.; Yao, B. Improvement in Thermostability of an *Achaetomium* sp. strain Xz8 Endopolygalacturonase via the Optimization of Charge-Charge Interactions. *Appl. Environ. Micrbiol.* **2015,** *81*(19), 6938–6944.

102. de Vries, R. Accessory Enzymes from *Aspergillus* Involved in Xylan and Pectin Degradation. Ph.D. Thesis, The Wageningen Agricultural University, Wageningen, Netherlands, 1999.

103. Yadav, S.; Anand, G.; Dubey, A. K.; Yadav, D. Purification and Characterization of an Exo-Polygalacturonase Secreted by *Rhizopus oryzae* MTCC1987 and its Role in Retting of *Crotolaria juncea* Fibre. *Biologia* **2012,** *67*(6), 1069–1074.

104. Yamada, H.; Hirano, M.; Kiyohara, H. Partial Structure of an Anti-Ulcer Pectic Polysaccharide from the Roots of *Bulpleurum falcatum L. Carbohydr. Res.* **1991,** *219,* 173–192.

105. Yan, H.; Liou, R. Cloning and Analysis of pppg1, an Inducible Endopolygalacturonase Gene from the Oomycete Plant Pathogen *Phytophthora parasitica. Fungal Genet. Biol.* **2005,** *42,* 339–350.

106. Yuan, P.; Meng K.; Wang, Y.; et al. A Protease-Resistant Exo-Polygalacturonase from *Klebsiella* sp. Y1 with Good Activity and Stability Over a Wide pH Range in the Digestive Tract. *Bioresour. Technol.* **2012,** *123,* 171–176.

107. Yuping, A.; Siwen, S.; Hui, H.; Chunping, X. Production, Purification and Characterization of an Exo-Polygalacturonase from *Penicillium janthinellum* sw09. *An. Acad. Bras. Ciênc.* **2016,** *88,* 479–487.

108. Zaslona, H.; Trusek- Holownia, A. Enhanced Production of Polygalacturonase in Solid-State Fermentation: Selection of the Process Conditions, Isolation and Partial Characterization of the Enzyme. *Acta Biochim. Pol.* **2015,** *62*(4), 651–657.

109. Zhang, J.; Bruton, B. D.; Biles, C. L. *Fusarium solani* Endo-Polygalacturonase from Decayed Muskmelon Fruit: Purification and Characterization. *Physiol. Mol. Plant Pathol.* **1999,** *54,* 171–186.

110. Zhang, J.; Henriksson, H.; Szabo, I. J.; et al. The Active Component in the Flax-Retting System of the Zygomycete *Rhizopus oryzae* sb is a Family 28 Polygalacturonase. *J. Ind. Microbiol. Biotechnol.* **2005,** *32,* 431–438.

111. Zheng, Z.; Shetty, K. Solid State Production of Polygalacturonase by *Lentinus edodes* Using Fruit Processing Wastes. *Process Biochem.* **2000,** *35,* 825–828.

112. Zhong, R.; Ye, Z.-H. Unraveling the Functions of Glycosyltransferase Family 47 in Plants. *Trends Plant Sci.* **2003,** *8,* 565–568.

113. Zhong, R.; Ye, Z.-H. Regulation of Cell Wall Biosynthesis. *Curr. Opin. Plant Biol.* **2007,** *10,* 564–572.

114. Zhou, H.; Li, X.; Guo, M.; Xu, Q.; Cao, Y.; Qiao, D.; Cao, Y.; Xu, H. Secretory Expression and Characterization of an Acidic Endo-Polygalacturonase from *Aspergillus niger* SC323 in *Saccharomyces cerevisiae. J. Microbiol. Biotechnol.* **2015,** *25*(7), 999–1006.

115. Zhu, C.; Ganguly, A.; Baskin, T. I.; McClosky, D. D.; Anderson, C. T.; Foster C.; et al. The Fragile Fiber1 Kinesin Contributes to Cortical Microtubule-Mediated Trafficking of Cell Wall Components. *Plant Physiol.* **2015,** *167,* 780–792.

CHAPTER 17

NATURAL AGENTS AGAINST FILARIASIS: A REVIEW

VISHAL KUMAR SONI*

*Department of Medical and Health, National Filaria Control Program Unit, Civil lines, Deoria, Uttar Pradesh 274001, India, *E-mail: vishalkumars@gmail.com*

CONTENTS

ABSTRACT

Lymphatic filariasis is a mosquito-borne neglected tropical disease caused by a group of nematodes (round worms), *Wuchereria bancrofti, Brugia malayi* and *Brugia timori,* which are transmitted by mosquito vectors. According to the recent fact sheet of World Health Organization, around 947 million people belonging to 54 countries are threatened by lymphatic filariasis infection. In the year 2000, over 120 million people were infected, with about 40 million disfigured and incapacitated by the disease globally. The disease results into a heavy socio-economic loss to the developing nations. The elimination of lymphatic filariasis relies on at least five effective annual rounds of mass drug administration program implemented in larger communities of endemic areas. It is done using oral medication of three drugs currently available for treatment: diethylcarbamazine, albendazole and ivermectin or a combination of albendazole and ivermectin. However, development of drug resistance is a reported phenomenon in nematodes. Also, these drugs act principally on larval stages (microfilariae) of the parasites and with limited or no effect on adult worms (macrofilariae). Thus, the biggest challenge is to combat adult parasites with a new and safe alternative which can be achieved only by developing newer chemotherapeutic agents against adult worms, a vaccine adjuvant or a prophylactic agent from natural flora. There are a number of plant species that are traditionally being used against many nematode infections; however the curative properties of these have not been explored much. The present review is an effort to collect the research outputs of various studies and approaches in the elimination of filariasis and gives detailed information about some of the major achievements in this area focusing on the development of prophylactic and chemotherapeutic agents from plants.

17.1   INTRODUCTION

Lymphatic filariasis (LF), commonly known as elephantiasis is a mosquito-borne neglected tropical disease. The disease ranks second in causing long-term disability in the individuals of the tropical and subtropical countries and results in a heavy socio-economic loss to the developing nations. World Health Organization has reported that 947 million people in 54 countries are living in areas at high risk of spread of infection. A global baseline estimate of persons affected by lymphatic filariasis reveals that there are around 25 million men with hydrocele and over 15 million people with lymphedema. It is a painful and profoundly disfiguring disease caused by three species of lymphatic-dwelling filariform nematode worms, known as filariae; *Wuchereria bancrofti*, *Brugia malayi* and *Brugia timori*. A major drawback of this disease is that the vast majority of infected people remain asymptomatic and serve to silently transmit the infection.[23,9,31]

Filarial nematodes are long, thread-like, tissue-dwelling obligate parasites which form nests by invading and blocking the lymphatic system of infected persons. They are able to persist in the infected host for many years (adult life span 5–12 years) by downregulating the immune response of infected individuals. The inability of the immune system to mount parasite-specific T cell responses causes filarial parasites to persist in the host body for a long time. This is attributed to parasite stimulation of endogenous host immunosuppressive controls.[26,1] Blockage of lymphatics causes an altered lymphatic system and the abnormal enlargement or lymphedema of the limbs, breasts and genitals causing hydrocele, chylocele and swelling of the scrotum and penis due to excessive accumulation of tissue fluid. The recurrent acute attacks are painful and accompanied by fever and chills. Once lymphedema is established there is no radical cure, however the presently available drugs prevent the further progression of the edema.[9,10]

Control of LF is limited to treatment with diethylcarbamazine (DEC) or ivermectin, which principally act on the circulating microfilariae (Mf) resulting into reappearance of microfilaraemia after a certain period of withdrawal of the drug.[17] WHO recommends at least five effective rounds of the annual mass drug administration (MDA) program targeted at the detection of infected individuals through night blood screening of the disease in the endemic population of an area followed by treatment with a single dose of combination of the two antifilarial drugs, DEC (6 mg/kg) and albendazole (400 mg/kg) or ivermectin (150–200 mg/kg) through oral medication. A combination of albendazole (400 mg/kg) and ivermectin is used in areas where onchocerciasis (river blindness) is prevalent. Antihistamines and steroids are useful to reduce swelling and hypersensitivity while antibiotics are a must to treat the opportunistic secondary infections.[9] Recent clinical trials with above combinations have been quite successful in decreasing Mf density[4,20] although, the treatment needs to be repeated for many years in the endemic population to impair Mf transmission through the vector. After the withdrawal of MDA (prophylaxis), chances of the reappearance of infection remain which is the biggest challenge. Furthermore, the threat of development of resistance to mainstay drugs especially ivermectin and albendazole in veterinary settings advocates sincere efforts to develop new macrofilaricidal (adulticidal) agent/s for long-term effects and reduction of filarial pathology. This can be achieved only by newer prophylactic or chemotherapeutic agents against microfilariae/adult worms, and an effective vaccine that can prevent future infections. Thus, there is an unequivocal call for the development of new and safe drugs, especially from natural resources with a novel mode of action and a short period of dose regimen to overcome the problems associated with the current treatment.

The characteristic impairment in the Th1 immune response of host during active filarial infection suggests novel methods of management of the disease by applying immunoprophylactic and immunotherapeutic methods. Therefore, development of protective immunity by boosting the immune system through immunostimulants may be an effective alternative to overcome the parasite burden and to enhance the antifilarial efficacy of existing antifilarial drugs when used as an adjunct, as some antiparasitic drugs act through the immune system of the host.

Plants are considered as the key reservoirs of natural entities having tremendous medicinal value. It is estimated that around 250,000 flowering plant species are found on the earth, of which around 125,000 are found in the tropical forests, and around 2000 species of higher plants are used for medicinal purposes throughout the world. However, till date only about 1% of tropical species have been exploited for their pharmaceutical potentials.[25,3,12] Ayurveda and other traditional literature mention about 1500 plants with medicinal uses and around 800 of these have been used as ethnomedicine in the treatment of various human ailments.[5,29,18] More than 70% of the Indian population still relies on these natural product-derived medicines.[8] Large numbers of tropical plants have not yet been exploited in detail for their pharmacological properties or chemical constituents. Usage of plants was recognized for centuries to resist and recover from various ailments without any untoward side effects. The emergence of drug resistance in pathogens against commonly used drugs has warranted an urgent need to develop new and safe drugs with an alternate mode of action. Despite the introduction of allopathic medicines, plant-based medicines are much in demand. The compounds isolated from medicinal plants may serve as lead molecules for the development of safe and better drugs against various diseases or may enhance the efficacy of the presently available synthetic drugs when used in combination. Some of the important medicinal uses of these plants are discussed below.

17.2   REVIEW OF NATURAL ANTIFILARIALS

Helminth infections are among the most common infections in humans, affecting a large proportion of population all over the world and pose a large threat to public health. LF, a mosquito-borne disease, is one of the most prevalent helminth infections in the tropical and subtropical countries, and is accompanied by a number of pathological conditions. Due to the lack of an adulticidal drug or vaccine against the adult filarial parasite, the disease has got an uncontrolled prevalence in tropical countries.[24] Hence, it is essential to develop an effective antifilarial drug which could either kill or permanently sterilize the adult filarial parasites and also produce less toxicity or side effects. WHO has recognized the potential of plant-derived products as natural medicine.[30]

There are a number of reliable plant species known for their antiparasitic efficacy which lies in the secondary metabolites produced by these plants, constituting a rich source of bioactive substances. Secondary metabolites have recently gained attention of the scientific community for the research on new drugs from plant origin. These metabolites maybe categorized as alkaloids, saponins, glycosides, tannins and so forth.[2] Till date, majority of the drugs available are in fact derived from the medicinal plants or prepared as analogs of the active entities (leads). Not only the pure compounds, but also the extracts confer protection as certain activity appears in combination of two or more metabolites. Like the extracts derived from *Piper betle* L. leaves demonstrated in vitro antifilarial as well as antileishmanial activity against *B. malayi* and *L. donovani*.[28] Crude extract prepared from the stem portion of *Lantana camara* revealed significant antifilarial activity against *B. malayi* and sterilized surviving female worms in vivo in animal models.[15] The extracts from three plant species that is *Xylocarpus granatum* (Family: Meliaceae), *Tinospora crispa* (Family: Menispermaceae), *Andrographis paniculata* (Family: Acanthaceae) showed in vitro activity against adult worms of *B. malayi*. The effect of dried husks (HX), dried seeds (SX), dried leaves (LX) of *X. granatum*, dried stem (ST) of *T. crispa* and dried leaves (LA) of *A. paniculata* was tested in vitro on adult *B. malayi* parasites for antifilarial activity, of which, SX extract of *X. granatum* demonstrated the best activity, followed by the LA, ST, HX and LX extract.[32] The glycosides asperoside and strebloside isolated from *Streblus asper* have been observed to interfere with the glutathione metabolism of the adult filarial parasite *Setaria cervi in vitro* and ultimately resulted in the death of the parasites.[21] The extracts from parts of *Carapa procera*, *Polyalthia suaveolens* and *Pachypodanthium staudtii* also exhibited microfilaricidal activity on *Onchocerca volvulus*.[27]

Mathew et al., (2007)[13] have screened the methanolic extracts of 20 medicinal plants at 1–10 mg/ml for in vitro macrofilaricidal activity by worm motility assay against the adult cattle filarial worm, *Setaria digitata*.

Four plant extracts out of these showed macrofilaricidal activity by worm motility at concentrations below 4 mg/ml. Complete inhibition of worm motility and subsequent mortality was observed at 3, 2, 1 and 1 mg/ml concentrations of the extracts prepared from *Centratherum anthelminticum, Cedrus deodara, Sphaeranthus indicus* and *Ricinus communis,* respectively. 3-[4,5-dimethylthiazol-2-yl]-2,5-diphenyltetrazolium bromide (MTT) reduction assay was carried out at 1 mg/ml over a 4 h exposure, and the results showed that *C. deodara, R. communis, S. indicus* and *C. anthelminticum* exhibited 86.56, 72.39, 61.20 and 43.15% inhibition, respectively in MTT reduction (formazan formation) as compared to that of control. Methanolic extracts from *C. anthelminticum, C. deodara, S. indicus* and *R. communis* had macrofilaricidal activities and completely inhibited worm motility in vitro (3–1 mg/ml) when exposed for 100 min. Mathew et al., (2008)[14] also identified a lead molecule against the adult bovine filarial worm *S. digitata* from the methanolic extract of fruits of *Trachyspermum ammi* (Family: Apiaceae) in vitro. The crude extract and the active fraction showed significant activity against the adult *S. digitata* both by worm motility and MTT reduction assays. The isolated active principle was chemically characterized by infrared (IR), proton nuclear magnetic resonanace (1H-NMR) and mass spectrometry (MS) analysis and identified as a phenolic monoterpene. It was further screened in vivo against *B. malayi* in a rodent model *Mastomys coucha* and showed profound macrofilaricidal activity apart from female worm sterility.

Singh et al., (2009)[22] have reported antifilarial activity in the plant *P. betle* Linn. (Family: Piperaceae). Based on the immunostimulatory findings, the antifilarial activity of crude methanol extract, chloroform and n-hexane fractions of *P. betle* was evaluated against *B. malayi* in *M. coucha*. The crude MeOH extract was reported to suppress microfilaraemia, causing significant female worm sterilization and had moderate action on adult worms (noticeably at 100 mg/kg) as compared to the untreated control groups. However, both the fractions led to suppressed microfilaraemia at 30 mg/kg dose on day 90. Further, oral administration of the *P. betle* extract and both the fractions (n-hexane and chloroform) at 100 mg/kg for five consecutive days caused significant increase in Nitric oxide (NO) production correlating well with the reduced blood microfilarial density. The extract and fractions also led to increased T- and B-cell proliferation in the presence of mitogens as well as a filaria-specific antigen and increased IgG antibody production. Gaur et al.,

(2008)[7] evaluated the antifilarial activity of *Caesalpinia bonducella* seed kernel crude extract and its various fractions against rodent filarial parasite, *Litomosoides carinii* in cotton rats and against human filarial parasite *B. malayi* in M. coucha. The crude extract and fractions showed microfilaricidal, macrofilaricidal and female-sterilizing efficacy against *L. sigmodontis* and microfilaricidal as well as female-sterilizing efficacy against *B. malayi* in animals. A study has reported that prolonged treatment with oral or topical coumarin or flavonoids can shrink the lymphedema caused due to lymphatic blockage.[6] Sahare et al., (2008)[19] have reported significant inhibition in the motility of microfilariae of *B. malayi* after in vitro incubation for 48 h with aqueous/methanol extracts (in the concentration range of 20–100 ng/ml) prepared from the roots of *Vitex negundo* L. (Family: Verbenaceae), roots and leaves of *Butea monosperma* L. (Family: Fabaceae) and leaves of *Aegle marmelos* Corr. (Family: Rutaceae).

Misra et al., (2011)[16] have reported the in vitro and in vivo antifilarial activity in the fruit extract of *X. granatum,* a mangrove plant from Andaman, India. The various extracts, fractions and pure molecules of *X. granatum* fruits were evaluated in vitro on adult worms and microfilariae of *B. malayi,* and the active samples were further evaluated in vivo in the primary and secondary in vivo screens. The antifilarial activity was primarily localized in the ethyl acetate-soluble fraction which revealed the inhibition concentration (IC50) of 8.5 and 6.9 µg/ml in adult and Mf, respectively. This fraction possessed moderate adulticidal and embryostatic action in vivo in mastomys. Out of eight pure molecules isolated from the active fraction, two compounds, gedunin and photogedunin at 5×100 mg/kg by subcutaneous route revealed excellent adulticidal efficacy resulting in the death of 80% and 70% transplanted adult *B. malayi,* respectively in the peritoneal cavity of birds in addition to noticeable microfilaricidal action on the day of autopsy. A study using root extracts of *Withania somnifera* chemotypes NMITLI-101, NMITLI-118, NMITLI-128 and pure compounds withanolide, withaferin A demonstrated potential immunoprophylactic efficacies against *B. malayi* infection in rodents. The administration of aqueous ethanolic extracts (10 mg/kg) of NMITLI-101, NMITLI-118, NMITLI-128 and withaferin A (0.3 mg/kg) for 7 days before and after interaction with human filarial parasite *B. malayii* offered differential protection in rodent *M. coucha;* the chemotype 101R conferred the best protection (53.57%) as compared to other chemotypes. Findings also demonstrated that establishment of *B. malayi* larvae was adversely affected by pretreatment with withaferin A as evidenced by 63.6% reduction in adult worm establishment. A large percentage (66.2%) of the established adult female worms also showed defective embryogenesis.[11] Thus, a large number of floras have been reported to possess antifilarial activities with some on larvae and others on adult parasites. The points to be noted are that not only the different plant parts but the geographical conditions too contribute to a variation in metabolites and, hence the pharmacological properties.

17.3   CONCLUSION

Thus, the review demonstrates the importance of natural products especially plant-derived test substances in the treatment of filarial infection. In spite of rich biodiversity and traditional knowledge, the pharmacological exploitation of these medicinal plants has been limited. Though some of the plants have been extensively investigated, others await thorough investigation. Not only this, but the scientific validation of the traditional value of the medicinal plants also needs in-depth study. It is well-known that the active constituents also vary in the plant during different seasons, and geographical variation also has an impact on the compound composition. Thus, the pharmacological profile of the plant chemotypes becomes worth investigating. The natural compounds also find use as anti-infectives as well as immunoprophylactants and can be used in combination with chemotherapeutic agents to enhance their anti-infective activity. Since filarial nematodes take advantage of the suppressed immune response of host for their prolonged stay in the host body, it would be of great interest to have a compound that confers immune stimulating activity together with adulticidal efficacy.

KEYWORDS

- lymphatic filariasis
- nematodes
- extract
- fraction
- natural compounds
- prophylactic
- chemotherapeutic
- mass drug administration
- diethylcarbamazine
- albendazole
- ivermectin

REFERENCES

1. Babu, S.; Blauvelt, C. P.; Kumaraswami, V.; Nutman, T. B. Regulatory Networks Induced by Live Parasites Impair both Th1 and Th2 Pathways in Patent Lymphatic Filariasis: Implications for Parasite Persistence. *J. Immunol.* **2006,** *176,* 3248–3256.
2. Bauri, R. K.; Tigga, M. N.; Kullu, S. S. A Review on Use of Medicinal Plants to Control Parasites. *Indian J. Nat. Prod. Resour.* **2015,** *6*(4), 268–277.
3. Carneiro, A. L. B.; Teixeira, M. F. S.; de Oliveira, V.M.A.; Fernandes, O.C.C.; Cauper, G.S. de B.; Pohlit, A.M. Screening of Amazonian plants from the Adolpho Ducke forest reserve, Manaus, state of Amazonas, Brazil, for antimicrobial activity. *Mem. Inst. Oswaldo Cruz.* **2008,** *103*(1), 31–38.
4. Chauhan, P. M. S. Current and Potential Filaricides. *Drugs Future* **2000,** *25*(5), 481–488.
5. Chopra, A. Ayurvedic Medicine and Arthritis. *Clin. Rheum. Dis.* **2000,** *26,* 133–144.
6. Das, L. K.; Reddy, G. S.; Pani, S. P. Some Observations on the Effect of Daflon (Micronized Purified Flavonoid Fraction of *Rutaceae aurantiae*) in Bancroftian Filarial Lymphoedema. *Filaria J.* **2003,** *2,* 5.
7. Gaur, R. L.; Sahoo, M. K.; Dixit, S.; Fatma, N.; Rastogi, S.; Kulshreshtha, D. K.; Chatterjee, R. K.; Murthy, P. K. Antifilarial Activity of *Caesalpinia bonducella* against Experimental Filarial Infections. *Indian J. Med. Res.* **2008,** *128*(1), 65–70.
8. Jain, J. A. T.; Devasagayam, T. P. A. Cardioprotective and Other Beneficial Effects of Some Indian Medicinal Plants. *J. Clin. Biochem. Nutr.* **2006,** *38,* 9–18.
9. Kalyanasundaram, R. Lymphatic Filariasis: The Current Status of Elimination Using Chemotherapy and the Need for a Vaccine. *Top. Med. Chem.* **2016.** DOI: 10.1007/7355_2015_5002.
10. Khatoon, S. S.; Rehman, M.; Rehman, A. The Role of Natural Products in Tropical Diseases. *Nat Prod Chem.* **2016,** *2,* 179–228.
11. Kushwaha, S.; Soni, V. K.; Singh, P. K.; Bano, N.; Kumar, A.; Sangwan, R. S.; Misra-Bhattacharya, S. *Withania somnifera* Chemotypes NMITLI 101R, NMITLI 118R, NMITLI 128R and Withaferin A Protect *Mastomys coucha* from *Brugia malayi* Infection. *Parasite Immunol.* **2012,** *34,* 1–11.
12. Mahesh, B.; Satish, S. Antimicrobial Activity of Some Important Medicinal Plant against Plant and Human Pathogens. *World J. Agric. Sci.* **2008,** *4*(S), 839–843.
13. Mathew, N.; Kalyanasundaram, M.; Paily, K. P.; Abidha, V. P.; Balaraman, K. In Vitro Screening of Medicinal Plant Extracts for Macrofilaricidal Activity. *Parasitol. Res.* **2007,** *100,* 575–579.
14. Mathew, N.; Bhattacharya, S. M.; Perumal, V.; Muthuswamy, K. Antifilarial Lead Molecules Isolated from *Trachyspermum ammi*. *Molecules* **2008,** *13,* 2156–2168.
15. Misra, N.; Sharma, M.; Raj, K.; Misra-Bhattacharya, S. Chemical Constituents and Antifilarial Activity of *Lantana camara* against Human Lymphatic Filariid *Brugia malayi* and Rodent Filariid *Acanthocheilonema viteae* Maintained in Rodent Hosts. *Parasitol. Res.* **2007,** *100*(3), 439–448.
16. Misra, S.; Verma, M.; Mishra, S. K.; Srivastava, S.; Lakshmi, V.; Misra-Bhattacharya, S. Gedunin and Photogedunin of *Xylocarpus granatum* Possess Antifilarial Activity against Human Lymphatic Filarial Parasite *Brugia malayi* in Experimental Rodent Host. *Parasitol. Res.* **2011,** *109*(5), 1351–1360.

17. Ottesen, E. A.; Ramachandran, C. D. Lymphatic Filariasis. Infection and Disease: Global Morbidity Attributable to Intestinal Nematode Infections. *Parasitol.* **1995,** *109*, 373–387.

18. Rao, N. S.; Das, S. K. Herbal Gardens of India: A Statistical Analysis Report. *Afr. J. Biotechnol.* **2011,** *10*(31), 5861–5868.

19. Sahare, K. N.; Anandhraman, V.; Meshram, V. G.; Meshram, S. U.; Reddy, M. V.; Tumane, P. M.; Goswami, K. Anti-Microfilarial Activities of Methanoloic Extract of *Vitex negundo* and *Aegle marmelos* and their Phytochemical Analysis. *Indian J. Exp. Biol.* **2008,** *46*, 128–131.

20. Sharma, D. C. New Goals Set for Filariasis Elimination in India. *Lancet Infect. Dis.* **2002,** *2*(7), 389.

21. Singh, S.N.; Raina, D.; Chatterjee, R.K.; Srivastava, A. K. Antifilarial Glycosides of Streblus Asper: Effect on Metabolism of Adult Setaria Cervi Females. *Helminthology* (Bratislava) **1998,** *35*(4), 173–177.

22. Singh, M.; Shakya, S.; Soni, V. K.; Dangi, A.; Kumar, N.; Bhattacharya, S. The N-hexane and Chloroform Fractions of *Piper betle* L. Trigger Different Arms of Immune Responses in BALB/c Mice and Exhibit Antifilarial Activity against Human Lymphatic Filarid *Brugia malayi. Int. Immunopharmacol.* **2009,** *9*, 716–728.

23. Singh, G.; Sharma, P. K.; Dudhe, R.; Singh, S. Biological Activities of *Withania somnifera. Ann. Biol. Res.* **2010,** *1*(3), 56–63.

24. Singh, P. K.; Ajay, A.; Kushwaha, S.; Tripathi, R. P.; Misra-Bhattacharya, S. Towards Novel Antifilarial Drugs: Challenges and Recent Developments. *Future Med. Chem.* **2010a,** *2*(2), 251–283.

25. Tag, H.; Das, A. K.; Loyi, H. Anti-inflammatory Plants Used by the Khamti Tribe of Lohit District in Eastern Himachal Pradesh. *Indian J. Nat. Prod. Resour.* **2007,** *6*(4), 334–340.

26. Taylor, M.; Le Goff, L.; Harris, A.; Malone, E.; Allen, J. E.; Maizels, R. M. Removal of Regulatory T Cell Activity Reverses Hyporesponsiveness and Leads to Filarial Parasite Clearance In Vivo. *J. Immunol.* **2005,** *174*, 4924–4933.

27. Titanji, V. P. K.; Evehe, M. S.; Ayafor, J. F.; Kimbu, S. F. Novel *Onchocerca volvulus* Filaricides from *Carapa procera, Polyalthia suaveolens* and *Pachypodanthium staudtii. Acta Leiden.* **1990,** *59*(1–2), 377–382.

28. Tripathi, S. Singh, N.; Shakya, S. Dangi, A. Misra-Bhattacharya, S. Dube, A.; Kumar, N. Landrace/Gender-Based Differences in Phenol and Thiocyanate Contents and Biological Activity in *Piper betle* L. *Curr. Sci.* **2006,** *91*(6), 746–749.

29. Vayalil, P. K.; Kuttan, G.; Kuttan, R. Rasayanas: Evidence for the Concept of Prevention of Diseases. *Am. J. Chin. Med.* **2002,** *30*(1), 155–171.

30. WHO (World Health Organization). Report of Informal Consultation on Surgical Approaches to the Urogenital Manifestations of Lymphatic Filariasis. World Health Organization, Geneva, 2002, 15–16. http://www.who.int/iris/handle/10665/67600 (Accessed March 30, 2017).

31. WHO (World Health Organization). Fact Sheet. http://www.who.int/mediacentre/factsheets/fs102/en/ (Accessed March 26, 2017).

32. Zaridah, M. Z.; Ididb, S. Z.; Omarc, A. W.; Khozirah, S. In Vitro Antifilarial Effects of Three Plant Species against Adult Worms of Subperiodic *Brugia malayi. J. Ethnopharmacol.* **2001,** *78*(1), 79–84.

CHAPTER 18

MAXIMIZING EXPRESSION AND YIELD OF HUMAN RECOMBINANT PROTEINS FROM BACTERIAL CELL FACTORIES FOR BIOMEDICAL APPLICATIONS

MANASH P. BORGOHAIN[1,†], GLORIA NARAYAN[2,†],
H. KRISHNA KUMAR[3], CHANDRIMA DEY[4], and
RAJKUMAR P. THUMMER[5,*]

[1]*Laboratory for Stem Cell Engineering and Regenerative Medicine (SCERM), Department of Biosciences and Bioengineering, Indian Institute of Technology Guwahati (IITG), Guwahati, Assam 781039, India, Mobile: +919854850277, E-mail: m.borgohain@iitg.ernet.in*

[2]*Laboratory for Stem Cell Engineering and Regenerative Medicine (SCERM), Department of Biosciences and Bioengineering, Indian Institute of Technology Guwahati (IITG), Guwahati, Assam 781039, India, Mobile: +919957966856, E-mail: gloria@iitg.ernet.in*

[3]*Laboratory for Stem Cell Engineering and Regenerative Medicine (SCERM), Department of Biosciences and Bioengineering, Indian Institute of Technology Guwahati (IITG), Guwahati, Assam 781039, India, Mobile: +919962180535, E-mail: hk.kumar@iitg.ernet.in*

[4]*Laboratory for Stem Cell Engineering and Regenerative Medicine (SCERM), Department of Biosciences and Bioengineering, Indian Institute of Technology Guwahati (IITG), Guwahati, Assam 781039, India, Mobile: +919163303801, E-mail: dey.chandrima@iitg.ernet.in*

[5]*Laboratory for Stem Cell Engineering and Regenerative Medicine (SCERM), Department of Biosciences and Bioengineering, Indian Institute of Technology Guwahati (IITG), Guwahati, Assam 781039, India, Mobile: +917086867025, *E-mail: rthu@iitg.ernet.in*

† These authors contributed equally to this work.

CONTENTS

ABSTRACT

The field of recombinant protein production has attained widespread popularity due to the enormous applications of recombinant proteins in therapeutics, diagnostics or to understand basic biological questions. Several factors hamper efficient production of recombinant human proteins in a foreign host. Such aspects are needed to be addressed meticulously in order to increase the expression and yield of the protein while retaining its biological activity. Choice of host system is one of the most critical criterion. *Escherichia coli*, a Gram-negative bacterium, is a commonly used bacterial host for the production of recombinant proteins. Its use as a cell factory is well-established due to the vast knowledge of its genetics and physiology. Other features include faster doubling time, low-maintenance costs and ability to express proteins in abundance. Hence, several molecular tools and approaches have been developed for enhanced expression and production of heterologous proteins, such as codon optimization softwares, a vast catalogue of expression plasmids and engineered strains, and various purification and production strategies. In this review, we discuss these essential prerequisite factors for enhanced expression and yield of human proteins in *E. coli*, and highlight the biomedical applications of such recombinant proteins in this ever-growing field.

18.1 INTRODUCTION

Human body produces several proteins that act as signaling molecules, catalytic agents, potent hormones, antibodies, cytokines and structural components and also perform many other functions. However, dysfunction in a specific protein results in the development of severe pathological conditions such as diabetes, cystic fibrosis, dwarfism, thalassemia, blood clotting disorders, and so forth. Due to a dearth of standardized gene therapy treatments that would genetically restore the dysfunctional protein within the patient, protein therapy has emerged as a promising alternative. This therapy involves delivery of properly folded biologically active protein at optimal doses to replenish the deficient or absent protein, so as to reach normal physiological levels to perform its function. These therapeutic proteins are generated in biological systems, which assure restoration of protein function, cost-effective industrial production and the absence of hazardous contaminants. In addition, protein therapy needs to be adhered to stringent quality standards, which consequently defines the choice of recombinant expression hosts and strains, purification protocols and production strategies. Recombinant protein production has revolutionized the field of biotechnology as several purified proteins have reached the market in a short span of time. Currently, there are over 400 marketed recombinant products and other 1300 are under development as antibodies, enzymes, hormones, diagnostic kits, and many more.[111]

Large scale production of recombinant proteins employs a wide range of expression systems. Mammalian expression system is a suitable choice for proteins that require appropriate post-translational modifications for their function. For other proteins, the high production time and costs along with slow growth kinetics and complex nutritional requirements associated with mammalian systems renders them less preferred than bacterial systems.[131] Therefore, for such proteins *Escherichia coli* became an obvious choice for large-scale production due to their well-established genetic makeup and knowledge of its physiology, reasonable costs and unparalleled growth rate.[63,108] In this chapter, we discuss the recent advances in several aspects that are required to be carefully addressed that enhance heterologous protein expression and yield (Figs. 18.1 and 18.2), and the diverse applications of the recombinant proteins produced using this bacterial cell factory *E. coli*.

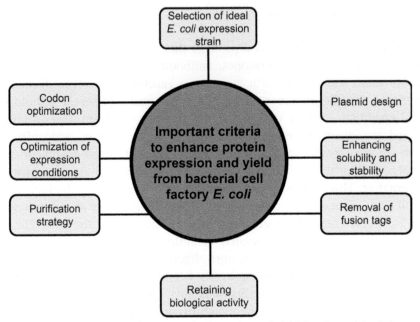

FIGURE 18.1   Criteria to enhance protein expression and yield from bacterial cell factory *Escherichia coli*. Important criteria listed in the figure has to be addressed for maximizing protein expression and yield along with considering the nature of the protein of interest, required for augmenting the recombinant protein production.

18.2   STRATEGIES FOR ENHANCING HETEROLOGOUS PROTEIN EXPRESSION

In recent times, many strategies have been developed to increase the expression of proteins of foreign origin in an expression system. The strategies mainly deal with the criteria of selection of expression systems, vector design and also how to overcome the problems associated with rare codons.

18.2.1   SELECTION OF EXPRESSION SYSTEM

Selection of a suitable expression system is one of the key steps in heterologous production of recombinant proteins and it depends on the nature of the protein to be produced. Among many expression systems developed in recent times, bacteria, yeast and mammalian cell-based expression systems

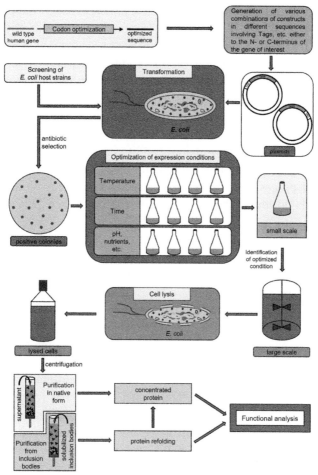

FIGURE 18.2 Generation of biologically active recombinant human proteins from bacterial cell factory *E. coli.* The desired wild type DNA sequence of a human gene is subjected to codon optimization for obtaining the most suitable codon combination leading to maximum expression of recombinant proteins. Various combinations of gene constructs involving tag(s) and other features are generated following codon optimization. Plasmid with optimized sequence of the gene of interest is transformed into *E. coli* and screened for positive colonies. Bacterial cultures of these positive colonies are then screened to determine standardized conditions on the basis of time, temperature, pH, nutrients, and so forth for optimal expression of the desired protein. Ideal conditions determined for small scale are then transferred to large scale for procuring maximum yield of the desired protein. Further, cells are lysed and subjected to centrifugation. Depending on the solubility of the protein, purification is done in either of the two ways: native or from inclusion bodies. Purified proteins are concentrated using various approaches, refolded in its active form and examined for its functionality.

offer better advantages in context of the characteristics of the protein and applications of the produced protein. Mammalian expression systems are preferable for those proteins that require post-translational modifications, as bacterial systems lack the modification machinery.[131] In mammalian systems, chinese hamster ovary (CHO) cells and murine myeloma cell lines (NS0 and SP2/0) are predominantly used for the production of recombinant proteins.[23] Another mammalian expression system derived from human embryonic kidney cells, HEK293 with two variants HEK293-T and HEK293-EBNA has got applicability because of its ease of culture, amenability of transfection with high efficiency, faithful post-translational folding and processing, and high efficiency of protein production. Till date, there is no report showing any difference between the recombinant proteins derived from the HEK293 and CHO cells. However, CHO cells find better industrial utility, because of its well-established purification and production strategies.[7,127]

Among microbial expression systems, two yeast-based expression systems namely, *Saccharomyces cerevisiae* and *Pichia pastoris* are also used for heterologous protein expression.[16,42] They offer advantages of easy genetic manipulation, higher production efficiency, majority of post-translational processing capability, low production costs, and so forth. It has been shown that these systems are very efficient in producing various recombinant proteins of human origin like chitinase, insulin, and so forth.[3] However, lack of certain cofactors essential for protein function and/or complex assembly and limited post-translational modification capacity render these systems unsuitable for widespread applications.[94]

Bacterial expression systems remain the most attractive because of their rapid growth, excellent productivity, ease of maintenance and low costs. Out of many bacterial systems, the Gram-negative *E. coli* bacterium is the most widely used expression system (Table 18.1). Apart from cost-effective production and rapid growth rate, widespread applicability of *E. coli* as a host is supported by well-characterized genetics and availability of many cloning vectors and genetically modified host strains.[63,108] Nevertheless, *E. coli*-based expression systems lack post-translational machinery and produce inactive proteins due to their accumulation into inclusion bodies. Hence, for those proteins whose biological activity is profoundly dependent on post-translational modifications, alternative systems like yeast, mammalian cell-based system, and so forth, as mentioned earlier should be used. However, in recent times a number of protocols have been developed describing strategies for expression of these proteins as soluble and active forms. Another disadvantage of *E. coli* derived recombinant proteins, especially related to therapeutic use is endotoxin (LPS) accumulation and therefore, a purification step should be incorporated in the production protocol.[91] Generally, in *E. coli*, the periplasm and cytoplasm are the two principal sites for accumulation of heterologous proteins. Cytoplasm is the site of choice for the expression of these proteins in view of higher yield. It is important to note that this will not assure that a recombinant gene product will accrue in *E. coli* at high levels as a full-length protein. It is also likely that it is unable to fold these products into their proper conformation resulting in a biologically inactive protein. However, a substantial amount of work has been carried out on enhancing the versatility and performance of this workhorse microorganism.

TABLE 18.1 Some of the Commonly Used *E. coli* Strains for Heterologous Protein Expression.

E. coli strain	Background	Features	Supplier
AD494	K-12	thioredoxin reductase (*trxB*) mutant; facilitates cytoplasmic disulphide bond formation; enhances proper folding of proteins	Novagen
BL21	B834	*lon* and *ompT* proteases deficient; increases yield of recombinant protein	Novagen
BL21-AI	BL21	has a chromosomal insertion T7 RNA polymerase gene into *araB* locus of *araBAD* operon, placing regulation of T7 polymerase under the control of the arabinose-inducible *araBAD* promoter; useful in the expression of the genes that may be toxic to other BL21 strains	Invitrogen
BL21*trxB*	BL21	has similar features as the AD494 but derived from BL21 background	Novagen
BLR	BL21	*recA* mutant; *lon* and *ompT* proteases deficient; stabilizes tandem repeats and increases yield of recombinant protein	Novagen
C41	BL21	has at least one uncharacterized mutation; Mutant designed for expression of toxic proteins like membrane proteins, nucleases and some cytoplasmic proteins	Lucigen

TABLE 18.1 *(Continued)*

E. coli strain	Background	Features	Supplier
C43	BL21	double mutant; derived from C41 by selecting for resistance to a different toxic protein and can express a different set of toxic proteins to C41	Lucigen
HMS174	K-12	has similar features as the BLR but derived from K-12 background; stabilizes certain gene product of which may cause loss of DE3 prophage	Novagen
JM83	K-12	enables periplasmic secretion of recombinant proteins	ATCC
NovaBlue	K-12	has *recA* and *endA* mutations; high transformation efficiency and blue-white screening capabilities; yields excellent quality plasmid D	Novagen
Origami	K-12	thioredoxin reductase and glutathione reductase (*trxB /gor*) mutant; greatly enhances cytoplasmic disulphide bond formation	Novagen
Origami B	BL21	has similar characteristics as Origami except they are derived from BL21 with *lacZY* mutation	Novagen
Rosetta	BL21	derived from BL21 strain with *lacZY* mutation; supplies tRNAs for the rare *E. coli* codons AUA, AGG, AGA, CUA, CCC, and GGA; enhances expression of eukaryotic proteins containing these rare codons	Novagen
Rosetta Blue	K-12	has high transformation efficiency and *recA* and *endA* mutations; supplies tRNAs for rare *E. coli* codons AGG, AGA, AUA, CUA, CCC and GGA; enhances expression of eukaryotic proteins containing these rare codons	Novagen
Rosetta-gami	K-12	*trxB /gor* mutant; supplies tRNAs for codon rarely used in *E. coli* AGG, AGA, AUA, CUA, CCC and GGA; enhances the expression of proteins containing these rare *E. coli* codons	Novagen
Turner	BL21	*lacZY* deletion mutant; enable adjustable levels of protein expression throughout all cells in a culture; by fine-tuning the IPTG concentration, expression can be controlled from very low levels up to the robust, completely induced levels	Novagen

The production efficiency and bioactivity of the produced recombinant proteins are greatly influenced by the strain of *E. coli* used. In routine practice, BL21 and K12 and their mutant stains are most widely used (Table 18.1). BL21 cells belong to B-strain (B834) and deficient in *lon* and *Omp T* proteases. These properties facilitate protein accumulation at a higher rate and reduce degradation of the expressed proteins during purification. In addition, these cells do not carry the gene encoding T7 RNA polymerase. In contrast to BL21 cells, BL21 (DE) mutants carry λDE3 lysogen with gene T7 RNA polymerase under the control of lac UV5 promoter. Presence of T7 RNA polymerase in these cells induces specific expression of gene(s) cloned downstream to a T7 promoter at a higher rate. Likewise, other two mutants BL21 CodonPlus-RIL/BL21 CodonPlus(DE3)-RIL and BL21 CodonPlus-RP (Stratagene) enhance the production of eukaryotic proteins containing rare *E. coli* codons AGG, AGA, AUA, CUA and AGG, AGA, CCC respectively. BL21 CodonPlus-RIL/BL21 CodonPlus(DE3)-RIL (Stratagene) carries additional copies of *argU, ileY* and *leuW* genes that encode tRNA recognizing rare *E. coli* codons for arginine, isoleucine and leucine. The variant BL21 CodonPlus-RP (Stratagene) carries extra copies of *argU* and *proL* genes. These genes encode tRNAs that read rare codons for arginine and proline. The variants C41 and C43 (Lucigen) of BL21 are mainly suitable for the expression of toxic proteins like membrane proteins, nucleases, few cytoplasmic proteins, and so forth, as they prevent cell death during expression. A *recA* mutant of BL21 cells, BLR (Novagen), has been designed to improve plasmid monomer yield and also stabilizes tandem repeats. The *recA* mutant of K12 background, HMS174 (Novagen), like BLR stabilizes several target genes whose product may result in loss of DE3 prophage. Origami and Origami B (Novagen) are the two *trxB/gor* (thioredoxin reductase) mutants of K12 and BL21 respectively. These cells show greatly enhanced formation of cytoplasmic disulphide bonds. Rosetta (Novagen) is a single variant form of BL21 with properties of both BL21 CodonPlus-RIL and BL21 CodonPlus-RP strains; whereas Rosetta-gami (Novagen) has additional properties of Origami strain. Likewise, a number of variants of BL21 and K12 have been developed to enhance heterologous expression of proteins in *E. coli*. An improved version of BL21 strain, BL21-Gold (Stratagene) cells have high transformation efficiency and produce improved miniprep DNA. These cells utilize T7 RNA polymerase promoter to direct protein expression to a higher level. The BL21-Gold derived strain with the Hte phenotype greatly enhances the transformation efficiency up to $> 1 \times 10^8$cfu/ μg of pUC18 DNA. Moreover, in these cells the gene that encodes endonuclease I (*endA*) is inactivated, thereby protecting the plasmid DNA from degradation. These *E. coli* variants are also available as DE3 (as described earlier) and pLysS strains. pLysS strain carries gene for T7 lysozyme, the expression of which leads to the inhibition of basal expression of T7 RNA polymerase prior to induction and stabilizes pET recombinants that encode target proteins affecting cell growth and viability. Another strain of *E. coli,* DH5 and its derivative DH5 alpha (Novagen) are most often preferred for DNA transformation. The DH5 alpha cells have *recA* mutation and free of heterologous recombination, thus ensuring enhanced insert stability. Additionally, these cells have *endA1* mutation and are devoid of endonucleases responsible for degrading plasmids during isolation process. Further, DH5 alpha cells (Novagen) are competent for blue-white screening. On the other hand, BL21 and K12 derivatives are especially designed for recombinant protein expression and lack certain proteases that digest the over-expressed proteins. Although specific post-translational modifications (e.g. glycosylation) will mostly remain beyond the reach of *E. coli*, robust engineered strains apt for economical production of numerous complex eukaryotic proteins should become attainable in the near future.

18.2.2   CODON OPTIMIZATION

Codon optimization is an important step to be carried out in heterologous expression of recombinant proteins. By definition, codon optimization is a technique which involves alteration of nucleotide sequence of DNA of one species into nucleotide sequence of DNA of another species without changing the amino acid sequence of the protein. This will maximize the efficiency of heterologous expression of a protein of interest because rare codons are replaced with codons that are more abundant in the genes of the host organism. Taking *E. coli* as an expression system has several advantages as described earlier, but the desired efficiency of heterologous expression is generally not achieved. However, before discussing the needs of codon optimization in detail, it is very important to highlight the concepts of 'codon bias usage' and 'rare codons'. Majority of the amino acids have more than one codon and selection of these synonymous codons varies significantly from organism to organism. Some codons (common codons) are very frequently used by some organisms while some (rare codons) are not. In *E. coli,* for example, the usage frequency of codons AGG for Arginine is 2.1%, which is approximately five times less (i.e. 11.2%) than the usage frequency of mammalian cells. This phenomenon is known as codon bias usage. Explaining reasons behind it, existing speculations have linked codon bias usage to genome GC content and growth conditions of an organism.[4,64] However, it is becoming increasingly evident that codon bias usage has a negative impact on the heterologous protein expression. Since there is a positive correlation existing between common codons and the abundancy of cognate tRNAs, it leads to the preferential utilization of common codons for enhanced translational efficiency and vice versa. Moreover, high level misincorporation and frameshift events are associated with the heterologous expression of rare codon genes. Frameshift events are generally seen due to errors in codon reading and processing. These mistranslational events ultimately give rise to undesired products.[51,140] The negative impact of codon bias on heterologous protein production can be overcome by overexpressing the rare tRNA encoding genes. For *E. coli,* Zhang and coworkers have identified eight least used codons; AGG and AGA for Arginine (Arg), AUA for Isoleucine (Ile), CUA and CUG for Leucine (Leu) and CCC and CCU for Proline (Pro).[142] Therefore, the primary strategy to enhance human protein production would be overexpression of the genes that encode rare tRNAs.[51] Even though this strategy somehow manages to fulfill the requirements of intracellular cognate tRNA pool, there are many overwhelming factors. It has been reported that tRNA molecules undergo extensive processing and modification before entering into the translation process. These modifications regulate gene expression by bringing conformation and dynamics for codon recognition, reading frame maintenance and by reducing frameshifts. These modifications are also important for the recognition of tRNAs by some aminoacyl synthetases.[52] Therefore, it remains broadly unclear whether forcibly increased rare tRNA pool meets requirements of tRNA modifications and aminoacylation and thereby the composition of heterologously expressed proteins. In addition, some metabolic processes are also believed to be involved in controlling tRNA concentration within a cell.[51]

Apart from these, many foreign genes or recombinant proteins severely retard the growth and survival of *E. coli* cells. These genes when expressed in the bacterial cells yield toxic products or proteins that significantly interfere with the normal cellular functions and homeostasis. Most importantly, some of these proteins affect cellular viability when they are overexpressed. This could be due to over-accumulation of some toxic moieties during the over-expression of heterologous proteins.[32,68,109] Similarly, the stability and half-life of heterologous mRNA also significantly affects the foreign protein production in *E. coli*. As we mentioned earlier, in many organisms there is a strong correlation between genomic GC contents and frequency of codon usage. Most organisms prefer decreased mRNA stability near the initiator codon and the codons with low GC contents increase mRNA stability probably allowing it to fold to a stable secondary structure.[48]

On the other hand, mRNA turnover is also found to have a critical role in heterologous protein expression. In general, mRNAs with longer half-lives yield more proteins as they are available for translation for a longer period of time than those with shorter half-lives.[19,29] Previously, it was believed that thermodynamic stability greatly influences the mRNA half-lives. However, it is becoming increasingly evident that the codon optimality, rather than thermodynamic stability, has greater influence on mRNA half-lives.[10,95] Advanced studies revealed that mRNAs harbouring optimal/common codons in their open reading frame (ORF) regions are more stable or have longer half-lives and the opposite is applicable to unstable or short-lived mRNAs. In addition, it was observed that substitution of optimal codons with synonymous nonoptimal/rare results in dramatic reduction of mRNA stability and vice versa.[10,95] Therefore, codon optimization is critical for upscale production of foreign/human proteins in *E. coli* or any other expression systems.

Describing its positive impact, Burgess-Brown et al.,[17] have analysed 30 human genes in *E. coli* and showed that codon optimization remarkably improved the yield and quality of the expressed proteins. This approach however, should be employed with caution so that it does not affect the optimum codon levels necessary for translation rate and mRNA stability and thereby proper folding of proteins.

Codon optimization facilitates heterologous production of recombinant proteins providing several advantages.[50] They are:

- Ensure proper folding by opportunity to match the codon frequencies in target and host,
- Reduce tandem repeat codons impairing gene construction in turn of expression,

- Insert or remove restriction sites to/from the gene construct,
- Enhance mRNA stability or prevent secondary structures,
- Allow modification in ribosome binding and mRNA degradation sites,
- Ensure proper transcriptional and translational control,
- Facilitate addition, removal and/or shuffling of protein domains,
- Allow control over translational rate for proper folding of proteins,
- Insert or delete additional sites for protein trafficking,
- Alter the proteins for their post translational modification sites, etc.

Development of various software (Table 18.2) makes the codon optimization easier. Most of these software are designed for the optimization of a gene sequence based on the information of a highly expressed gene of an organism. In addition, some of them also offer expertise in insertion and deletion of restriction sites, restriction site prediction and identification, T_m and stem loop determination, design verification or error checking, codon adapter indexing, codon maximization, and so forth. It has been claimed that software enabled codon optimization can increase the heterologous expression of protein up to 100- to 500-fold.[50]

TABLE 18.2 Commonly Used Codon Optimization Software and Their Features.

Software	Feature(s)	Reference
MEGA (Molecular Evolutionary Genetics Analysis)	Calculates codon frequencies	[67]
Codon Optimizer	Optimizes codon and finds type II restriction sites in the sequence	[38]
INCA (Interactive Codon Usage Analysis)	Computes codon frequencies and provides graphical display of values	[125]
UpGene	Multilevel optimization of gene sequence	[40]
GeMS	Apart from optimizing codon, it also predicts restriction sites, T_m and stem–loops. Also, breaks long sequences into small synthesizable parts.	[60]
Synthetic Gene Design	Optimizes codon and traces branches of life that evolved due to different deviations from standard genetic code.	[139]
GASCO (Genetic Algorithm Simulation for Codon Optimization)	This algorithm decreases search space and is used for designing DNA vaccines.	[112]

TABLE 18.2   *(Continued)*

Software	Feature(s)	Reference
CAI (Codon Adaptation Index)	It quantifies the similarity between codon usage in a gene and its frequency in a reference set.	[96]
QPSOBT	Optimization software works on Quantum-behaved Particle Swarm Optimization (QPSO) algorithm	[18]
COStar	D-star Lite-based dynamic search algorithm for optimization of codon.	[73]

18.2.3   PLASMID DESIGN

Efficiency of gene transfer greatly depends upon the nature of the plasmid construct. Majority of the plasmid DNA vectors used in current practice are a combination of replicons, promoters, selectable markers and multiple cloning sites (Fig. 18.3). Most frequently, a fusion protein and/or its removal strategy is incorporated.

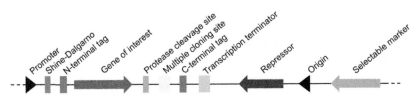

FIGURE 18.3   Schematic representation (not drawn to scale) of *E. coli* protein expression vector. The figure shows the important features present in *E. coli* protein expression vectors. The affinity tags and coding sequences for the removal are positioned randomly.

18.2.3.1   REPLICON

The origin of replication and the sequences located in the vicinity that regulate plasmid copy number are together called replicon. Copy number is an important factor to be considered while designing a plasmid construct, and the control of copy number per cell resides in the replicon.[30] Plasmid copy number decides the gene dosage available for expression and therefore, high copy number generally results in high yield. However, high genome dosage may sometimes exert metabolic burden affecting the growth rate of bacteria and thereby, productivity of recombinant proteins.[9,55] Therefore, this strategy of increasing plasmid copy number is not helpful in increasing the expression level of proteins. pET vector series is used most often, carrying pMB1 origin with 15–60 plasmid copies/cell and in pUC series the pMB1 origin has been mutated and carries 500–700 plasmid copies/cell.[12,81] A novel expression vector having the capability of regulating both the plasmid copy number and gene expression has also been developed. The simplest version of the vector is developed copy control pBAC/OriV plasmid[137,138] with trfA up-mutant gene controlled by the L-arabinose *para* promoter. These vectors are maintained as a single copy in the absence of an inducer and thus, tightly regulating any residual expression. Thus, induction by L-arabinose leads to the amplification of plasmid and subsequently the expression of the cloned gene which can be increased up to 50,000-fold. The vector exhibits very low expression in the absence of an inducer, making it suitable for maintenance of toxic genes.[138] pACYC and pBAD are expression systems with p15A origin, which can be used for the double expression of recombinant proteins.[20,53] Likewise, triple expression can also be obtained by using pSC101 plasmid. Plasmids containing this replicon offer advantages when high dose of cloned gene becomes detrimental to cells.[124,132]

18.2.3.2 PROMOTER

Perhaps, the promoter of an expression vector is the key regulatory unit determining the duration and strength of transgene expression. This is the site where RNA polymerase binds along with associated factors and they initiate transcription. An ideal promoter should favour high degree of protein synthesis. Several commonly used promoters have been identified, however, each of these promoters have their own advantages and limitations (Table 18.3). It must be strong enough to support protein accumulation up to 10–30% or more. The promoter should also exhibit low basal expression that is, it should be tightly regulated. In case of toxic proteins, tight regulation of promoter is very important when protein of interest is detrimental to host cell itself. Very recently, a novel tightly inducible expression system (pyP_{fhuA}) with strong iron-regulated promoter, P_{fhuA} has been developed. Using tightly regulated promoter the authors successfully expressed toxic genes E, gef and mazf in *E. coli*.[49] This strategy does not always give fruitful result as a small amount of severely toxic gene product may also drastically affect bacterial growth and survival.[141] Apart from these two properties, a promoter must be simple, cost-effective, compatible with other *E. coli* strains and also should not be affected by commonly used culture media components.

TABLE 18.3 Some of the Commonly Used Promoter Systems, Their Advantages and Disadvantages.

Promoter	Induction	Regulation	Advantages	Disadvantages
lac	IPTG, thermal	lacI, lacI^q	Strong promoter allows efficient transcription in the presence or absence of the complex between cyclic AMP and its receptor protein (CRP)	Lower level of expression compared to other systems; leaky expression
trp	Trp starvation, IAA		Well characterized, control is well-understood, powerful and promotes highly efficient transcription	Leaky expression
tac	IPTG, thermal	lacI, lacI^q	Strong and efficient transcription	Leaky expression
T7	IPTG, thermal	λcIts 857	Selective and higher rate of transcription initiation and processive transcription elongation rate	Abrupt transcription reduces cell densities, leaky expression
λpL	Thermal	λcIts 857	Highly productive, finely regulated, easy scalability and eliminates use of chemical inducers	Difficult to induce at low temperature, partial induction cannot be achieved
RecA	Nalidixic acid	lexA	Strong and efficient transcription, selective and simple induction	Not titratable
PhoA	Phosphate starvation	phoB+, phoR$^-$	Tightly controlled and selective induction, induction without intervention of operator	Limited media availability, not titratable
Ara	L-arabinose	araC	Highly efficient, rapid induction rate and high level expression	Limited vector availability, catabolite repressed by glucose
Cad	pH	cadR	Specific and highly efficient	Not well characterized, limited vector availability

A well-known mechanism of *E. coli* is lactose utilization. Based on *lac*-derived regulatory element many promoters were designed. However, lac promoter is not strong and rarely used for large scale production of recombinant proteins. But its use is desirable for the production of membrane proteins or the toxic gene products because of its leakiness. The promoters of synthetic origin, *tac*[11] and *trc*[14] induced by isopropyl-β-D-galactopyranoside (IPTG) are strong and have widespread utility. These promoters are more efficient and confer protein accumulation up to 15–30% of total cellular proteins. However, the use of IPTG for large scale production of human recombinant proteins is restricted because of its toxic effects towards host cells,[115] and high costs. Moreover, the strength of these promoters may cause problems when proteins toxic to the host cells are expressed. Most importantly, it may lead to degradation when membrane proteins are overexpressed using strong *tac* promoter.[97] Alternatively, thermal induction of these promoters is now possible because of the development of mutant gene *lacI*(Ts) encoding a thermosensitive lac repressor.[5,57,58]

An extremely popular promoter system for recombinant protein production is the T7 promoter.[82] This promoter system is generally available in the pET vector series. In this system, the target gene is cloned next to the promoter which is recognized by T7 RNA polymerase. By using this promoter system, accumulation of more than 50% of target protein can be achieved.[47] But tight repression is very important to prevent leakiness of this promoter. Multiple approaches have been developed to regulate expression of this active T7 polymerase. An IPTG inducible system has also been described where it is placed in an *E. coli* BL21 (DE3) genome under the control of L8-UV *lac*. The activity of T7 RNA polymerase of non-induced cells can be reduced by using specific inhibitor such as T7 lysozyme.[82] Similarly, a reversibly attenuated T7 RNA expression cassette has also been described based on λ p_L regulation.[80]

Temperature sensitive phage λ p_L promoter has a moderate expression level which is regulated by a *cI* repressor. This promoter exhibits highest activity at 20°C because cI857 repressor becomes fully functional at 29°C and higher, and turns off the expression.[76] But higher temperature, that is at 42°C the repressor becomes inactivated inducing gene expression.[34,75] Many derivatives of p_L with higher efficiency at a temperature below 20°C, have been recently described.[46] The strategy behind the use of temperature sensitive promoters is to provide enough time for protein refolding, so that biologically active proteins are produced, minimizing the formation of inclusion bodies that is, inactive proteins. This is based on the fact that at lower temperatures that is, 15–20°C, rate of protein folding is affected slightly whereas, rates of transcription and translation are affected significantly.[46]

In recent times, many other promoters with attractive features have been described. The rRNA promoters P1 and P2, for example, are extraordinarily strong. However, its application for high level expression of recombinant proteins is hampered by regulation difficulties. Similarly, a pH sensitive promoter has very good strength, and the expression level of up to 35% of total cell protein can be achieved. But the level of expression may vary depending upon the gene of interest because protein production depends not only on the promoter strength, but also on the translational efficiency.

18.2.3.3   TRANSLATION INITIATION SEQUENCE

Translation initiation sequence or Shine–Dalgarno sequence is a sequence with purine-rich nucleotides complementary to the 3′ end of 16S RNA, located 5–13 bases 5′ to an initiator ATG in prokaryotic mRNA, which targets the RNA to a small ribosomal subunit for translation. This sequence is in close vicinity to an initiator methionine[117] and is required for ribosomal binding and protein synthesis to occur in prokaryotic cells. It is also known as the ribosome binding site. Ribosomal binding site elements typically used in expression vectors are derived from well-translated *E. coli*.

18.2.3.4   SELECTABLE MARKER

Selectable markers are sequence elements added to the vector construct to make the transformed cells distinct from non-transformed cells. In general, a selectable marker is an antibiotic resistant gene. For example, cells become resistant to ampicillin because of the presence of *bla* gene and its product β-lactamase (resistance enzyme), an enzyme that splits β-lactam antibiotics. However, the selection becomes problematic when constitutive high level expression of β-lactamase leads to nearly complete depletion of ampicillin in a couple of hours.[65] Under this condition non-transformed cells also increase their numbers. To avoid this, an antibiotic like tetracycline should be selected, the resistance of which is due to active efflux of antibiotic from the cells.[65,100] Moreover, it additionally imposes a metabolic burden on the bacterial cells and has risks of contamination in the final product. Around two-fold increase in the efficiency of foreign gene expression has been reported on the removal of *bla* gene from a therapeutic plasmid. This was applicable to other vector sequences as well.[56] Apart from the stability of antibiotic during cultivation and other associated problems, high cost is a major drawback in the large scale production of recombinant proteins in *E. coli*.

Alternatively, antibiotic-free strategies have been developed for the maintenance of plasmids in an expression system. Most of these systems are based on the principle of plasmid addiction, which implies a phenomenon where plasmid-free cells cannot grow or survive.[92] The systems that have been developed on the principle of plasmid addiction include—(i) auxotrophy complementation,[13,106] (ii) Toxin/antitoxin-based system[106,134] and (iii) operator-repressor titration-based system[25, 26, 41] Although this technology has proved its applicability for industrial scale, it is still not widely used.

18.2.3.5   FUSION PROTEIN

The development of versatile fusion proteins or fusion partners significantly revolutionized high level production and purification of recombinant proteins. Coupling of fusion tags performs three important functions—(i) acts as an affinity tag to facilitate purification, (ii) improves folding by acting as solubilizing partner and (iii) easy detection along with its expression, as well as purification scheme. For example, Staphylococcal protein A and its derivatives of synthetic origin are not only responsible for the ease of purification and folding, but also cause secretion of the proteins into culture medium.[84,85,110] Alternatively, streptococcal protein G has been developed for affinity tagging and has been successfully employed for purifying cDNA-encoded proteins.[69] Furthermore, a combination of protein A and G resulted in a tripartite fusion protein with providing additional protection to the expressed proteins from different proteases.[33,86] Some other examples of widely-used fusion proteins are poly-His-, poly-Arg-, FLAG-, c-Myc, and so forth. Availability of specific antibodies for these fusion peptides facilitates their detection (through western blot analysis) even though they are expressed in a very small quantity. Development of tag-specific resins that bind tightly to it, also enable single-step affinity purification. His-tagged recombinant proteins, for example, can be purified by Ni^{2+}–bound NTA (nitrilotriacetic acid)–agarose resin-based affinity chromatography.[93] A small ubiquitous protein, thioredoxin when used as a fusion partner remarkably enhances the solubility of heterologous proteins in *E. coli* preventing inclusion body formation.[71]

A multipartite affinity system utilizes seven different affinity tags to provide conditional flexibility for binding and elution of target proteins, and further provides advantages of a variety of molecular interactions of the seven affinity domains for detection, purification and solubilization.[87]

18.2.3.6   FUSION PROTEIN REMOVAL STRATEGY

After successful recovery of recombinant proteins, it is desirable to remove the fusion tags as the effects of these are unpredictable and may interfere with the biological activity[36] as well as structure[21,122] of the protein. Tag removal is preferably done by enzymatic cleavage, but chemical mediated removal can also be employed. Therefore, it is very important to incorporate a sequence encoding a protease cleavage site into the plasmid downstream to the gene that codes for fusion partner.

A large number of endoproteases have been described as tag removing enzymes. Among them enterokinase, factor Xa, thrombin and tobacco etch virus (TEV) protease are most successfully employed.[1,61,79,133] The use of endoproteases like enterokinase and factor Xa resulted in completely tag-free native proteins. In contrast, endoproteases such as TEV protease and thrombin give rise to proteins with one or two residual amino acids at the N-terminus. Therefore, effects of these residual amino acid(s) must be analyzed as they may alter the activity as well as structure of the processed protein. Recently, an endoprotease based on the human granzyme B has been developed for this purpose.[74] However, broad substrate specificity limits its use because of the risk of non-specific degradation. Further, the endoproteases have common limitations like usage of high enzyme to protein ratio, longer incubation time, and so forth, and in these conditions, sometimes non-specific cleavage occurs at cryptic sites of the protein affecting the product yield.[6,133] Exoproteases are not enzymes of choice for human recombinant protein production as they have a risk of exogenous contamination and other associated problems. However, a mentionable exoprotease-based tag removal system is TAGzyme system.[90] The system exploits the aminopeptidase dipeptidyl peptidase I (DPPI) with or without the combination of glutamyl cyclotransferase (GCT) and proglutamyl aminopeptidase (PGAP) to remove the protein tag. For a protein with a stop position at N-terminus, the system employs DPPI alone whereas for all other proteins with the combination of two enzymes acting on N-terminal amino acids that is, GCT and PGAP are employed. The system offers advantages of low enzyme to protein (tagged) ratio and complete removal of tags within a shorter incubation time.[70,90] However, an alternative promising strategy for large scale production of recombinant proteins comprises of a column affinity purification and tag removal using immobilized metal affinity chromatography (IMAC),[44,113] provided the issues related to process validation are addressed properly.

18.3 STRATEGIES FOR ENHANCING THE FINAL YIELD

Recombinant proteins when expressed in bacterial systems are secreted extracellularly or are found intracellularly in solubilized form that facilitates easy and effortless isolation, or in the form of insoluble aggregates called inclusion bodies that require elaborate processing to obtain a bioactive protein. If the protein of interest is not secreted extracellularly, the first step of each purification process involves cell lysis. Based on the fragility of the protein as well as stability of the cells, several approaches can be used for cell lysis: (i) sonication, (ii) repeated cycles of freezing and thawing, (iii) homogenization by grinding (bead beating), (iv) homogenization by high pressure (French press), and (v) permeabilization by enzymes (e.g. lysozyme) and detergents (e.g. Triton X-100).[116] When the proteins are not expressed intracellularly in inclusion bodies, the soluble proteins can be easily released into the solution by cell lysis (lysis buffer containing protease inhibitors and preferably DNAse), and further purified using chromatographic techniques in the presence of fusion tags.

Protein purification mostly employs three characteristics to separate proteins. (i) Proteins can be separated based on their molecular weight or size using sodium dodecyl sulphate-polyacrylamide gel electrophoresis (SDS-PAGE) or size exclusion chromatography. Proteins are also purified using two-dimensional (2D)-PAGE and are subsequently examined by peptide mass fingerprinting to confirm the identity of the protein. (ii) Proteins can be purified based on their isoelectric points by running them through an ion exchange column (ion exchange chromatography) or a pH graded gel (free-flow-electrophoresis). (iii) Proteins may be separated by hydrophobicity / polarity via hydrophobic interaction chromatography, reversed-phase chromatography or high performance liquid chromatography.

In addition, proteins are also commonly separated based upon molecular conformation, which often exploits application specific resins (affinity chromatography). Affinity chromatography is further categorized into: (i) metal binding (polyhistidine tag that binds strongly to divalent metal ions such as cobalt and nickel), (ii) immunoaffinity chromatography (precise binding of an antibody-antigen to selectively purify the protein of interest), (iii) purification of a tagged protein (engineering an antigen peptide tag against the protein, and further purify the protein by incubating with a resin coated with an immobilized antibody or on a column. This technique is called immunoprecipitation).

However, when heterologous proteins get entrapped in inclusion bodies, the purification process becomes quite tedious. Nevertheless, this has been productive for proteins that are challenging to produce in the cytoplasm of *E. coli* as soluble and bioactive proteins, including numerous growth factors,[22] recombinant Fab fragments,[121] and receptors.[39]

18.3.1 PURIFICATION OF RECOMBINANT PROTEINS FROM INCLUSION BODIES

Inclusion bodies are densely packed intracellular insoluble protein aggregates which are formed when a gene of interest is overexpressed in the cytoplasm of *E. coli*. Inclusion body formation is advantageous: (i) it helps in higher yield of protein in a pure form, (ii) protects the protein from intracellular proteases, (iii) homogeneity of protein reduces the purification step, and (iv) entrapped protein can be easily isolated based on its size and density.[119] Since, these inclusion bodies are resistant to proteolysis, they contain a large amount of relatively pure protein of interest.[8] However, these bodies are sites of misfolded proteins. Formation of these inclusion bodies is mainly due to the usage of strong promoters, high inducer concentrations, inability to form correct, or any, intra- or intermolecular disulphide bonds in the reducing intracellular environment, failure of bacteria to provide all post-translational modifications that a protein requires to fold, imbalance between in vitro protein solubilization and aggregation, and so forth.[8] This can be minimized by controlling parameters such as temperature, expression and host metabolism. Rise in the temperature (around 42°C) is directly proportional to aggregation.[72] Thus, reducing the temperature from the usual 37°C in *E. coli* resulted in increased production of native protein and decreased aggregation.[114] Addition of carbon sources like deoxyglucose reduces the rate of metabolism and aggregation.[72] A number of factors such as the rate at which the polypeptide chain is formed, rate of aggregation and protein synthesis also determine the formation of inclusion bodies.[6] Slow folding due to presence of disulphide bridges often results in inclusion body formation.[8]

Isolation of inclusion bodies can be achieved by a treatment with lysozyme before cell homogenization to enable cell disruption.[118] Inclusion bodies are isolated by low speed centrifugation of bacterial cells that have been mechanically ruptured either by high pressure homogenization or by sonication.[102] Frequently, bacterial outer membrane or envelop proteins may co-precipitate with the insoluble fractions as contaminants. These contaminants can easily be removed by treatment with high salt concentration or detergents such as Triton X-100[89] or low concentrations of chaotropic compounds.

18.3.2   ENHANCING SOLUBILIZATION OF ENTRAPPED PROTEIN

Solubilization of inclusion bodies is performed by treatment with different concentrations of chaotropic denaturants such as urea or guanidinium hydrochloride. The latter is preferred more due to its superior chaotropic and strong denaturant characteristics. Moreover, urea may contain isocyanate causing carbamylation of free amino acids.[105] A pH shock is also known to facilitate solubilization along with chaotrophic agents.[119] Fusion tags are also used to solubilize proteins. Sachdev and Chirgwin compared the solubility of two mammalian aspartic proteinases (procathepsin D and pepsinogen) by tagging them to maltose binding protein (MBP) and thioredoxin (TRX) at C-termini of the proteinases placing factor Xa between the tags.[107] These tags enhanced solubility, however only the tagged pepsinogen was enzymatically active. Fused procathepsin D requires glycosylation which is carried out in endoplasmic reticulum. When the solubility of MBP and TRX was estimated, MBP was better in resolubilizing the protein.[107] In addition to MBP and TRX, different *E. coli* proteins like NusA, GrpE, Skp and BFR have been used to solubilize proteins. NusA when fused to human interleukin-3 (hIL-3) enhanced the solubility to 97%.[28] NusA tagging also facilitated easy purification using affinity chromatography. In 2012, Singh et al., used 6 M of different alkyl alcohols to solubilize inclusion bodies formed during the expression of recombinant human growth hormone and found *n*-propanol with 2M urea to be effective.[120] This composition improved solubility without compromising the native structure of the protein. Alkyl alcohols are known to unfold proteins due to their hydrophobic interactions and reducing dielectric constant of medium. They also have helix stabilizing ability and these features make them apt for solubilization. *n*-Propanol has detergent-like characteristics at high concentrations due to the presence of -OH group and protects the secondary structure by inducing helix formation, whereas urea disrupts hydrophobic interactions within the polypeptide.[120] Moreover, inclusion body proteins solubilized under mild denatured conditions are easy to refold and regain their biological activity.

18.3.3   REFOLDING OF PROTEIN TO REGAIN BIOACTIVITY

Various strategies involving solubilization of inclusion bodies can lead to non-native conformation of the expressed protein. This can be solved by proper protein refolding techniques of a target protein at low denaturant concentrations. High concentration of the misfolded protein often results in decreased refolding, irrespective of the refolding method. Therefore, it is necessary to maintain low concentration of the initial misfolded protein for proper refolding.

18.3.3.1   PROTEIN REFOLDING APPROACHES

Refolding of protein is the process of restoring the native structure of the protein completely or partially, using different traditional methods like direct dilution, pulse renaturation, or high-end methods like chromatographic techniques. These techniques use the physical and chemical properties of the protein to bring about its refolding.

18.3.3.1.1   Direct Dilution

Direct dilution of protein is the process of diluting the protein concentration to 1–10 µg/ml in a proper folding buffer that allows the formation of native structure. Often dialysis is carried out after dilution. This method gets complicated when the protein requires disulphide bond formation leading to higher dilution for pairing of cysteine residues.[135] Although this process finds use in experimental purpose, huge vessels and additional expenses restrict its use for industrial purposes.[130]

18.3.3.1.2   Pulse Renaturation

Pulse renaturation is a direct implementation of direct dilution method in which the denatured solubilized protein is added continuously to the refolding buffer.[62] This process enhances the final active protein concentration but requires extensive knowledge of kinetics of the target protein. Katoh and co-workers showed that the rate of addition of protein solution should not exceed the rate determining folding step.[62] Reports stated that 80% of the protein must have attained its native conformation before the second batch of pulse is added. Moreover, the amount of pulse must also be optimized so that residual amount of protein does not exceed.[24]

18.3.3.1.3   Membrane Dependent Refolding

Membrane dependent refolding makes use of dialysis and diafiltration for the removal of denaturants. This process often causes greater amount of aggregation and non-specific adsorption to the membrane, however, it efficiently works with a few proteins like refolding of bovine carbonic anhydrase using hollow fibre ultrafiltration.[136] Streptavidin was also refolded using dialysis and yielded twice the normal quantity when the dilution rate and protein feed was controlled.[123]

18.3.3.1.4   Role of Aiding Agents in Protein Folding

Chaperones play a major role as aiding agents. These are proteins that bind to unfolded or denatured polypeptides, suppress aggregation pathway and initiate proper folding. Chaperones are generally categorized based on the source: natural or artificial. Among the natural ones, GroEL finds the maximum use. It assists the folding of 10% of polypeptides in vivo and can reach up to 30% when under stress.[35] Apart from GroEL, DnaK was also reported to increase the production of folded protein, when added to B3 (Fv)-PE38KDEL which is a recombinant immunotoxin. In 2009, an *E. coli* chaperone, Skp, was reported to stabilize human Nanog and Sox2 transcription factors.[54] DNA binding assay showed that co-expression of Skp with recombinant proteins did not perturb their DNA recognizing capacity.[55] But high cost and concentration of molecular chaperone and additional steps for purification of protein often limits the use of chaperone for refolding.[15]

Artificial chaperones address the problems associated to natural chaperones by a two-step method. In the first step, detergent conjugates to the protein molecule preventing aggregation and in the second step, cyclodextrin cleaves the detergent facilitating folding of protein to its native conformation. This method was used to concentrate a number of proteins like lysozyme, carbonic anhydrase B[104] and citrate synthase.[27] Instead of cyclodextrin, Machida et. al., used cycloamylose as a cleaving agent that stripped off the attached detergent moiety and enhanced protein folding. It proved to work equally well for both oligomeric and monomeric protein.[77] Artificial chaperone system requires the complete elimination of surfactants and since they are difficult to remove, Nomura et al., used amphiphilic nanogel system to assist in refolding. Nanogels had heat shock protein (HSP)-like characteristics and were used to refold carbonic anhydrase and citrate synthase.[88]

18.3.3.1.5   Physical and Chemical Features Enhancing Refolding Yield

Change in the physical conditions like lowering down the refolding temperature often results in decrease in the rate of aggregation reaction as hydrophobic interaction is repressed. Ideal temperature of 15°C to initiate the reaction is often helpful.[24] However, further lowering in temperature results in slowing down of the rate, and therefore increases the time required. Pressure also plays an important role in productive folding. High pressure disfavours aggregation reactions and proteins are known to retain their native structure. Slow removal of pressure favours native folding even at high concentrations.[101]

Among the chemical assistants, L-arginine is used more since it impedes aggregation by concealing the hydrophobic patches. In addition, low molecular weight chemicals like glycerol and high molecular weight chemicals like polyethylene glycols (PEGs) also act as enhancing aids.[24] Zwitterionic agents like sulphobetaines, aminocyclohexanes and pyrroles are also employed.[130] Moreover, different polymers are applied, but further optimization is required.

18.3.4 PURIFICATION TO MAXIMIZE YIELD

Protein purification is carried out to remove major contaminants and enzymes interfering with the biological activity of the protein. Moreover, high recovery at every step is required to maximize the final yield. Some key points that must be considered before designing a purification protocol are choice of assays prior to purification, choice of starting material, precautionary measures to avoid inactivation and choice of fractionation step. Purification is carried out using several approaches as discussed earlier. A number of studies describe different novel ways to purify recombinant proteins. Mojsin et al.,[83] carried out purification of human RXRα protein expressed in *E. coli* strain and tagged it to Glutathione S-transferase (GST) using three different lysis methods: sonication, freeze/thaw and bead beating. The study compared the solubility, yield and functionality of the protein purified using three methods. The data showed that purified proteins obtained from all three approaches were soluble, with the highest protein yield achieved via the freeze/thaw lysis approach.[83] Functional analysis data showed that the protein purified using the sonication and freeze/thaw approaches displayed comparable DNA binding affinity, whereas the proteins purified by bead beating were functional too, but with reduced DNA binding affinity.[83] Tao et al.,[126] used sarkosyl for solubilizing entrapped protein . Initial report suggested 0.3–2% of sarkosyl were sufficient to solubilize when the protein is expressed in Luria–Bertani (LB) media.[37] But the reported concentration of sarkosyl was insufficient to solubilize GST and MBP fused proteins. When higher concentration (10% sarkosyl) was used along with Tris, NaCl and β-mercaptoethanol for 6-24 h, more than 95% of protein got solubilized. With more than 10% sarkosyl, viscosity increased and this involved laborious purification steps.[126] The protein of interest is often tagged to MBP for easy solubilization. Recently, Raghava et al.[98] reported that precipitation of HIV protein gp120 with cationic starch combined with PEG and CaCl$_2$ facilitated simple purification . Smart polymers were also known to assist refolding and purification. These are pseudochaperonins which react to various stimuli such as change in pH, temperature and chemical species. One can optimize these stimuli to bring about precipitation, thus causing purification. Initially smart polymers were only used for chemically dena-tured proteins, but its use for active proteins is also reported.[43] pH sensi-tive methyl methacrylate polymer was used to isolate cell division or death B (CcdB) protein.[43] In a different study, denaturant along with reducing agent (β-mercaptoethanol) and detergent (Triton-X 100) were used to isolate the recombinant outer membrane protein (rOmp28) of *Brucella melitensis*. Addition of Triton-X 100 enhanced lysis and solubilization. β-mercaptoethanol cleaved the disulphide linkages that might have been formed with contaminants. Moreover, 10% glycerol was also added to the wash buffer to aid solubilization.[66]

18.4   APPLICATIONS OF RECOMBINANT PROTEINS

Recombinant proteins can be used for both preparative and analytical applications. The goal of preparative purifications is to produce a large quantity of purified proteins for commercial products such as nutritional proteins (e.g. soy protein isolate), enzymes (e.g. lactase) and certain biopharmaceuticals (e.g. insulin). Analytical purification aims to produce a relatively small quantity of proteins for various analytical or research applications (e.g. restriction enzymes).

Since the production of somatostatin in 1977 by Itakura et al., recombi-nant protein production has become an important tool for treating diseases as well as for research purposes. Somatostatin is a growth hormone with a known sequence that inhibits secretion of other hormones. Low somatostatin levels often cause hormonal imbalance and the patients require recombi-nant somatostatin externally. In order to produce native somatostatin, the gene for somatostatin was fused to β-galactosidase gene in pBR322 plasmid. Transformation in *E. coli* resulted in active polypeptide that could be cleaved with cyanogen bromide. However, the somatostatin yield was low and further optimization of purification conditions are required.[59] In 1979, two different bacterial strains were used to produce A chain and B chain of insulin. Insulin deficiency causes type 1 diabetes mellitus. Administration of insulin hormone is required to maintain normal glucose levels. The peptide chain of insulin was fused to β-galactosidase gene with a methionine residue. The chains were held together by disulphide bridges. When the chains were purified, 100% recovery was obtained in case of A chain, whereas B chain had inefficient purification. The final yields of A chain were quite high, around 10mg of protein from 24 g of cells.[45] Rosenberg and co-workers produced bioactive human interleukin-2 (IL2) protein in *E. coli*. IL-2 protein acts as a helper molecule in B- and T-responses, and thus plays an important role in immunological responses. They reported large scale production of the protein from *E. coli* as an expression system. The engineered IL-2 showed similar bioactivity as that of native IL-2.[103]

Recombinant protein production has also been extensively used to generate various growth factors. Epidermal growth factors (EGFs) trigger cell growth and proliferation, however, their deficiency may cause diabetic foot ulcer. Recombinant production of EGFs that can be used for treatment of this disease was carried out using *E. coli* HB-101 strain. Media composition, quantity of inoculum and the time of IPTG induction was optimized to increase the yield, and eventually a final yield of 94% was achieved.[129] A critical growth factor, fibroblast growth factor 20 (FGF20) is responsible for several biological processes such as neuroprotection and development of neural tissues. Clinical application of FGF20 requires abundant production of active recombinant protein. Optimization of inducing conditions produced a large quantity of insoluble protein that upon solubilization, refolding and purification with HiTrap affinity chromatography yielded, 218 mg of rFGF20 for 100 g of wet cells.[128] Human insulin-like growth factor I (hIGF-I) has got significant clinical applications (somatomedin C). It plays a crucial role in growth and has an anabolic effect. Large-scale production of this growth factor takes into consideration these three factors: culture condition, amount of inducer and exposure time of inducer, which assist in over production of this protein. When bacteria were cultured in LB, terrific broth (TB) and 32y media, 32y culture medium at 32°C and induction with 0.05 mM IPTG were found to be optimal. Moreover, the addition of 10 g/L of glucose gave best results.[99] A recombinant form of human growth hormone called somatropin is used as a drug to treat growth disorders in children and growth hormone deficiency in adults. Large-scale production of somatropin can be performed using *E. coli* expression system for the prevention of such growth-related disorders.

Human epidermal growth factor receptor (HER2) dimerizes with other HER receptors activating signalling pathways that promote cell growth and proliferation. Pertuzumab, a recombinant monoclonal antibody inhibits dimerization by specifically blocking the dimerization of HER2 with other HER receptors thus, preventing metastatic breast cancer. For this reason, it is also called as "HER dimerization inhibitor".[2] In addition, the production of Trastuzumab, a recombinant purified monoclonal antibody inhibits the growth of cells overexpressing HER2 protein. Trastuzumab in combination with paclitaxel and carboplatin checks the growth of tumour and prevents its progression.

Recombinant proteins can also be used for stem cell research.[31] Colony-stimulating factors (CSF) are primarily glycoproteins that stimulate haematopoietic stem cells to differentiate to a specific cell lineage. Multi CSFs are also called interleukin-3 (IL3) and they trigger the differentiation of progenitor cells to haematopoietic lineage.[78] Techniques for the production of CSF recombinant protein were also employed for its large-scale production. One of the most recent developments in this field has been the generation of induced pluripotent stem (iPS) cells with the help of stem cell-specific recombinant proteins. Recently, Zhou et al. demonstrated that stem cell-specific transcription factors Oct4, Sox2, Klf4 and c-Myc can be easily purified from E. coli.[143] These recombinant proteins were applied to fibroblast cells to reprogram them to iPS cells in the presence of a small molecule valproic acid.[143] These cells have tremendous biomedical potential in drug discovery, disease modeling, regenerative medicine as well as understanding human development.

Therefore, recombinant proteins can be used for the treatment of various diseases in which a protein is dysfunctional (growth hormone and insulin), obstructing a biological process (antibodies that block blood supply to tumours), biopharmaceuticals, signalling modulators, stem cell research and for other numerous applications.

18.5   CONCLUSION

Recombinant proteins have found tremendous biomedical applications on a large-scale in therapeutics, diagnostics and understanding biology. Several recombinant proteins have been generated for the treatment of various diseases and alleviate the symptoms. E. coli as a bacterial expression is the most common expression system in practice for the production of recombinant proteins. A wise choice of the engineered *E. coli* strain which expresses the protein of interest with high efficiency is the key to successful protein production. In addition, extensive research devoted to gene expression in *E. coli* has delivered several choices for transcriptional and translational regulatory elements that can be useful for the expression of foreign genes. *E. coli* is one of the most potent and versatile expression systems because of its rapid growth and protein production rates in combination with easy manipulation, cost-effective cultivation, ample physiological understanding and advanced genetic tools (increasingly large number of expression vectors and mutant host strains). Furthermore, the right selection of the engineered host, codon optimization, expression plasmids and various purification and production strategies are equally vital for the successful production of high quality recombinant human proteins from these bacterial cell factories. In summary, *E. coli* remains one of the most attractive expression hosts for small and large-scale production of heterologous proteins for a variety of applications.

18.6 ACKNOWLEDGMENTS

We thank all the members of Laboratory for Stem Cell Engineering and Regenerative Medicine (SCERM) for their excellent support. This work was supported by grants funded by North Eastern Region—Biotechnology Programme Management Cell (NER-BPMC), Department of Biotechnology, Government of India (BT/PR16655/NER/95/132/2015) and Science and Engineering Research Board (SERB), Department of Science and Technology, Government of India (Early Career Research Award; ECR/2015/000193).

KEYWORDS

- **bacterial cell factory**
- ***E. coli***
- **recombinant proteins**
- **expression and purification**
- **protein yield**
- **inclusion bodies**
- **protein refolding**

REFERENCES

1. Abdullah, N.; Chase, H. A. Removal of Poly-Histidine Fusion Tags from Recombinant Proteins Purified by Expanded Bed Adsorption. *Biotechnol. Bioeng.* **2005,** *92*(4), 501–513.
2. Adams, C. W.; Allison, D. E.; Flagella, K.; Presta, L.; Clarke, J.; Dybdal, N.; McKeever, K.; Sliwkowski, M. X. Humanization of a Recombinant Monoclonal Antibody to Produce a Therapeutic HER Dimerization Inhibitor, Pertuzumab. *Cancer Immunol. Immunother.* **2006,** *55*, 717–727.
3. Andersen, D. C.; Krummen, L. Recombinant Protein Expression for Therapeutic Applications. *Curr. Opin. Biotechnol.* **2002,** *13*(2), 117–123.
4. Andersson, G. E.; Kurland, C. G. An Extreme Codon Preference Strategy: Codon Reassignment. *Mol. Biol. Evol.* **1991,** *8*(4), 530–544.
5. Andrews, B.; Adari, H., et al . A Tightly Regulated High Level Expression Vector that Utilizes a Thermosensitive Lac Repressor: Production of the Human T Cell Receptor Vβ5.3 in *Escherichia coli*. *Gene* **1996,** *182*(1), 101–109.
6. Arnau, J.; Lauritzen, C.; Petersen G. E.; Pedersen J. Current Strategies for the use of Affinity Tags and Tag Removal for the Purification of Recombinant Proteins. *Protein Expression Purifi.* **2006,** *48*(1), 1–13.
7. Baldi, L., et al. Recombinant Protein Production by Large-Scale Transient Gene Expression in Mammalian Cells: State of the Art and Future Perspectives. *Biotechnol. Lett.* **2007,** *29*(5), 677–684.
8. Baneyx, F. O.; Mujacic, M. Recombinant Protein Folding and Misfolding in *Escherichia coli*. *Nat. Biotechnol.* **2004,** *22*, 1399–1408.
9. Bentley, W. E.; Mirjalili, N.; Andersen, D. C.; Davis R. H.; Kompala D. S. Plasmid-Encoded Protein: the Principal Factor in the "metabolic burden" Associated with Recombinant Bacteria. *Biotechnol. Bioeng.* **1990,** *35*(7), 668–681.
10. Boël, G.; Letso, R. et al. Codon Influence on Protein Expression in *E. coli* Correlates with mRNA Levels. *Nature* **2016,** *529*(7586), 358–363.
11. de Boer, H. A.; Comstock, L. J.; Vasser, M. The Tac Promoter: a Functional Hybrid Derived from the Trp and Lac Promoters. *Proc. Natl. Acad. Sci. U. S. A.* **1983,** *80*(1), 21–25.
12. Bolivar, F.; Betlach, M. C.; Heyneker, H. L.; Shine, J.; Rodriguez R. L.; Boyer H. W. Origin of Replication of pBR345 Plasmid DNA. *Proc. Natl. Acad. Sci.* **1977,** *74*(12), 5265–5269.
13. Borsuk, S.; Mendum, T. A.; Fagundes, M. Q.; Michelon, M.; Cunha, C. W.; McFadden, J.; Dellagostin O. A. Auxotrophic Complementation as a Selectable Marker for Stable Expression of Foreign Antigens in *Mycobacterium bovis* BCG. *Tuberculosis*, **2007,** *87*(6), 474–480.
14. Brosius, J.; Erfle, M.; Storella, J. Spacing of the-10 and-35 Regions in the Tac Promoter. Effect on Its in Vivo Activity. *J. Biol. Chem.* **1985,** *260*(6), 3539–3541.
15. Buchner, J.; Brinkmann, U.; Pastan, I. Renaturation of a Single–Chain Immunotoxin Facilitated by Chaperones and Protein Disulfide Isomerase. *Nat. Biotechnol.* **1992,** *10*(6), 682–685.

16. Buckholz, R.G.; Gleeson, M. A. G. Yeast Systems for the Commercial Production of Heterologous Proteins. *Nat. Biotechnol.* **1991,** *9*(11), 1067–1072.

17. Burgess-Brown, N.A.; Sharma, S. F.; Sobott, F.; Loenarz, C.; Oppermann U.; Gileadi O. Codon Optimization Can Improve Expression of Human Genes in *Escherichia coli*: A Multi-Gene Study. *Protein expression and purification,* **2008,** *59*(1), 94–102.

18. Cai, Y.; Sun, J.; Wang, J.; Ding, Y.; Liao, X.; Xu, W. QPSOBT: One Codon Usage Optimization Software for Protein Heterologous Expression. *J. Bioinform. Seq. Anal.* **2010,** *2*(2), 25–29.

19. Caponigro, G.; Muhlrad, D.; Parker, R. A Small Segment of the MAT Alpha 1 Transcript Promotes mRNA Decay in *Saccharomyces cerevisiae*: a Stimulatory Role for Rare Codons. *Mol. Cell. Biol.* **1993,** *13*(9) 5141–5148.

20. Chang, A.C.;Cohen, S.N. Construction and Characterization of Amplifiable Multicopy DNA Cloning Vehicles Derived from the P15A Cryptic Miniplasmid. *J. Bacteriol.* **1978,** *134*(3), 1141–1156.

21. Chant, A.; Kraemer-Pecore, C. M.; Watkin R.; Kneale G. G. Attachment of a Histidine Tag to the Minimal Zinc Finger Protein of the *Aspergillus nidulans* Gene Regulatory Protein AreA Causes a Conformational Change at the DNA-Binding Site. *Protein Expression Purif.* **2005,** *39*(2), 152–159.

22. Cheah, K.-C.; Harrison, S.; King, R.; Crocker, L.; Wells, J. R. E.; Robins, A. Secretion of Eukaryotic Growth Hormones in *Escherichia coli* is Influenced by the Sequence of the Mature Proteins. *Gene* **1994,** *138*, 9–15.

23. Chu, L.; Robinson, D. K. Industrial Choices for Protein Production by Large-Scale Cell Culture. *Curr. Opin. Biotechnol.* **2001,** *12*(2), 180–187.

24. Clark, E.D.B.; Schwarz, E. & Rudolph, R. Inhibition of Aggregation Side Reactions During in Vitro Protein Folding. *Methods Enzymol.* **1999,** *309*, 217–236.

25. Cranenburgh, R. M.; Hanak, J. A. J.; Williams S. G.; Sherratt D. J. *Escherichia coli* Strains that Allow Antibiotic-Free Plasmid Selection and Maintenance by Repressor Titration. *Nucleic Acids Res.* **2001,** *29*(5), e26–e26.

26. Cranenburgh, R. M.; Lewis, K. S.; Hanak, J. A. J. Effect of Plasmid Copy Number and Lac Operator Sequence on Antibiotic-Free Plasmid Selection by Operator-Repressor Titration in *Escherichia coli*. *J. Mol. Microbiol. Biotechnol.* **2004,** *7*(4), 197–203.

27. Daugherty, D. L.; Rozema, D.; Hanson P. E.; Gellman S. H. Artificial Chaperone-Assisted Refolding of Citrate Synthase. *J. Biol. Chem.* **1998,** *273*(51), 33961–33971.

28. Davis, G. D.; Elisee, C.; Newham, D. M.; Harrison, R.G. New Fusion Protein Systems Designed to Give Soluble Expression in *Escherichia coli*. *Biotechnol. Bioeng.* **1999,** *65*, 382–388.

29. Deana, A.; Belasco, J. G. Lost in Translation: The Influence of Ribosomes on Bacterial mRNA Decay. *Genes Dev.* **2005,** *19*(21), 2526–2533.

30. Del Solar, G.; Espinosa, M. Plasmid Copy Number Control: An Ever-Growing Story. *Mol. Microbiol.* **2000,** *37*(3), 492–500.

31. Dey, C.; Narayan, G.; Krishna, K. H.; Borgohain, M. P.; Lenka, N.; Tummer, P. R. Cell-Penetrating Peptides as a Tool to Deliver Biologically Active Recombinant Proteins to Generate Transgene-Free Induced Pluripotent Stem Cells. *Global J. Stem Cell Biol. Transplant.* **2017,** *3*, 006–015.

32. Doherty, A. J.; Connolly, B. A.; Worrall, A. F. Overproduction of the Toxic Protein, Bovine Pancreatic DNaseI, in *Escherichia coli* Using a Tightly Controlled T7-Promoter-Based Vector. *Gene* **1993**, *136*(1), 337–340.

33. Eliasson, M.; Olsson, A.; Palmcrantz, E.; Wiberg, K.; Inganäs, M.; Guss, B.; Lindberg, M.; Uhlen, M. Chimeric IgG-Binding Receptors Engineered from Staphylococcal Protein A and Streptococcal Protein G. *J. Biol. Chem.* **1988**, *263*(9), 4323–4327.

34. Elvin, C.M.; Thompson, P. R.; Argall, M. E.; Hendr, N. P.; Stamford, P. J.; Lilley P. E.; Dixon N. E. Modified Bacteriophage Lambda Promoter Vectors for Overproduction of Proteins in *Escherichia coli*. *Gene* **1990**, *87*(1), 123–126.

35. Ewalt, K.L.; Hendrick, J. P.; Houry W. A.; Hartl, F. U. In Vivo Observation of Polypeptide Flux Through the Bacterial Chaperonin System. *Cell* **1997**, *90*(3), 491–500.

36. Fonda, I.; Kenig, M.; Gaberc-Porekar, V.; Pristovaek P.; Menart V. Attachment of Histidine Tags to Recombinant Tumor Necrosis Factor-Alpha Drastically Changes its Properties. *Sci. World J.* **2002**, *2*, 1312–1325.

37. Frangioni, J. V.; Neel, B. G. Solubilization and Purification of Enzymatically Active Glutathione S-Transferase (pGEX) Fusion Proteins. *Anal. Biochem.* **1993**, *210*(1), 179–187.

38. Fuglsang, A. Codon Optimizer: A Freeware Tool for Codon Optimization. *Protein Expression Purif.* **2003**, *31*(2), 247–249.

39. Fuh, G.; Cunningham, B. C.; Fukunaga, R.; Nagata, S.; Goeddel, D. V.; Wells, J. A. Rational Design of Potent Antagonists to the Human Growth Hormone Receptor. *Science* **1992**, *256*, 1677–1680.

40. Gao, W.; Rzewski, A.; Sun, H.; Robbins, P. D.; Gambotto, A. UpGene: Application of a Web-based DNA Codon Optimization Algorithm. *Biotechnol. Progr.* **2004**, *20*(2), 443–448.

41. Garmory, H. S.; Leckenby M. W.; et al. Antibiotic-Free Plasmid Stabilization by Operator-Repressor Titration for Vaccine Delivery by Using Live *Salmonella enterica Serovar typhimurium*. *Infect. Immun.* **2005**, *73*(4), 2005–2011.

42. Gasser, B.; Prielhofer R.; et al. *Pichia pastoris*: Protein Production Host and Model Organism for Biomedical Research. *Future Microbiol.* **2013**, *8*(2), 191–208.

43. Gautam, S.; Dubey P.; Singh P.; Kesavardhana S.; Varadarajan R.; Gupta M. N. Smart Polymer Mediated Purification and Recovery of Active Proteins from Inclusion Bodies. *J. Chromatogr. A*, **2012**, *1235*, 10–25.

44. Glynou, K.; Ioannou, P. C.; Christopoulos, T. K. One-Step Purification and Refolding of Recombinant Photoprotein Aequorin by Immobilized Metal-Ion Affinity Chromatography. *Protein Expression Purif.* **2003**, *27*, 384–390.

45. Goeddel, D. V.; Kleid D. G. et al. Expression in *Escherichia coli* of Chemically Synthesized Genes for Human Insulin. *Proc. Natl. Acad.Sci.* **1979**, *76*(1), 106–110.

46. Goldenberg, D.; Azar, I.; Oppenheim, A. B.; Brandi, A.; Pon C. L.; Gualerzi, C. O. Role of *Escherichia coli* cspA Promoter Sequences and Adaptation of Translational Apparatus in the Cold Shock Response. *Mol. Gen. Genet. MGG* **1997**, *256*(3), 282–290.

47. Graumann, K.; Premstaller, A. Manufacturing of Recombinant Therapeutic Proteins in Microbial Systems. *Biotechnol. J.* **2006**, *1*(2), 164–186.

48. Gu, W.; Zhou, T.; Wilke, C. O. A Universal Trend of Reduced mRNA Stability Near the Translation-Initiation Site in Prokaryotes and Eukaryotes. *PLoS Comput. Biol.* **2010,** *6*(2), e1000664.

49. Guan, L.; Liu, Q.; Li, C.; Zhang, Y. Development of a Fur-Dependent and Tightly Regulated Expression System in *Escherichia coli* for Toxic Protein Synthesis. *BMC Biotechnol.* **2013,** *13*(1), 25.

50. Gupta, S.; Joshi, L. Codon Optimization. Computational Bioscience. Arizona State University, **2003**.

51. Gustafsson, C.; Govindarajan, S.; Minshull, J. Codon Bias and Heterologous Protein Expression. *Trends Biotechnol.* **2004,** *22*(7), 346–353.

52. Gustilo, E. M.; Vendeix, F. A. P.; Agris, P. F. tRNA's Modifications Bring order to Gene Expression. *Curr. Opin. Microbiol.* **2008,** *11*(2), 134–140.

53. Guzman, L.-M.; Belin, D.; Carson M. J.; Beckwith J. O. N. Tight Regulation, Modulation, and High-Level Expression by Vectors Containing the Arabinose PBAD Promoter. *J. Bacteriol.* **1995,** *177*(14), 4121–4130.

54. Ha, S. C.; Pereira, J. H.; Jeong, J. H.; Huh J. H.; Kim S.-H. Purification of Human Transcription Factors Nanog and Sox2, Each in Complex with Skp, an *Escherichia coli* Periplasmic Chaperone. *Protein Expression Purif.* **2009,** *67*(2), 164–168.

55. Hägg, P.; De Pohl, J. W.; Abdulkarim, F.; Isaksson, L. A. A Host/Plasmid System That is not Dependent on Antibiotics and Antibiotic Resistance Genes for Stable Plasmid Maintenance in *Escherichia coli.* *J. Biotechnol.* **2004b,** *111*, 17–30.

56. Hartikka, J.; Sawdey, M., et al. An Improved Plasmid DNA Expression Vector for Direct Injection into Skeletal Muscle. *Hum. Gene Ther.* **1996,** *7*(10), 1205–1217.

57. Hasan, N.; Szybalski, W. Construction of Laclts and Lacl q ts Expression Plasmids and Evaluation of the Thermosensitive Lac Repressor. *Gene* **1995,** *163*(1), 35–40.

58. Honeyman, A. L.; Cote, C. K.; Curtiss, R. Construction of Transcriptional and Translational LacZ Gene Reporter Plasmids for Use in *Streptococcus mutans*. *J. Microbiol. Methods* **2002,** *49*(2), 163–171.

59. Itakura, K.; Hirose, T.; Crea, R.; Riggs, A. D.; Heyneker, H. L.; Bolivar F.; Boyer H. W. Expression in *Escherichia coli* of a Chemically Synthesized Gene for the Hormone Somatostatin. *Science* **1977,** *198*(4321), 1056–1063.

60. Jayaraj, S.; Reid, R.; Santi, D. V. GeMS: An Advanced Software Package for Designing Synthetic Genes. *Nucleic Acids Res.* **2005,** *33*(9), 3011–3016.

61. Jenny, R. J.; Mann, K.G.; Lundblad, R. L. A Critical Review of the Methods for Cleavage of Fusion Proteins with Thrombin and Factor Xa. *Protein Expression Purif.* **2003,** *31*(1), 1–11.

62. Katoh, S.; Katoh, Y. Continuous Refolding of Lysozyme with Fed-Batch Addition of Denatured Protein Solution. *Process Biochemi.* **2000,** *35*(10), 1119–1124.

63. Khow, O.; Suntrarachun, S. Strategies for Production of Active Eukaryotic Proteins in Bacterial Expression System. *Asian Pac. J. Trop. Biomed.* **2012,** *2*(2), 159–162.

64. Knight, R. D.; Freeland, S.J.; Landweber, L. F. A Simple Model Based on Mutation and Selection Explains Trends in Codon and Amino-Acid Usage and GC Composition within and Across Genomes. *Genome Biol.* **2001,** *2*(4), research0010-1.

65. Korpimäki, T.; Kurittu, J.; Karp, M. Surprisingly Fast Disappearance of β-Lactam Selection Pressure in Cultivation as Detected with Novel Biosensing Approaches. *J. Microbiol. Methods* **2003,** *53*(1), 37–42.

66. Kumar, A.; Tiwari, S.; Thavaselvam, D.; Sathyaseelan, K.; Prakash, A.; Barua, A.; Arora, S.; Rao, M. K. Optimization and Efficient Purification of Recombinant Omp28 Protein of *Brucella melitensis* Using Triton X-100 and β-Mercaptoethanol. *Protein Expression Purif.* **2012,** *83*(2), 226–232.

67. Kumar, S.; Tamura, K.; Nei, M. MEGA3: Integrated Software for Molecular Evolutionary Genetics Analysis and Sequence Alignment. *Brief. Bioinform.* **2004,** *5*(2), 150–163.

68. Kurland, C. G.; Dong, H. Bacterial Growth Inhibition by Overproduction of Protein. *Mol. Microbiol.* **1996,** *21*(1), 1–4.

69. Larsson, M.; Brundell, E.; Nordfors, L.; Höög, C.; Uhlén M.; Ståhl S. A General Bacterial Expression System for Functional Analysis of cDNA-Encoded Proteins. *Protein Expression Purif.* **1996,** *7*(4), 447–457.

70. Lauritzen, C.; Pedersen, J.; Madsen, M. T.; Justesen, J.; Martensen P. M.; Dahl S. W. Active recombinant rat dipeptidyl aminopeptidase I (cathepsin C) produced using the baculovirus expression system. *Protein Expression Purif.* **1998,** *14*(3), 434–442.

71. LaVallie, E.R.; DiBlasio, E. A.; Kovacic, S.; Grant, K. L.; Schendel P. F.; McCoy, J. M. A Thioredoxin Gene Fusion Expression System that Circumvents Inclusion Body Formation in the *E. coli* Cytoplasm. *Biotechnology* **1993,** *11*, 187–193.

72. Lilie, H.; Schwarz, E.; Rudolph, R. Advances in Refolding of Proteins Produced in *E. coli*. *Curr. Opinion Biotechnol.* **1998,** *9*(5), 497–501.

73. Liu, X., Deng, R.; Wang, J.; Wang, X. COStar: A D-star Lite-based Dynamic Search Algorithm for Codon Optimization. *J. Theor. Biol.* **2014,** *344,* 19–30.

74. Lorentsen, R.H.; Fynbo, C. H. Thøgersen, H. Etzerodt M.; Holtet T. L. Expression, Refolding, and Purification of Recombinant Human Granzyme B. *Protein Expression Purif.* **2005,** *39*(1), 18–26.

75. Love, C. A.; Lilley, P. E.; Dixon, N. E. Stable High-Copy-Number Bacteriophage λ Promoter Vectors for Overproduction of Proteins in *Escherichia coli. Gene* **1996,** *176*(1), 49–53.

76. Lowman, H. B.; Bina, M. Temperature-Mediated Regulation and Downstream Inducible Selection for Controlling Gene Expression from the Bacteriophage λp L Promoter. *Gene* **1990,** *96*(1), 133–136.

77. Machida, S.; Ogawa, S.; Xiaohua, S.; Takaha, T.; Fujii K.; Hayashi, K. Cycloamylose as an Efficient Artificial Chaperone for Protein Refolding. *FEBS Lett.* **2000,** *486*(2), 131–135.

78. Maiorella, B.; Inlow, D.; Shauger A.; Harano, D. Large-Scale Insect Cell-Culture for Recombinant Protein Production. *Nat. Biotechnol.* **1988,** *6*(12), 1406–1410.

79. de Marco, A.; de Marco, V. Bacteria Co-Transformed with Recombinant Proteins and Chaperones Cloned in Independent Plasmids are Suitable for Expression Tuning. *J. Biotechnol.* **2004,** *109*(1), 45–52.

80. Mertens, N.; Remaut, E.; Fiers, W. Versatile, Multi-Featured Plasmids for High-Level Expression of Heterologous Genes in *Escherichia coli*: Overproduction of Human and Murine Cytokines. *Gene* **1995,** *164*(1), 9–15.

81. Minton, N. P. Improved Plasmid Vectors for the Isolation of Translational Lac Gene Fusions. *Gene* **1984**, *31*(1), 269–273.
82. Moffatt, B. A.; Studier, F. W. T7 Lysozyme Inhibits Transcription by T7 RNA Polymerase. *Cell* **1987**, *49*(2), 221–227.
83. Mojsin, M.; Nikčević, G. T.; Kovačević-Grujičić, N. R.; Savić, T.; Petrović, I.; Stevanović, M. Purification and Functional Analysis of the Recombinant Protein Isolated from *E. coli* by Employing Three Different Methods of Bacterial Lysis. *J. Serb. Chem. Soc.* **2005**, *70*(7), 943–950.
84. Moks, T.; Lars, A., et al. Large-Scale Affinity Purification of Human Insulin-Like Growth Factor I from Culture Medium of *Escherichia coli*. *Nat. Biotechnol.* **1987**, *5*(4), 379–382.
85. Moore, J. T.; Uppal, A.; Maley F.; Maley, G. F. Overcoming Inclusion Body Formation in a High-Level Expression System. *Protein Expression Purif.* **1993**, *4*(2), 160–163.
86. Murby, M.; Cedergren, L.; Nilsson, J.; Nygren, P. A.; Hammarberg, B.; Nilsson, B.; Enfors S. O.; Uhlen M. Stabilization of Recombinant Proteins from Proteolytic Degradation in *Escherichia coli* Using a Dual Affinity Fusion Strategy. *Biotechnol. Appl. Biochem.* **1991**, *14*(3), 336–346.
87. Nilsson, J.; Larsson, M.; Stefan, S.; Per-Åke N.; Mathias U. Multiple Affinity Domains for the Detection, Purification and Immobilization of Recombinant Proteins. *J. Mol. Recognit.* **1996**, *9*(5–6), 585–594.
88. Nomura, Y.; Ikeda, M.; Yamaguchi, N.; Aoyama Y.; Akiyoshi, K. Protein Refolding Assisted by Self-Assembled Nanogels as Novel Artificial Molecular Chaperone. *FEBS Lett.* **2003**, *553*(3), 271–276.
89. Palmer, I.; Wingfield, P. T. Preparation and Extraction of Insoluble (Inclusion-Body) Proteins from *Escherichia coli*. In *Current Protocols in Protein Science*; Chapter 6, Unit 6.3, John Wiley and Sons: Hoboken, NJ, **2004.**
90. Pedersen, J.; Lauritzen, C. Madsen M. T.; Dahl S. R. W. Removal of N-Terminal Polyhistidine Tags from Recombinant Proteins Using Engineered Aminopeptidases. *Protein Expression Purif.* **1999**, *15*(3), 389–400.
91. Petsch, D.; Anspach, F. B. Endotoxin Removal from Protein Solutions. *J. Biotechnol.* **2000**, *76*(2) 97–119.
92. Peubez, I.; Chaudet N., et al. Antibiotic-free Selection in *E. coli*: New Considerations for Optimal Design and Improved Production. *Microb. Cell Fact.* **2010**, *9*(1), 65.
93. Porath, J.; Olin, B. Immobilized Metal Affinity Adsorption and Immobilized Metal Affinity Chromatography of Biomaterials. Serum Protein Affinities for Gel-Immobilized Iron and Nickel Ions. *Biochemistry* **1983**, *22*(7), 1621–1630.
94. Portolano, N.; Watson, P. J.; Fairall, L.; Millard, C. J.; Milano, C. P.; Song, Y.; Cowley S. M.; Schwabe J. W. R. Recombinant Protein Expression for Structural Biology in HEK 293F Suspension Cells: A Novel and Accessible Approach. *J. Visualized Exp.* **2014**, (92), e51897–e51897.
95. Presnyak, V.; Alhusaini N.; et al. Codon Optimality is a Major Determinant of mRNA Stability. *Cell* **2015**, *160*(6), 1111–1124.
96. Puigbò, P.; Bravo, I. G.; Garcia-Vallvé, S. E-CAI: A Novel Server to Estimate an Expected Value of Codon Adaptation Index (eCAI). *BMC Bioinformatics* **2008**, *9*(1), 65.

97. Quick, M.; Wright, E. M. Employing *Escherichia coli* to Functionally Express, Purify, and Characterize a Human Transporter. *Proc. Natl. Acad. Sci.* **2002,** *99*(13), 8597–8601.

98. Raghava, S.; Aquil, S.; Bhattacharyya, S.; Varadarajan R.; Gupta, M. N. Strategy for Purifying Maltose Binding Protein Fusion Proteins by Affinity Precipitation. *Journal of Chromatography A,* **2008,** *1194*(1), 90–95.

99. Ranjbari, J.; Babaeipour, V.; Vahidi, H.; Moghimi, H.; Mofid, M. R.; Namvaran M. M.; Jafari, S. Enhanced Production of Insulin-Like Growth Factor I Protein in *Escherichia coli* by Optimization of Five Key Factors. *Iran. J. Pharm. Res.* **2015,** *14*(3), 907.

100. Roberts, M. C. Tetracycline Resistance Determinants: Mechanisms of Action, Regulation of Expression, Genetic Mobility, and Distribution. *FEMS Microbiol. Rev.* **1996,** *19*(1), 1–24.

101. Robinson, C. R.; Sligar, S. G. Hydrostatic and Osmotic Pressure as Tools to Study Macromolecular Recognition. *Methods Enzymol.* **1995,** *259*, 395–427.

102. Rodríguez-Carmona, E.; Cano-Garrido, O.; Seras-Franzoso, J.; Villaverde, A.; García-Fruitós, E. Isolation of Cell-Free Bacterial Inclusion Bodies. *Microb. Cell Fact.* **2010,** *9,* 71.

103. Rosenberg, S. A.; Grimm, E. A.; McGrogan, M.; Doyle, M.; Kawasaki, E.; Koths K.; Mark, D. F. Biological Activity of recombinant Human Interleukin-2 Produced in *Escherichia coli. Science* **1984,** *223*, 1412–1416.

104. Rozema, D.; Gellman, S. H. Artificial Chaperone-Assisted Refolding of Denatured-Reduced Lysozyme: Modulation of the Competition Between Renaturation and Aggregation. *Biochemistry* **1996,** *35*(49), 15760–15771.

105. Rudolph, R.; Lilie, H. In Vitro Folding of Inclusion Body Proteins. *FASEB J.* **1996,** *10*(1), 49–56.

106. Ryan, E. T.; Crean, T. I.; Kochi, S. K.; John, M.; Luciano, A. A.; Killeen, K. P.; Klose K. E.; Calderwood, S. B. Development of a ∆glnA balanced lethal plasmid system for expression of heterologous antigens by attenuated vaccine vector strains of Vibrio cholerae. *Infect. Immun.* **2000,** *68*(1), 221–226.

107. Sachdev, D.; Chirgwin, J. M. Solubility of Proteins Isolated from Inclusion Bodies is Enhanced by Fusion to Maltose-Binding Protein or Thioredoxin. *Protein Expression Purif.* **1998,** *12*(1), 122–132.

108. Sahdev, S.; Khattar, S. K.; Saini, K. S. Production of Active Eukaryotic Proteins Through Bacterial Expression Systems: A Review of the Existing Biotechnology Strategies. *Mol. Cell. Biochem.* **2008,** *307*(1–2), 249–264.

109. Saida, F. Overview on the Expression of Toxic Gene Products in *Escherichia coli. Current Protoc. Protein Sci.* **2007,** 5.19. 1–5.19. 13.

110. Samuelsson, E.; Wadensten, H.; Hartmanis, M.; Moks T.; Uhlén M. Facilitated in Vitro Refolding of Human Recombinant Insulin-Like Growth Factor I Using a Solubilizing Fusion Partner. *Nat. Biotechnol.* **1991,** *9*(4), 363–366.

111. Sanchez-Garcia, L.; MartÃn, L.; Mangues, R.; Ferrer-Miralles, N.; VÃ¡zquez, E. & Villaverde, A. Recombinant Pharmaceuticals from Microbial Cells: a 2015 Update. *Microb. Cell Fact.* **2016,** *15*, 33.

112. Sandhu, K. S.; Pandey, S.; Maiti, S.; Pillai, B. GASCO: Genetic Algorithm Simulation for Codon Optimization. *In Silico Biol.* **2008**, *8*(2), 187–192.

113. Schauer, S.; Lüer, C.; Moser, J. Large Scale Production of Biologically Active *Escherichia coli* Glutamyl-tRNA Reductase from Inclusion Bodies. *Protein Expression Purif.* **2003**, *31*, 271–275.

114. Schein, C. H.; Noteborn, M. H. M. Formation of Soluble Recombinant Proteins in *Escherichia coli* is Favored by Lower Growth Temperature. *Nat. Biotechnol.* **1988**, *6*(3), 291–294.

115. Schügerl, K.; Zeng, A. P. *Tools and Applications of Biochemical Engineering Science*; Springer: Berlin, **2002**; Vol. 74.

116. Scopes, R. K. *Protein Purification: Principles and Practice*; Springer Science & Business Media: New York, **2013**.

117. Shine, J.; Dalgarno, L. The 3′-Terminal Sequence of *Escherichia coli* 16S Ribosomal RNA: Complementarity to Nonsense Triplets and Ribosome Binding sites. *Proc. Natl. Acad.Sci.* **1974**, *71*, 1342–1346.

118. Singh, A.; Upadhyay, V.; Upadhyay, A. K.; Singh, S. M.; Panda, A. K. Protein Recovery from Inclusion Bodies of *Escherichia coli* Using mild Solubilization Process. *Microb. Cell Fact.* **2015**, *14*, 41.

119. Singh, S. M.; Panda, A. K. Solubilization and Refolding of Bacterial Inclusion body Proteins. *J. Biosci. Bioeng.* **2005**, *99*(4), 303–310.

120. Singh, S. M.; Sharma, A.; Upadhyay, A. K.; Singh, A.; Garg L. C.; Panda A. K. Solubilization of Inclusion Body Proteins Using n-Propanol and its Refolding into Bioactive form. *Protein Expression Purif.* **2012**, *81*(1), 75–82.

121. Skerra, A. Use of the Tetracycline Promoter for the Tightly Regulated Production of a Murine Antibody Fragment in *Escherichia coli*. *Gene* **1994**, *151*, 131–135.

122. Smyth, D. R.; Mrozkiewicz, M. K.; McGrath, W. J.; Listwan P.; Kobe B. Crystal Structures of Fusion Proteins with Large-Affinity Tags. *Protein Sci.* **2003**, *12*(7), 1313–1322.

123. Sørensen, H. P.; Sperling-Petersen, H. U.; Mortensen, K. K. Dialysis strategies for Protein Refolding: Preparative Streptavidin Production. *Protein Expression Purif.* **2003**, *31*(1), 149–154.

124. Stoker, N. G.; Fairweathe, N. F.; Spratt, B. G. Versatile Low-Copy-Number Plasmid Vectors for Cloning in *Escherichia coli*. *Gene* **1982**, *18*(3), 335–341.

125. Supek, F.; Vlahoviček, K. INCA: Synonymous Codon Usage Analysis and Clustering by Means of Self-organizing Map. *Bioinformatics* **2004**, *20*(14), 2329–2330.

126. Tao, H.; Liu, W.; Simmons, B. N.; Harris, H. K.; Cox T. C.; Massiah, M. A. Purifying Natively Folded Proteins from Inclusion Bodies Using Sarkosyl, Triton X-100, and CHAPS. *Biotechniques* **2010**, *48*(1), 61–64.

127. Thomas, P.; Smart, T. G. HEK293 Cell line: A Vehicle for the Expression of Recombinant Proteins. *J. Pharmacol. Toxicol. Methods* **2005**, *51*(3), 187–200.

128. Tian, H.; Zhao Y., et al. High Production in *E. coli* of Biologically Active Recombinant Human Fibroblast Growth Factor 20 and its Neuroprotective Effects. *App. Microbiol. Biotechnol.* **2016**, *100*(7), 3023–3034.

129. Tong, W. Y.; Yao, S. J.; Zhu, Z. Q.; Yu, J. An Improved Procedure for Production of Human Epidermal Growth Factor from Recombinant *E. coli*. *App. Microbiol. Biotechnol.* **2001,** *57*(5), 674–679.

130. Vallejo, L. F.; Rinas, U. Strategies for the Recovery of Active Proteins Through Refolding of Bacterial Inclusion Body Proteins. *Microb. Cell Fact.* **2004,** *3*(1), 11.

131. Walsh, G.; Jefferis, R. Post-Translational Modifications in the Context of Therapeutic Proteins. *Nat. Biotechnol.* **2006,** *24*(10), 1241–1252.

132. Wang, R. F.; Kushner, S. R. Construction of Versatile Low-Copy-Number Vectors for Cloning, Sequencing and Gene Expression in *Escherichia coli*. *Gene* **1991,** *100*, 195–199.

133. Waugh, D. S. An Overview of Enzymatic Reagents for the Removal of Affinity Tags. *Protein Expression Purif.* **2011,** *80*(2), 283–293.

134. Weaver, K. E.; Reddy, S. G.; Brinkman, C. L.; Patel, S.; Bayles K. W.; Endres, J. L. Identification and Characterization of a Family of Toxin-Antitoxin Systems Related to the *Enterococcus faecalis* Plasmid pAD1 Par Addiction Module. *Microbiology* **2009,** *155*(9), 2930–2940.

135. Werner, M. H.; Clore, G. M.; Gronenborn, A. M.; Kondoh, A.; Fisher, R. J. Refolding Proteins by Gel Filtration Chromatography. *FEBS Lett.* **1994,** *345*, 125–130.

136. West, S. M.; Chaudhuri, J. B.; Howell, J. A. Improved Protein Refolding Using Hollow-Fibre Membrane Dialysis. *Biotechnol. Bioeng.* **1998,** *57*(5), 590–599.

137. Wild, J.; Hradecna, Z.; Szybalski, W. Conditionally Amplifiable BACs: Switching from Single-Copy to High-Copy Vectors and Genomic Clones. *Genome Res.* **2002,** *12*(9), 1434–1444.

138. Wild, J.; Szybalski, W. Copy-Control Tightly Regulated Expression Vectors Based on pBAC/oriV. In *Methods in Molecular Biology, Recombinant Gene Expression: Reviews and Protocols;* Balbás, P., Lorence, A., Walker, J. M., Eds.; Chapter 11, Humana Press: Totowa, NJ, 2004, Vol. 267; pp 155–167.

139. Wu, G.; Bashir-Bello, N.; Freeland, S. J. The Synthetic Gene Designer: A Flexible Web Platform to Explore Sequence Manipulation for Heterologous Expression. *Protein Expression Purif.* **2006,** *47*(2), 441–445.

140. Wu, X.; Jörnvall, H.; Berndt K. D.; Oppermann, U. Codon Optimization Reveals Critical Factors for High Level Expression of Two Rare Codon Genes in *Escherichia coli*: RNA Stability and Secondary Structure but not tRNA Abundance. *Biochem. Biophys. Res. Commun.* **2004,** *313*(1), 89–96.

141. Yike, I.; Zhang, Y.; Ye, J.; Dearborn, D. G. Expression in *Escherichia coli* of Cytoplasmic Portions of the Cystic Fibrosis Transmembrane Conductance Regulator: Apparent Bacterial Toxicity of Peptides Containing R-Domain Sequences. *Protein Expression Purif.* **1996,** *7*(1), 45–50.

142. Zhang, S.; Zubay, G.; Goldman, E. Low-Usage Codons in *Escherichia coli*, Yeast, Fruit Fly and Primates. *Gene* **1991,** *105*(1), 61–72.

143. Zhou, H.; Wu, S.; Joo, J. Y.; Zhu, S.; Han, D. W.; Lin, T.; Trauger, S.; Bien, G.; Yao, S.; Zhu, Y. Generation of Induced Pluripotent Stem Cells Using Recombinant Proteins. *Cell Stem Cell* **2009,** *4*, 381–384.

PART IV
Microbes in Nanotechnology

CHAPTER 19

MICROBES AND NANOTECHNOLOGY: A POTENT CONFLUENCE

REECHA SAHU and LATA S. B. UPADHYAY*

*Department of Biotechnology, National Institute of Technology, Raipur, Chattisgarh 492010, India, *E-mail: lupadhyay.bt@nitrr.ac.in*

CONTENTS

ABSTRACT

Microbes or microscopic living organism can serve as both boon and bane to humankind. Microorganisms whether single-celled or multicellular can either be pathogenic or non-pathogenic. Pathogenic microbes can invade the body and cause infection, while certain species of nonpathogenic microbes even constitute the microbiota of human digestive system. They participate in potent health-promoting activities such as synthesis of vital components (vitamins) required by the body. They also create a barrier in the prevention of infections caused by pathogenic microbes.

A possible co-relationship can be established between microbiology and nanotechnologies; the microbial systems are nano-machineries and nanotechnology deals with the particles of nano-metric dimensions. Nanotechnology involves creating and modifying nano-sized particles and exploiting their wide range of properties acquired for various applications right from material sciences to the health sector. The convergence of nanotechnology and microbiology has not only addressed various issues related to health care but also directs us to usher revolution in this regard.

Study of microbes and nanotechnology are correlative scientific fields. Use of microorganisms provides an eco-friendly technique for the synthesis of stable and well-defined nanoparticles (NPs) with desired specific properties. It is not only ecologically safe but also economical, stable and easily scalable. Apart from the use of microbes in NP synthesis, NPs are also used as an antimicrobial agent for various microbial systems such as bacteria, fungi and even viruses and, thus have been used for combating drug resistance. Microbial-engineered NPs have found to be useful for microbial sensing, in-vivo imaging, drug delivery, nano-peptide delivery and development of nano-biosensors for environmental pollutants. This chapter discusses various applications where nanotechnology and microbiology have coalesced to bring out the beauty of interdisciplinary sciences.

19.1   INTRODUCTION

Nanoparticle (NP) development and its research are getting attention in the scientific community due to their unusual and typical properties and advantages they offer. Materials with size ranging from 1–1000 nm are considered to be the NPs. The unique property exhibited by the NP is associated with their size and surface characteristics and is significantly different from their bulk counterparts. NPs have given rise to a plethora of possibilities of developments in various industries and research ranging from optical sensing to health sector.

The term nanotechnology comes from the Greek word 'nanos' which means dwarf; this refers to the materials of one-billionth (10^{-9} m) in size. Richard Feymann is known as the father of nanotechnology, who gave the concept of nanotechnology in a lecture at the American Institute of Technology in 1959 entitled 'There is a plenty of room at the bottom'. The term 'nanotechnology' was coined by Prof. Norio Taniguchi from Tokyo Science University.

The characteristic property NPs exhibit is a measure of their size, shape, crystallinity and surface morphology. Nanoparticle s possess high surface to volume ratio which results in the dominance of the surface atoms over those in the interior. They have a great surface energy and large electron density on the surface atom. Metallic NPs have a characteristic surface plasmon resonance (SPR) unlike their bulk counterparts and also an enhanced Rayleigh and surface-enhanced Raman scattering (SERS). NPs impart supermagnetism in magnetic nano-materials while quantum size effect in semiconductors. These properties have made NPs an excellent matrix and carrier for various chemical and biochemical reactions and applications. Nanoparticles easily interact with larger particles; this property makes them very useful in various areas like permanent magnets, magnetic fluids[68] magnetic recording media, solar energy absorption, optical, electronics and catalytic materials.[42] The NPs synthesized from metals are also referred as magnetic NPs because they can be manipulated using magnetic field gradient.[119] Their magnetic applications include tunable colloidal photonic crystals[48] microfluidics[60] magnetic resonance imaging[34] data storage[54] nano fluids[90] and so forth. Catalytic applications include nano-material-based catalysts, biomedicines and tissue-specific targeting.

A wide range of NPs have been documented to date. They can be classified based on different parameters such as their origin (natural or anthropogenic), chemical composition (organic and inorganic), formation (biogenic, geogenic, anthropogenic, and atmospheric), size, shape and characteristic.[17] The classification can even be based on their application in research and industries. Usually, NPs are classified on the basis of their chemical composition, and on the basis of this criterion NPs can be classified as follows:

Organic NPs: Micelles, reverse micelles, liposomes, dendrimers, lipid NPs, nanogels and polymeric NPs.

Inorganic NPs: Metal NPs, quantum dots, carbon nanotubes and NPs.

19.2 ROLE OF MICROORGANISMS IN NANOPARTICLE (NP) BIOSYNTHESIS

The synthesis of mono-dispersed NPs with a well definite shape and size is challenging and there are different types of methodology available which are broadly categorized as physical, chemical and biological methods of NPs synthesis. The most common approach for the synthesis of nano materials and fabrication of nano structures are:

Top-down approach: In the top-down approach, bulk materials are broken down or reduced to nano-sized material either by physical or chemical means.

Bottom-up approach: On the other hand, the buildup of a material from the bottom: joining atom by atom, molecule by molecule or cluster by cluster is the principle behind bottom-up approach. Figure 19.1 summarizes various approaches to NPs synthesis.

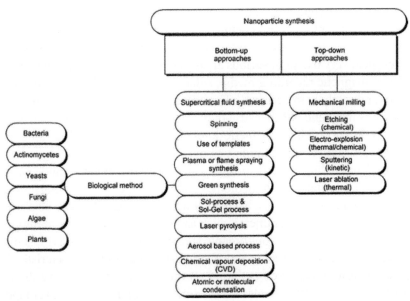

FIGURE 19.1   Various physical, chemical and biological methods of nanoparticle (NP) synthesis.

Source: Reprinted with permission Iravani, S., Green synthesis of metal nanoparticles using plants. *Green Chemistry* 2011, 13(10), 2638. © 2011 Royal Society of Chemistry.

As we can see there is an array of physical, chemical and biological methods used for NP synthesis. In spite of the popularity of above-mentioned methods, NP synthesized through physical and chemical means have limitations in biomedical and clinical applications due to poor surface morphology and nonuniform size. Furthermore, in such methodologies, the synthesis condition involves the role of high temperature pressure and use of expensive and toxic chemical. The byproducts released are harmful and pollute the environment. So, there is a demand for the development of economical, ecological, non-toxic and reliable methods for NP synthesis to give a new dimension to its applications especially in the biomedical field. One of the feasible solutions for NP biosynthesis is using microorganism, it is a convergence of nanotechnology and microbiological science. Micro-organisms act as nanofactories that have a potential not only to synthesize but also to control the size and shape of the NPs.[85]

Microbial-mediated NP synthesis comes under bottom-up approach and can also be termed 'green' route for NP formation. NPs synthesis employing microbes are not only environment-friendly but also economical and less energy intensive as compared to chemical and physical approaches for the same which require harsh conditions like high temperature, pressure, a longer duration, expensive chemicals and equipment[63,137,80,65,85,133,72,52] and also the byproducts released are toxic and harmful to environment. These drawbacks coerce for the development of more 'green' routes for NPs synthesis. Looking into increasing the demand of NPs owing to the properties and applications 'green' route for synthesis will be a better option for future.

As ever technique has its own advantages and disadvantages/draw-backs/limitation, similarly apart from advantages offered by microbial NPs synthesis, there are a few drawbacks also associated with it. Synthesis of NPs with defined composition, mono-dispersity, size and shape are concerns. Since the rate of synthesis is slow large-scale production of NP is difficult. There are a number of ways recognized in past few years to address these challenges which include (i) optimizing the physiochemical growth parameters and downstream processing, (ii) detailed study of the overall mechanism of NP synthesis at the molecular, cellular and biochemical level, (iii) exploring tools of recombinant DNA technology.

Microorganisms synthesize NPs either extracellularly or intracellularly depending upon the site of synthesis. There are a number of examples that have exploited various bacteria, fungi, actinomycetes, yeast and viruses for the synthesis of different metal NPs, bimetallic/alloy NPs, oxides NPs and quantum dots.

19.2.1   *BACTERIAL NPs SYNTHESIS*

Within the diverse array of the microorganism exploited for the NP synthesis, bacteria are most suited candidates among prokaryotes. Bacteria culturing is a cost-effective process, they have high growth rate and their genome can easily be manipulated if required, thereby making them the best candidate for NP synthesis. The site of the reductive component in the cell decides the localization of NP synthesis, whether it is extracellular or intracellular. Extracellular-synthesized NPs have a much-extended range of applications than intracellular-synthesized NPs such as in biosensing, sensor development, optoelectronics, semiconductors, automobiles and missiles, bioimaging, gene therapy and different biomedical applications. Table 19.1 and 19.2 summarizes bacterial intracellular and extracellular synthesis, respectively.

TABLE 19.1   Bacterial Intracellular NP Synthesis.

Microorganism used	Mechanism	Nanoparticle formed	References
Pedomicrobium sp.	Bioaccumulation	Gold	[85]
Sulphate-reducing bacteria		Gold	[128]
Bacillus subtilis	Reduction	Octahedral Au NP	[126]
Geobacter ferrireducans	Precipitation	Au NP	[85]
Shewanella algae	Reduction	Au 10–20 nm periplasmic surface 50–200 nm bacterial surface	[86]
Plectonema boryanum UTEX485	Precipitation	1–10 nm inside cell Octahedral Au np 1–10 μm in solution	[56]
Escherichia coli DH5α	Bioreduction	Au NP mostly spherical at the cell surface with a few triangles and quasi-hexagons	[29, 85, 56]
Rhodobacter capsulatus	Bioreduction and biosorption	Au NP both extracellular as well as intracellular	[37]
Pseudomonas stutzeri AG259		Triangular, hexagonal Ag NP of around 200 nm at periplasmic space.	[56]
Lactobacillus sp.	Biosorption and bioreduction	Ag NP on cell surface	[26, 56, 43]

TABLE 19.1 *(Continued)*

Microorganism used	Mechanism	Nanoparticle formed	References
Corynebacterium sp. SHO9		10–15 nm Ag NPs on the cell wall	[40]
Bacillus sp.	Bioaccumulation and bioreduction	Ag NPs 5–15 nm	[85, 40]
Lactobacillus sp. from butter milk	Bioreduction	Hexagonal and triangular Au NP, Ag NP as well as clusters and bimetallic alloys of Ag and Au.	[40]
Stenotrophomonas maltophilia SELTE02 *Enterobacter cloacae* SLD1a-1 *Rhodospirillum rubrum*	Reduction and accumulation	Spherical or granular Selenium NPs both extracellular as well as extracellular.	[85, 56]
Desulfovibrio desulfuricans *D. desulfuricans*			
S. algae	Reduction	Platinum NP of around 5 nm size	[85]
D. desulfuricans NCIMB 8307 *Shewanella oneidensis* MR-1	Bioreduction and biocrystallization	Palladium NPs	[85, 56]

TABLE 19.2 Bacterial Extracellular NPs Synthesis.

Microorganism used	Mechanism	Nanoparticle formed	Reference
Rhodopseudomonas capsulate	Reduction	Au NP spherical 10–20 nm triangular 50–400 nm size and shape were pH-dependent	[85]
R. capsulate	Reduction and biocapping	10–20 nm spherical Au NPs	[85, 115]
Pyrobaculum islandicum, Thermotoga maritime, S. algae, Geobacter sulfurreducens and *Pyrococcus furiosus*	Bioreduction	Au NPs 5–40 nm	[56]

TABLE 19.2 *(Continued)*

Microorganism used	Mechanism	Nanoparticle formed	Reference
Pseudomonas aeruginosa	Bioreduction	Au NP	[56]
B. megatherium D01	Reduction and capping	Monodispersed Au NP	[139]
Geobacter metallireducens GS-15	Reduction	10–50 nm magnetite particles Fe_2O_3	[10, 56]
Klebsiella pneumoniae, E. coli and *E. cloacae*	Reduction by nitroreductase enzyme	Ag nanoparticles	[85, 56, 40]
Bacillus licheniformis	Reduction by extracellular enzyme	Ag nanoparticles ~50 nm Au nanoparticle	[114]
P. islandicum, T. maritime, S. algae *G. sulfurreducens* *P. furiosus*	Reduction in the presence of hydrogen as electron donor by reductase	Au NP	[85, 56]
Sulfurospirillum barnesii, B. selenitireducens, Selenihalanaerobacter shriftii *Streptomyces* sp. ES2–5	Dissimilatory reduction	~300 nm selenium nanospheres	[85, 122]
B. selenireducens *L. plantarum*	Reduction	10×200 nm tellurium nanorods forming clusters	[85, 79]
S. barnesii	Reduction	<50 nm irregular shaped tellurium nanoparticles	[56]
Lactobacillus sp.	Reduction by culture supernatant	40–60 nm spherical titanium NP aggregates	[85, 56]
P. boryanum UTEX 485	Reduction	30–300 nm platinum NP of different shapes	[65, 85]
G. metallireducens GS-15	Dissimilatory reduction of ferric oxide	10–15 NP magnetite NP crystals	[63, 65, 85]
G. metallireducens thermophilic bacteria TOR-39 *Actinobacteria* sp.	Reduction	10–50 nm octahedral and spherical magnetite NP	[101, 85]

TABLE 19.2 *(Continued)*

Microorganism used	Mechanism	Nanoparticle formed	Reference
Micrococcus lacti-lyticus, Alteromonas putrefaciens, G. metal-lireducens GS-15	Reduction of U (VI) to U (IV) by the cell free extract.	Uranium NPs Uranium NPs	[65, 85, 56]
	Reduction of U (VI) to U (IV) in the presence of an electron donor		
Rhodopseudomonas palustris	Reduction	CdS nanocrystals	[85]
Rhodobacter sphaeroides	Reduction by immobilized cells	ZnS nanospheres	[85, 56]
R. sphaeroides	Reduction by immobilized cells	monodispersed spherical PbS nanoparticles	[85]

19.2.2 NP SYNTHESIS BY FUNGI

Bacterial-mediated NP synthesis is well characterized and has been studied for a variety of NPs synthesis especially gold and silver but in the past few years the attention has shifted and centered on the use of eukaryotic organisms, especially fungi for such synthesis. The possible reason might be (i) simplicity of scaling up- and downstream processing, (ii) presence of mycelia structure affording an increased surface area provide a key advantage. Morphology spectrum of the fungal-synthesized NPs is broad. It results in the formation of NPs of a wide range of shapes and sizes. Nevertheless, understanding of fungi-mediated synthesis is limited and required extensive research and studies in the area. Fungi are mostly reported to produce NPs extracellularly as the enzymes or protein involved in the synthesis are secreted in the extracellular environment helping in reduction and capping of NPs. The NPs, thus, formed are dispersed and are of well-defined morphology (Table 19.3).

Other microbes like actinomycetes and viruses have also been attempted for NPs synthesis and the reports are compiled in Table 19.4. The frequency and number is less.

TABLE 19.3   NP Synthesis by Fungi.

Location	Microorganism	Nanoparticle formed	Mechanism	References
Intracellular	Fusarium oxysporum by photobiological method	Ag NP		[85]
	Verticillium sp.	20 nm Au NP in the cytoplasmic membrane and on the surface	Reduction and accumulation	[72, 62, 109]
	Trichothecium sp.	Au NP	Accumulation	[80, 65, 52]
	V. Luteoalbum	Au NP of <10 nm size and of different shapes depending upon the pH	Reduction	[85]
	Phoma gardeniae	Ag NP	Accumulation	[95]
	Aspergillus flavus	Ag NP	Accumulation	[80, 38, 65, 72, 52, 111, 17, 93, 107]
	Rhizopus oryzae	Au nanoparticles	Biomineralization	[23]
extracellular	Colletotrichum sp.	8–40 nm spherical and rod-shaped Au NP	Reduction and capping	[80, 65, 25, 17, 107]
	Trichothecium sp.	5–200 nm Au NPs, spherical/rod-shaped and polydispersed	Reduction and capping	[80, 65, 52]
	Aspergillus fumigatus	5–30 nm monodispersed Ag NP	Reduction and capping in 10 min	[16]
	Phanerochaete chrysosporium	50–200 nm Ag NP	Reduction by reductase enzymes	[85]
	Fusarium semitectum	10–60 nm spherical Ag NPs	Reduction	[15]

TABLE 19.3 (Continued)

Location	Microorganism	Nanoparticle formed	Mechanism	References
	Aspergillus niger	Ag NP	Reduction and stabilization by nitrate reductase enzyme and quinine	[65, 52, 17, 11, 107]
	Penicillium brevicompactum, Rhizophora annamalayana, P. brevicompactum WA2315, Penicillium fellutanum	Ag NP	Reduction	[85, 17, 11, 107]
	V. volvacea	Ag NP, Au NP, Ag-Au NP	Reduction	[65, 52, 107]
extracellular	*P. aeruginosa* SNT1	Spherical Se nanoparticles	Reduction	[85, 56]
	Verticillium sp.	10–400 nm iron oxide NP	The secretion of cationic proteins with the molecular weight of 55 KDa and 13 KDa were found to be responsible for the hydrolysis of magnetite precursors and/or capping the magnetite nanoparticles	[65, 85, 17, 11]
extracellular and intercellular	*F. oxysporum* f. sp. *Lycopersici*	10–100 nm Pt. NP	Reduction and accumulation	[65, 17, 5, 107]
extracellular as well as intracellular	*F. oxysporum*	gold, silver, bimetallic Au–Ag alloy, silica, titania, zirconia, quantum dots, magnetite, strontianite, Bi and barium titanate NPs	Reduction, capping or bioaccumulation	[80, 148, 53, 93]

TABLE 19.4 NP Synthesized by Yeast, Actinomycetes and Virus.

Location	Microorganism	Nanoparticle	Mechanism	References
Intracellular	*Rhodococcus* sp.	10–15 nm Au NP	Reduction and accumulation by reductase enzyme	[80, 65, 93]
	Pichia jadinii	Au NP	Accumulation	[17, 107]
Extracellular	*Thermomonospora* sp.	Au NP	Reduction and capping	[80, 65, 93]
	Torulopsis sp.	PbS NP	Reduction	[65]
	S. cerevisiae	Sb_2O_3	Reduction	[85]
	Tobacco mosaic virus (TMV)	SiO, CdS, PbS and Fe_2O_3	Oxidative hydrolysis with the help of Glutamate and aspartate on the surface	[93]
	M13 bacteriophages	ZnS, PbS	Template peptides A7/J140-pviiim13 are involved	[93]
NA	*Candida glabrata*, *Schizosaccharomyces pombe*	CdS NP	Reduction	[52, 8, 93]

19.2.3 BASIC MECHANISM OF MICROBIAL NPs SYNTHESIS

Microbes respond to stress condition in different ways. For example, microorganisms capable of growing in the presence of heavy metal ions exhibit a specific resistance against the metal ion present in the niche they are growing. Such resistance is due to the presence of genes especially in the extrachromosomal DNA content such as plasmid. These genes encode for specific enzyme or protein taking part in the mechanism to deal with metal ions present in the environment. They respond either by transporting the heavy metals across the membrane, intracellular bioaccumulation, precipitation, complexation or by altering the solubility, changing the redox state of metal ions and thus reduce their toxicity.[63,137,80,65,83,36,85,133,72,11]

So, basically, microbial-metal interaction in the form of detoxification mechanism is employed for the synthesis of NPs by them. The same can be understood in detail by taking few reported examples (i) silver metal resistance by microorganism, silver NPs are synthesized as a means to protect the cell from silver toxicity. The detoxification is done by a periplasmic protein which can bind silver; it binds silver at the cell surface and stops the incoming of metal ions further by efflux pumps. Metal resistance can either be achieved by blocking efflux of metal ions or by chelation of metal by meal bind protein or by both the mechanisms. Silver resistance is a result of both active efflux and metal ion binding by microbial proteins. Hence, it can be concluded that the binding protein present in the cell matrix gives amino acid moieties that acts as a site for nucleation for NP synthesis. Figure 19.2 shows Au biomineralization mechanism by *Rhizopus oryzae*. The Au biomineralization mechanism is nicotinamide adenine dinucleotide phosphate (NADP)-based nitrate reductase-dependent.

Apart from proteins, there are a number of biomolecules that are involved in the synthesis of NPs either by reduction or by any other mechanism. In reduction process, proteins, enzymes, carbohydrates and different byproducts are involved, the process is mostly energy intensive (NADPH-dependent) or in few cases energy-independent. For example, nitrate reductase in case of Ag NP synthesis by bacteria and fungi, sulphite reductase in case of Au NP synthesis is NADPH-dependent. There are a group of proteins collectively called gold shape directing protein (GSP, Mol mass of 25–55 KDa) responsible for the size and shape of gold NP formed by the extract of *Chlorella vulgaris*.[65,125,52,93] There are various capping peptides and molecules, too that are involved at the redox centres apart from enzymes like quinine (naphthoquinones and anthraquinones), glutathione, phytochelatin, hydrogenase, aspartate and so forth participating during NP synthesis. In case of bacteria, NADPH-dependent enzyme of 50–90 KDa is responsible for intracellular NP synthesis while extracellular synthesis is achieved by means of nitroreductase.[52,107] Actinomycete *Thermomonospora* sp. secretes the extracellular protein of 10–80 KDa which reduces and stabilizes protein, Fourier-transform infrared spectroscopy (FTIR) analysis shows that NPs are capped with proteins. Studies have shown that Mms6 is a dominant protein that tightly associates with the surface of bacterial magnetites in *Magnetospirillum magneticum* AMB-1.[72]

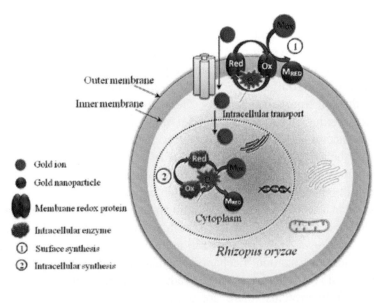

FIGURE 19.2 Mechanism of biomineralization of Au NP by *Rhizopus oryzae.*
Source: Reprinted with permission from Das, S. K.; Liang, J.; Schmidt, M.; Laffir, F.; Marsili, E. Biomineralization Mechanism of Gold by Zygomycete Fungi *Rhizopous oryzae*. *ACS Nano.* **2012,** *6*(7), 6165–6173. Copyright © 2012, American Chemical Society.

19.3 NPs AS ANTIMICROBIAL AGENT

The word 'antimicrobial' was derived from the Greek words *anti* (against), *mikros* (little) and *bios* (life) and refers to all agents that act against microorganisms. This is not synonymous with antibiotics, all antibiotics are antimicrobials but not all antimicrobials are antibiotics. An antimicrobial agent is a substance that kills or inhibits the growth of microorganisms (bacteria, fungi, viruses and protozoa) without causing damage to the host cell. The antimicrobial agent can be natural, semisynthetic or synthetic in origin.

Until the beginning of 20th century, microbial infections were the leading cause of deaths annually worldwide. With the advent of various antimicrobials agents especially antibiotics, the mortality due to infection was exponentially reduced or controlled. However, usage of traditional antimicrobial agents is facing a major setback due to the development of antibiotic resistance. This situation has aroused due to an excessive, repeated and improper use of antibiotics and, thus there is an exponential rise in the multidrug-resistant microorganisms. In spite of an in-depth knowledge of microbial pathogenesis and prophylaxis, the problem of antibiotic resistance cannot be effectively controlled. Mostly to combat this issue, antibiotics are administered at higher doses and the frequency of doses has been increased which results in toxicity and side effects. This calls for either development of new and better antibiotics or employing an alternative agent or nontraditional antimicrobial agents for a cure. Development of new antibiotics requires a huge monetary investment, it is time-consuming and labour intensive. Because of this, the discovery and release of new antibiotics in past decades have dramatically decreased. Figure 19.3 represents antibiotics approved by Food and Drug Administration (FDA) from 1983–2012.

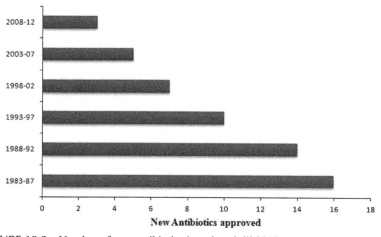

FIGURE 19.3 Number of new antibiotics introduced till 2012.

Also, it is difficult to cope up with the rate with which the microorganisms develop resistance to a range of antibiotics. With the continuous evolution, antibiotic resistance is posing a serious threat to humankind and demands a permanent solution to this ever-growing issue. Nontraditional antimicrobial agent combats antibiotic resistance and also works on the cells which normally are inaccessible, therefore use of nontraditional antimicrobial agents are clinching attention over traditional antibiotics [51]. Nanoparticles can serve as one of such antimicrobial agent. Nanoparticles as antimicrobial agent reduce toxicity, impart a better immunization, economical and also overcome the issue of acquired resistance. On the other hand, NPs can also be used in drug delivery, diagnosis, nano-therapeutics and theranostics[7, 19, 39, 51]. Antibiotics when administered along with NPs the overall pharmacokinetics of antibiotics is improved as well as the body retention time is enhanced. [51] Free NPs come under the category of broad-spectrum antimicrobial agent and the mode of action of NPs depends on the size and type of NP under consideration. The mode of actions of antimicrobial NPs includes cidal destruction of cell membranes, blockage of enzyme pathways, alterations of the microbial cell wall and microbial DNA synthesis. [144] The various modes of action of antimicrobial NPs have schematically represented in Figure 19.4.

Antimicrobial property of a metal is quite evident from decades. In ancient times, meals were consumed in silver or gold utensils; preferably water storage utensils were also made up of metals like silver, gold and copper and assumed to be effective to impart antimicrobial property to food and water. Thus, in present time, silver is introduced widely to treat acute infections, wounds and burns. In an era of antibiotic resistance, when a potent and traditional antibiotic fails to work NPs have emerged as a powerful tool to act against microbial infections. As quoted earlier, nanoscience has not only brought together material science and microbiology at the same page as an antimicrobial but also in various fields like diagnosis, prophylaxis, sensing, industries, textile, packaging and so forth, to name a few. The current gaga over NPs by the industries and research fraternities is because of the bewildering properties they offer which are already been discussed in details. The antimicrobial property of NP is due to its small size and large surface area to volume ratio. The nano dimension increases its surface area and also the interaction with the microbes to carry out its antimicrobial activity.[84]

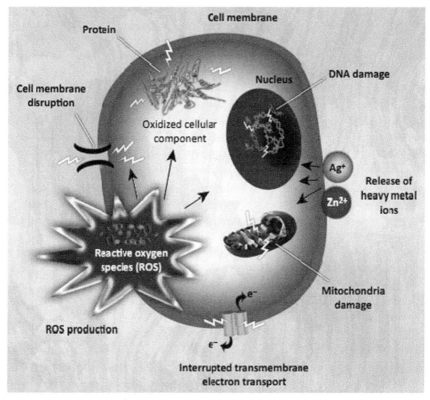

FIGURE 19.4   Mechanisms of toxicity of NPs against bacteria. NPs and their ions (for example silver and zinc) can produce free radicals, resulting in induction of oxidative stress (that is reactive oxygen species; ROS). The produced ROS can irreversibly damage bacteria (for example their membrane, DNA and mitochondria), resulting in bacterial death.

Source: Reprinted with permission from Hajipour, M. J.; Fromm, K. M.; Ashkarran, A. A.; Aberasturi, D. J. De; Larramendi, I. R. De; Rojo, T.; Serpooshan, V.; Parak, W. J.; Mahmoudi, M. Antibacterial Properties of Nanoparticles. *Trends Biotechnol.* **2012,** *30*(10), 499–511. © 2012 Elsevier.

    Different metal and metal oxide NPs have been studied for their antimicrobial activity. Ag, Au, ZnO, MgO, TiO_2, CuO, aluminium oxide, ferric oxide, zirconia and silicon NPs are some such metallic NP reported to exhibit antimicrobial property. Further, NP-polymer conjugates have been reported to exhibit an increased antimicrobial property in comparison to the bare counterparts. Table 19.5 enlists different NPs which exhibit antimicrobial property along with their possible mechanism.

TABLE 19.5   Summary of Antimicrobial NPs.

Nanoparticle	Forms	Applications	Targeted microorganism	Reference
Silver (Ag) nanoparticle	Bare and functionalized Ag NP, chitosan-Ag NP, Fe-Oxide NP & Ag NP, polyvinylpyrrolidone (PVP)-coated silver nanoparticles, mercaptoethane sulfonate polysaccharide coated Ag NP, Ag-NPs-coated PUC, MWCNT-AgNPs, Ag–Fe/SWCNTs, alloy of gold and silver NP, Ag₂O NP	Bare, functionalized and conjugated Ag nanoparticle works as an antimicrobial agent for a broad range of microorganisms, it includes gram positive and gram negative bacteria, pathogenic strains of fungi, viruses, protozoa and parasites. They have been used for drug delivery. Antimicrobial properties of sole Ag NP and Ag NPs conjugated with polymers and polysaccharides have made it an excellent matrix for wound healing and dressing as it is biodegradable and non-toxic to damaged and surrounding tissues. Combats antibiotic resistance and has come up as an effective nanomedicine for pathogenic strains that have developed multiple drug resistance.	Bacteria: *E. coli, B. subtilis, Staphylococcus aureus,* methicillin-resistant coagulase-negative staphylococci, vancomycin-resistant *Enterococcus faecium,* ESBL-positive *K. pneumoniae, S. typhi, Vibrio cholerae, S. aureus, Streptococcus agalactiae, S. dysgalactiae* virus: HIV-1, Influenza virus, Herpes Simplex virus, Respiratory syncytial virus, Monkey pox virus Parasite: *A. stephensi, A. aegypti, and C. quinquefasciatus,* etc. fungi: *Trichophyton mentagrophytes and Candida species, A. niger, A. flavus, Penicillium* spp.,	[118, 127, 138, 7, 14, 36, 67, 77, 51, 103, 108, 84, 47, 92, 44, 88, 99, 27, 49, 111, 95, 144]

TABLE 19.5 (Continued)

Nanoparticle	Forms	Applications	Targeted microorganism	Reference
Gold (Au) nanoparticles	Bare and core shell functionalized Au nps, mercaptoethane sulfonate (MES)-coated gold nanoparticles, Sialic-acid functionalized gold nanoparticles, *Chrysosporium tropicum*-mediated gold nanoparticles,	Gold NPs apart from being antimicrobial, they are majorly exploited in nanomedicine field because of its biocompatibility and inertness.	Bacteria: MRSA, VRE, *Pseudomonas aeruginosa, B. Anthracis spores, Streptococcus mutans, Listeria monocytogenes, E. Coli, Salmonella typhimurium, Vibrio parahaemolyticus, Aspergillus fumigates,*	[39, 18, 138, 32, 58, 51, 84, 108, 47, 70, 88, 136, 27, 31, 49, 95, 144, 19]
		Au NPs are widely used for chemotherapy via photothermal heating, photodynamic antimicrobial chemotherapy.	*Aspergillus niger,*	
		They are used for photothermal killing of potent and antibiotic-resistant pathogenic microbes.	Viruses: Herpes simplex virus type 1 (HSV-1), HIV virus, various influenza virus.	
		Currently being used for wound dressing and healing.	Fungi: *Penicillium and Candida* sp.	
			Parasites: *Aedes aegypti larvae.*	
Copper Np	bare and functionalized Cu NP, CuO NP, CuCl$_2$ NP, Copper (II) nanohybrid solids,	Have a number of important physical and chemical properties which makes it useful for many industries.	*B. subtilis, Pseudomonas aeruginosa and Proteus* sp.,	[7, 84, 27, 55, 144]
	LCu (CH$_3$COO)$_2$ and LCuCl$_2$	Anti-parasitic and anti-bacterial in nature	*Plasmodium falciparum* (MRC 2)	

TABLE 19.5 *(Continued)*

Nanoparticle	Forms	Applications	Targeted microorganism	Reference
Cobalt NP	Co NP	Antiparasitic in nature	malaria vector *Anopheles subpictus* and dengue vector *Aedes aegypti* (*Diptera: Culicidae*).	[144]
metal oxide NP	ZnO, TiO₂, Al oxide NP, SiO₂, ZrO₂, AP-MgO NP	ZnO NPs have a broad range microcidal property, they are stable and highly toxic to microbes but causes no or low toxicity to healthy cells. They are used as cosmetics, for agricultural applications and food packaging. AP-MgO (synthesized by aerogel procedure) NP are used as a potent disinfectant, they have high surface area and capability to carry higher amount of active halogens which brings out micro-bial cidal action. TiO₂ NP used as a potent disinfectant agent and prevents biofilm formation. Acts against bacteria and virus and are used for food packaging. Al oxide NP: wide range of applications in agricultural and personal care products	Bacteria: *E. coli, P. aeruginosa, B. subtilis, Campylobacter jejuni, Shigella dysenteriae, Salmonella typhimurium, S. aureus,* Fungi: *Candida albicans, Aspergillus niger* Parasites: *Phthiraptera: Pediculidae,* *Acari: Ixodidae* *Diptera:Culicidae* *Rhipicephalus* (*Boophilus) microplus, Canestrini* (*Acari: Ixodidae,)* *R. microplus, Culex quinquefasciatus*	[59, 7, 87, 57, 76, 84, 88, 124, 27, 49, 144]

19.3.1 ANTIMICROBIAL MECHANISM OF METAL AND METAL OXIDE NPs

The mechanism or mode of action as an antimicrobial agent is almost similar to both traditional antibiotics as well as antimicrobial NPs/conjugates. Advantages offered by NPs over antibiotics are (i) microorganisms do not develop resistance against these NPs, (ii) NP exhibits a broad range of antimicrobial activity, (iii) they are non-toxic with reduced side effects, (iv) NPs can easily pass bloodbrain barrier (BBB), (v) they also sustain the antimicrobial activity for a longer time duration[1,6,7,9,27,39,47,49,51,84,88,144].

Antimicrobial NPs act by blocking critical metabolic pathways, interference with cell-wall synthesis, protein denaturation, DNA denaturation and hampered DNA replication. The mechanism of microbial growth inhibition depends on the type of NP being used as an antimicrobial agent.

In order to have a clear understanding of antimicrobial property of NPs, let us discuss some of the specific mechanism reported by a researcher in support of antimicrobial activity of different NP preparation.

19.3.1.1 ANTIMICROBIAL PROPERTY OF SILVER NPs

Silver ions are being used as an antimicrobial agent in a variety of applications. The applications include antimicrobial coating on medical and surgical devices, dental resin composite formulations, antimicrobial sprays, as an antimicrobial agent in air sanitizers, laundries used in hospitals, respirators, wound dressing, in personal care products like soaps, shampoo, detergents, wet wipes, socks and so forth.[92] The mechanism of antimicrobial action of silver NP or silver ions has not been rightly apprehended and is a topic of debate among researchers. Morones et al. claim that silver ions have a high affinity towards sulphur and phosphorus and thus it gets attached to sulphur and phosphorous containing bio-molecules. Binding of silver NP to sulphur-containing proteins of cell membrane and thiol enzymes leads to protein denaturation which ultimately leads to cell lysis.[81,84] Similarly, attachment of silver NPs to DNA and nucleic acids component of the cell through its phosphate-sugar backbone causes damage to DNA and interference in DNA replication, which also lead to cell death. Inhibition of respiratory enzymes by silver ions generates reactive oxygen species (ROS) and free radicals. Free radicals, thus generated attack the cell membrane lipids enhancing the cell porosity and ultimately death.[81,92] A synergistic effect of Ag NPs on the mode of action of certain antibiotics such as ampicillin, kanamycin, erythromycin and chloramphenicol has been reported by Fayaz et al., when used as a combined formulation of antibiotic and silver NP. The antibacterial efficacy of these antibiotics was reported to get enhanced in the presence of AgNPs against both gram-positive and gram-negative strains.[36] Bare Ag NPs are less effective antibacterial agents against gram-positive bacteria than gram-negative bacteria. This may be due to the structural difference in cell wall composition of gram-positive and gram-negative bacteria. However, a recent study has proved that the toxicity mechanism of Ag NPs on bacteria is by induction of apoptosis and inhibition of newborn DNA synthesis and not cell wall composition. Apoptosis and DNA synthesis inhibition are directly proportional to the concentration of Ag NPs used.[12] Minimum Inhibitory Concentration (MIC) of Ag NPs for different pathogens as reported by Balakumaran et al. is 3.125 µg/ml, 6.25 µg/ml and 12.5 µg/ml for *Escherichia coli* (*E. faecalis*, methicillin-resistant *Staphylococcus aureus*), *Candida albicans* (*P. aeruginosa*, *Klebsiella pneumoniae*, *Bacillus subtilis*) and *S. aureus*, respectively.[11]

19.3.1.2 ANTIMICROBIAL PROPERTY OF GOLD NPs

The role of gold in therapeutic application can be traced back to 2500 BC. The colloidal solution of gold was in practice in the Indian Ayurvedic medicine for rejuvenation and revitalization to control ageing process under the name of Swarna Bhasma ('Swarna' meaning gold, 'Bhasma' meaning ash).[84] Au NPs were exploited for antimicrobial studies against human pathogenic microbes due to its biocompatibility and inert nature. However, functionalized and engineered Au NP possesses photothermal activity and therefore is extensively used in chemotherapy as photodynamic antimicrobial agent.[84] Initially, the antimicrobial property of gold NPs was thought to be size-dependent but recent studies have shown that antimicrobial action of gold NP is due to its physical interaction with bacterial membrane modulating its membrane potential, decreases in ATP level and inhibition of tRNA binding to the ribosome.[21,47,27,11,106] MIC of Au NP for few microbes is reported, for example, Au NP of size 7–34 nm have MIC of 2.93 µg/ml, 7.56 µg/ml, 3.92 µg/ml and 3.15 µg/ml for *E. coli*, *B. subtilis*, *S. aureus* and *K. pneumonia*, respectively, whereas the MIC values of gold NPs of 20–40 nm particle size are 2.96 µg/ml, 8.61 µg/ml, 3.98 µg/ml and 3.3 µg/ml for *E. coli*, *B. subtilis*, *S. aureus* and *K. pneumonia*, respectively.[106] Au NPs have been proved to be an excellent therapeutic agent. Apart from antimicrobial applications, they are being used for chemotherapy, biosensing, bioimaging, biodetection and so forth. Au NPs are inert and the mechanism of antimicrobial activity is ROS-independent, this makes them safer nanomaterial as a therapeutic agent for human than any other metallic or metal oxide NPs. Furthermore, high stability of functionalization gold NP makes them ideal nanomaterials to be applied as targeted antimicrobial agents.[27]

19.3.1.3 ANTIMICROBIAL PROPERTY OF METAL OXIDE NPs

The spectra of antimicrobial property of ZnO NP is broad, it works against both gram-positive and gram-negative bacteria as well as on spores (spores are highly resistant to high temperature and antibiotic). ZnO NPs are also nontoxic and compatible with human skin which makes it an excellent antimicrobial agent in clinical dermal applications. That is why it is extensively exploited in various textile and cosmetics industry, medical sunscreen lotions and so forth. The mechanism of action is concentration and size-dependent. Antimicrobial activity increases with increase in the concentration of NPs while lesser the size of NP better is its antimicrobial property. Size-dependent antimicrobial effect is a measure of increase in the surface area and the electron cloud over it. In presence of UV light, ZnO NP generates ROS on the NP surface. The ROS generated, thus, helps in formation of hydrogen peroxide which convert the ZnO into zinc ions. Zinc ion released, thus, creates membrane dysfunction and facilitates its own internalization causing cell lysis and eventual cell death,[27,59,87,76,124]. However the standard mechanism of antimicrobial action of ZnO NP is still unknown. Other metal oxide NP being studied includes CuO, TiO_2, aluminium oxide, SiO_2, iron oxide NP and so forth. Among these metal oxides, NPs studies ZnO NPs have been reported to show highest antimicrobial activity. CuO is found to have bactericidal activity against *Bacillus subtilis*, mechanism of action can be attributed to the high affinity of copper towards amino and carboxyl group present in the bacterial cell membranes. TiO_2 NP requires UV light for photo-killing of bacteria and viruses. The metal oxide NPs other than ZnO can act as a potent disinfectant rather than antimicrobial agent. For potential antimicrobial action, a higher concentration of these NPs is required[27,84,47,95].

19.3.1.4   ANTIVIRAL MECHANISM OF NPs

Copper iodide, gold NP, iron oxide NP and silver are found to have antiviral activity. Both bare and NPs coated with polyvinylpyrrolidone (PVP), mercaptoethane sulphonate and so forth exhibit antiviral activity[14,18,27,51,84,96,144]. It has been studied and proved that in comparison to other metal NPs, Ag NPs have a better and strong antiviral potential. Detailed study and understanding of the exact site of interaction and mode of action of Ag NPs has led researcher to modify the surface property of Ag NPs to enhance antiviral activity. Functionalization and modification of the surface properties of NPs can make the virucidal action more effective. Ag NPs attach directly to the viral surface glycoproteins and enter the cell preventing genome (DNA or RNA) replication, ultimately blocking the viral multiplication. Ag NPs also block the viral invasion into host cell by preventing the fusion of the viral envelope with host cell[14,33,67,78,96,100,113,118,142].

19.3.2   NPs TO DEFY ANTIBIOTIC RESISTANCE

Antibiotic resistance is the potency exhibited by microorganisms that help them to defy effects of an antibiotic. Antimicrobial resistance is a natural phenomenon acquired by the microorganism with time usually through genome modification. Misuse and overuse of antibiotics in human and animals without professional oversight is in regular practice. Such practices are accelerating resistance against antibiotics in human system. Antibiotic resistance evolves naturally via natural selection through random mutation, but it could also be developed by applying an evolutionary stress on a population. Once such antibiotic-resistant gene is generated, bacteria can then transfer the genetic information in a horizontal fashion (between individuals) by plasmid exchange. A bacterium carrying multiple resistance genes, exhibit multi-antibiotic/drug resistance and such organisms are informally also referred as a superbug. According to the recent report by WHO,[140] 480,000 people are developing multidrug-resistant tuberculosis (TB) annually worldwide. Such scenarios are complicating the fight against HIV and malaria. WHO report also suggests that microorganisms are developing new resistance mechanisms and spreading globally. Thus, developed resistance mechanisms are threatening our existing ability to treat many common infectious diseases. This is ultimately leading to prolonged illness, disability, and death. In lack of effective antibiotics for prevention and treatment of infectious diseases medical procedures such as organ transplantation, cancer chemotherapy, diabetes management and major surgery are at high risk. Figure 19.5 illustrates the antibiotics that have developed resistance and deployed till 2012. A study by WHO was performed globally to check the current global scenario of antibiotic resistance. According to their report[140] following observations were documented.

1. Antibiotic resistance against pathogenic bacteria is present worldwide. Even though worldwide carbapenem antibiotics is used as last resort treatment for *K. pneumoniae* infection. Carbapenem resistance in *K. pneumoniae* (common intestinal bacteria which causes most of the deadly infections in hospitals like pneumonia, infections in bloodstream etc.) have been reported in some part of the world. Most widely used medicines for the treatment of urinary tract infections caused by *E. coli* are fluoroquinolone antibiotics; there are reports from developing countries regarding acquired resistance towards this drug. There are similar acquired drug resistance reports in case of *S. aureus, Neisseria gonorrhoeae, Enterobacteriaceae* and so forth infection too.

2. Among new TB cases reported worldwide in 2014, an estimated 3.3% were the multidrug-resistant type (in clinical reference known as MDR-TB). Extensively drug-resistant tuberculosis (XDR-TB) is a form of tuberculosis that is resistant to minimum four of the core anti-TB drugs. Such cases have been identified in around 105 countries. According to the WHO report, estimated 9.7% of people suffering from MDR-TB being converted to XDR-TB state.

3. Cases of antibiotics resistance have also been reported in case of malaria, HIV and influenza.[140]

Antibiotic deployment

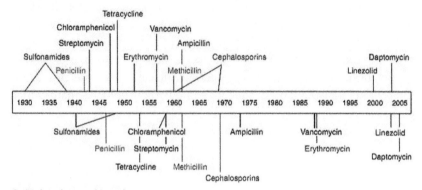

Antibiotic resistance observed

FIGURE 19.5   Number of antibiotics that have been deployed till 2005 due to the development of antibiotic resistance against them.

Source: Reprinted with permission from [152] Clatworthy A.E.; Pierson, E.; Hung, D.T. Targeting virulence: a new paradigm for antimicrobial therapy. *Nature Chemical Biology,* **2007** *3*(9), 514-548. © 2007 Nature Publishing Group.

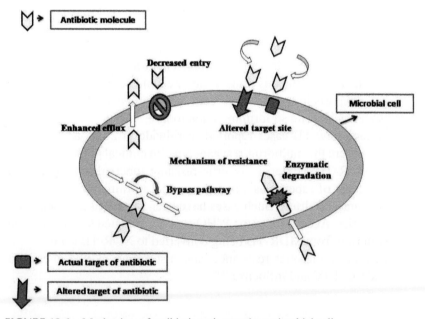

FIGURE 19.6   Mechanism of antibiotic resistance in a microbial cell.

Mechanism of resistance development depends on factors like the class of antibiotics for which resistance has been developed, mechanism of action of the antibiotic and microbe genera under consideration. Antibiotic resistance development is a genetic and biochemical phenomenon and resistance might be intrinsic or acquired. Possible mechanisms reported are by genetic polymorphism, overexpression of efflux pumps, altering the antibiotics targeted microbial cell component, enhanced and rapid hydrolysis of antibiotic, enzymatic destruction of antibiotic, reprogramming peptidoglycan synthesis and acetylation of antibiotic resulting into loss of its biological potential.[24] Figure 19.6 schematically illustrates the mechanism of antibiotic resistance by microorganism.

Nanoparticle serves as an alternative to conventional traditional antibiotics. Usage of NP alone or in conjugation with the antibiotics for antimicrobial therapy combats drug resistance development, improves the solubility of existing drug and reduces the side effects of the drug. For example, vancomycin-capped Au NPs (Au@Van NPs) exhibited 64-fold improved efficacy against vancomycin-resistant *E. coli* (VRE) strains and gram-negative bacteria such as *E. coli* in comparison with vancomycin alone [51]. Microbes are very well adapted to meet the challenges faced by current antibiotics as it is being considered as a foreign particle by the microbial cell. While using NPs as an antimicrobial agent relies on the fact that NPs even though being a foreign molecule for the microbial cell is not recognized so. Therefore, the chances of the development of resistance against NP are minimized. These nanostructures should, therefore, be capable of depositing drugs within individual microbial cells thus bypassing the problem of resistance. It would not be wrong if it is said that NP-based antimicrobial agents are the future of antimicrobial therapy. NP-based antimicrobial therapy will improve treatment against bacterial infections, especially caused by multidrug-resistant *S. aureus* (MRSA), vancomycin-resistant *E. coli* (VRE), vancomycin-resistant *S. aureus* (VRSA) and multidrug-resistant *P. aeruginosa* [51]. One of the complications of external injuries, open wounds and burns are the formation of biofilms on the surface. Such biofilms are usually nonresponsive towards the use of an antimicrobial agent. NP-based treatments are also helpful in preventing biofilm formation. There are reports of using NP of silver, gold, iron and ZnO to prevent the biofilm formation by *P. aeruginosa*, *S. aureus* and various gram-negative and gram-positive bacteria [45,51,69,104] However, there have been some reports of resistance against NPs, too.

Cu-doped TiO$_2$ NPs are bactericidal for many pathogenic strains but does not work against *Shewanella oneidensis* MR-1 (–). The resistance of *Shewanella oneidensis* MR-1 (–) against Cu doped TiO$_2$ NP is due to the production of extracellular polysaccharide (EPS) under stress which resists the bactericidal activity of Cu^{2+} ions released by the NPs. Similarly, *E. coli, P. putida, S typhimurium, P. aeruginosa* have shown resistance against certain NPs.[47]

19.3.3   NPs AS INTRACELLULAR DRUG DELIVERY VEHICLES

The intracellular location of bacterial infection provides a preferential micro-environment for bacterial growth since they are not only protected by host defence mechanism but also from antimicrobial therapy. It has been reported that some of the antibiotic families such as β-lactams and amino-glycosides owing to their high hydrophilicity are being prevented from cellular penetration. While antibiotics families like fluoroquinolones and macrolides have no solubility issue and can diffuse well inside the cells cytoplasm but possess low intracellular retention. In addition, the subcellular distribution of antibiotics is not uniform, so different antibiotics show different distribution and activity profiles.[130] The activity of antibiotics might also be influenced by factors such as pH, enzymatic inactivation within the cell. Therefore, at the intracellular level the active concentration of antibiotics is often subtherapeutic resulting in low effectiveness against intracellular pathogens and, hence the emergence of antibiotic resistance. Moreover, to circumvent the problems of drug resistance high doses of antibiotics are often given in such case, generating more side effects and toxicity.[1]

Nanocarriers with optimized physicochemical and biological properties are successfully being used as a carrier for drug molecules since they can easily be taken up by the cells. Liposomes, solid lipids NPs, dendrimers, biodegradable polymers such as polylactic acid coglycolic acid (PLGA), polylactic acid (PLA), chitosan, gelatin, polycaprolactone and poly-alkyl-cyanoacrylates, silicon or carbon nanotube, metal NPs and magnetic NPs are the examples of nanocarriers that have been tested for drug delivery systems. They have been introduced in the therapeutics as nanocarriers to improve the pharmacokinetics of the conventional drug molecule. NPs impart better solubility and improve the overall therapeutic efficacy of the drug. The NPs in the size range of 10–100 nm are found to be most suitable for drug delivery. NP of size less than 10 nm size face renal clearance and particle size more than 100 nm causes opsonization and thus removed from the bloodstream.[141] Thus, the advantages of NP-based antimicrobial drug delivery can be summarized as; (i) it improves solubility of poorly water-soluble drugs, (ii) imparts prolonged drug half-life and systemic circulation time, (iii) results in sustained and stimuli-responsive drug release, which eventually lowers administration frequency and dose, (iv) lowers the side effects of the drug [51] (v) last but not the least; it overcomes the structural barriers (BBB) and delivers drugs to intracellular targets where a conventional carrier fails to deliver the drug molecules.

Blood–brain barrier has made the treatment of neuronal disorders, neural infections, tumours and so forth a challenging task, but with the advent of nanocarriers, a ray of hope for treatments of such deadly diseases have made their way.[131] However, a lot still needs to be studied for development of a nanocarrier for successful delivery of drugs to such tissues without therapeutic failure and toxicity. The toxicity quotient in case of intracellular delivery and delivery in neuronal cells by NPs can be lowered down by utilization of biodegradable polymers like polylactic acid, polycaprolactone, poly (lactic-co-glycolic acid), poly (fumaric-co-sebacic) anhydride chitosan, modified chitosan, liposome, dendrimers and solid lipid for the synthesis of such NPs. Application of NPs in targeted drug delivery depends not only on the biocompatibility of NP but also on shape, size and surface property of the NP for a non-toxic efficient delivery of drug molecules in tough and sensitive targets[1,141,143] (Jung and Jik 2011, Upadhyay 2014).

19.3.4   NP IN WOUND HEALING

It is well evident that certain metal, metal oxide, polymeric NPs possesses antimicrobial property which makes them an excellent agent to be used as a disinfectant and bactericidal agents for external tissue applications. The recent development in this regard is the employment of NPs for wound healing and dressing. It has been reported that Au/collagen nanocomposite can be used for wound healing; this nanocomposite has imparted high wound closure percentage, improved elastic modulus of the healed epidermis, reduced inflammatory response, increased neovascularization and granulation tissue formation.[4] If more of such biocompatible nano-composites are made it can change the face of skin tissue engineering. Silver NPs have been used in many studies for wound dressing. Application of silver NP-based wound dressing also facilitated the wound healing process in terms of formation of healed dermis and epidermis without any sign of necrosis and apoptosis in-vitro.[99] In a similar in-vitro study, silver-chitin composites are used which also shows bactericidal action and has a superior blood clotting ability. However, the composite was cytotoxic on prolonging use.[77] Thrombin-conjugated γ-Fe$_2$O$_3$ NPs were found to be an effective wound healing material for incisional wounds, the results were better than free thrombin and untreated wounds. This treatment imparted a relatively higher skin tensile strength. The successful in vitro studies are positive indicators for novel thrombin conjugates usage for treatment and it may reduce complications in surgery such as wound dehiscence.[150] Such studies will also open an avenue to improve adhesion and take of skin grafts. There will be reduction in the number of sutures required in cosmetic and reconstructive surgery. It will also permit earlier removal of stitches thereby reducing stitch-induced scar formation.

19.4   NANOTECHNOLOGY-BASED DETECTION AND BIOSENSING OF MICROORGANISM

Food and water-borne diseases are epidemic since decades and reason for an increased mortality and morbidity, especially in the developing countries. WHO has reported more than 200 types of diseases caused by water and foodborne pathogens. More than 2 million people worldwide die annually due to consumption of contaminated water and food.

The demands a rapid and sensitive detection tools for the determination of pathogens and infectious agents in food, potable water, biological (tissue and fluids) and environmental samples (soil, water and air). NPs have been used and implemented as a detector and sensor molecule in optical, electrochemical or magnetic biosensors. NP-based biosensors are effective with respect to old and conventional techniques like PCR, immunoassays and so forth in terms of detection time, high throughput screening, selectivity, reliability, cost-effectiveness (per sample-based), sensitivity and infield application.

The biosensors used in microorganism detection are usually optical, electrochemical, fluorescent or magnetic.[102] Figure 19.3 illustrates a diagrammatic representation of NP-based biosensor/detection system. Optical biosensors are based on SPR which allow label-free qualitative and quantitative measurements of microbe in the test sample.[102] However, SPR-based techniques are less sensitive. Au NPs-based optical biosensor has been fabricated for detection of *E. coli*. The working principle of the biosensor was change in SPR peak due to sensitive detection of 16s rRNA (of *E coli*) using peptide-nucleic acid probe.[102] A hybrid Au–Ag NP array was used for detection of *Staphylococcus aureus* enterotoxin B. The detection was based on local SPR (LSPR) sensing technique which is more sensitive compared to SPR technique. Magnetic biosensors (an immunomagnetic approach) are widely being used for the detection of *E. coli* and *S. aureus*. NP-based magnetic biosensors make the immunomagnetic detection much more sensitive and rapid. Table 19.6 illustrates few examples of NPs-based detection of pathogens along with the detection principle (adopted and modified from[121]) (Figure 19.7).

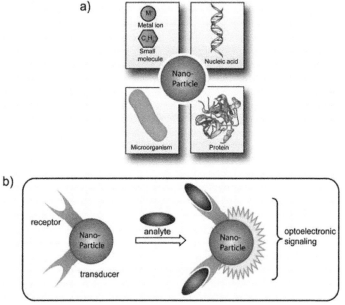

FIGURE 19.7   a) Examples of targets for nanoparticle-based detection and b) Schematic depiction of a representative nanoparticle-based detection system.

Source: Reprinted with permission from [151] Agasti, S.S; Rana, S.; Park, M-H; Kyu, C.; You, C-C; Rotello, V.M. Nanoparticles for detection and diagnosis, *Advanced Drug Delivery Reviews*, **2010,** *62*(3). © 2010 Elsevier.

TABLE 19.6 NPs-Based Detection of Pathogens.

Pathogen	Nanomaterial	Recognition	Detection method	detection limit	Reference
E. coli O157:H7	Qdots	Biotinylated antibody	Fluorescence microscopy	2 orders more sensitive than conventional dyes	[46]
	Qdots and magnetic NPs	Antibody	Fluorometry	100 times more sensitive than FITC	[116]
	Qdots	Fim-H mannose-specific lectin	Fluorometry	10^4 bacteria/ml	[82]
	Dye-doped silica NPs	Antibody	Plate counting/flow cytometry	1–400 E. coli within 20 min	[149, 132]
			Surface plating	$1.6 \times 10^1 – 7.2 \times 10^7$ CFU/ml	
	Magnetic NPs	Antibody	IR spectroscopy	$10^4 – 10^5$ CFU/ml	[98, 20]
			ATP bioluminescence	20 CFU/ml	
	Au NPs	surface Plasmon resonance (SPR)	optical	NA	[102]
B. anthracis	Eu-doped silica NPs	Ca dipicolinate	Fluorescence spectra	0.2 nM in 2 min	[3]
	Eu doped polystyrene NPs	Antibody	Fluoroimmunoassay	0.01–100 ng/ml	[123]
	Au NPs	Antibody	Microscope and visual	10 ng	[73]
M. tuberculosis	Dye-doped silica NPs	Antibody	Fluorescence microscopy	Amplified signal within 4 h	[94, 13]
	Au NPs	Oligonucleotide	UV–Vis spectroscopy	Visual detection	

TABLE 19.6 (*Continued*)

Pathogen	Nanomaterial	Recognition	Detection method	detection limit	Reference
Listeria monocytogenes	Magnetic NPs	Antibody	PCR	226 CFU/.5 ml	[145]
M. pneumonia	Ag nanorod array	–	SERS	NA	[28]
S. saprophyticus	Magnetic NPs	Vancomycin	MALDI-MS	7×10^4 CFU/ml in urine	[74]
S. aureus	Magnetic NPs	Vancomycin antibody	MALDI-MS test strip	7×10^4 CFU/ml in urine	[74, 117, 50]
	CCMV				
	Au NPs	Antibody	I. C. assay test device	Visual detection 100% Sensitivity	
S. aureus entero-toxin B	Au–Ag	LSPR	optical	–	[102]
Cholera toxin	Liposomes	Gangliosides thio-lated lactose	Fluoroimmunoassay test strip	1 nM	[110, 2, 105]
	Au NPs		visual and UV-Vis	10 fg/ml in 20 min 10 min	
Salmonella	Au NPs	Antibody Test	test strip	Red dots appearance in 2 h	[35]
Salmonella enteritidis	Au and magnetic NPs	DNA assay	Fluorescence	1 ng/ml	[147]
H. pylori	Au NPs	Antibody	SEM	10 ng	[22]
Staphylococcus	Magnetic NPs	Antibody	TEM	10 CFU/ml in human blood	[41]

TABLE 19.6 (Continued)

Pathogen	Nanomaterial	Recognition	Detection method	detection limit	Reference
E. faecalis, S. epidermidis	Magnetic NPs	Vancomycin	Plate counting	–	[61]
P.fimbriated E. coli	Magnetic NPs	Pigeon ovalbumin	MALDI-MS	~9.6 × 10⁴ CFU/0.5 ml	[75]
E. coli, Salmonella, S. aureus	Dye-doped silica NPs	Antibody	Luminophore immunoassay	within 20 min	[135]
RSV	Qdots	Antibody	Colour change	Single step/short time	[129]
	Ag nanorod array	–	SERS	–	[28]
Rotavirus	Ag nanorod array	–	SERS		[28]
HIV-1 DNA	Au NPs	Probe DNA	NLO properties	100 pM	[22]
HIV	Ag nanorod array	–	SERS	–	[28]
H5N1	Magnetic NPs	Antibody	MRI	5 viral particles in 10 µl	[89, 146]
			SQUID	5 pg/ml	
HBV, HCV, HIV	Qdots	Antibody	Fluorescence	100 µl sample/1 h/50 times more sensitive	[64, 112]
	Au nanowire barcoded Coated with silica	Thiolated DNA	Fluorescence imaging		
HBV, HCV	Au NPs/Ag staining	Protein A	Protein chip assay	Visual detection	[134, 30]
Giardia lamblia	Au NPs/Ag staining	Antibody	UV–Vis spectroscopy	1.088 × 10³ cells/ml	[71]
P. falciparum	Eu doped polystyrene NPs	Antibody	Fluoroimmunoassay	–	[91]

Advantages of NP-based biosensors:

1. NP enhances sensitivity and enables simultaneous detection of multiple targets.
2. Due to small size and large surface-volume ratio of NP, the amount of biomolecules immobilized onto the surface of NPs in process of designing the biological element of the sensor is more, which results in amplified biomolecular recognition events.
3. NP imparts an enormous signal enhancement when coupled to a highly sensitive optical or electrochemical transducers component setting the basis for ultrasensitive detection.
4. Enhances the sensitivity of detection to several folds and makes the process rapid, economical and eco-friendly.

19.5   NANOTOXICITY: A LIMITATION OR CHALLENGE TO BE ADDRESSED

Biological and physicochemical properties NPs allow its extensive application in the field of heath care as antimicrobial agent and drug delivery vehicle. The physicochemical properties and advantages of NPs have already been discussed in details earlier in this chapter. The effect of NPs on tissues and toxicity is a topic of debate and concern among researchers. However, the mechanism of toxicity and its effect on human tissues and cells are poorly understood and reported. A number of in-vivo and in-vitro tests have been performed to access the toxicity of NPs. These studies have shown NPs to be a potent carcinogen, cytotoxic and genotoxic compounds that inhibit cell proliferation. The NP toxicity depends on the chemical composition, size, shape and surface morphology of NPs. Most of the reported toxicity study is on Ag and Au NPs as they have been extensively used in therapeutic applications. The study suggested that both Au and Ag NPs are non-toxic and inert to cells.[49] There are even few reports suggesting toxicity of silver NPs. Ag NPs coated with mercaptoethane sulphonate have antiviral activity but it also possesses toxicity against mammalian cells.[14] The toxic effect of NPs in human cells is at cellular, subcellular and protein level. Metal NPs except gold imparts toxicity by generation of ROS and thus damaging the tissues. The various methodologies used for NP toxicity detection are (i) cell viability assay, (ii) apoptosis assay, (iii) necrosis assay, (iv) oxidative stress assay, (v) inflammatory assay.[120] SERS a latest developed technique is modified cell viability assay. It was used to study the toxicity if Au and ZnO NPs. The change in the SERS was observed due to protein denaturation and DNA damage.[66]

19.6   CONCLUSION

Nanoscience and microbiology together have emerged as a powerful alliance. Microorganisms facilitate the biosynthesis of NPs and on the other hand NPs act as an antimicrobial agent against pathogenic microorganisms. NP biosynthesis by microbes intend to make the synthesis process 'green' thereby reducing the harmful impact of physical and chemical processes on environment. Not only this, NPs thus synthesized has proved to exhibit better properties and have less toxicity when used for therapeutic applications like nano-antibiotics, drug delivery and wound dressing. Microorganism used for NP biosynthesis can be bacteria, fungi and even viruses. The possible mechanisms of NP biosynthesis reported are reduction, bioaccumulation, biomineralization, biosorption, precipitation and crystallization. The site of synthesis of NPs by microorganism can be either extracellular or intracellular. It is governed by the location of the active component responsible for NP biosynthesis, for example enzymes, peptides and other microbial components. Nano-antibiotics (NPs alone and NP-antibiotic conjugate) have enhanced antimicrobial efficacy with reduced side effects. Among the various advantages nano-antibiotics offer, combating antibiotic resistance is the most salient feature. An efficient drug delivery is yet another beneficial attribute of NPs. NP-based drug delivery system facilitates effective intracellular drug delivery. Generally a drug reaches at it subtherapeutic level on administration on the other hand NPs can cross BBB. The sensitivity and efficacy of NPs have also been exploited in detection of pathogenic microorganism in food, water and environmental samples. However, combating nanotoxicity is challenge and a lot of study has been going in this direction to infer it in a better way. Coalition of NPs and microbiology has given an array of applications and advantages. With recent progress and ongoing efforts in meeting up the challenges and improving the properties of NPs, it is hopeful that NP-based techniques would be the future of therapeutics, detection and imaging.

KEYWORDS

- nanotechnology
- microbial-engineered nanoparticles
- interdisciplinary approach

REFERENCES

1. Abed, N.; Couvreur, P. Nanocarriers for Antibiotics: A Promising Solution to Treat Intracellular Bacterial Infections. *Int. J. Antimicrob. Agents* **2014,** *43*(6), 485–496.
2. Ahn-Yoon, S.; DeCory, T. R.; Baeumner, A. J.; Durst, R. A. Ganglioside-Liposome Immunoassay for the Ultrasensitive Detection of Cholera Toxin. *Anal. Chem.* **2003,** *75*(10), 2256–2261.
3. Ai, K.; Zhang, B.; Lehui, L. Europium-Based Fluorescence Nanoparticle Sensor for Rapid and Ultrasensitive Detection of an Anthrax Biomarker. *Angew. Chem.* **2009,** *48*(2), 304–308.
4. Akturk, O.; Kismet, K.; Yasti, A. C.; Kuru, S.; Duymus, M. E.; Kaya, F.; Caydere, M.; Hucumenoglu, S.; Keskin, D. Collagen/gold Nanoparticle Nanocomposites: A Potential Skin Wound Healing Biomaterial. *J. Biomater. Appl.* **2016,** *31*(2), 283–301.
5. Alghuthaymi, A. M.; Almoammar, H.; Rai, M.; Galiev, E. S.; Abd-Elsalam, K. A. Myconanoparticles: Synthesis and Their Role in Phytopathogens Management. *Biotechnol. Biotechnol. Equip.* **2015,** *29*(2), 221–236.
6. Allahverdiyev, A. M.; Kon, K. V.; Abamor, E. S.; Bagirova, M.; Rafailovich, M. Coping with Antibiotic Resistance: Combining Nanoparticles with Antibiotics and Other Antimicrobial Agents. *Expert. Rev. Anti. Infect. Ther.* **2011,** *9*(11), 1035–1052.
7. Allaker, R. P.; Ren, G. Potential Impact of Nanotechnology on the Control of Infectious Diseases. *Trans R Soc Trop Med Hyg.* **2008,** *102*(1), 1–2.
8. Al-Shalabi, Z.; Doran, P. M. Biosynthesis of Fluorescent CdS Nanocrystals with Semiconductor Properties: Comparison of Microbial and Plant Production Systems. *J. Biotechnol.* **2016,** *223,* 13–23.
9. Arora, D.; Sharma, S.; Sharma, V.; Abrol, V.; Shankar, R.; Jaglan, S. An Update on Polysaccharide-Based Nanomaterials for Antimicrobial Applications. *Appl. Microbiol. Biotechnol.* **2016,** *100*(6) 1–13.
10. Bharde, A.; Wani, A.; Shouche, Y.; Joy, P. A.; Prasad, B. L.; Sastry, M. Bacterial Aerobic Synthesis of Nanocrystalline Magnetite. *J. Am. Chem. Soc.* **2005,** *127*(26), 9326–9327.
11. Balakumaran, M. D.; Ramachandran, R.; Balashanmugam, P.; Mukeshkumar, D. J.; Kalaichelvan, P. T. Mycosynthesis of Silver and Gold Nanoparticles: Optimization, Characterization and Antimicrobial Activity against Human Pathogens. *Microbiol. Res.* **2016,** *182,* 8–20.

12. Bao, H.; Yu, X.; Xu, C.; Li, X.; Li, Z.; Wei, D.; Liu, Y. New Toxicity Mechanism of Silver Nanoparticles: Promoting Apoptosis and Inhibiting Proliferation. *PLoS One* **2015,** *10*(3), e0122535.

13. Baptista, P. V.; Koziol-Montewka, M.; Paluch-Oles, J.; Doria, G.; Franco, R. Gold-Nanoparticle-Probe–Based Assay for Rapid and Direct Detection of *Mycobacterium tuberculosis* DNA in Clinical Samples. *Clin. Chem.* **2006,** *52*(7), 1433–1434.

14. Baram-pinto, D.; Shukla, S.; Perkas, N.; Gedanken, A.; Sarid, R. Inhibition of Herpes Simplex Virus Type 1 Infection by Silver Nanoparticles Capped with Mercaptoethane Sulfonate. *Bioconjug. Chem.* **2009,** *20*(8), 1497–1502.

15. Basavaraja, S.; Balaji, S. D.; Lagashetty, A. K.; Rajasab, A. H.; Venkataraman, A. Extracellular Biosynthesis of Silver Nanoparticles Using the Fungus *Fusarium semitectum. Mater. Res. Bull.* **2008,** *43*(5), 1164–1170.

16. Bhainsa, K. C.; D'Souza, S.F. Extracellular Biosynthesis of Silver Nanoparticles Using the Fungus *Aspergillus fumigatus*. Colloids. *Surf. B. Biointerfaces* **2006,** *47*(2), 160–164.

17. Borouman, M. A.; Namvar, F.; Moniri, M.; Paridah, Md. T.; Azizi, S.; Mohamad, R. Nanoparticles Biosynthesized by Fungi and Yeast: A Review of Their Preparation, Properties, and Medical Applications. *Molecules* **2015,** *11*(20), 16540–16565.

18. Bowman, M. C.; Ballard, T. E.; Ackerson, C. J.; Feldheim, D. L.; Margolis, D. M.; Melander, C. Inhibition of HIV Fusion with Multivalent Gold Nanoparticles. *J. Am. Chem. Soc.* **2008,** *130*(22), 6896–6897.

19. Chen, G.; Roy, I.; Yang, C.; Prasad, P. N. Nanochemistry and Nanomedicine for Nanoparticle-Based Diagnostics and Therapy. *Chem. Rev.* **2016,** *116* (5), 2826–2885.

20. Cheng, Y.; Liu, Y.; Huang, J.; Li, K.; Zhang, W.; Xian, Y.; Jin, L. Combining Biofunctional Magnetic Nanoparticles and ATP Bioluminescence for Rapid Detection of *Escherichia coli. Talanta* **2009,** *77*(4), 1332–1336.

21. Cui, Y.; Zhao, Y.; Tian, Y.; Zhang, W.; Lü, X.; Jiang, X. The Molecular Mechanism of Action of Bactericidal Gold Nanoparticles on *Escherichia coli. Biomaterials* **2012,** *33*(7), 2327–2333.

22. Darbha, G. K.; Rai, U. S.; Singh, A. K.; Ray, P. C. Gold-Nanorod-Based Sensing of Sequence Specific HIV-1 Virus DNA by Using Hyper-Rayleigh Scattering Spectroscopy. *Chemistry* **2008,** *14*(13), 3896–3903.

23. Das, S. K.; Liang, J.; Schmidt, M.; Laffir, F.; Marsili, E. Biomineralization Mechanism of Gold by Zygomycete Fungi *Rhizopous oryzae. ACS Nano.* **2012,** *6*(7), 6165–6173.

24. Davies, J.; Davies, D. Origins and Evolution of Antibiotic Resistance. *Microbiol. Mol. Biol. Rev.* **2010,** *74*(3), 417–433.

25. Dhanya, K. C.; Nair, K. C. Nanobiotechnology: Relevance in Microbiology Nanobiotechnology: Relevance in Microbiology. *Ann. Basic Appl. Sci.* **2011,** *2*(1), 53–58.

26. Dhoondia, Z. H.; Chakraborty, H. Lactobacillus Mediated Synthesis of Silver Oxide Nanoparticles. *Nanomater. Nanotechno.* 2012, 2(15).

27. Dizaj, S. M.; Lotfipour, F.; Jalali, M. B.; Zarrintan, M. H.; Adibkia, K. Antimicrobial Activity of the Metals and Metal Oxide Nanoparticles. *Mater. Sci. Eng. C Mater. Biol. Appl.* **2014,** 44, 278–284.

28. Driskell, J. D.; Shanmukh, S.; Liu, Y. J.; Hennigan, S.; Jones, L.; Zhao, Y. P.; Dluhy, R. A.; Krause, D. C.; Tripp, R. A. Infectious Agent Detection With SERS-Active Silver Nanorod Arrays Prepared by Oblique Angle Deposition. *IEEE Sens. J.* **2008**, 8(6), 863–870.

29. Du, L.; Jiang, H.; Liu, X.; Wang, E. Biosynthesis of Gold Nanoparticles Assisted by Escherichia coli DH5α and Its Application on Direct Electrochemistry of Hemoglobin. *Electrochem. Commun.* **2007**, 9(5), 1165–1170.

30. Duan, L.; Wang, Y.; Li, S. S.; Wan, Z.; Zhai, J. Rapid and Simultaneous Detection of Human Hepatitis B Virus and Hepatitis C Virus Antibodies Based on a Protein Chip Assay Using Nano-Gold Immunological Amplification and Silver Staining Method. *BMC Infect. Dis.* **2005**, 5, 53.

31. Duncan, T. V. Applications of Nanotechnology in Food Packaging and Food Safety: Barrier Materials, Antimicrobials and Sensors. *J. Colloid. Interface. Sci.* **2011**, 363, 1, 1–24.

32. Eid, K. A. M.; Salem, H. F.; Zikry, A. A. F.; El-Sayed, A. F. M.; Sharaf, M. A. Antifungal Effects of Colloidally Stabilized Gold Nanoparticles: Screening by Microplate Assay. *Nat. Sci.* **2011**, 9(2), 29–33.

33. Elechiguerra, J. L.; Burt, J. L.; Morones, J. R.; Camacho-Bragado, A.; Gao, X.; Lara, H. H.; Yacaman, M. J. Interaction of Silver Nanoparticles with HIV-1. *J Nanobiotechnol* **2005**, 3(6).

34. Estelrich, J.; Martín, M. J. S.; Busquets, M. A. Nanoparticles in Magnetic Resonance Imaging: From Simple to Dual Contrast Agents. *Int. J. Nanomed.* **2015**, 10, 1727–1741.

35. Fang, S. B.; Tseng, W. Y.; Lee, H. C.; Tsai, C. K.; Huang, J. T.; Hu, S. Y. Identification of Salmonella Using Colony-Print and Detection with Antibody-Coated Gold Nanoparticles. *J. Microbiol. Methods* **2009**, 77(2), 225–228.

36. Fayaz, A. M.; Balaji, K.; Girilal, M.; Yadav, R., Kalaichelvan, P. T.; Venketesan, R. Biogenic Synthesis of Silver Nanoparticles and Their Synergistic Effect with Antibiotics: A Study against Gram-Positive and Gram-Negative Bacteria. *Nanomedicine* **2010**, 6(1), 103–109.

37. Feng, Y.; Lin, X.; Wang, Y.; Wang, Y.; Hua, J. Diversity of Aurum Bioreduction by *Rhodobacter capsulatus*. *Mater. Lett.* **2008**, 62(37), 4299–4302.

38. Ferrer, M. N.; Domingo, E. J.; Corchero, J. L.; Vázquez, E.; Villaverde, A. Microbial Factories for Recombinant Pharmaceuticals. *Microb. Cell. Fact.* **2009**, 8, 17.

39. Filipponi, L.; Duncan, S. Applications of Nanotechnology: MEDICINE (Part 1). *NanoCap* **2007**, 1–10. http://nanopinion.archiv.zsi.at/en/reading-material/factsheet-5-application-nanotechnology-medicine-part-1.html.

40. Firdhouse, M. J.; Lalitha, P. Biosynthesis of Silver Nanoparticles and Its Applications. *J. Nanotech.* **2015**, 2015, 1–18.

41. Gao, J.; Li, L.; Ho, P. L.; Mak, G. C.; Gu, H.; Xu, B. Combining Fluorescent Probes and Biofunctional Magnetic Nanoparticles for Rapid Detection of Bacteria in Human Blood. *Adv. Mater.* **2006**, 18(23), 3145–3148.

42. Gawande, M. B.; Goswami, A.; Felpin, F. X.; Asefa, T.; Huang, X.; Silva, R.; Zou, X.; Zboril, R.; Varma, R. S. Cu and Cu-Based Nanoparticles: Synthesis and Applications in Catalysis. *Chem. Revs.* **2016**, 116(6), 3722–3811.

43. Gowramma, B.; Keerthi, U.; Rafi, M.; Rao, D. M. Biogenic Silver Nanoparticles Production and Characterization from Native Stain of Corynebacterium Species and Its Antimicrobial Activity. *3 Biotech* **2015,** *5*(2), 195–201.

44. Guidelli, E. J.; Kinoshita, A.; Ramos, A. P.; Baffa, O. Silver Nanoparticles Delivery System Based on Natural Rubber Latex Membranes. *J. Nanopart. Res.* **2013,** *15*(4), 1536.

45. Gurunathan, S.; Han, J. W.; Kwon, D. N.; Kim, J. H. Enhanced Antibacterial and Anti-Biofilm Activities of Silver Nanoparticles against Gram-Negative and Gram-Positive Bacteria. *Nanoscale Res. Lett.* **2014,** *9*(1), 373.

46. Hahn, M. A.; Tabb, J. S.; Krauss, T. D. Detection of Single Bacterial Pathogens with Semiconductor Quantum Dots. *Anal. Chem.* **2005,** *77*(15), 4861–4869.

47. Hajipour, M. J.; Fromm, K. M.; Ashkarran, A. A.; Aberasturi, D. J. De; Larramendi, I. R. De; Rojo, T.; Serpooshan, V.; Parak, W. J.; Mahmoudi, M. Antibacterial Properties of Nanoparticles. *Trends Biotechnol.* **2012,** *30*(10), 499–511.

48. He, L.; Wang, M.; Ge, J.; Yin, Y. Magnetic Assembly Route to Colloidal Responsive Photonic Nanostructures. *Acc. Chem. Res.* **2012,** *45*(9), 1431–1440.

49. Herman, A.; Herman, A. P. Nanoparticles as Antimicrobial Agents: Their Toxicity and Mechanisms of Action. *J. Nanosci. Nanotechnol.* **2014,** *14*(1), 946–957.

50. Huang, S. H. Gold Nanoparticle-Based Immunochromatographic Test for Identification of *Staphylococcus aureus* from Clinical Specimens. *Clin. Chim. Acta.* **2006,** *373*(1–2), 139–143.

51. Huh, A. J.; Kwon, Y. J. "Nanoantibiotics": A New Paradigm for Treating Infectious Diseases Using Nanomaterials in the Antibiotics Resistant Era." *J. Control Release* **2011,** *156*(2), 128–145.

52. Hulkoti, N. I.; Taranath, T. C. Biointerfaces Biosynthesis of Nanoparticles Using Microbes—A Review. *Colloids Surf. B* **2014,** *121*, 474–483.

53. Husseiny, S. M.; Salah, T. A.; Anter, H. A. Biosynthesis of Size Controlled Silver Nanoparticles by *Fusarium oxysporum*, Their Antibacterial and Antitumor Activities. *Beni-Seuf Univ. J. Appl. Sci.* **2015,** *4*(3), 225–231.

54. Hyeon, T. Chemical Synthesis of Magnetic Nanoparticles. *Chem. Commun.* **2003,** *8*, 927–934.

55. Ingle, A. P.; Duran, N.; Rai, M. Bioactivity, Mechanism of Action, and Cytotoxicity of Copper-Based Nanoparticles: A Review. *Appl. Microbiol. Biotechnol.* **2014,** *98*(3), 1001–1009.

56. Iravani, S. Bacteria in Nanoparticle Synthesis: Current Status and Future Prospects. *Int. Sch. Res. Notices* **2014,** *2014*, 359316.

57. Jin, T.; Sun, D.; Su, J. Y.; Zhang, H.; Sue, H. J. Antimicrobial Efficacy of Zinc Oxide Quantum Dots against *Listeria monocytogenes*, *Salmonella enteritidis*, and *Escherichia coli* O157: H7. *J. Food. Sci.* **2009,** *74*(1), 46–52.

58. John, St. J.; Moro, D. J. Hydrogel Wound Dressing and Biomaterials Formed in situ and Their Uses. U.S. Patent 7,910,135, 2011, B2.

59. Jones, N.; Ray, B.; Ranjit, K. T.; Manna, A. C. Antibacterial Activity of ZnO Nanoparticle Suspensions on a Broad Spectrum of Microorganisms. *FEMS Microbiol. Lett.* **2008,** *279*(1), 71–76.

60. Kavre, I.; Kostevc, G.; Kralj, S.; Vilfan, A.; Babič, D. Fabrication of Magneto-Responsive Microgears Based on Magnetic Nanoparticle Embedded PDMS. *RSC Adv.* **2014**, *4*(72), 38316–38322.

61. Kell, A. J.; Stewart, G.; Ryan, S.; Peytavi, R.; Boissinot, M.; Huletsky, A.; Bergeron, M. G.; Benoit, S. Vancomycin-Modified Nanoparticles for Efficient Targeting and Preconcentration of Gram-Positive and Gram-Negative Bacteria. *ACS Nano.* **2008**, *2*(9), 1777–17788.

62. Kitching, M.; Ramani, M.; Marsili, E. Fungal Biosynthesis of Gold Nanoparticles: Mechanism and Scale Up. *Microb. Biotechnol.* **2015**, *8*(6), 904–917.

63. Klaus, J. T.; Joerger, R.; Olsson, E.; Granqvist, C. Bacteria as Workers in the Living Factory: Metal-Accumulating Bacteria and Their Potential for Materials Science. *Trends Biotechnol.* **2001**, *19*(1), 15–20.

64. Klostranec, J. M.; Xiang, Q.; Farcas, G. A.; Lee, J. A.; Rhee, A.; Lafferty, E. I.; Perrault, S. D.; Kain, K. C.; Chan, W. C. Convergence of Quantum Dot Barcodes with Microfluidics and Signal Processing for Multiplexed High-Throughput Infectious Disease Diagnostics. *Nano. Lett.* **2007**, *7*(9), 2812–2818.

65. Korbekandi, H.; Iravani, S.; Abbasi, S. Production of Nanoparticles Using Organisms. *Crit. Rev. Biotechnol.* **2009**, *29*(4), 279–306.

66. Kuku, G.; Saricam, M.; Akhatova, F.; Danilushkina, A.; Fakhrullin, R.; Culha, M. Surface-Enhanced Raman Scattering to Evaluate Nanomaterial Cytotoxicity on Living Cells. *Anal. Chem.* **2016**, *88*(19), 9813–9820.

67. Lara, H. H.; Ayala-Nuñez, N. V.; Ixtepan-turrent, L.; Rodriguez-padilla, C. Mode of Antiviral Action of Silver Nanoparticles against HIV-1. *J. Nanobiotechnol.* **2010**, *8*(1).

68. Latorre, M.; Rinaldi, C. Applications of Magnetic Nanoparticles in Medicine: Magnetic Fluid Hyperthermia. *P R Health Sci. J.* **2009**, *28*(3), 227–238.

69. Lee, J. H.; Kim, Y. G.; Cho, M. H.; Lee, J. ZnO Nanoparticles Inhibit Pseudomonas Aeruginosa Biofilm Formation and Virulence Factor Production. *Microbiol. Res.* **2014**, *169*(12), 888–896.

70. Leu, J. G.; Chen, S. A.; Chen, H. M.; Wu, M. M.; Hung, C. F.; Yao, Y. D.; Tu, C. S.; Liang, Y. J. The Effects of Gold Nanoparticles in Wound Healing with Antioxidant Epigallocatechin Gallate and α-Lipoic Acid. *Nanomedicine* **2012**, *8*(5), 767–775.

71. Li, X. X.; Cao, C.; Han, S. J.; Sim, S. J. Detection of Pathogen Based on the Catalytic Growth of Gold Nanocrystals. *Water Res.* **2009**, *43*(5), 1425–1431.

72. Li, X.; Xu, H.; Chen, Z. S.; Chen, G. Biosynthesis of Nanoparticles by Microorganisms and Their Applications. *J. Nanomater.* **2011**, *2011*.

73. Lin, F. Y. H.; Sabri, M.; Alirezaie, J.; Li, D.; Sherman, P. M. Development of a Nanoparticle-Labeled Microfluidic Immunoassay for Detection of Pathogenic Microorganisms. *Clin. Diagn. Lab. Immunol.* **2005**, *12*(3), 418–425.

74. Lin, Y. S.; Tsai, P. J.; Weng, M. F.; Chen, Y. C. Affinity Capture Using Vancomycin-Bound Magnetic Nanoparticles for the MALDI-MS Analysis of Bacteria. *Anal. Chem.* **2005**, *77*(6), 1753–1760.

75. Liu, J. C.; Tsai, P. J.; Lee, Y. C.; Chen, Y. C. Affinity Capture of Uropathogenic Escherichia coli Using Pigeon Ovalbumin-Bound $Fe_3O_4@Al_2O_3$ Magnetic Nanoparticles. *Anal. Chem.* **2008**, *80*(14), 5425–5432.

76. Liu, Y.; He, L.; Mustapha, A.; Li, H.; Hu, Z. Q.; Lin, M. Antibacterial Activities of Zinc Oxide Nanoparticles Against *Escherichia coli* O157: H7. *J. Appl. Microbiol.* **2009,** *107*(4), 1193–1201.

77. Madhumathi, K.; Sudheesh Kumar, P. T.; Abhilash, S.; Sreeja, V.; Tamura, H.; Manzoor, K.; Nair, S. V.; Jayakumar, R. Development of Novel Chitin/nanosilver Composite Scaffolds for Wound Dressing Applications. *J. Mater. Sci. Mater. Med.* **2010,** *21*(2), 807–813.

78. Mehrbod, P.; Motamed, N.; Tabatabaian, M.; Soleimani Estyar, R.; Amini, E.; Shahidi, M.; Kheiri, M. T. In Vitro Antiviral Effect of "Nanosilver" on Influenza Virus. *J. Pharm. Sci.* **2009,** *17*(2), 88–93.

79. Mirjani, R.; Faramarzi, M. A.; Sharifzadeh, M.; Setayesh, N.; Khoshayand, M. R.; Shahverdi, A. R. Biosynthesis of Tellurium Nanoparticles by *Lactobacillus plantarum* and the Effect of Nanoparticle-Enriched Probiotics on the Lipid Profiles of Mice. *IET Nanobiotechnol.* **2015,** *9*(5), 300–305.

80. Mohanpuria, P.; Rana, N. K.; Yadav, S. K. Biosynthesis of Nanoparticles: Technological Concepts and Future Applications. *J. Nanopart. Res.* **2008,** *10*(3), 507–517.

81. Morones, J. R.; Elechiguerra, J. L.; Camacho, A.; Holt, K.; Kouri, J. B.; Ramirez, J. T.; Yacaman, M. J. The Bactericidal Effect of Silver Nanoparticles. *Nanotechnology* **2005,** *16*(10), 2346–2353.

82. Mukhopadhyay, B.; Martins, B. M.; Karamanska, R.; Russell, D. A.; Field, R. A. Bacterial Detection Using Carbohydrate-Functionalised Cds Quantum Dots: A Model Study Exploiting *E. coli* Recognition of Mannosides. *Tetrahedron. Lett.* **2009,** *50*(8), 886–889.

83. Nanda, A.; Saravanan, M. Biosynthesis of Silver Nanoparticles from Staphylococcus aureus and Its Antimicrobial Activity against MRSA and MRSE. *Nanomedicine* **2009,** *5*(4), 452–456.

84. Nanoparticles and Their Potential Application as Antimicrobials; Rai, R.; Bai, J; Science Against Microbial Pathogens: Communicating Current Research and Technological Advances (Microbiology Book Series- Number 3 Vol. 1; Formatex Research Center, 2011.

85. Narayanan, K. B.; Sakthivel, N. Biological Synthesis of Metal Nanoparticles by Microbes. *Adv. Colloid Interface Sci.* **2010,** *156*(1–2), 1–13.

86. Ogi, T.; Saitoh, N.; Nomura, T.; Konishi, Y. Room-Temperature Synthesis of Gold Nanoparticles and Nanoplates Using Shewanella Algae Cell Extract. *J. Nanopart Res.* **2010,** *12*(7), 2531–2539.

87. Padmavathy, N.; Vijayaraghavan, R. Enhanced Bioactivity of ZnO Nanoparticles— an Antimicrobial Study. *Sci. Technol. Adv. Mater.* **2008,** *9*(3).

88. Pelgrift, R. Y.; Friedman, A. J. Nanotechnology As a Therapeutic Tool to Combat Microbial Resistance. *Adv. Drug Deliv. Rev.* **2013,** *65*(13–14), 1803–1815.

89. Perez, J. M.; Josephson, L.; Loughlin, T. O.; Högemann, D.; Weissleder, R. Magnetic Relaxation Switches Capable of Sensing Molecular Interactions. *Nat. Biotechnol.* **2002,** *20*(8), 816–820.

90. Philip, J.; Shima, P. D.; Raj, B. Nanofluid with Tunable Thermal Properties. *Appl. Phys. Lett.* **2008,** *92*(4).

91. Polpanich, D.; Tangboriboonrat, P.; Elaissari, A.; Udomsangpetch, R. Detection of Malaria Infection via Latex Agglutination Assay. *Anal. Chem.* **2007,** *79*(12), 4690–4695.

92. Prabhu, S.; Poulose, E. K. Silver Nanoparticles: Mechanism of Antimicrobial Action, Synthesis, Medical Applications, and Toxicity Effects. *Int. Nano. Lett.* **2012,** *2*(1), 32.

93. Prasad, R.; Pandey, R.; Barman, I. Engineering Tailored Nanoparticles with Microbes: Quo Vadis?. *Wliey Interdiscip. Rev. Nanomed. Nanobiotechnol.* **2016,** *8*(2), 316–330.

94. Qin, D.; He, X.; Wang, K.; Zhao, X. J.; Tan, W.; Chen, J. Fluorescent Nanoparticle-Based Indirect Immunofluorescence Microscopy for Detection of *Mycobacterium tuberculosis*. *J. Biomed. Biotechnol.* **2007,** *2007*(7), 89364. DOI: 10.1155/2007/89364.

95. Rai, M.; Ingle, A. P.; Gade, A.; Duran, N. Synthesis of Silver Nanoparticles by *Phoma gardeniae* and in Vitro Evaluation of Their Efficacy Against Human Disease-Causing Bacteria and Fungi. *IET Nanobiotechnol.* **2015,** *9*(2), 71–75.

96. Rai, M.; Deshmukh, S. D.; Ingle, A. P.; Gupta, I. R.; Galdiero, M.; Galdiero, S. Metal Nanoparticles: The Protective Nanoshield Against Virus Infection. *Crit. Rev. Microbiol.* **2016a,** *42*(1), 42–56.

97. Rai, M.; Ingle, A. P.; Birla, S.; Yadav, A.; Santos, C. A. Strategic Role of Selected Noble Metal Nanoparticles in Medicine. *Crit. Rev. Microbiol.* **2016b,** *42*(5), 696–719.

98. Ravindranath, S. P.; Mauer, L. J.; Deb-Roy, C.; Irudayaraj, J. Biofunctionalized Magnetic Nanoparticle Integrated Mid-Infrared Pathogen Sensor for Food Matrixes. *Anal. Chem.* **2009,** *81*(8), 2840–2846.

99. Rigo, C.; Ferroni, L.; Tocco, I.; Roman, M.; Munivrana, I.; Gardin, C.; Cairns, W. R. L.; Vindigni, V.; Azzena, B.; Barbante, C.; Zavan, B. Active Silver Nanoparticles for Wound Healing. *Int. J. Mol. Sci.* **2013,** *14*(3).

100. Rogers, J. V.; Christopher, V. P.; Choi, Y. W.; Speshock, J. L.; Hussain, S. M. A Preliminary Assessment of Silver Nanoparticle Inhibition of Monkeypox Virus Plaque Formation. *Nanoscale Res. Lett.* **2008,** *3*(4), 129–133.

101. Roh, Y.; Vali, H.; Phelps, T. J.; Moon, J. W. Extracellular Synthesis of Magnetite and Metal-Substituted Magnetite Nanoparticles. *J. Nanosci. Nanotechnol.* **2006,** *6*(11), 3517–3520.

102. Sanvicens, N.; Pastells, C.; Pascual, N.; Marco, M. P. Nanoparticle-Based Biosensors for Detection of Pathogenic Bacteria. TrAC, *Trends Anal. Chem.* **2008,** *28*(11), 1243–1252.

103. Saravanan, M.; Vemu, A. K.; Barik, S. K. Biointerfaces Rapid Biosynthesis of Silver Nanoparticles from *Bacillus megaterium* (NCIM 2326) and Their Antibacterial Activity on Multi Drug Resistant Clinical Pathogens. *Colloids Surf. B Biointerfaces* **2011,** *88*(1), 325–331.

104. Sathyanarayanan, M. B.; Balachandranath, R.; Srinivasulu, Y. G.; Kannaiyan, S. K.; Subbiahdoss, G. The Effect of Gold and Iron-Oxide Nanoparticles on Biofilm-Forming Pathogens. *ISRN Microbiol.* **2013,** *2013*.

105. Schofield, C. L.; Field, R. A.; Russell, D. A. Glyconanoparticles for the Colorimetric Detection of Cholera Toxin. *Anal. Chem.* **2007,** *79*(4), 1356–1361.

106. Shamaila, S.; Zafar, N.; Riaz, S.; Sharif, R.; Nazir, J.; Naseem, S. Gold Nanoparticles: An Efficient Antimicrobial Agent Against Enteric Bacterial Human Pathogen. *Nanomaterials* **2016,** *6*(4).

107. Siddiqi, K. S.; Husen, A. Fabrication of Metal Nanoparticles from Fungi and Metal Salts: Scope and Application. *Nanoscale Res. Lett.* **2016**, *11*(1).

108. Singh, R.; Nalwa, H. S. Medical Applications of Nanoparticles in Biological Imaging, Cell Labeling, Antimicrobial Agents, and Anticancer Nanodrugs. *J. Biomed. Nanotechnol.* **2011**, *7*(4), 489–503.

109. Singh, A. K.; Srivastava, O. N. One-Step Green Synthesis of Gold Nanoparticles Using Black Cardamom and Effect of pH on Its Synthesis. *Nanoscale Res. Lett.* **2015**, *10*(1).

110. Singh, A. K.; Harrison, S. H.; Schoeniger, J. S. Gangliosides As Receptors for Biological Toxins: Development of Sensitive Fluoroimmunoassays Using Ganglioside-Bearing Liposomes. *Anal. Chem.* **2000**, *72*(24), 6019–6024.

111. Singh, D.; Rathod, V.; Ninganagouda, S.; Hiremath, J.; Singh, A. K.; Mathew, J. Optimization and Characterization of Silver Nanoparticle by Endophytic Fungi *Penicillium sp.* Isolated from Curcuma Longa (Turmeric) and Application Studies Against MDR *E. coli* and *S. aureus*. *Bioinorg. Chem. Appl.* **2014**, *2014*.

112. Sioss, J. A.; Stoermer, R. L.; Sha, M. Y.; Keating, C. D. Silica-Coated, Au / Ag Striped Nanowires for Bioanalysis. *Langmuir* **2007**, *23*(22), 11334–11241.

113. Speshock, J. L.; Murdock, R. C.; Braydich-stolle, L. K.; Schrand, A. M.; Hussain, S. M. Interaction of Silver Nanoparticles with Tacaribe Virus. *J. Biotechnol.* **2010**, *8*.

114. Sriram, M. I.; Kalishwaralal, K.; Gurunathan, S. Biosynthesis of Silver and Gold Nanoparticles Using Bacillus licheniformis. *Methods Mol. Bio.* **2012**, *906*, 33–43.

115. Srivastava, S. K.; Yamada, R.; Ogino, C.; Kondo, A. Biogenic Synthesis and Characterization of Gold Nanoparticles by *Escherichia coli* K12 and Its Heterogeneous Catalysis in Degradation of 4-Nitrophenol. *Nanoscale Res. Lett.* **2013**, *8*(1).

116. Su, X. L.; Li, Y. Quantum Dot Biolabeling Coupled with Immunomagnetic Separation for Detection of *Escherichia coli* O157:H7. *Anal. Chem.* **2004**, *76*(16), 4806–4810.

117. Suci, P. A.; Berglund, D. L.; Liepold, L.; Brumfield, S.; Pitts, B.; Davison, W.; Oltrogge, L.; Hoyt, K. O.; Codd, S.; Stewart, P. S.; Young, M.; Douglas, T. High-Density Targeting of a Viral Multifunctional Nanoplatform to a Pathogenic, Biofilm-Forming Bacterium. *Chem. Bio.* **2007**, *14*(4), 387–398.

118. Sun, R. W.; Chen, R.; Chung, N. P.; Ho, C. M.; Lin, C. L.; Che, C. M. Silver Nanoparticles Fabricated in Hepes Buffer Exhibit Cytoprotective Activities toward HIV-1 Infected Cells. *Chem. Commun.* (Camb). **2005**, (40), 5059–5061.

119. Tadic, M.; Kralj, S.; Jagodic, M.; Hanzel, D.; Makovec, D. Magnetic Properties of Novel Superparamagnetic Iron Oxide Nanoclusters and Their Peculiarity under Annealing Treatment. *Appl. Surf. Sci.* **2014**, *322*, 255–264.

120. Takhar, P.; Mahant, S. In Vitro Methods for Nanotoxicity Assessment: Advantages and Applications. *Arch. Appl. Sci. Res.* **2011**, *3*(2), 389–403.

121. Tallury, P.; Malhotra, A.; Byrne, L. M.; Santra, S. Nanobioimaging and Sensing of Infectious Diseases. *Adv. Drug Deliv. Rev.* **2010**, *62*(4–5), 424–437.

122. Tan, Y.; Yao, R.; Wang, R.; Wang, D.; Wang, G.; Zheng, S. Reduction of Selenite to Se(0) Nanoparticles by Filamentous Bacterium Streptomyces Sp. ES2-5 Isolated from a Selenium Mining Soil. *Microb. Cell Fact.* **2016**, *15*(1).

123. Tang, S.; Moayeri, M.; Chen, Z.; Harma, H.; Zhao, J.; Hu, H.; Purcell, R. H.; Leppla, S. H.; Hewlett, I. K. Detection of Anthrax Toxin by an Ultrasensitive Immunoassay Using Europium Nanoparticles. *Clin. Vaccine Immunol.* **2009**, *16*(3), 408–413.

124. Shi, L. E.; Li, Z. H.; Zheng, W.; Zhao, Y. F.; Jin, Y. F.; Tang, Z. X. Synthesis, Antibacterial Activity, Antibacterial Mechanism and Food Applications of ZnO Nanoparticles: A Review. *Food Addit. Contam.* **2014**, *31*(2), 173–186.

125. Thakkar, K. N.; Mhatre, S. S.; Parikh, R. Y. Biological Synthesis of Metallic Nanoparticles. *Nanomedicine* **2010**, *6*(2), 257–262.

126. Thirumurugan, A.; Ramachandran, S.; Tomy, N. A.; Jiflin, G. J.; Rajagomathi, G. Biological Synthesis of Gold Nanoparticles by *Bacillus subtilis* and Evaluation of Increased Antimicrobial Activity Against Clinical Isolates. *Korean J. Chem. Eng.* **2012**, *29*(12), 1761–1765.

127. Tian, J.; Wong, K. K.; Ho, C. M.; Lok, C. N.; Yu, W. Y.; Che, C. M.; Chiu, J. F.; Tam, P. K. Topical Delivery of Silver Nanoparticles Promotes Wound Healing. ChemMedChem. **2007**, *2*(1), 129–136.

128. Tikariha, S.; Singh, S.; Banerjee, S.; Vidyarthi, A. S. Biosynthesis of Gold Nanoparticles, Scope and Application: A Review. *IJPSR.* **2012**, 1603–1615.

129. Tripp, R. A.; Alvarez, R.; Anderson, B.; Jones, L.; Weeks, C.; Chen, W. Bioconjugated Nanoparticle Detection of Respiratory Syncytial Virus Infection. *Int. J. Nanomedicine* **2007**, *2*(1), 117–124.

130. Tulkens, P. M. Intracellular Distribution and Activity of Antibiotics. *Eur. J. Clin. Microbiol. Infect. Dis.* **1991**, *10*(2), 100–106.

131. Upadhyay, R. K. Drug Delivery Systems, CNS Protection, and the Blood Brain Barrier. *Biomed. Res. Int.* **2014**, *2014*.

132. Varshney, M.; Yang, L.; Su, X. L.; Li, Y. Magnetic Nanoparticle-Antibody Conjugates for the Separation of *Escherichia Coli* O157:H7 in Ground Beef. *J. Food Prot.* **2005**, *68*(9), 1804–1811.

133. Villaverde, A. Nanotechnology, Bionanotechnology and Microbial Cell Factories. *Microb. Cell Fact.* **2010**, *9*(1), 53.

134. Wang, Y. F.; Pang, D. W.; Zhang, Z. L.; Zheng, H. Z.; Cao, J. P.; Shen, J. T. Visual Gene Diagnosis of HBV and HCV Based on Nanoparticle Probe Amplification and Silver Staining Enhancement. *J. Med. Virol.* **2003**, *70*(2), 205–211.

135. Wang, L.; Zhao, W.; O'Donoghue M. B.; Tan W Fluorescent Nanoparticles for Multiplexed Bacteria Monitoring. *Bioconjugate Chem.* **2007**, *18*(2), 297–301.

136. Wani, I. A.; Ahmad, T. Size and Shape Dependant Antifungal Activity of Gold Nanoparticles: A Case Study of Candida. *Colloids Surf. B* **2013**, *101*, 162–170.

137. Weiner, S.; Dove, P. M. An Overview of Biomineralization Processes and the Problem of the Vital Effect. *Rev. Mineral. Geochem.* **2001**, *54*(1), 1–29.

138. Weir, E.; Lawlor, A.; Whelan, A.; Regan, F. The Use of Nanoparticles in Anti-Microbial Materials and Their Characterization. *Analyst* **2008**, *133*(7), 835–845.

139. Wen, L.; Lin, Z.; Gu, P.; Zhou, J.; Yao, B.; Chen, G.; Fu, J. Extracellular Biosynthesis of Monodispersed Gold Nanoparticles by a SAM Capping Route. *J. Nanoparticle Res.* **2009**, *11*(2), 279–288.

140. WHO Fact sheet: Media Centre Antimicrobial Resistance, September 2016.

141. Wilczewska, A. Z.; Niemirowicz, K.; Markiewicz, K. H.; Car, H. Nanoparticles as Drug Delivery Systems. *Pharmacol. Rep.* **2012,** *64*(5), 1020–1037.

142. Xiang, D. X.; Chen, Q.; Pang, L.; Zheng, C. L. Inhibitory Effects of Silver Nanoparticles on H1N1 Influenza A Virus in Vitro. *J. Virol. Methods.* **2011,** *178* (1–2) 137–142.

143. Xie, S.; Tao, Y.; Pan, Y.; Qu, W.; Cheng, G.; Huang, L.; Chen, D.; Wang, X.; Liu, Z.; Yuan, Z. Biodegradable Nanoparticles for Intracellular Delivery of Antimicrobial Agents. *J. Control* **2014,** *187*, 101–117.

144. Yah, C. S.; Simate, G. S. Nanoparticles as Potential New Generation Broad Spectrum Antimicrobial Agents. *Daru: J. Pharm. Sci.* **2015,** *23*.

145. Yang, H.; Qu, L.; Lin, Y.; Jiang, X.; Sun, Y. P. Detection of *Listeria monocytogenes* in Biofilms Using Immunonanoparticles. *J. Biomed. Nanotechnol.* **2007,** *3*(2), 131–138.

146. Yang, S. Y.; Cheih, J. J.; Wang, W. C.; Yu, C. Y.; Lan, C. B.; Chen, J. H.; Horng, H. E.; Hong C. Y.; Yang, H. C.; Huang W. Ultra-Highly Sensitive and Wash-Free Bio-Detection of H5N1 Virus by Immunomagnetic Reduction Assays. *J. Virol. Methods* **2008,** *153*(2), 250–252.

147. Zhang, D.; Carr, D. J.; Alocilja, E. C. Fluorescent Bio-Barcode DNA Assay for the Detection of *Salmonella enterica* Serovar Enteritidis. *Biosens. Bioelectron.* **2009,** *24*(5), 1377–1381.

148. Zhang, X.; Yan, S.; Tyagi, R. D.; Surampalli, R. Y. Synthesis of Nanoparticles by Microorganisms and Their Application in Enhancing Microbiological Reaction Rates. *Chemosphere* **2011,** *82*(4), 489–494.

149. Zhao, X.; Hilliard, L. R.; Mechery, S. J.; Wang, Y.; Bagwe, R. P.; Jin, S.; Tan, W. A Rapid Bioassay for Single Bacterial Cell Quantitation Using Bioconjugated Nanoparticles. *Proc. Natl. Acad. Sci. USA* **2004,** *101*(42), 15027–15032.

150. Ziv, P. O.; Topaz, M.; Brosh, T.; Margel, S. Enhancement of Incisional Wound Healing by Thrombin Conjugated Iron Oxide Nanoparticles. *Biomaterial* **2010,** *31*(4), 741–747.

151. Agasti, S. S., Rana, S., Park, M. H., Kim, C. K., You, C. C., & Rotello, V. M. Nanoparticles for detection and diagnosis. *Adv. Drug Deliv. Rev.* **2010,** *62*(3), 316–328.

152. Clatworthy, A. E., Pierson, E., & Hung, D. T. Targeting virulence: A new paradigm for antimicrobial therapy. *Nat. Chem. Biol.* **2007,** *3*(9), 541–548.

153. Iravani, S. Green synthesis of metal nanoparticles using plants. *Green Chem.* **2011,** *13*(10), 2638

CHAPTER 20

POTENTIAL APPLICATIONS OF NANOTECHNOLOGY: AN INTERFACE IN MODERN SCIENCE

ANJAN KUMAR PRADHAN[1,*] and SUSHIL KUMAR SAHU[2]

[1]Virginia Commonwealth University, Richmond, Virginia, USA,
*E-mail: pradhan.anjan@gmail.com

[2]John Hopkins University, Baltimore, Maryland, USA

There's plenty of room at the bottom.

Richard Feynman

CONTENTS

ABSTRACT

Scientific exploration can open new perspectives in almost every field of the world. This has been shown earlier and is currently going on leading to newer discoveries and inventions ultimately improving human life and health. In this context nanotechnology has emerged as an interdisciplinary science created by an eclectic band of physicists, chemists, engineers pooling their talents and the application of it to improve human health is nanobiology or nanomedicine. Nanotechnolgy has been applied to almost every field of biology and here the topic of interest is microbiology that is, the study of small organisms. The part of science that studies small stuff is nanotechnology and the part that studies small organisms is microbiology. At a number of levels, microbiology and nanoscience are related. Bacterial entities, formation of hyphae in fungi, viral capsids can be compared with the nanoscale self-assembly. Present day molecular biology techniques are already applied to nanotechnology. Imaging of single molecules, nanoscale molecules and determination of spatial organization can be studied using nanotools and technology. Food contaminants (chemical and biological) can be studied through nanotechnology. Infectious diseases can be studied and targeted through nanotechnology. This chapter describes the nanoscale objects in the study of microorganisms for the betterment of society and health science.

20.1 INTRODUCTION

The concept of nanobiology came from the applications of nanotechnology in the biochemical and biological fields. The term refers to the link between nanotechnology and biology.[1] The term nanotechnology was first coined by Norio Taniguchi, a professor at Tokyo Science University, in 1974. It is an interdisciplinary science which uses the principles of chemistry, mathematics, physics, electronics and biology. The areas that come under the umbrella include nanomaterials, nanodevices and nanoparticles in the first part and biological systems in the second part.[2] Nanometer is one thousands of micrometer which is one millionth of a meter. This explains the size of materials in the nano-range (Table 20.1) and their strength. Funded by billions of dollars around the world, a lot of people are working to use the technology to develop new approaches towards the welfare of human kind. This approach allows scientists to design systems used for plant and animal health care research.

TABLE 20.1   Sizes of Natural Nano-size Objects.

Objects	Approximate size
Atom	0.2 nM
Carbon nanotube	1 nM
DNA	2 nM
Protein	50 nM
Ribosome	25 nM
Virus	100 nM
Mitochondria	500 nM
Bacteria	1000 nM
WBC	10000 nM
RBC	7 μM
Hair	80 μM
Pin head	1 mM

Microbiology can be related to nanoscience at several levels.[3] Many bacterial entities can be compared to nano-machines in nature, that is, molecular motors like pili or flagella.[4] The formation of hyphae by fungi is directed by the controlled assembly of building blocks. Formation of capsids in a virus is a process of molecular and cellular recognition and self-assembly at the nanometer scale.[5]

Nanotechnology involves the design, manufacture, and manipulation of particulate matter at the nanoscale.[6] It provides the means for manipulative nanomaterial with diverse chemical, physical, and biological/biochemical properties.[7–9] Nanotechnology has been applied in the genetic modification of crops with the present day molecular biology tools and techniques.[10,11] Material scientists and engineers have developed newer technologies to synthesize nanomaterials. Technology of preparation depends on the type of material. They are getting more important day by day and their preparation costlier. This makes the use of nanoparticles less affordable.

Nanotechnology can be a tool for a better future to improve the food value, quality, safety and food protection[12,13] with a combinatorial approach involving western technologies, eastern traditional knowledge and ethical standards (Fig. 20.1). Nanotechnology promises that the materials have self-assembling and self-healing properties. It can be applied in almost every field and aspect of science.

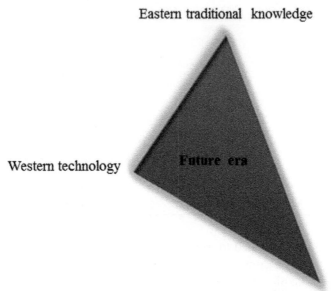

FIGURE 20.1   Future of medicine lies in a combinatorial force of eastern traditional knowledge and western technology with a focus to ethics and safety measures.

This chapter describes the use and aspects of nanotechnology in the development of human health, food science, water purification and pollution control and so forth. After 20 years of nanoscience research and a focused last decade research, nanotechnology is delivering in expected and unexpected areas to benefit the society. This field and era is improving and revolutionizing the technology sector, the environmental sector, medicine and food industry, energy, transportation and security.

20.2   TERMINOLOGIES

Nanomedicine: The application of nanotechnology in medicine is nanomedicine. It has a wide range from nanobiological devices to nanobiosensors.[14]

Nanobiotechnology: The application of nanotechnology in biology is nanobiotechnology.[15]

Bionanotechnology: Bionanotechnology is the branch of nanotechnology that uses biological tools and techniques, including biological materials or design.[16]

Nanobiomechanics: Nanobiomechanics or bionanomechanics is the combination of biomechanics using nanomechanics in biology.[17] Nanobiomechanics can measure the mechanical characteristics of individual living cells.

Nanosubmarine: Nanosubs or nanosubmarines are microscopic devices that navigate and perform specific tasks within the human body. Nanosubmarines use a lot of methods to navigate through the body.[18]

Colloidal gold: Colloidal gold is a colloidal suspension of fluidic gold nanoparticles.[19]

Nanobeacons: Nanobeacons are functionalized nanoparticles with a fluorophore-labelled hairpin-DNA. This can follow synthesis of RNA or DNA replication.[20]

BioNEMS: Biofunctionalized nanoelectromechanical systems.[21]

Nanofabrication is the design and manufacture of devices measured in nanometers.[22,23]

Nanolithography is the science of writing and printing at the nanoscale level.[24]

20.3   USE OF NANOMICROBIOLOGY

Nanotechnology is applied in almost every field that is, sports goods, fabrics, electronics and technology, computer and camera displays, health and medicine, food, air and water pollution, automobile industry, petroleum industry, energy sector, environmental remediation (Fig. 20.2). This chapter focuses mostly on the food industry, health and medicine and environmental remediation (air and water).

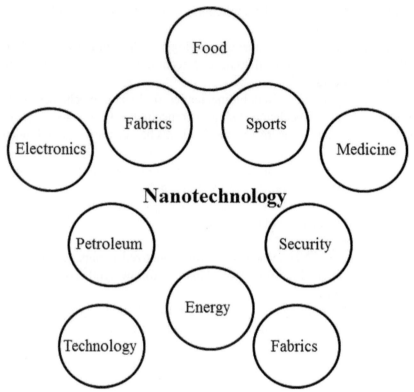

FIGURE 20.2   Applications of Nanotechnology.

20.3.1   FOOD INDUSTRY

Nanoparticles can enter the human body due to their small size through various usual or unusual routes.[25] They can easily pass through the cell membranes or the blood-brain barrier.[26] Food industry is the most growing area in today's era and nanotechnology can be applied at different stages in food industry: food production, processing, packaging, transporting, tracing, and keeping the quality of food product and to extend the product shelf life. This can overcome food wastage, water filtration, removal of undesirable tastes, flavours or allergens from food products and nanobiosensors for food safety. The possibility of combining biology and nanotechnology in the design and construction of sensors holds the potential of increased sensitivity and, therefore reduced time consumption for potential problems. Based on nanobiosensors, and due to the low infectious doses,[27] rapid and sensitive detection methods are developed for pathogens *Escherichia coli* strains, *Bacillus subtilis, Listeria monocytogenes* and yeast.[28] Nanosensors to detect biofilm formation by bacteria are coming up.[29] Detection and inhibition of bacterial biofilms is a leading problem and a developing area which utilizes the concepts and advantage of nanomaterials that have antimicrobial properties (such as silver, gold, copper, titanium, zinc, magnesium, cadmium and alumina).[30] Employees, using sensors with either light or magnetic materials, can detect minute quantities of harmful pathogens. The advantage of such a system is they can enter any place due to their small sizes.

Detection of a small minute chemical or microbiological (bacterial or viral or other pathogens) contaminant in food systems is a potential application of nanotechnology.[31] Combination of nanotechnology and biology can be used in the sensors that can detect contaminations in food.[32]

Researchers have developed nanosensors that can be modified to fluoresce or are magnetically modified.[33] These can selectively attach to food pathogens which can be detected by infrared light or a magnetic field.[34] A lot of nanoparticles can be made into single nanosensors to accurately detect bacteria and other pathogens.[35] Nanosensors can gain access to any small place where pathogens often hide, where other systems cannot pass. [36]

Bioanalytical Microsystems are other detection systems which use the principle of RNA/DNA recognition or RNA-DNA hybridization or liposome amplification.[37] The bioanalytical microsystems use the principles of molecular biology, which is RNA recognition via RNA/DNA hybridization and liposome amplification. These are used in the rapid detection of pathogens in water testing and analysis of food contaminants.

20.3.2 MEDICINE

Nanomedicine is the most studied area in nanotechnology. Drugs can be tagged to nanoparticles which increase the specificity, bioavailability and targeted delivery.[38] Biomedicine is the most lucrative field that has been explored for the possibility to use nanotechnology.[39] One can visualize the effect of drugs on cell surfaces.[40] Early investigations have shown the ability of microscopy to visualize the drug-induced alterations in *E. coli, Helicobacter pylori* and *Staphylococcus aureus*.[41] More recently, drug-induced alterations were also demonstrated on *Mycobacteria*.[42] The applications of nanotechnology in medical application are well studied in literature and clinics.

Efficient and targeted delivery of drugs is an important challenge in medicine. Use of nanotechnology has a promising future in the field of biomedicine covering a wide range of areas including genomics and proteomics (Fig. 20.3). Many drugs are hydrophobic due to which they cannot reach their destination. Most promising use of nanotechnology is that nanotransporters can overcome this effect due to their small size, enhanced delivery and protective coatings releasing the drugs at the target. Eliminating tumour cells without major collateral damage is a significant interest in nanomedicine research.

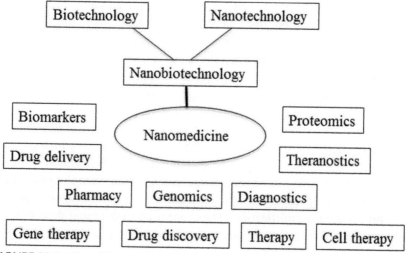

FIGURE 20.3   Use of Nanomedicine in different aspects of medicine.

20.3.2.1   *RELEASE OF DRUGS AT THE TARGET CAN BE MODULATED IN DIFFERENT WAYS*

- Hydrophobicity induced drug release: Hydrophobic forces can be applied to the nanoparticles-drug conjugate.
- pH induced drug release: Drug-nanoparticle conjugates can be modulated to release the drug according to the pH of the microenvironment where the drug is targeted.

- Temperature mediated drug delivery
- Biological agents mediated drug release: Drugs can be conjugated to ligands specific to the receptors and expressed in the organs where they need to be released. Also enzymatic release and chemical reduction-based release approaches are used for nanotechnology-mediated drug delivery.

20.3.2.2   THERANOSTICS

Biomedical applications or nanomedicine has both the wings that is, therapeutics and diagnostics (Fig. 20.4) with indirect or targeted delivery of therapeutics. Nowadays, systems are being developed which can detect the disease both in the initial stage and progression by imaging and act as therapeutics. Much of the literature has shown a combined use of diagnostics and therapy which has laid the concept of the field of theranostics. Due to their unique properties, the field is growing full of potential.

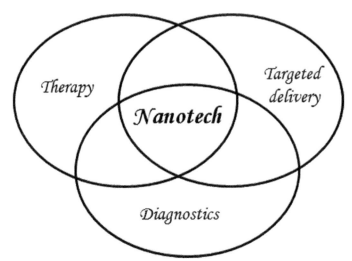

FIGURE 20.4   Application of Nanotechnology in therapy, diagnostics, and targeted delivery.

Modification of nanoparticles can help in delivering drugs through the blood–brain barrier. Hence the system can be used to treat brain diseases. Doxorubicin cannot cross the blood–brain barrier, combining it with polysorbate 80 increases the delivery to the brain significantly. Nanoparticles can penetrate into tissues and cells can absorb them efficiently. Pharmaco-therapy has a new widened future with this development.

20.3.2.3 TYPES OF NANOPARTICLES USED IN MEDICINE

Different forms and types of nanoparticles used in research and clinics are quantum dots, carbon nanotubes, liposomes, dendrimers, metal nanopar-ticles; peptide-based nanoparticles, polymeric nanoparticles and inorganic nanoparticles (Fig. 20.5).

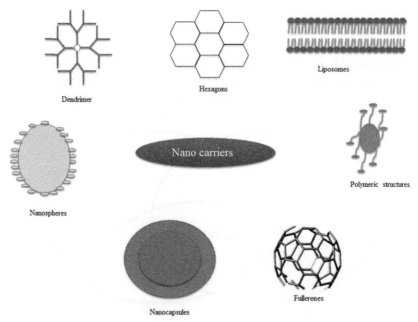

FIGURE 20.5   Types of Nanoparticles.

20.3.2.4 APPLICATION OF NANOTECHNOLOGY FOR CURING VARIOUS DISEASES

20.3.2.4.1 Diabetes

Nanotechnology has a substantial potential for improving the diabetes patients. Early diagnosis of type I diabetes is possible by nanoparticles-based contrast agents. Glucose nanosensors enable real time tracking of blood glucose levels. Diabetes mellitus is a metabolic disorder which is caused by increased blood glucose level. Incidence of diabetes is rising world-wide. Type I diabetes, also known as juvenile diabetes, is due to the deficiency of insulin. Insulin, produced by Islets of Langerhans, regulates blood glucose levels. Prolonged diabetes can lead to multiple organ failure including eye, kidney, nerves and heart. Molecular and biomolecular imaging can create new opportunities to monitor the two types of diseases. Nanoprobes with beta cells of Islet of Langerhans specificity, can detect the status and condition of the insulin producing cells. Glucose sensors can detect the blood glucose levels more accurately. For therapeutic purposes, nanoparticle-based insulin delivery can work better rather than the traditional insulin injections due to obvious reasons as described earlier.

20.3.2.4.2   Cancer

Cancer nanomedicine is the use of nanotechnology for monitoring, detection, diagnosis, treatment and prevention of cancer. This is expected to revolutionize the way of chemotherapy using pharmacokinetics and molecular targeting to deliver to a wide range of diseases with increased efficacy, enhanced specificity and heightened safety (Fig. 20.5).

Immunotherapy is a newer promising tool for the treatment of cancer. Several clinical trials are ongoing and some have been   accomplished successfully. Nanoparticles have been shown to target immune cells within lymph nodes or mucosal tissues so that they can induce immune responses in cancer treatment.

Examples of nanoparticle-mediated drugs are Doxil (pegylated doxorubicin), Abraxane (Albumin-paclitaxel nanoparticles), Onivyde, liposome-encapsulated irinotecan, C-dots (Cornell dots).

20.3.2.4.3   Infectious Diseases

Rapid and specific detection of pathogens can be done as described above. Bioconjugated nanoparticle-based diagnostics can accurately detect bacteria in very less time.[43] Most of the conventional diagnostic methods lack accuracy and sensitivity, which can be overcome by nanotech-based tools. Detection of single molecule has been possible using hybridization method using oligonucleotide-based nanoprobes.[44] Rapid detection of viruses by nanotechnology-based techniques is easy, specific and cost effective.[45] Nanoparticles as labels or tags can detect infectious agents in small volume and are specific, rapid, sensitive and accurate leading to fast and prompt treatment.[46]

Quantum dot technology is another recently employed technology in medicine.[47] Cantilever technology uses the DNA or protein-based microarrays used in proteomics, genomics and molecular diagnostics.[48]

Bioterrorism is a leading threat in today's generation. It can empty the whole world in an hour. Nanosensors can be used to detect the microbes, mostly deadly viruses. This can also be used in clinical settings.[49]

20.3.3   POLLUTION

Unwanted release of waste to nature resulting from the excess resource production and consumption is causing pollution. Most of the industrial, technological and household waste cannot be reintegrated into the environment or else, the methodologies are extremely costly due to which the industry management avoids it. Extraction of petroleum and coal results in photochemical smog and acid rain. Biological systems, on the contrary, oxidize fuel through molecular nanoscale mechanisms without altering the chemical energy. Nanofabrication can be a potential therapy for pollution control.[50] It has a major disadvantage of being expensive, preventing it from commercialization.

20.3.3.1   AIR QUALITY

Air pollution can be controlled and remediation can be done using nanotechnology. Nano-catalysts with an increased surface area can be one of the methods for such gaseous reactions.[51] Nano-catalysts work by modulating the chemical reactions which transform the harmful gases from cars and industrial plants to harmless vapours. Currently, nano-catalysts made of manganese oxide are in use that remove harmful volatile compounds from industrial smokestacks.[52]

The other approach uses the nanostructured membranes having small pores to separate or carbon dioxide or methane from the exhaust.[53] Conventional membranes can separate one particle from the other, while nanomembranes can separate a huge amount of particulate matter from air. Research is going on to develop technologies which can separate large volumes of gas effectively. The substances left out still hold a disposal problem.

20.3.3.2   WATER POLLUTION

The major prerequisite for a healthy life is safe drinking water. Water-borne diseases are a major threat to human as well as the flora and fauna. There is a growing health concern about newer diseases worldwide which are occurring by a complex blend of socio-economic and ecological factors. Biodiagnostics or microbial tests are time-consuming, costly with specific culture methods and antibiotic selection. Harmful pollutants in water can be converted to harmless compounds by applying the principles of chemistry and various chemical reactions. Trichloroethane, a common water pollutant can be effectively catalyzed by nanoparticles.[50]

So the important challenge is specific, sensitive and rapid detection of pathogens. Newer methods are being developed based on enzymatic reactions, genetic methods or immunological methods. Nanotechnology and nanoscience-based methods and devices have been introduced for the detection pathogens in water which can lead to a better healthy life.

Nanotechnology can be the most useful in water cleaning process because addition of nanoparticles into ground water is a cheaper and efficient insertion process.[54] The deionization methods are also cheaper and energy-efficient. Traditional filtration systems use membranes for reverse osmosis or electro dialysis. The pore size of the membrane can be decreased to nanometer range to increase the selectivity of the molecules allowed to pass through.[55] Membranes to filter out viruses or other harmful pathogens are now available.

20.4   CHALLENGES AND ISSUES

20.4.1   CYTOTOXICITY, CLEARANCE, EFFICACY

20.4.1.1   TOXICOLOGICAL HAZARDS

Using nanotechnology in medicine has drawn a lot of criticism due to safety and toxicological issues. The dose needed for efficient therapeutics sometimes has adverse effects. The nanoparticles which are given intentionally can cause health issues and also have polluting effects on nature.

Since the surface area is more, they have more exposure to body tissues. These can be toxic to human health. Although literature claims that they are nontoxic, the long term effect of these nanoparticles is unknown. Also they can alter the body homeostasis by triggering different vital processes like blood coagulation. So a basic understanding of chemical and biological behaviour of the nanoparticles has to be studied before their use. So the risk has to be assessed at a later stage. Nanoparticle accumulation can cause pulmonary inflammation and fibrosis. Quantum dots and carbon nanoparticles can enter the skin. Manganese oxide or titanium oxide can enter the brain through olfactory neurons which can have severe health hazards.

20.4.1.2   CLEARANCE ISSUES

Research has shown that tissues retained the nanoparticles with the drugs even after 24 h.

20.4.1.3   EFFICACY

Gold nanoparticles can enter other organs including the central nervous system. Efficiency is not as much as claimed and as expected in some cases. Also the kinetics of drug release varies with the type of nanomaterial.

Famous Nanotechnology companies: Several companies deal with nanotechnology-based products, some of them are:

Nancor: Nanocomposite barrier to gases

Nano Science Diagnostics: Detect food contaminants

Nanomedia: Targeted drug delivery

Oxonica: Gold nanoparticles as diagnostics

Invitrogen: Quantum dots in medical imaging

CytImmune: Targeted delivery of drugs to tumours

$NanoH_2O$: Nanotechnology membranes for desalination of water

American elements: Ground water treatment by iron nanoparticles

20.5   CONCLUSIONS

The potential use of nanoparticles in different fields continues to make it more exciting and fruitful. An understanding of physical properties to modulate their efficacy is critical to overcome the issues associated with nanotechnology. These systems are rapidly emerging as the focus of new era in science and technology. Nanomedicine is a rapidly growing field in translational medicine. Effective therapeutics in different diseases require appropriate temporal resolution. Development of nanomaterials is a crucial driver towards the progress in the field. Better understanding of the fundamental molecular processes is critical towards the therapy of different diseases. Early diagnosis can help the treatment at an early stage. Innovative research is needed to treat metastatic tumours. Biocompatibility, safety and toxicity issues are important in the design and application of nanomedicine. Too many speculations have occurred bringing much more investments, making nanotechnology a magical buzz. Companies that invested in this field, though not all, went through a loss, decreasing their stocks. The applications of nanotechnology, thus come with a risk of productiveness and investment costs.

KEYWORDS

- **nanotechnology**
- **microbiology**
- **medicine**
- **cells**
- **single molecule**
- **pollution**

REFERENCES

1. Salata, O. V Applications of Nanoparticles in Biology and Medicine. *J. Nanobio-technol.* **2004,** *2*(1), 3.
2. Jha, R. K; Jha, P. K.; Chaudhury, K.; Rana, S. V. S.; Guha, S. K. An Emerging Interface Between Life Science and Nanotechnology: Present Status and Prospects of Reproductive Healthcare Aided by Nano-Biotechnology. *Nano Rev.* **2014,** *5*(10), 22762.
3. Nasr, N. F. Applications of Nanotechnology in Food Microbiology. *Int. J. Curr. Microbiol. Appl. Sci.* **2015,** *4*(4), 846–853.

4. Magdum, S. S. Functions and Future Applications F1 ATPase as Nanobioengine Powering the Nanoworld. *Nano Hybrids* **2013**, *5*, 33–53.

5. Stephanopoulos, N.; Liu, M.,Tong, G. J.; Li, Z.; Liu, Y.; Yan, H.; Francis, M. B. Immobilization and One-Dimensional Arrangement of Virus Capsids with Nanoscale Precision Using DNA Origami. *Nano Lett.* **2010**, *10*(7), 2714–2720.

6. Silva, G. A. Introduction to Nanotechnology and its Applications to Medicine. *Surg. Neurol.* **2004**, *61*(3), 216–20.

7. Nakatsuka, N.; Barnaby, S. N.; Fath, K. R.; Banerjee, I. A. Fabrication of Collagen-Elastin-Bound Peptide Microtubes for Mammalian Cell Attachment. *Biomater Sci. Polym. Ed.* **2012**, *23*(14), 1843–62.

8. Mitragotri, S.; Lahann, J. Physical Approaches to Biomaterial Design. *Nat. Mater.* **2009**, *8*, 15–23.

9. Navya, P. N.; Daima, H. K. Rational Engineering of Physicochemical Properties of Nanomaterials for Biomedical Applications with Nanotoxicological Perspectives. *Nano Convergence.* **2016**, *3*(1), 1–14.

10. Parisi, C.; Vigani, M.; Rodriguez-Cerezo, E. Agricultural Nanotechnologies: What are the Current Possibilities? *Nano Today* **2015**, *10*(2), 124–127.

11. Sekhon, B. S. Nanotechnology in Agri-Food Production: An Overview. *Nanotechnol. Sci. Appl.* **2014**, *7*, 31–53.

12. Cho, Y. J. Emerging Technologies for Food Quality and Food Safety Evaluation. *Contemporary Food Engineering Series,* Taylor and francis group, CRC press; 2011; pp 307–334.

13. Ravichandran, R. Nanotechnology Applications in Food and Food Processing: Innovative Green Approaches, Opportunities and Uncertainties for Global Market. *Int. J. Green Nanotechnol. Phys. Chem.* **2010**, *1*(2):72–96.

14. Barve, A.; Jain, A.; Liu, H.; Jin, W.; Cheng, K. An Enzyme-Responsive Conjugate Improves the Delivery of a PI3K Inhibitor to Prostate Cancer. *Nanomedicine* **2016**, *12*(8), 2373–2381.

15. Khan, I.; Khan, M.; Umar, M. N.; Oh, D. H. Nanobiotechnology and its Applications in Drug Delivery System: A Review. *IET Nanobiotechnol.* **2015**, *9*(6), 396–400.

16. Chan, W. C. Bionanotechnology Progress and Advances. *Biol. Blood Marrow. Transplant.* **2006**, *12*(1), 87–91.

17. Long, M.; Sato, M.; Lim, C. T.; Wu, J.; Adachi, T.; Inoue, Y. Advances in Experiments and Modeling in Micro- and Nano-Biomechanics: A Mini Review. *Cell. Mol. Bioeng.* **2011**, *4*, 327.

18. Guix, M.; Mayorga-Martinez, C. C.; Merkoçi, A. Nano/Micromotors in (Bio)chemical Science Applications. *Chem. Rev.* **2014**, *114*(12), 6285–6322.

19. Bendayan, M. A Review of the Potential and Versatility of Colloidal Gold Cytochemical Labeling for Molecular Morphology. *Biotech. Histochem.* **2000**, *75*(5), 203–42.

20. Conde, J.; Rosa, J.; Baptista, P. Gold-Nanobeacons as a Theranostic System for the Detection and Inhibition of Specific Genes. *Nat. Protoc. Exch.* **2013**, DOI: 10.1038/protex.2013.088.

21. Bhushan, B. Nanotribology and Materials Characterization of MEMS/NEMS and BioMEMS/BioNEMS Materials and Devices. In *Springer Handbook of Nanotechnology;* Springer Berlin Heidelberg, 2017; pp 1575–1638.

22. Yang, L., Akhatov, I., Mahinfalah, M., Jang, B. Z. Nano-Fabrication: A Review. *J. Clin. Inst. Eng.* **2007**, *30*(3), 441–446.

23. Biswas, A.; Bayer, I. S.; Biris, A. S.; Wang, T.; Dervishi, E.; Faupel, F. Advances in Top-Down and Bottom-up Surface Nanofabrication: Techniques, Applications & Future Prospects. *Adv. Colloid Interface Sci.* **2012**, *170*(1–2):2–27.

24. Xie Z.; Yu, W.; Wang, T.; Zhang, H.; Fu, Y.; Liu, H.; Li, F.; Lu, Z.; Sun, Q. Plasmonic Nanolithography: A Review. *Plasmonics* **2011**, *6*, 565.

25. Agarwal, M.; Murugan, M. S.; Sharma, A.; Rai, R.; Kamboj, A. Nanoparticles and its Toxic Effects: A Review. *Int. J. Curr. Microbiol. App. Sci.* **2013**, *2*(10), 76–82.

26. Saraiva, C.; Praça, C.; Ferreira, R.; Santos, T.; Ferreira, L.; Bernardino, L. Nanoparticle-Mediated Brain Drug Delivery: Overcoming Blood–Brain Barrier to Treat Neurodegenerative Diseases. *J. Controlled Release.* **2016**, *235*, 34–47.

27. Singh, A.; Poshtiban, S.; Evoy, S. Recent Advances in Bacteriophage Based Biosensors for Food-Borne Pathogen Detection. *Sensors (Basel)* **2013**, *13*(2), 1763–1786.

28. Koedrith, P.; Thasiphu, T.; Tuitemwong, K. Recent Advances in Potential Nanoparticles and Nanotechnology for Sensing Food-Borne Pathogens and Their Toxins in Foods and Crops: Current Technologies and Limitations. *Sens. Mater.* **2014**, *26*(10), 711–736.

29. Li, X.; Kong, H.; Mout, R.; Saha, K.; Moyano, D. F.; Robinson, S. M.; Rana, S.; Zhang, X.; Riley, M. A.;Vincent, M.; Rotello, V. M. Rapid Identification of Bacterial Biofilms and Biofilm Wound Models Using a Multichannel Nanosensor. *ACS Nano.* **2014**, *8*(12), 12014–12019.

30. Azamal Husen, A.; Siddiqi, K. S. Phytosynthesis of Nanoparticles: Concept, Controversy and Application. *Nanoscale Res. Lett.* **2014**, *9*(1), 229.

31. Liang, P.; Park, T. S.; Yoon, J. Y. Rapid and Reagentless Detection of Microbial Contamination within Meat Utilizing a Smartphone-Based Biosensor. *Sci. Rep.* **2014**, *4*, 1–8.

32. Nikalje, A. P. Nanotechnology and its Applications in Medicine. *Med. Chem.* **2015**, *5*, 81–89.

33. Desai, A. S.; Chauhan, V. M.; Johnston, A. P. R.; Esler, T.; Aylott, J. W. Fluorescent Nanosensors for Intracellular Measurements: Synthesis, Characterization, Calibration, and Measurement. *Front Physiol.* **2013**, *4*, 401.

34. Inbaraj, B. S., Chen, B. H. Nanomaterial-Based Sensors for Detection of Foodborne Bacterial Pathogens and Toxins as well as Pork Adulteration in Meat Products. *J. Food Drug Anal.* **2016**, *24*(1), 15–28.

35. Chung, H. J.; Castro, C. M.; Im, H.; Lee, H.; Weissleder, R. A Magneto-DNA Nanoparticle System for Rapid Detection and Phenotyping of Bacteria. *Nat. Nanotechnol.* **2013**, *8*, 369–375.

36. Anker, J. N.; Hall, W. P.; Lyandres, O.; Shah, N. C.; Zhao, J.; Van Duyne, R. P. Biosensing with Plasmonicnanosensors. *Nat. Mater.* **2008**, *7*, 442–453.

37. Lagally, E. T.; Scherer, J. R.; Blazej, R. G.; Toriello, N. M.; Diep, B. A.; Ramchandani, M.; Sensabaugh, G. F.; Riley, L. W.; Mathies, R. A. Integrated Portable Genetic Analysis Microsystem for Pathogen/Infectious Disease Detection. *Anal. Chem.* **2004**, *76*(11), 3162–3170.

38. Samarasinghe, R. M.; Kanwar, R. K.; Kanwar, J. R. The Role of Nanomedicine in Cell Based Therapeutics in Cancer and Inflammation. *Int. J. Mol. Cell. Med.* **2012**, *1*(3),133–144.

39. Kulkarni, R. P. Nano-Bio-Genesis: Tracing the Rise of Nanotechnology and Nano-biotechnology as "big science". *J. Biomed Discov Collab.* **2007**, *2*, 3.

40. Li, M.; Liu, L.; Xi, N.; Wang Y. Nanoscale Monitoring of Drug Actions on Cell Membrane Using Atomic Force Microscopy. *Acta Pharmacol. Sin.* **2015**, *36*(7), 769–782.

41. Alsteens, D.; Dague, E.; Verbelen, C.; Andre, G.; Francius, G.; Dufrene, Y. F. Nano-microbiology. *Nanoscale Res. Lett.* **2007**, *2*(8), 365–372.

42. Alsteens, D.; Verbelen, C.; Dague, E.; Raze, D.; Baulard, A. R.; Dufrêne, Y. F. Organization of the Mycobacterial Cell Wall: A Nanoscale View. *Pflugers Arch.* **2008**, *456*(1), 117–25.

43. Zhao, X.; Hilliard, L. R.; Mechery, S. J.; Wang, Y.; Bagwe, R. P.; Jin, S.; Tan, W. A Rapid Bioassay for Single Bacterial Cell Quantitation Using Bioconjugated Nanoparticles. *Proc. Natl. Acad. Sci. U. S. A.* **2004**, *101*(42), 15027–15032.

44. Tsang, M. K.; Ye, W.; Wang, G.; Li, J.; Yang, M.; Hao, J. Ultrasensitive Detection of Ebola Virus Oligonucleotide Based on Upconversion Nanoprobe/Nanoporous Membrane System. *ACS Nano.* **2016**, *10*(1), 598–605.

45. Kaittanis, C.; Santra, S.; Perez, J. M. Emerging Nanotechnology-Based Strategies for the Identification of Microbial Pathogenesis. *Adv. Drug Deliv. Rev.* **2010**, *62*(4–5), 408–423.

46. Abraham, A. M.; Kannangai, R.; Sridharan, G. Nanotechnology: A New Frontier in Virus Detection in Clinical Practice. *Ind. J. Med. Microbiol.* **2008**, *26*(4), 297–301.

47. Maiti, A.; Bhattacharyya, S. Quantum Dots and Application in Medical Science. *Int. J. Chem. Chem. Eng.* **2013**, *3*(2), 37–42.

48. Xu, S.; Mutharasan, R. Cantilever Biosensors in Drug Discovery. *Expert Opin. Drug Discov.* **2009**, *4*(12), 1237–51.

49. Wang, X.; Lou, X.; Wanga, Y.; Guo, Q.; Fang, Z.; Zhong, X.; Mao, H.; Jin, Q.; Wu, L.; Zhao, H.; Zhao, J. QDs-DNA Nanosensor for the Detection of Hepatitis B Virus DNA and the Single-Base Mutants. *Biosens. Bioelectron.* **2010**, *25*(8), 1934–1940.

50. YunusI, S., Harwin Kurniawan, A., Adityawarman, D., Indarto, A. Nanotechnologies in Water and Air Pollution Treatment. *Environ. Technol. Rev.* **2012**, *1*(1), 136–148.

51. Chaturvedi, S.; Dave, P. N.; Shah, N. K. Applications of Nano-Catalyst in New Era. *J. Saudi Chem. Soc.* **2012**, *16*(3), 307–325.

52. Chirag, P. N. Nanotechnology: Future of Environmental Pollution Control. *Int. J. Recent Innovation Trends Comput. Commun.* **2015**, *3*(2), 146–166.

53. Nwogu, N. C.; Gobina, E.; Kajama, M. N. Improved Carbon Dioxide Capture Using Nanostructured Ceramic Membranes. *Low Carbon Economy.* **2013**, *4*, 125–128.

54. Gehrke, I. v. Geiser, A.; Somborn-Schulz, A. Innovations in Nanotechnology for Water Treatment. *Nanotechnol. Sci. Appl.* **2015**, *8*, 1–17.

55. Abidi, A.; Gherraf, N.; Ladjel, S.; Rabiller-Baudry, M.; Bouchami, T. Effect of Operating Parameters on the Selectivity of Nanofiltration Phosphates Transfer Through a Nanomax-50 Membrane. *Arabian J. Chem.* **2011**, *9*, S334–S341.

CHAPTER 21

A SYSTEMATIC STUDY ON PHYTO-SYNTHESIZED SILVER NANOPARTICLES AND THEIR ANTIMICROBIAL MODE OF ACTION

ANUPRIYA BARANWAL[1], ANANYA SRIVASTAVA[2], and PRANJAL CHANDRA[1,*]

[1]Department of Biosciences and Bioengineering, Indian Institute of Technology Guwahati, Guwahati, Assam 781039, India, Tel.: + 91 (0)-361–258-3207, *E-mail: pranjalmicro13@gmail.com, pchandra13@iitg.ernet.in, Web: www.iitg.ac.in/biotech/P.Chandra.html

[2]Department of Chemistry, Indian Institute of Technology Delhi, New Delhi 110016, India

CONTENTS

ABSTRACT

Nanotechnology in past few decades witnessed tremendous advancement in terms of its application and still continues to charm the scientific community. The unique physicochemical properties exhibited by metallic nanoparticles enable them to engender new hopes in terms of improving different aspects of human life. Silver nanoparticles, amongst other noble metallic nanoparticles, have gained enormous attention in different fields starting from material science and technology, engineering to nanomedicine and nanobiotechnology. The serious threat to the contemporary world is the development of drug resistance in microbial strains rendering them unstoppable. Therefore, development of a drug which could mitigate this threat from the face of the earth is the need of the hour. Owing to its nobility and other unique properties (petite size, high surface area, stability, etc.), silver nanoparticle in recent years has enormously attracted the scientific community to overcome the aforementioned threats. Silver nanoparticles synthesis via conventional routes, chemical reduction, for instance has raised human health and environmental safety concerns and to mitigate these issues green synthetic approaches have come into picture. Amongst different green synthetic approaches, plant extract mediated synthesis serves the purpose of being an inexpensive, facile, clean and environmentally benign technique to its best. The phytochemicals present in plant extract assist in silver salt reduction to give away silver nanoparticles and also provide stability by acting as capping agent. Albeit, this approach is advantageous in several ways, however not completely devoid of shortcomings like difficulty to control dispersity of nanoparticles, control over the composition of plant extract in terms of phytochemical concentration, and so forth. Herein, we explore the vast plant diversity that has been utilized for the rapid and facile synthesis of silver nanoparticle giving an insight into the mechanism behind its synthesis and its probable antimicrobial mode of action.

21.1   INTRODUCTION

Advancement in nanotechnology has revolutionized almost every field of science and technology by leaving a great impact on every aspect of human lives. Nanotechnology being an interdisciplinary branch of science involves enormous applications of nanoparticles. The word 'nano' is derived from Greek word '*nanos*' which denotes a parameter in the range of one billionth (10^{-9}) of a meter and nanoparticles are nano-sized particles having their size in the range of 1–100 nm. Nanotechnology can be ascribed as a technology which enables nanomaterial synthesis by using different approaches and eventually using these materials for an anticipated purpose. It was an American physicist Richard Feynman who brought in the very concept of nanotechnology and its importance into focus in his lecture 'There's plenty of room at the bottom' at American Physical Society on 29th December 1959, however, the term nanotechnology was coined by N. Taniguchi in 1974 at an international conference on industrial production in Tokyo, Japan.[49] He defines nanotechnology as 'the processing of separation, consolidation, and deformation of material by one atom or one molecule'.

The concept of nanotechnology is recent, however; generation and usage of nano-objects have been there since BC without any knowledge of the synthesis process and nature of these objects. For instance, cultivation and fabrication of natural fabrics like cotton, silk, wool and so forth. were known to people across different civilizations thousands of years ago. Presence of beautiful nanoporous networks of size >20 nm impart these fabrics with spectacular properties, quick swelling and drying after absorbing sweat which makes them stand out from other varieties of fabrics. Fermented product synthesis (wine, bread, beer and so forth.), where nano level fermentation is of extreme importance, is another such example that was known to people since ancient times. The Lycurgus cup which now has been kept in British museum represents a 4th century old Roman glass object. It is known to possess bizarre optical properties depending on whether light is passing through it or not. The fragmental analysis of this cup has shown it to be composed of soda lime quartz glass along with manganese, nanosilver, and nanogold in small fraction and it is these nanoparticles which contribute to the unusual colour change of the cup.[49]

Until past few decades, application of nanotechnology was only limited to electronics which has now slowly spread its claws to several other fields, such as nanomedicine, nanobiotechnology and various other fields of material science. Applicability of nanotechnology in different fields has only been possible due to the unique properties exhibited by nanoparticles which substantially differ from those exhibited by the same material at macroscopic scale.[7] There are several examples in nature where nanostructures have shown inexplicable properties and allowed the plant or organism to sustain in peculiar conditions. For instance, the nanoscale architecture present on the lotus leaf surface makes it ultra-hydrophobic and causes 'lotus effect' where, dirt particles are washed away with the droplets of water resulting in self-cleaning of the leaves. This effect inspired people and helped in developing commercial self-cleaning windows.[35] Likewise, other existing examples of natural nano-structures encouraged the scientific community to mimic these nanoforms and use them for widespread applications.

Fabrication of nanoparticles can be brought about by either of the two approaches, top-down approach or bottom-up approach;[9] however, bottom-up approach is preferred more. Top-down approach generates nanoparticles through the reduction of appropriate macroscopic material into nanorange by employing different physical or chemical techniques. Bulk material is cut, milled, and shaped using external tools to give away nanoparticles of anticipated shape and size.[31] Top-down approach has seen limited usage for generating nanoparticles due to a major drawback, introducing surface imperfections in nanoparticles, which greatly influences its surface chemistry and physical properties. Consumption of high energy during nanoparticle production in order to maintain high tempera-ture and pressure makes this approach highly expensive which thus, limits its usage at commercial scale.[50] Bottom-up is known to be a self-assembly approach where atoms, molecules, or clusters are assembled to create nanoparticles. These initially created nanoparticles are then assembled together eventually to generate ultimate material of the desired conforma-tion by using the chemical or biological technique. Bottom-up approach has gained more attention for generating nanoparticles of uniform surface chemistry and uniform composition for much less cost when compared with top-down approach. This approach is known to deliver a wet synthesis of nanoparticles by employing chemical, photochemical, electrochemical etc. reduction procedures.[52] These procedures are known to be inex-pensive for commercial nanoparticle production, however; use of toxic reluctant, stabilizing and capping agents; and hydrophobic solvents in the production methodology restricts their implementation in the biomedical and clinical field. The excessive usage of lethal chemicals in the afore-mentioned synthetic procedures has raised pronounced concern for the environmental safety and well-being of human race.[7] Therefore, develop-ment of non-hazardous, biocompatible, facile and environmentally benign synthetic procedure is the need hour. To overcome the aforementioned drawbacks of wet synthetic procedures, researchers shifted their attention toward biosynthetic approaches. There has been enormous work on 'green chemistry' based synthetic procedures in past few decades. Researchers exploited various unicellular (bacteria, cyanobacteria and yeast) and multicellular (fungus and plant systems) living entities to bring out inexpensive nanoparticles synthesis at bulk scale. Albeit microbe mediated nanoparticle synthesis is environmentally benign, however it has not gained required attention because of exclusive necessities, such as maintenance of sterile conditions, strain isolation and purification, culture preparation, downstream processing, and so forth. Employing different plant systems and their extracts for nanoparticle synthesis is considered a better method of choice due to several explicable reasons viz. ease of extract preparation, ease of availability and no special requirements unlike for microbe mediated synthesis. Apart from this, it usually requires atmospheric temperature and pressure which prevents huge capital and energy expenditure. Phytofabrication of nanoparticles is based on the capability of plants to uptake, accumulate and exploit metal ions by reducing them.[11]

Ever since nanotechnology has witnessed advancement, a wide range of metallic nanoparticles have been synthesized, however noble metallic nanoparticles (nanogold, nanosilver etc.) continue to top the list when it comes to their applications, as these noble nanoparticles are biocompatible, chemically inert, catalytically active, and least or non-corrosive.[8,33] Silver has extensively been used for different purposes across civilization. In ancient time, silver found its way through different societies where it was used as jewellery, cutlery and storage wares and believed to offer health benefits to its users. It has a very ancient history of being used as a coating material in milk bottles and as wound healing agent to discourage microbial infections.[5] These olden applications of silver metal gave silver nanoparticle its much seeking attention which eventually led to the development of different synthetic procedures. However, a revolution in its synthesis was seen when it was found to eradicate metal ion resistant and drug-resistant pathogens.[45]

21.2   BIOSYNTHESIS OF SILVER NANOPARTICLES

Several physicochemical and biological methodologies have been exploited to achieve silver nanoparticles of different shape and size, however phytofabrication tops them all. A widespread range of plant sources, *Alternanthera dentata, Argyreia nervosa, Tea extract, Cocos nucifera, Brassica rapa, Abutilon indicum, Pistacia atlantica, Acalypha indica, Ziziphora tenuior, Cymbopogon citratus, Premna herbacea, Calotropis procera, Coccinia indica, Vitex negundo, Melia dubia, Pogostemon benghalensis, Swietenia mahogany, Musa paradisiacal, Moringa oleifera, Garcinia mangostana, Eclipta prostrata, Nelumbo nucifera, Acalypha indica, Allium sativum,* and so forth.[4] have been utilized to synthesize silver nanoparticles and subsequently used them for diverse applications. Investigation of different plant extracts that have been used for reduction revealed the presence of phytochemicals (secondary metabolites) such as flavonoids, alkaloids, polysaccharides, amino acids, tannins, oximes and so forth. which can act as reducing agent for the silver metal ion. Different plant tissues (leaf, stem, flower, fruit, bark, root, peel etc.) have successfully facilitated silver nanoparticle synthesis in recent years. The general protocol of synthesizing silver nanoparticles using plant tissue (Fig. 21.1) involves following steps and pre-requisites to accomplish this protocol are silver salt and a phyto-reductant.

- Prepare aqueous silver metal ion salt solution of specific concentration
- Select a medicinal plant rich in secondary metabolites, choose a desired plant tissue and prepare its extract solution.
- Add silver salt solution and plant extract solution in optimized ratio under controlled reaction conditions (temperature, pH, incubation time, reactant concentration, etc.)
- Observe the colour change from transparent or pale white to yellow or brown.
- Analyse the reaction mixture in UV-Visible spectrophotometer for change in surface plasmon resonance (SPR).

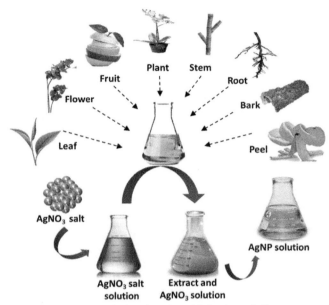

FIGURE 21.1   General protocol showing phytofabrication of silver nanoparticles using different plant tissues. Plant tissue extract is mixed with aqueous silver salt solution under optimized condition to give away silver nanoparticles. Colour change from transparent to deep yellow or yellowish brown gives preliminary confirmation of silver nanoparticle synthesis.

21.2.1   MECHANISM OF SILVER NANOPARTICLE FABRICATION

Biosynthesis of silver nanoparticle is performed by a simple chemical reduction reaction where phytochemicals like alkaloids, polyphenols, flavonoids, polysaccharides, proteins, terpenoids, etc. donate an electron to Ag^+ ion in silver salt solution and give away Ag^0. Dehydrogenation of alcohols and acid, tautomeric transition causing keto-enol conversions of sugars, flavonoids etc. are some of the plausible reasons causing nanotransformation of silver metal ions. The entire mechanism (Fig. 21.2) of phytofabrication of the silver nanoparticle can be divided into three important steps: first is the stage of activation during which the ions of silver salt (Ag^+) obtain an electron from reducing agent, phytochemical in this case and get reduced to silver atoms (Ag^0). The reduction is followed by nucleation step to initiate the process of nanoparticle formation. Second is the stage of growth, during which aggregation of small particle nucleations formed in the first stage occurs spontaneously. The third is the stage of termination, the final step during newly aggregated silver nanoparticles from stage second are stabilized by means of capping agents present in plant extract.[7]

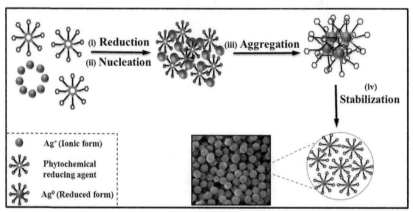

FIGURE 21.2   Schematic representation of the general mechanism behind biosynthesis of silver nanoparticles using plant tissue extract.

21.2.2   PARAMETERS AFFECTING FABRICATION OF SILVER NANOPARTICLES

Reactant concentration, reaction temperature, pH of the reaction mixture, incubation time and metal ion concentration are the key parameters affecting synthesis of silver nanoparticles. Variable phytochemical concentration in plant extracts obtained from different plant systems influences the morphology of silver nanoparticles. In 2006, Chandran et al. studied the influence of increasing *Aloe vera* extract concentration and reported the change in nanoparticle shape from triangular to sphere.[10] Reports from different scientific communities showed high reaction temperature generally leads to the formation of spherical nanoparticles, however low reaction temperature helps in synthesizing nanoparticles of a particular shape, nanotriangles for instance.[26,44] Moreover, high temperature also results in the formation of smaller size nanoparticles by increasing the rate of particle formation.[46] Reaction mixture pH is also known to play a crucial role in nanoparticle fabrication. Variation in hydronium ion concentration can extensively influence texture and morphology of nanoparticles. The basic nature of reaction mixture is reportedly suitable for generating silver nanoparticles of high stability. The additional benefits imparted by basic pH of the reaction mixture are monodispersity, increased growth rate and better yield of nanoparticles, however highly basic pH (> 11) can lead to the formation of agglomerated and unstable silver nanoparticles.[13,48]

21.3   EXTRACTION AND CHARACTERIZATION OF SILVER NANOPARTICLES

The primitive indication of silver nanoparticle synthesis observed by most researchers is the colour change from colourless to yellow or slightly brownish yellow. UV-visible spectroscopy (UV-Vis) analysis of reaction mixtures reveal the SPR peak to lie within the range of 400–450 nm, a range very specific for silver nanoparticles.[41] UV-Visible spectral analysis has widely been used to investigate the impact of different chemical and physical parameters over nanoparticle formation in the reaction mixture, thus does not require separation of nanoparticles. Dynamic light scattering (DLS) is another spectroscopy technique which provides information on size distribution; hydrodynamic radius, surface charge and quality of prepared silver nanoparticles in hydrated form. Nonetheless, other charac-terization techniques (X-ray diffraction (XRD), Fourier-transform infrared spectroscopy (FTIR), Transmission Electron Microscope (TEM), Atomic Force Microscopy (AFM), Scanning Electron Microscope (SEM), Energy Dispersive X-Ray Analyser (EDX) etc.) require nanoparticle separation from reaction mixture that is in powder form, post completion of nanopar-ticle synthesis. Centrifugation is the technique which enables separation of nanoparticle in the form of a pellet which is later oven dried or lyophi-lized to get the ultimate nanopowder form.[39] XRD analysis plays a very crucial role to provide confirmation of nanoparticle synthesis. It assists in phase identification and crystallite size determination of nanomaterial. Analysis of the functional groups attached to the surface of nanoparticles as a result of phytofabrication and surface chemistry analysis is medi-ated by FTIR. TEM and SEM are the two most commonly used electron microscopy techniques to divulge the morphology and surface topology, respectively of prepared nanoparticles. EDX reveals the elemental compo-sition of nanoparticles when integrated with SEM and AFM enables one to study three-dimensional morphology of nanomaterial under realistic conditions.[7]

21.4   APPLICATION OF SILVER NANOPARTICLES TO MITIGATE THE EFFECT OF PATHOGENIC MICROBES

Today emergence of multi-drug resistant microbial strains, high cost and side effects of available drugs pose a serious matter of concern. Commonly available drugs fail to curtail this rising concern and, therefore development of a drug which is inexpensive, targets multi-crucial pathways and poses least or no side effect is inevitable. Reports published in last few decades have revealed the potential of silver nanoparticles as antimicrobial agent.[4] Their ability to curb the pathogenic effect of microbes depends on several unique properties viz. nobility, biocompatibility, minuscule size and large surface area. Impact of silver nanoparticles on different microbial strains is described in Table 21.1 and the following sections elaborate bactericidal, fungicidal, virucidal and antiparasitic properties of silver nanoparticles. The generalized mechanism of antimicrobial activity of silver nanoparticles is shown in Figure 21.3.

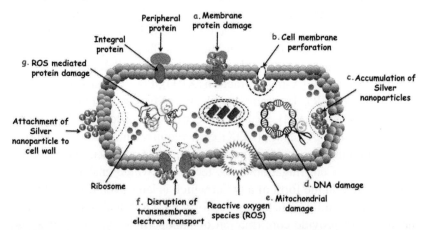

FIGURE 21.3   Generalized illustration of plausible modes exhibited by phytofabricated silver nanoparticles for antimicrobial action. a) Membrane protein damage is caused due to binding of silver nanoparticles with thiol group in membrane proteins leading to their denaturation, b) reactive oxygen species (ROS) formation disrupts cell membrane permeability by forming pits, c) accumulation of silver nanoparticle is the result of disruption of selective permeability of cell membrane, d) intercalation of silver nanoparticle between nitrogenous base pairs results in hindrance in DNA replication that eventually causes DNA damage, e) ROS generation in eukaryotic microbes causes oxidative damage to mitochondrial membrane which leads to mitochondrial damage, f) disruption of membrane protein channels hinder transmembrane electron transport, g) generation of ROS species causes oxidative damage to crucial microbial proteins.

TABLE 21.1 Antimicrobial Properties of Biogenic Silver Nanoparticles.

S. no.	Plant extracts	Nanoparticle size range	Test pathogens	Impact	Reference
(A) Antibacterial activity					
1.	*Acalypha indica*	20–30 nm	*E. coli* and *V. cholerae*	Alteration in membrane permeability and respiration led to pathogen death	[20]
2.	*Sesuvium portulacastrum L.*	5–20 nm	*P. aeruginosa, S. aureus, L. monocytogenes, M. luteus* and *K. pneumoniae*	Maximum inhibition of *S. aureus* and minimum in the case of *M. luteus* was observed	[28]
3.	*G. mangostana*	35 nm	*E. coli* and *S. aureus*	Inhibition of test pathogens	[51]
4.	*Ocimum sanctum* (*Tulsi*)	18 nm	*S. aureus* and *E. coli*	High antibacterial activity	[38]
5.	*S. cumini, C. sinensis, S. tricobatum and C. asiatica*	53, 41, 52 and 42 nm respectively	*S. aureus, P. aeruginosa, K. pneumoniae* and *E. coli*	Maximum antibacterial activity observed for nanoparticles synthesized from *C. sinensis* and *C. asiatica* against *P. aeruginosa*	[24]
6.	*Chenopodium murale*	30–50 nm	*S. aureus*	Effective inhibition of test pathogens	[1]
7.	*Chrysanthemum indicum*	37.71–71.99 nm	*K. pneumonia, E. coli* and *P. aeruginosa*	Significant antibacterial activity against test pathogens	[6]
8.	*Trianthema decandra*	36–74 nm	*S. faecalis, P. vulgaris, E. faecalis, S. aureus, B. subtilis, E. coli, Y. enterocolitica* and *P. aeruginosa*	Effective inhibition of all test pathogens	[15]
9.	*Adansonia digitata*	44 nm	*E. coli, B. subtilis, S. typhimurium, K. pneumoniae, P. vulgaris, P. aeruginosa* and *S. aureus*	Highest inhibition was observed in *E. coli* followed by *S. typhimurium, P. vulgaris, K. pneumoniae, P. aeruginosa, B. subtilis* and *S. aureus*	[21]

TABLE 21.1 (Continued)

S. no.	Plant extracts	Nanoparticle size range	Test pathogens	Impact	Reference
(B) Antifungal Activity					
1.	*Sesuvium portulacas-trum* L.	5–20 nm	*A. alternata, P. italicum, F. euisetii* and *C. albicans*	Highest inhibition of *Penicillium species* and lowest in the case of *C. albicans*	[28]
2.	*Sinapis arvensis*	<100 nm	*Neofusiccum parvum*	Significant antimicrobial activity against test pathogen	[19]
3.	*Adansonia digitata*	44 nm	*Alternaria solani, Penicillium chrysogenum, Aspergillus flavus, Trichoderma harzianum, A. niger*	Effective inhibition of test pathogens with highest inhibition of *A. flavus*	[21]
4.	*Brassica rapa* L.	16.14 nm	*Gloeophyllum abietinum, G. trabeum, Chaetomium globosum* and *Phanerochaete sordida*	Prominent inhibition of test pathogens	[30]
(C) Antiparasitic and Antiviral Activity					
1.	*Nelumbo nucifera*	45 nm	*A. subpictus* and *C. quinquefasciatus*	High inhibition of pathogenic larvae	[40]
2.	*Cinnamomum cassia*	42 nm	Avian influenza virus subtype H7N3	Effective antiviral activity of synthesized NPs	[14]
3.	*Catharanthus roseus*	35–55 nm	*Plasmodium falciparum*	Effective against test microbe	[36]
4.	*Eclipta prostrata*	35–60 nm	*C. quinquefasciatus* and *A. subticus*	Effective inhibition of test pathogens	[37]
5.	*Azadirachta indica, Saraca asoca*	5–20 nm	*Plasmodium falciparum*	Inhibition of *P. falciparum* growth	[25]
6.	*Isatis tinctoria*	10–20 nm	*Leishmania tropica*	Inhibition of test parasite and enhanced inhibition when bound with Amphotericin B	[3]
7.	*Sargentodoxa cuneata*	15–30 nm	*Leishmania tropica*	Significant antileishmanial activity	[2]

21.4.1 BACTERICIDAL ACTION OF SILVER NANOPARTICLES

Usage of the silver nanoparticle as bactericidal agent depends on the fact that silver metal has been known to eradicate bacterial infections since ancient time. It is reported to offer compelling bactericidal effects toward gram positive bacterial strains, *B. cereus*,[12] *Streptomyces* sp.,[55] *S. aureus*,[29] *L. fermentum*,[53] and so forth, and gram-negative bacterial strains, *Enterobacteria*,[42] *E. coli*[34] and so forth. The exact mechanism behind its bactericidal action is not yet well understood, however several possible mechanisms have been reported in the literature. Sondi et al. anticipated silver nanoparticles to have cell wall anchoring capability to bring out variations in bacterial cell structure and cause perforation which eventually leads to bacterial cell death. Another valid reason predicted by this group is an agglomeration of nanoparticles underneath bacterial cell surface due to their unhindered transport across perforations in cell wall.[43] Silver nanoparticles are furthermore, known to generate reactive oxygen species (ROS) which cause oxidative damage to crucial proteins and cellular membrane damage by forming 'pits' in it. It has been anticipated that these particles after being internalized release Ag^+ ions, which interact with –SH (thiol) group of various enzymes and proteins and inactivates them. Reports also suggest intercalation of silver nanoparticles between nitrogenous base pairs which prevent DNA from hydrogen bond formation and terminate DNA replication machinery.[17] The antibacterial effect of silver nanoparticle is found to be dose dependent, that is high concentration of silver nanoparticle is reported to enable better and increased permeability of bacterial cell membrane and, thus increased rate of bacterial cell death.[7]

21.4.2 FUNGICIDAL EFFECT OF SILVER NANOPARTICLES

In recent years, fungal infection has seen hike due to the incompetence of conventional medications to curb drug-resistant fungal strains. These conventional drugs often result in serious side effects on different body organs rendering their application restrained.[7] Therefore, development of potent antifungal agent which is inexpensive, imparts least or no side effects and could target crucial pathways is the need of the hour. Phytofabricated silver nanoparticles are reportedly capable of exhibiting spectacular fungicidal activity when compared with fluconazole and amphotericin, available commercial antifungal drug,[32] however not much work has been reported. The exact mechanism behind the antifungal mode of action of silver nanoparticles in not yet well understood but, certain assumptions have been made. Silver nanoparticle disrupts ionic gradient across fungal cytoplasmic membrane hindering their ability to maintain membrane potential which eventually results in loss of membrane conformity and fungal cell death.[32] Zhao et al. related disruption of membrane conformation and disturbance in ionic gradient across the membrane as the reason for ceasing of yeast budding.[54] Furthermore, silver nanoparticles are also reported to prevent the growth of sporulating fungus by causing serious damage to it.[23]

21.4.3   ANTIPARASITIC AND VIRUCIDAL ACTIVITY OF METALLIC NANOPARTICLES

Parasitic diseases such as filariasis, malaria, leishmaniasis, trypanosomiasis, yellow fever, Japanese encephalitis, dengue fever and chikungunya have endangered human health across the globe. A very large population around the world has fallen prey to this disease as these infections can't be treated with normal medications due to some of the following reasons: one, emergence of drug resistance in parasite, second, lack of conspicuous immune response excitation which prevents development of effective vaccinations, and third, lack of potent medications to mitigate drug-resistant parasites, which leaves antiparasitic chemotherapy as the only choice.[36] Keeping every problem in consideration, a need for better, efficacious and affordable drug to eradicate parasitic infection from the face of the earth comes to mind. People have put tremendous effort in search of better medications which would possess all the aforementioned features. Silver nanoparticles synthesized via. biosynthetic routes using plant extract has been reported to possess spectacular properties as a potent antiparasitic drug. For instance, anti-larval activity against several malarial vectors, *Aedes aegypti*[22] and *A. subpictus;*[37] dengue fever vectors, *Culex quinquefasciatus*[27] and *Aedes aegypt;*[47] and filariasis vector, *C. quinquefasciatus.*[40] Ahmad and group studied the impact of silver nanoparticle for its antileishmanial activity and reported that these phytofabricated nanoparticles could be a better alternative in many ways; effective, safe and inexpensive antileishmanial agent.[2,3] Not much endeavour has been put forth to unveil the exact mechanism behind silver nanoparticle's anti-parasitic mode of action. However, binding of silver nanoparticles with a phosphate group and thiol group result in denaturation of DNA and protein molecules, respectively, which is believed to impart nanoparticles their anti-larval effect.[16]

The attachment and entry of virus inside host cell involves multivalent biological interactions between surface components of virus and host cell membrane receptors. These interactions provide viral strain with enhanced specificity, efficiency and strength that allow it to easily hijack the host cell. The ability of viral strains to evade immune system either, by means of antigenic drift or antigenic shift mechanism has made viral infections, a global threat, rendering conventional vaccination programmes hope-less against them. Not only this, insensitive, expensive and harmful side effects of extended virucidal medications have brought in the need for the development of a better alternative over conventional drugs.[18] Phytofab-ricated silver nanoparticles have long been studied for their antimicro-bial properties and have proven to be good bactericidal agent against both Gram positive and Gram negative bacterial strains, however, their anti-viral activity still remains least explored. Chemical route mediated silver nanoparticles have been studied for their excellent antiviral activity. None-theless, similar antiviral activity was observed by Fatima et al. with phyto-fabricated silver nanoparticles revealing great potential they can offer in terms of a successful antiviral agent.[14] Scientific communities not yet, have a complete understanding of their antiviral mode of action, however, some predictions have been made. These nanoparticles attack the recog-nition system between viral surface components and host cell receptors, thus preventing the virus from entering host cells. The advantage silver nanoparticles have to offer as antiviral agent is that, it is unlikely for the virus to develop resistance against these nanoparticles because they attack varied range of targets on virus.[7]

21.5 CONCLUSION

The efficient and ingenious ability of nature to create minuscule material particles and to use them for numerous applications has greatly inspired scientific community and brought in the very concept of nanotechnology.

Different physiochemical approaches have been developed in order to successfully achieve nanoparticle synthesis; however, shortcomings (carcinogenicity, environmental toxicity, high cost, etc.) associated with these approaches have limited their use in the biomedical field. Increased awareness amongst people for health and environmental safety has led to the development of several greener approaches viz. clean, rapid, facile and environment-friendly. Silver nanoparticle synthesis using different plant extracts has revolutionized the world of green nanotechnology in terms of the benefits it has to offer. It allows inexpensive, less energy intensive, easily scalable and environmentally viable approach to synthesize safe nanoparticles with least generation of waste. Albeit it offers several benefits over other traditional approaches, however, it is not completely devoid of shortcomings. The composition of phytochemicals varies largely when analysed for same species of plant collected from different geographical regions all over the world. Difficulty in the generation of monodisperse silver nanoparticles is another shortcoming of this technique. The need to evade these shortcomings is imperative in order to make phytofabrication, a better method of choice. Phytofabricated silver nanoparticles uphold an entire world of applications starting from sensors, catalysis, wound dressings, medical implants, theranostics to cosmetics and personal care products and the ability of these nanoparticles to act as antimicrobial agent has been most analysed property, so far. However, the exact mechanism behind their antimicrobial mode of action is not yet well understood, therefore, a detailed study of their mode of action is required to make silver nanoparticles a better alternative to traditional drugs.

21.6   ACKNOWLEDGEMENT

This work is supported by DST Ramanujan Fellowship (SB/S2/RJN-042/2015) and DBT project no BT/PR16634/NER/95/225/2015, Government of India awarded to Dr. Pranjal Chandra.

21.7   CONFLICT OF INTEREST

Authors have no conflict of interest.

KEYWORDS

- phytofabrication
- silver nanoparticle
- plant extract
- antimicrobial activity

REFERENCES

1. Abdel-Aziz, M. S.; Shaheen, M. S.; El-Nekeety, A. A.; Abdel-Wahhab, M. A. Antioxidant and Antibacterial Activity of Silver Nanoparticles Biosynthesized Using Chenopodium Murale Leaf Extract. *J. Saudi Chem. Soc.* **2014,** *18,* 356–363.
2. Ahmad, A.; Syed, F.; Shah, A.; Khan, Z.; Tahir, K.; Khan, A. U.; Yuan, Q. Silver and Gold Nanoparticles from *Sargentodoxa cuneata*: Synthesis, Characterization and Antileishmanial Activity. *RSC Adv.* **2015,** *5,* 73793–73806.
3. Ahmad, A.; Wei, Y.; Syed, F.; Khan, S.; Khan, G. M.; Tahir, K.; Khan, A. U.; Raza, M.; Khan, F. U.; Yuan, Q. *Isatis tinctoria* Mediated Synthesis of Amphotericin B-Bound Silver Nanoparticles with Enhanced Photoinduced Antileishmanial Activity: A Novel Green Approach. *J. Photochem. Photobiol. B* **2016,** *161,* 17–24.
4. Ahmed, S.; Ahmad, M.; Swami, B. L.; Ikram, S. A Review on Plants Extract Mediated Synthesis of Silver Nanoparticles for Antimicrobial Applications: A Green Expertise. *J. Adv. Res.* **2016,** *7,* 17–28.
5. Ankanna, S.; Tnvkv, P.; Elumalai, E.; Savithramma, N. Production of Biogenic Silver Nanoparticles Using *Boswellia ovalifoliolata* Stem Bark. *Dig. J. Nanomater Biostructures* **2010,** *5,* 369–372.
6. Arokiyaraj, S.; Arasu, M. V.; Vincent, S.; Prakash, N. U.; Choi, S. H.; Oh, Y.-K.; Choi, K. C.; Kim, K. H. Rapid Green Synthesis of Silver Nanoparticles from *Chrysanthemum indicum* L and its Antibacterial and Cytotoxic Effects: an in vitro Study. *Int. J. Nanomed.* **2014,** *9,* 379.
7. Baranwal, A.; Mahato, K.; Srivastava, A.; Maurya, P. K.; Chandra, P. Phytofabricated Metallic Nanoparticles and Their Clinical Applications. *RSC Adv.* **2016,** *6,* 105996–106010.
8. Chandra, P.; Noh, H.-B.; Won, M.-S.; Shim, Y.-B. Detection of Daunomycin Using Phosphatidylserine and Aptamer Co-Immobilized on Au Nanoparticles Deposited Conducting Polymer. *Biosens. Bioelectron.* **2011,** *26,* 4442–4449.
9. Chandra, P.; Singh, J.; Singh, A.; Srivastava, A.; Shim, Y. B. Gold Nanoparticles and Nanocomposites in Clinical Diagnostics Using Electrochemical Methods. *J. Nanopart.* **2013,** *2013,* 1–12.
10. Chandran, S. P.; Chaudhary, M.; Pasricha, R.; Ahmad, A.; Sastry, M. Synthesis of Gold Nanotriangles and Silver Nanoparticles Using Aloevera Plant Extract. *Biotechnol. Prog.* **2006,** *22,* 577–583.

11. Dauthal, P.; Mukhopadhyay, M. Noble Metal Nanoparticles: Plant-Mediated Synthesis, Mechanistic Aspects of Synthesis, and Applications. *Ind. Eng. Chem. Res.* **2016**, *55*, 9557–9577.

12. Deepak, V.; Umamaheshwaran, P. S.; Guhan, K.; Nanthini, R. A.; Krithiga, B.; Jaithoon, N. M. H.; Gurunathan, S. Synthesis of Gold and Silver Nanoparticles Using Purified URAK. *Colloids Surf. B* **2011**, *86*, 353–358.

13. Edison, T. J. I.; Sethuraman, M. Instant Green Synthesis of Silver Nanoparticles Using *Terminalia chebula* Fruit Extract and Evaluation of Their Catalytic Activity on Reduction of Methylene Blue. *Process Biochem.* **2012**, *47*, 1351–1357.

14. Fatima, M.; Zaidi, N.; Amraiz, D.; Afzal, F.; In Vitro Antiviral Activity of *Cinnamomum cassia* and Its Nanoparticles Against H7N3 Influenza A Virus. *J. Microbiol. Biotechnol.* **2016**, *26*, 151.

15. Geethalakshmi, R.; Sarada, D. Gold and Silver Nanoparticles from *Trianthema decandra*: Synthesis, Characterization, and Antimicrobial Properties. *Int. J. Nanomed.* **2012**, *7*, 5375–5384.

16. Gnanadesigan, M.; Anand, M.; Ravikumar, S.; Maruthupandy, M.; Vijayakumar, V.; Selvam, S.; Dhineshkumar, M.; Kumaraguru, A. Biosynthesis of Silver Nanoparticles by Using Mangrove Plant Extract and Their Potential Mosquito Larvicidal Property. *Asian Pac. J. Trop. Med.* **2011**, *4*, 799–803.

17. Hatchett, D. W.; White, H. S. Electrochemistry of Sulfur Adlayers on the Low-Index Faces of Silver. *J. Phys. Chem.* **1996**, *100*, 9854–9859.

18. Hull, H. F.; Ward, N.; Hull, B. P.; Milstien, J.; de Quadros, C. Paralytic Poliomyelitis: Seasoned Strategies, Disappearing Disease. The Lancet **1994**, *343*, 1331–1337.

19. Khatami, M.; Pourseyedi, S.; Khatami, M.; Hamidi, H.; Zaeifi, M.; Soltani, L. Synthesis of Silver Nanoparticles Using Seed Exudates of *Sinapis arvensis* as a Novel Bioresource, and Evaluation of Their Antifungal Activity. *Bioresour. Bioprocess.* **2015**, *2*, 1.

20. Krishnaraj, C.; Jagan, E.; Rajasekar, S.; Selvakumar, P.; Kalaichelvan, P.; Mohan, N. Synthesis of Silver Nanoparticles Using *Acalypha indica* Leaf Extracts and its Antibacterial Activity Against Water Borne Pathogens. *Colloids Surf. B* **2010**, *76*, 50–56.

21. Kumar, C. M. K.; Yugandhar, P.; Savithramma, N. Adansonia Digitata Leaf Extract Mediated Synthesis of Silver Nanoparticles; Characterization and Antimicrobial Studies, *J. Appl. Pharm. Sci.* **2015**, *5*(8), 82–89.

22. Kumar, V.; Yadav, S. K. Synthesis of Stable, Polyshaped Silver, and Gold Nanoparticles Using Leaf Extract of *Lonicera japonica* L. *Int. J. Green Nanotechnol.* **2011**, *3*, 281–291.

23. Kuppusamy, P.; Yusoff, M. M.; Maniam, G. P.; Govindan, N. Biosynthesis of metallic nanoparticles using plant derivatives and their new avenues in pharmacological applications–An updated report. *Saudi Pharm. J.* **2014**, *4*, 473–484.

24. Logeswari, P.; Silambarasan, S.; Abraham, J. Ecofriendly Synthesis of Silver Nanoparticles from Commercially Available Plant Powders and Their Antibacterial Properties. *Sci. Iran.* **2013**, *20*, 1049–1054.

25. Mishra, A.; Kaushik, N. K.; Sardar, M.; Sahal, D. Evaluation of Antiplasmodial Activity of Green Synthesized Silver Nanoparticles. *Colloids Surf. B* **2013**, *111*, 713–718.

26. Mittal, J.; Batra, A.; Singh, A.; Sharma, M. M. Phytofabrication of Nanoparticles Through Plant as Nanofactories. *Ad. Nat. Sci. Nanosci. Nanotechnol.* **2014**, *5*, 043002.
27. Mondal, N. K.; Chowdhury, A.; Dey, U.; Mukhopadhya, P.; Chatterjee, S.; Das, K.; Datta, J. K. Green Synthesis of Silver Nanoparticles and its Application for Mosquito Control. *Asian Pac. J. Trop. Dis.* **2014**, *4*, S204–S210.
28. Nabikhan, A.; Kandasamy, K.; Raj, A.; Alikunhi, N. M. Synthesis of Antimicrobial Silver Nanoparticles by Callus and Leaf Extracts from Saltmarsh Plant, *Sesuvium portulacastrum L. Colloids Surf. B* **2010**, *79*, 488–493.
29. Nanda, A.; Saravanan, M. Biosynthesis of Silver Nanoparticles from *Staphylococcus aureus* and its Antimicrobial Activity Against MRSA and MRSE. *Nanomed. Nanotechnol. Biol. Med.* **2009**, *5*, 452–456.
30. Narayanan, K. B.; Park, H. H. Antifungal Activity of Silver Nanoparticles Synthesized Using Turnip Leaf Extract (*Brassica rapa L.*) Against Wood Rotting Pathogens. *Eur. J. plant pathol.* **2014**, *140*, 185–192.
31. Narayanan, K. B.; Sakthivel, N. Biological Synthesis of Metal Nanoparticles by Microbes. *Adv. Colloid Interface Sci.* **2010**, *156*, 1–13.
32. Nasrollahi, A.; Pourshamsian, K.; Mansourkiaee, P. Antifungal Activity of Silver Nanoparticles on Some of Fungi. *Int. J. Nano Dimens.* **2011**, *1*, 233–239.
33. Pallela, R.; Chandra, P.; Noh, H.-B. An Amperometric Nanobiosensor Using a Biocompatible Conjugate for Early Detection of Metastatic Cancer Cells in Biological Fluid. *Biosens. Bioelectron.* **2016**, *85*, 883–890.
34. Perni, S.; Hakala, V.; Prokopovich, P. Biogenic Synthesis of Antimicrobial Silver Nanoparticles Capped with l-Cysteine. *Colloids Surf. A* **2014**, *460*, 219–224.
35. Poinern, G. E. J. *A Laboratory Course in Nanoscience and Nanotechnology*, CRC Press: 2014, Boca Raton.
36. Ponarulselvam, S.; Panneerselvam, C.; Murugan, K.; Aarthi, N.; Kalimuthu, K.; Thangamani, S. Synthesis of Silver Nanoparticles Using Leaves of *Catharanthus roseus* Linn. G. Don and Their Antiplasmodial Activities. *Asian Pac. J. Trop. Biomed.* **2012**, *2*, 574–580.
37. Rajakumar, G.; Rahuman, A. A. Larvicidal Activity of Synthesized Silver Nanoparticles Using *Eclipta prostrata* Leaf Extract Against Filariasis and Malaria Vectors. *Acta Trop.* **2011**, *118*, 196–203.
38. Ramteke, C.; Chakrabarti, T.; Sarangi, B. K.; Pandey, R.-A. Synthesis of Silver Nanoparticles from the Aqueous Rxtract of Leaves of *Ocimum sanctum* for Enhanced Antibacterial Activity. *J. Chem.* **2012**, *2013*, 1–7.
39. Sadeghi, B.; Gholamhoseinpoor, F.; A Study on the Stability and Green Synthesis of Silver Nanoparticles Using *Ziziphora tenuior* (Zt) Extract at Room Temperature. *Spectrochim. Acta Part A* **2015**, *134*, 310–315.
40. Santhoshkumar, T.; Rahuman, A. A.; Rajakumar, G.; Marimuthu, S.; Bagavan, A.; Jayaseelan, C.; Zahir, A. A.; Elango, G.; Kamaraj, C. Synthesis of Silver Nanoparticles Using *Nelumbo nucifera* Leaf Extract and its Larvicidal Activity Against Malaria and Filariasis Vectors. *Parasitol Res* **2011**, *108*, 693–702.
41. Sastry, M.; Mayya, K.; Bandyopadhyay, K. pH Dependent Changes in the Optical Properties of Carboxylic Acid Derivatized Silver Colloidal Particles. *Colloids Surf. A* **1997**, *127*, 221–228.

42. Shahverdi, A. R.; Minaeian, S.; Shahverdi, H. R.; Jamalifar, H.; Nohi, A.-A. Rapid Synthesis of Silver Nanoparticles Using Culture Supernatants of Enterobacteria: A Novel Biological Approach. *Process Biochem.* **2007,** *42*, 919–923.

43. Sondi, I.; Salopek-Sondi, B. Silver Nanoparticles as Antimicrobial Agent: A Case Study on *E. coli* as a Model for Gram-Negative Bacteria. *J. Colloid Interface Sci.* **2004,** *275*, 177–182.

44. Soni, N.; Prakash, S. Factors Affecting the Geometry of Silver Nanoparticles Synthesis in *Chrysosporium tropicum* and *Fusarium oxysporum. Am. J. Nanotechnol.* **2011,** *2*, 112–121.

45. Sperling, R. A.; Gil, P. R.; Zhang, F.; Zanella, M.; Parak, W. J. Biological Applications of Gold Nanoparticles. *Chem. Soc. Rev.* **2008,** *37*, 1896–1908.

46. Srikar, S. K.; Giri, D. D.; Pal, D. B.; Mishra, P. K.; Upadhyay, S. N. Green Synthesis of Silver Nanoparticles: A Review. *Green Sustainable Chem.* **2016,** *6*, 34.

47. Suresh, G.; Gunasekar, P. H.; Kokila, D.; Prabhu, D.; Dinesh, D.; Ravichandran, N.; Ramesh, B.; Koodalingam, A.; Siva, G. V. Green Synthesis of Silver Nanoparticles Using *Delphinium denudatum* Root Extract Exhibits Antibacterial and Mosquito Larvicidal Activities. *Spectrochim. Acta Part A* **2014,** *127*, 61–66.

48. Tagad, C. K.; Dugasani, S. R.; Aiyer, R.; Park, S.; Kulkarni, A.; Sabharwal, S. Green Synthesis of Silver Nanoparticles and Their Application for the Development of Optical Fiber Based Hydrogen Peroxide Sensor. *Sens. Actuators B* **2013,** *183*, 144–149.

49. Tolochko, N. K. *History of nanotechnology. Nanosci. Nanotechnologies;* Eolss Publishers: Oxford, 2009.

50. Treguer, M.; de Cointet, C.; Remita, H.; Khatouri, J.; Mostafavi, M.; Amblard, J.; Belloni, J.; de Keyzer, R. Dose Rate Effects on Radiolytic Synthesis of Gold–Silver Bimetallic Clusters in Solution. *J. Phys. Chem. B* **1998,** *102*, 4310–4321.

51. Veerasamy, R.; Xin, T. Z.; Gunasagaran, S.; Xiang, T. F. W.; Yang, E. F. C.; Jeyakumar, N.; Dhanaraj, S. A. Biosynthesis of Silver Nanoparticles Using Mangosteen Leaf Extract and Evaluation of Their Antimicrobial Activities. *J. Saudi Chem. Soc.* **2011,** *15*, 113–120.

52. Won, S.-Y.; Chandra, P.; Hee, T. S.; Shim, Y.-B. Simultaneous Detection of Antibacterial Sulfonamides in a Microfluidic Device with Amperometry. *Biosens. Bioelectron.* **2013,** *39*, 204–209.

53. Zhang, M.; Zhang, K.; de Gusseme, B.; Verstraete, W.; Field, R. The Antibacterial and Anti-Biofouling Performance of Biogenic Silver Nanoparticles by *Lactobacillus fermentum. Biofouling* **2014,** *30*, 347–357.

54. Zhao, G.; Stevens, S. E., Jr. Multiple Parameters for the Comprehensive Evaluation of the Susceptibility of *Escherichia coli* to the Silver Ion. *Biometals* **1998,** *11*, 27–32.

55. Zonooz, N. F.; Salouti, M. Extracellular Biosynthesis of Silver Nanoparticles Using Cell Filtrate of *Streptomyces* sp. ERI-3. *Sci. Iran.* **2011,** *18*, 1631–1635.

INDEX

Printed and bound by CPI Group (UK) Ltd, Croydon, CR0 4YY

23/10/2024

01777702-0018